Proceedings in Technology Transfer

The book series "Proceedings in Technology Transfer" provides a valuable platform for researchers, practitioners, and policymakers to share their experiences and insights related to technology transfer. Our aim is to facilitate knowledge exchange between academic researchers and industry professionals, accelerating the introduction of new technologies and driving economic growth.

Our series publishes books and conference proceedings that focus on original research findings, best practice recommendations, emerging trends, and technology evaluations. The series covers all topics which explore the transfer of technology in various industry sectors that require innovations and economic growth.

Join us in advancing the technology transfer and its impact on innovation. Publish your proceedings with us and offer insights and expertise to professionals in your field.

Huaqun Guo · Ian McLoughlin ·
Umayal Lakshmanan · Eyasu Getahun Chekole ·
Weizhi Meng · Peng Cheng Wang · Jiqiang Lu ·
Nicholas Heng Loong Wong
Editors

Proceedings of the 11th IRC Conference on Science, Engineering and Technology

IRC-SET 2025, 16 August, Singapore

Springer

Editors
Huaqun Guo
Infocomm Technology Cluster
Singapore Institute of Technology
Singapore, Singapore

Ian McLoughlin
Infocomm Technology Cluster
Singapore Institute of Technology
Singapore, Singapore

Umayal Lakshmanan
Scientific Advisor and Business Development Lead
SenzeHub
Singapore, Singapore

Eyasu Getahun Chekole
Department of Information Systems Technology and Design
Singapore University of Technology and Design
Singapore, Singapore

Weizhi Meng
Department of Computing and Communications
Lancaster University
Lancaster, UK

Peng Cheng Wang
Engineering Cluster
Singapore Institute of Technology
Singapore, Singapore

Jiqiang Lu
School of Cyber Science and Technology
Beihang University
Beijing, China

Nicholas Heng Loong Wong
Infocomm Technology Cluster
Singapore Institute of Technology
Singapore, Singapore

ISSN 2948-2321 ISSN 2948-233X (electronic)
Proceedings in Technology Transfer
ISBN 978-981-95-9629-4 ISBN 978-981-95-9630-0 (eBook)
https://doi.org/10.1007/978-981-95-9630-0

This Springer imprint is published by the registered company Springer Nature Singapore Pte Ltd.
The registered company address is: 152 Beach Road, #21-01/04 Gateway East, Singapore 189721, Singapore

Preface

International Researchers Club (IRC) (www.irc.org.sg) was set up in 2001. The vision of IRC is to create a vibrant and innovative research community for Singapore and beyond, through contributions of technical specialties and occupational experiences from its members and fostering strong networking and social interactions of expatriates and new citizens with the local community.

With the vision of IRC, it is our great pleasure to organize IRC conference on Science, Engineering and Technology (IRC-SET, www.ircset.org) for the younger talents and researchers. IRC-SET 2015 was the inaugural conference of IRC and IRC-SET 2025 is now in its eleventh run. IRC-SET conference aims to provide a platform for young and talented researchers to share their research findings, to obtain comments and feedback from experts in the field, and exchange innovative ideas of leading-edge research in multi-disciplinary areas. Researchers from universities, junior colleges, polytechnics and secondary schools are warmly welcome to participate in this conference to showcase and present their research projects, results and findings. Unlike other academic conferences, this conference focuses specifically on Education and Youth development, and has officially been given technically sponsorship from six universities, namely National University of Singapore (NUS), Nanyang Technological University (NTU), Singapore Management University (SMU), Singapore University of Technology and Design (SUTD), Singapore Institute of Technology (SIT), and Singapore University of Social Sciences (SUSS). IRC-SET 2025 conference is also supported by IEEE Intelligent Transportation Systems Society (ITSS) Singapore Chapter, IEEE Broadcast Technology Society (BTS) Singapore Chapter, and IEEE Singapore Section Women in Engineering (WIE) Affinity Group.

The program of IRC-SET 2025 advocates the importance of innovative technology backed by the strong foundation of science and engineering education. Exposing our young participants to the key technology enablers will encourage more interest in the fields of science, engineering and technology. To select good quality papers, the IRC-SET 2025 Call for Papers was broadcast to universities, junior colleges, polytechnics and secondary schools. The researchers then submitted their technical papers to the conference online system. To meet the criteria of the conference, the submitted papers (which had to follow the required template) were peer reviewed by the conference technical program committee, which comprised of a few reviewers coming from the IRC researchers, professors, lecturers and teachers. Comments and recommendations based on novelty of the work, scientific, engineering and technology relevance, technical treatment plausible, and clarity in writing, tables, graphs and illustrations were then provided back to the authors. Through this rigorous review, the technical program committee has selected these papers to be presented in the IRC-SET 2025 conference, which are now published in this proceeding.

IRC-SET 2025 Conference was held online and consisted of Opening Speech by General Chair Prof. Ian McLoughlin (Professor in Singapore Institute of Technology), and nine parallel presentation sessions. The nine presentation sessions covered the key focus areas which included Life Sciences Session, Biomedical Sciences Session, Chemical Engineering Session, BTS Session, Computer Science Session, WIE session, ITSS Session, Math/Cyber Security Session, and Data Science/AI Session. In the closing ceremony, A/Prof. Huaqun Guo (President of IRC and Associate Professor in Singapore Institute of Technology) announced the winners of best paper awards and best presenter awards and concluded the conference with an introduction to the International Researchers Club and a closing speech.

This proceeding is dedicated to International Researchers Club and its members.

November 2025

Huaqun Guo
Ian McLoughlin
Umayal Lakshmanan
Eyasu Getahun Chekole
Weizhi Meng
Peng Cheng Wang
Jiqiang Lu
Nicholas Heng Loong Wong

Acknowledgments

We would like to extend our utmost gratitude to the people who have, in one way or other, inspired, aided and contributed to the successful completion of this book.

First, we would like to thank all members of the conference organizing committee who have contributed their time and professional knowledge to make the conference a big success.

Special thanks to all reviewers for their expertise, time, effort and timely response throughout the peer evaluation process.

Special thanks to all Session Chairs for their expertise, time and effort to make the presentations smooth and have the fruitful knowledge sharing at the conference day.

Finally, our greatest appreciation to all IRC members for their unity.

Organizing Committee

General Chair

Ian McLoughlin	Singapore Institute of Technology, Singapore

Technical Program Chair

Umayal Lakshmanan	SenzeHub Social Enterprise, Singapore
Eyasu Getahun Chekole	Singapore University of Technology and Design, Singapore
Aldy Gunawan	Singapore Management University, Singapore

Publicity Chair

Weizhi Meng	Lancaster University, UK
Noori Kim	Purdue University, USA

Publication Chair

Peng Cheng Wang	Singapore Institute of Technology, Singapore
Jiqiang Lu	Beihang University, China
Nicholas Heng Loong Wong	Singapore Institute of Technology, Singapore

ePoster Chair

Elena Lisova	Malardalen University, Sweden and Volvo Construction Equipment, Sweden
Rong Li	Shanghai University of Sport, China

Local Arrangement Chair

Kasthurirengan Sampath	Temasek Lifesciences Laboratory, Singapore

Communication Chairs

Yongqing Zhu	Singapore University of Social Sciences, Singapore
Yu Wang	A*STAR Research Entities, Singapore

Steering Committee

Huaqun Guo	Singapore Institute of Technology, Singapore
Anand Bhojan	National University of Singapore, Singapore

International Advisory Panel

Chip Hong Chang	Nanyang Technological University, Singapore
Takako Hashimoto	Chiba University of Commerce, Japan
Chau Yuen	Nanyang Technological University, Singapore
Jianying Zhou	Singapore University of Technology and Design, Singapore

TPC Members

Aldy Gunawan	Singapore Management University, Singapore
Boon Seng Chew	Singapore Polytechnic, Singapore
Chin Keong Ho	Sentinan, Singapore
Daniel Reijsbergen	Singapore University of Technology and Design, Singapore
Elena Lisova	Malardalen University and Volvo Construction Equipment, Sweden
Eyasu Getahun Chekole	Singapore University of Technology and Design, Singapore
Fang Liu	Singapore University of Social Sciences, Singapore
Foram Rajdev	Marwadi University, Rajkot Gujarat, India
Howard Halim	Singapore University of Technology and Design, Singapore
Huaqun Guo	Singapore Institute of Technology, Singapore
Judice Koh	National University of Singapore, Singapore
Michael Sileshi Mekbib	Centers for Disease Control and Prevention, USA

Muthugala Arachchige Viraj Jagathpriya Muthugala	Singapore University of Technology and Design, Singapore
Peng Cheng Wang	Singapore Institute of Technology, Singapore
Pramila Ghode	National University of Singapore, Singapore
Umayal Lakshmanan	SenzeHub Social Enterprise, Singapore
Weizhi Meng	Lancaster University, UK
Yi Zhou	Singapore Institute of Technology, Singapore
Yinebeb Mezgebu Dagnachew	Bahir Dar University, Ethiopia
Yongqing Zhu	Singapore University of Social Sciences, Singapore
Yvonne Chow	A*STAR Research Entities, Singapore

Contents

Plant Science

Life Sciences

Chemical Engineering

Biomedical

High-Throughput Transcriptome Profiling of Angiosarcoma for Discovery of Novel Gene Expression Signature Associated with Chemoresistance

Glenys Mai Shia Khor(✉)

Raffles Institution, 1 Raffles Institution Ln, Singapore 575954, Singapore
expoexperiment@yahoo.com

Abstract. Angiosarcoma is a rare yet highly aggressive soft-tissue sarcoma characterized by dismal prognosis, especially in Asian patients. Treatment is notoriously challenging, with resistance to chemotherapy posing a critical challenge to angiosarcoma treatment efforts thus necessitating the discovery of biomarkers and therapeutic targets to guide precision oncology strategies. This study focuses on identifying transcriptomic biomarkers of chemoresistance in angiosarcoma. Data from 72 cases of Asian angiosarcomas were examined, including 35 cases treated with palliative chemotherapy, integrating information from Nanostring gene expression profiling, whole transcriptome profiling, immunohistochemistry, spatial transcriptomic profiling, in vitro testing, and clinicopathological data. In the chemoresistant cohort (defined as stable disease or progression), significant overexpression of genes, such as *SPP1*, *CXCL13*, *CD48*, and *CLEC5A*, was accompanied by significant enrichment of myeloid compartment as well as cytokine and chemokine signaling pathways, while the most enriched immune cell types were neutrophils and macrophages. Validation via spatial transcriptomic profiling showed *SPP1* expression throughout tumor tissues, and the most enrichment in myeloid cells. Immunohistochemistry showed a significant correlation between *SPP1* protein and mRNA expression, while higher *SPP1* protein expression correlated with lower chemotherapeutic sensitivity in patient-derived angiosarcoma cell lines MOLAS and ISOHAS. Additionally, *SPP1* mRNA overexpression is positively correlated with epithelioid histology, higher tumor grade and poorer overall survival outcomes. This study has successfully described the transcriptomic profiling of angiosarcoma for discovery of a novel gene expression signature indicative of chemoresistance and survival outcomes, thus advancing ongoing efforts for targeted therapeutic strategies for precision medicine.

Keywords: angiosarcoma · whole transcriptome sequencing · chemoresistance · secreted phosphoprotein 1 · immune-oncology

1 Introduction

Angiosarcoma (AS), a clinically aggressive cancer of endothelial origin, is a rare subtype of soft-tissue sarcoma with dismal prognosis. AS tumors exhibit significant clinical and genetic heterogeneity and can originate in various anatomical sites, most commonly in

H. Guo et al. (Eds.): IRC-SET 2025, PTT, pp. 3–20, 2026.
https://doi.org/10.1007/978-981-95-9630-0_1

the head and neck, breasts, and extremities [1]. Notably, AS is more prevalent among Asians, making up 7% of all sarcoma diagnoses in Asia and significantly poorer prognosis with a mere 7.8-month median overall survival in Asian AS patients [2]. Given the dismal survival outcomes of Asian patients, it is essential to examine this disease in the Asian setting.

Next, the treatment of AS is notoriously challenging, with low survival rates despite multimodal approaches combining radical surgical resection with radiotherapy and chemotherapy. In particular, while adjuvant cytotoxic chemotherapy is currently the main treatment method for metastatic AS, yet it is often hindered by primary resistance or produces only transient responses followed by rapid development of acquired drug resistance [3]. Excluding clinical factors to aid in diagnosis, such as patient age, disease stage, and histopathological characteristics, there remains a lack of reliable markers to predict survival outcomes and chemotherapy response. Hence, there is a pressing need to identify biomarkers that could enhance patient selection for precision treatment strategies [4]. This is especially true in view of many AS patients' elderly age, which increases the risk of agent-related toxicity and comorbidities from chemotherapy.

At the molecular level, gene expression profiling has revealed three distinct AS clusters represented by the lack or enrichment of immune-related signaling and immune cells, as well as varying tumor mutation burden and tumor inflammation signature scores [5]. In addition, genomic profiling has uncovered multiple actionable mutations, underscoring the promise of precision medicine [4]. However, while multi-omic analyses and immune profiling studies have revealed the genomic and immune landscapes of AS, there is a gap in our understanding of the molecular mechanism underlying chemoresistance. Hence, this study addresses this critical gap by adopting a comprehensive transcriptomic approach to identify novel biomarkers that could enhance the administration of targeted therapies for AS patients [6].

The top candidate gene associated with chemoresistance was secreted phosphoprotein 1 (*SPP1*), also known as osteopontin, a small integrin-binding ligand, N-linked glycoprotein (SIBLING) located at 4q22.1. This highly acidic secreted phosphoglycoprotein has a diverse range of functions, including cell adhesion, migration and inflammation [7]. Multiple studies have described overexpression of *SPP1* to be associated with poor prognosis and immune escape in multiple cancer types, including breast cancer and hepatocellular carcinoma, documenting the key role of *SPP1* in invasion, metastasis, and immune suppression [8]. Additionally, in lung adenocarcinoma, the upregulation of *PD-L1* by *SPP1* mediates macrophage polarisation, facilitates immune escape [9] and acts as an immune checkpoint that induces host tumour immune tolerance by suppressing T cell activation [10]. However, the role of *SPP1* in AS has never been explored. Hence, this study explores the role of *SPP1* as a key candidate gene involved in regulating chemoresistance in angiosarcoma.

To achieve this goal, gene expression profiles of AS were derived and analysed, with spatial transcriptomic profiling, immunohistochemistry, and in vitro testing conducted. Correlations between expression of the top candidate gene and clinicopathological features, immuno-oncology pathways, and patient survival outcomes were also established. It is important to note that this study does not intend to fully elucidate the mechanisms of chemoresistance. Instead, the goal is to identify novel transcriptomic biomarkers of chemoresistance in angiosarcoma thus paving the way for developing precision oncology strategies for this rare but highly aggressive cancer.

2 Materials and Methods

2.1 Patient Cohort

This study included 72 patients with histologically confirmed angiosarcoma treated at Singapore General Hospital and National Cancer Centre Singapore. Certified pathologists reviewed all cases, with diagnoses corroborated by immunohistochemical staining for vascular markers such as CD31 and/or ERG. Cases of Kaposi sarcoma, epithelioid hemangioendothelioma, and intimal sarcoma were excluded. Clinicopathologic characteristics are provided in Table 1 (Appendix).

2.2 Nanostring Gene Expression Profiling

The NanoString PanCancer IO360 panel (NanoString Technologies, Seattle, WA, USA) was used to perform gene expression profiling on formalin-fixed, paraffin-embedded (FFPE) tissue on the nCounter platform, following the manufacturer's protocol. RNA extraction was conducted from five 10-μm sections of all samples containing sufficient tumor tissue, which were subsequently analysed with the 2100 Bioanalyzer (Agilent Technologies, Palo Alto, CA, USA). From the final cohort, a subset of patients who were treated with palliative chemotherapy ($n = 35$), including paclitaxel ($n = 28$), liposomal doxorubicin ($n = 2$), doxorubicin plus ifosfamide ($n = 2$), doxorubicin ($n = 1$), ifosfamide ($n = 1$), and doxorubicin and cisplatin plus paclitaxel ($n = 1$), was analysed on the nSolver 4.0 Advanced Analysis module using the default settings to derive DEGs, pathway scores, and cell-type scores associated with chemotherapy response. The final cohort ($n = 72$) was subsequently reanalyzed against expression levels of *SPP1*, which was the top candidate DEG identified in the initial analysis.

2.3 Immunohistochemical Staining

FFPE tissue samples were procured from the Department of Pathology, Singapore General Hospital. *SPP1* staining was performed using an anti-osteopontin antibody [RM1018] (ab283656, Abcam, Cambridge, UK). The ImmPRESS Universal PLUS Polymer Kit (catalogue #MP-7800; Vector Laboratories, CA, USA) was used to stain the slides according to the manufacturer's protocol. FFPE sections underwent deparaffinization and rehydration using limonene and ethanol, and high-temperature-induced epitope retrieval was performed using a citrate-based buffer (pH 6.0) through 5 min of pressurized heating at 120 °C in a pressure cooker. The slides were incubated with BLOXALL blocking solution for 10 min to quench endogenous peroxidase and alkaline phosphatase activity, followed by treatment with 2.5% prediluted horse serum for 20 min. The slides were then incubated overnight at 4 °C with the anti-osteopontin antibody diluted to 1:200. Staining with substrate–chromogen mix was performed on the slides for 2 min, followed by counterstaining with Hematoxylin QS Counterstain (Vector

Laboratories, CA, USA). *SPP1* expression was quantified by two independent readers blinded to each other and the NanoString and clinical data. Histopathologic appearances were located through images taken at 10 × objective (magnification of 100 ×), and a more detailed evaluation was conducted at 40 × objective (magnification of 400 ×). A histochemical score (H-score) was assigned according to the staining intensity observed in representative sections, with scores ranging from 0% to 100%.

2.4 Western Blotting

Cells were harvested and washed with PBS. Lysis buffer was added to the cell pellet and stored at − 20 °C overnight. Cell lysates were separated based on molecular weight using SDS-PAGE with 4–15% Mini-PROTEANTM TGX Stain-Free™ Protein Gels (Bio-Rad Laboratories, Hercules, CA, USA). The blot was transferred onto 0.2 μm PVDF membranes (Bio-Rad Laboratories, Hercules, CA, USA) and blocked with 5% non-fat dry milk (Bio-Rad Laboratories, Hercules, CA, USA) in TBST solution (50 mM Tris/HCl pH 7.4, 150 mM NaCl, 0.1% Tween-20) for 1 h. Subsequently, the blot was left to roll at 4 °C overnight in anti-osteopontin antibody [RM1018] (ab283656, Abcam, Cambridge, UK) at 1:1000 dilution and β-Actin antibody (#4970S, Cell Signaling Technology, Danvers, MA, USA). The blot was incubated in secondary antibodies for 1 h and subsequently exposed to chemiluminescence detection using SuperSignal Substrate Western Blotting Kit (Thermo Fisher Scientific, MA, USA). Images were taken using ChemiDoc™ XRS + System with image Lab™ Software (Bio-Rad Laboratories, Hercules, CA, USA).

2.5 Cell Lines and Cell Viability Assays

Two AS cell lines (MOLAS and ISOHAS) were obtained from the Cell Resource Center for Biomedical Research, Institute of Development, Aging and Cancer, Tohoku University, Japan, courtesy of Dr. Mikio Masuzawa. MOLAS was established from a patient with scalp lymphangiosarcoma metastatic to the pleura, while ISOHAS was established from a patient with primary scalp hemangiosarcoma. Both cell lines were maintained in a DMEM medium supplemented with 10% FBS and 1% penicillin/streptomycin. These cells were grown in a humidified chamber with 5% CO2 at 37 °C. To examine the response to chemotherapeutic agents, MOLAS and ISOHAS cultures were exposed to paclitaxel and doxorubicin (Selleck Chemicals, TX, USA) at concentrations of 10, 20, 50, and 100 ng/mL for 72 h. The Quick Cell Proliferation Assay Kit II (ab65475, Abcam, Cambridge, UK) was used to quantify the overall cell viability after treatment with the respective agents. Cell cultures at approximately 70% confluence were used for all experimental drug treatments unless otherwise stated.

2.6 10 x Genomics Visium Platform

Formalin-fixed, paraffin-embedded (FFPE) tissue blocks of two breast AS samples and two fresh frozen AS samples of head and neck origin were examined for *SPP1* expression pattern. The AS samples were processed on the Visium CytAssist (10 × Genomics, CA, USA) following previously established protocols [11].

2.7 Analysis of Spatial Sequencing Data

The 10 × Genomics Visium platform was used for spatial transcriptomic profiling following previously established protocols [11]. Reads were demultiplexed and aligned to the hg38 reference genome using 10 × Space Ranger v.1.3.1 (10 × Genomics, CA, USA), employing default parameters for automatic alignment. Spatial data were loaded into count matrices using Seurat v4.0, retaining spots with less than 10% of transcripts mapping to mitochondrial genes for scaling and normalization of gene expression measurements via the sctransform method. Selected genes were utilized to annotate immune cells, including PECAM1 and ERG for tumor cells; *FBLN1*, *FAP*, and *DES* for fibroblasts; *CD79A* and *CD79B* for B-cells; *CD8A* and *CD8B* for CD8 T cells; *KLRK1* and *KLRD1* for NK cells; *CD14* and *CD68* for macrophages; *CSF3R* and FPR1 for neutrophils; *CD209* and *CCL13* for dendritic cells; and *TPSAB1*, *MS4A2*, *CPA3* and *HDC* for mast cells.

2.8 Statistics

SPP1 expression levels were dichotomized into high and low categories using the median split method. The expression of SPP1 in primary tumor samples was compared with various clinicopathological data. Progression-free survival was determined to be the period elapsed from diagnosis to either disease progression or death from any cause. Overall survival was calculated from diagnosis to death or censored at the last follow-up date for surviving patients. Using Kaplan–Meier analysis and Cox proportional hazards models, survival analyses were conducted with censoring applied at the last follow-up date. Box-and-whisker plots were used to represent continuous data, and associations with categorical variables were analysed using the Mann–Whitney U or Kruskal–Wallis tests, as applicable. All statistical analyses were performed using MedCalc (Windows version 19.0.4), with the statistical significance threshold set at two-tailed $p < 0.05$.

3 Results

3.1 Transcriptome Profiling

The top four genes that were significantly overexpressed in non-responders to palliative chemotherapy were *SPP1* (log2foldchange 3.49, adj. $p = 0.0112$), *CXCL13* (log2foldchange 2.57, adj. $p = 0.0112$), *CD48* (log2foldchange 2.97, adj. $p = 0.0166$), and *CLEC5A* (log2foldchange 2.55, adj. $p = 0.0247$). Conversely, *TAF3* was significantly overexpressed (adj. $p < 0.05$) in chemosensitive tumors (Fig. 1A). To survey the transcriptomic landscape further on a global level, whole transcriptome RNA-seq data of a subset of samples in tumor (n = 12) and matched normal tissue where available (n = 6) was evaluated. The expression of the top candidate biomarker, *SPP1*, was significantly higher in tumor tissue compared to matched normal tissue ($p = 0.0008$) (Fig. 1B) in keeping with NanoString gene expression profiling. The top 10 DEGs can be found in Table 2, Appx.

Pathway analysis indicated upregulation of myeloid compartment ($p = 0.007$), cytokine and chemokine signaling ($p = 0.017$), as well as metastasis/matrix remodeling

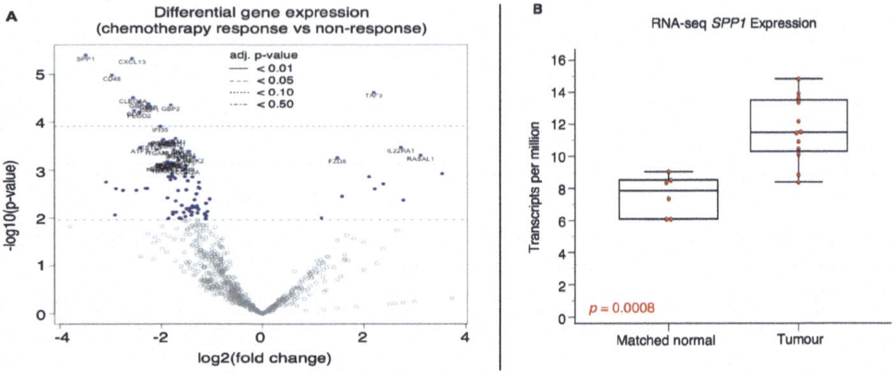

Fig. 1. **A.** Volcano plot of differential gene expression in chemoresistant versus chemosensitive AS reveals significant *SPP1* overexpression in chemoresistant AS. **B.** *SPP1* transcript levels from whole transcriptome sequencing data in tumor compared to matched normal tissue.

pathways in chemoresistant tumors. Conversely, Hedgehog, Notch, and Wnt signaling pathways were upregulated in chemosensitive tumors (Fig. 2A). Immune cell-type analysis suggested myeloid cells, including neutrophils ($p = 0.015$) and macrophages ($p = 0.039$), along with NK cells ($p = 0.149$), were enriched in chemoresistant tumors, whereas a greater population of mast cells, dendritic cells and regulatory T cells were observed in chemosensitive tumors (Fig. 2B). Statistical significance of upregulation of top 3 pathways and immune cell types in chemoresistant versus chemosensitive AS were then assessed using the Kruskal–Wallis test (Fig. 2C & 2D).

Further evaluation of all 72 samples, the DEGs between *SPP1*-high and *SPP1*-low tumors (n = 41 for $p < 0.05$, n = 23 for $p < 0.01$) were analysed with *p*-values adjusted for false discovery rate following Benjamini–Yekutieli procedure. Genes that were very significantly overexpressed ($p < 0.001$) included *SLC11A1, PLOD2, CXCL3, IL6, FSTL3, CXCL2, TREM1,* and *IL1β* (Fig. 3A). The top 10 DEGs can be found in Table 3, Appx.

A gene-specific analysis (GSA) revealed very significant overexpression ($p < 0.001$) of myeloid compartment pathway genes (*SLC11A1, CXCL3, CXCL2, TREM1, IL1β*), cytokine and chemokine signaling pathway genes (*CXCL3, IL6, CXCL2, IL1β*), and matrix remodeling and metastasis pathway genes (*PLOD2*) (Fig. 3B & 3C).

Additionally, cell-type analysis of *SPP1*-high tumors was suggestive of the enrichment of neutrophils, macrophages, and NK cells, similar to that in chemoresistant tumors (Fig. 3D).

3.2 Association of *SPP1* mRNA Expression with Clinicopathological Characteristics

SPP1 mRNA overexpression emerged as a significant candidate biomarker of chemoresistance, positively correlated to the presence of epithelioid features ($p = 0.007$) (Fig. 4A) and high FNCLCC histological tumor grade ($p = 0.006$) (Fig. 4B). Dichotomizing into high or low scores using the median *SPP1* expression value of 10.23, high *SPP1* expression was correlated with non-head and neck location ($p = 0.0576$), but no significant

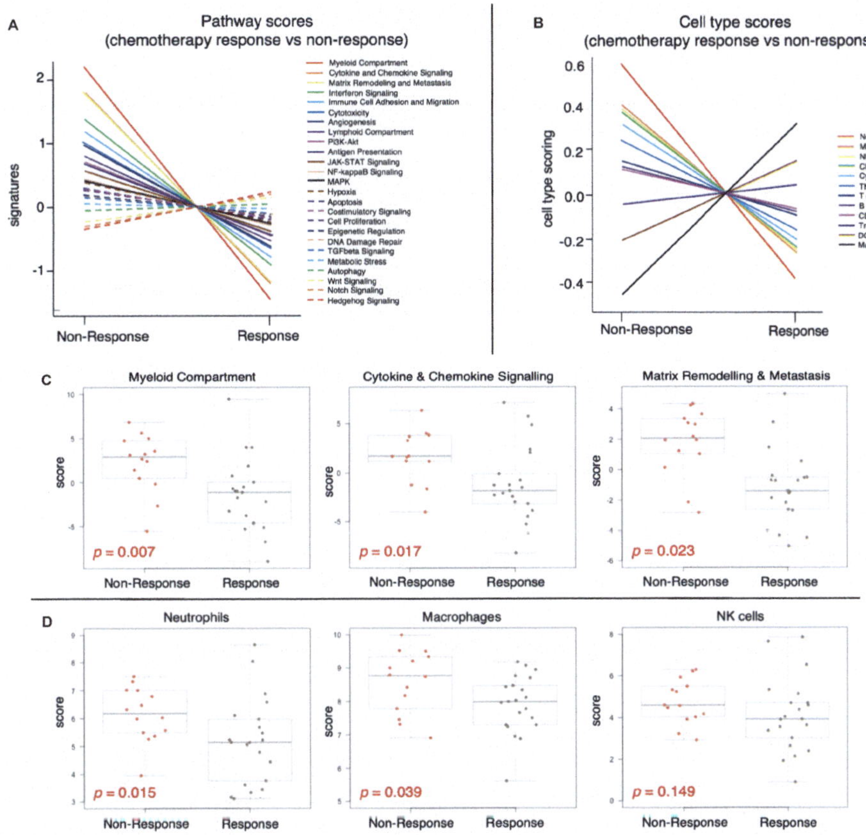

Fig. 2. **A.** Pathway enrichment scores of 25 cancer associated canonical pathways in chemoresistant vs chemosensitive tumors. **B.** Immune cell enrichment scores in chemoresistant vs chemosensitive tumors. **C.** Statistical significance of association of top three immuno-oncology pathways with chemotherapy non-response versus response in AS. **D.** Statistical significance of association of top three immune cell types with chemotherapy non-response versus response in AS.

association was found with TMB (Fig. 4C), TIS scores (Fig. 4D), the presence of human herpesvirus-7, ultraviolet exposure signature, or metastatic state at diagnosis (Table 1, Appx).

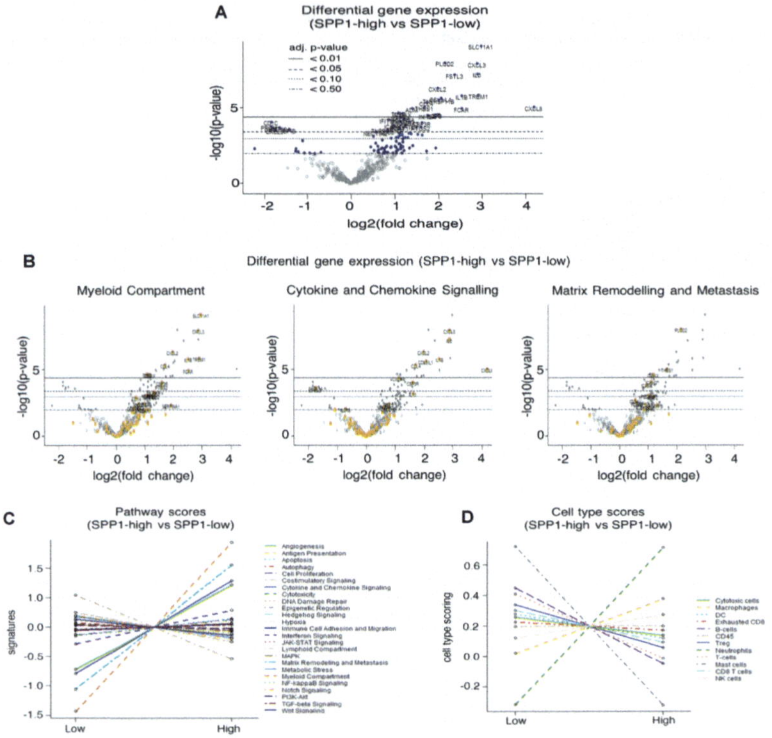

Fig. 3. **A.** Volcano plot of DEGs between *SPP1*-high and *SPP1*-low tumors. **B.** GSA volcano plots of differentially expressed genes between *SPP1*-high and *SPP1*-low tumors in the top 3 enriched pathways in chemoresistant AS. **C.** Pathway enrichment scores of 25 cancer-associated canonical pathways. **D.** Immune cell enrichment scores in *SPP1*-high vs *SPP1*-low tumors.

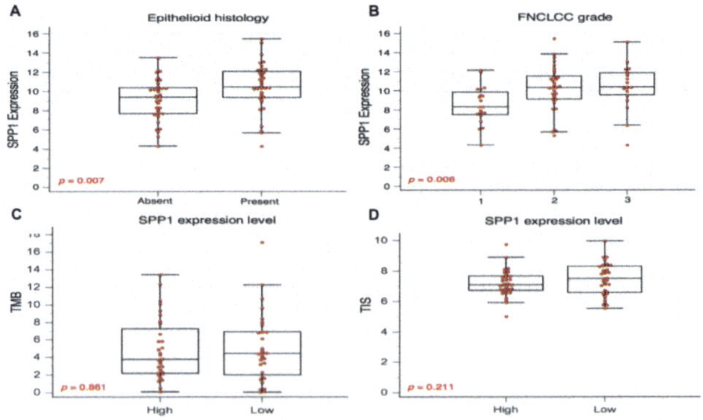

Fig. 4. Association between (**A**) presence of epithelioid histology, (**B**) higher FNCLCC grading, (**C**) tumor mutational burden (TMB) and (**D**) tumor inflammatory signature (TIS) scores with *SPP1* gene expression levels.

3.3 Spatial Analysis of *SPP1* Expression

The 10 × Genomics Visium spatial transcriptomics was used to analyze 2 AS samples of the breast and 2 of the head and neck. A total of 28,988 55-micron spots were analysed (primary breast AS, 13101; post-radiation therapy breast AS, 13600; HHV7-positive scalp AS, 748; HHV7-negative scalp AS, 1539). A spatial transcriptomics analysis showed *SPP1* expression displayed significant heterogeneity in spatial distribution in all samples, with focal clusters of cells expressing higher levels of *SPP1* scattered throughout the tumor tissues (Fig. 5A). *SPP1* expression was present across all cell types in AS, including stromal, immune, and tumor cells, and in corroboration with NanoString immune profiling, was most enriched in myeloid cells (Fig. 5B&5C).

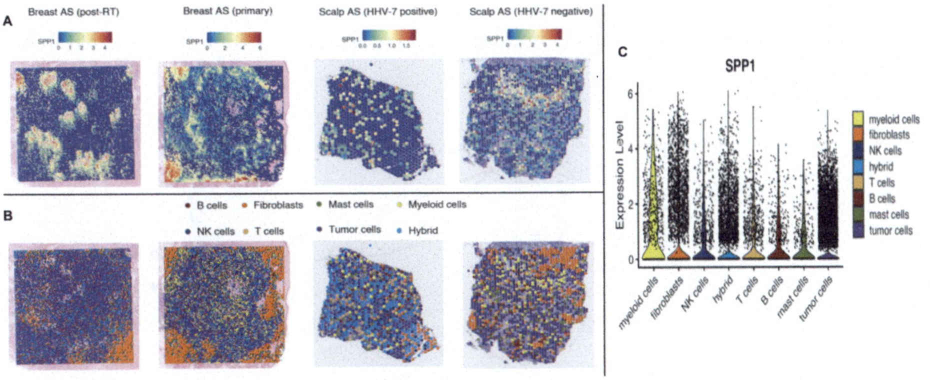

Fig. 5. **A.** Spatial transcriptomics analysis showed spatial distribution of *SPP1* expression in tumor tissues. **B.** Spatial distribution of various cell types in the tumor tissues. **C.** Violin plots showing *SPP1* expression across all cell types in AS.

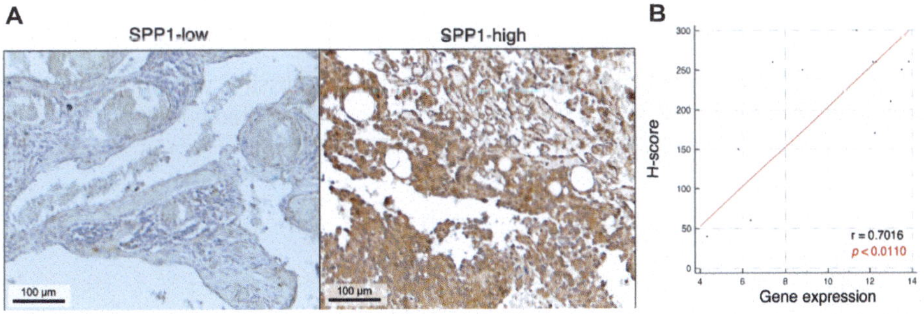

Fig. 6. **A.** IHC images at 400 × magnification (scale bar, 100 μM) of tissue sections stained with anti-osteopontin antibody (*SPP1*). Representative images for high and low *SPP1* H-scores are shown. **B.** Scatterplot showing correlation between *SPP1* expression levels from IHC H-scores and gene expression profiling.

3.4 Immunohistochemical Staining for *SPP1* Protein Expression

SPP1 expression was examined in the AS tumor archival FFPE samples using immunohistochemistry (Fig. 6A). A significant moderate positive correlation (r = 0.7016; $p < 0.0110$) was observed between *SPP1* expression levels obtained via immunohistochemistry and NanoString gene profiling (Fig. 6B).

3.5 *In Vitro* Response to Chemotherapeutic Drugs

Western blot analyses demonstrated higher protein expression of *SPP1* in ISOHAS compared to MOLAS (Fig. 7A). MOLAS and ISOHAS cell lines were exposed to increasing concentrations of paclitaxel and doxorubicin (10, 20, 50, and 100 ng/mL for 72h). Both drugs resulted in a dose-dependent reduction in cell viability in the AS cell lines, although ISOHAS displayed greater chemoresistance compared with MOLAS (Fig. 7B).

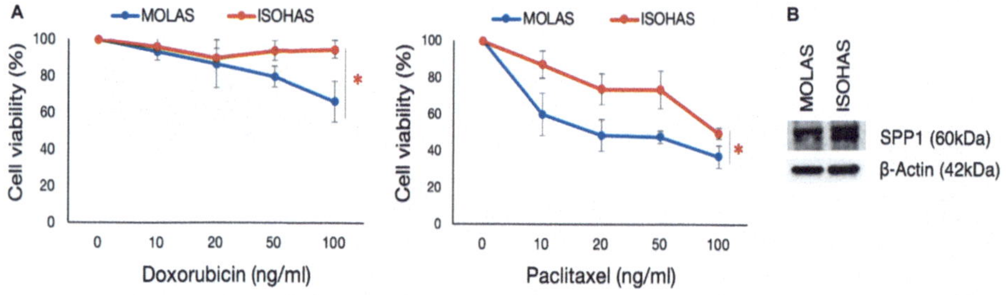

Fig. 7. A. Western blot of *SPP1* protein expression in ISOHAS and MOLAS cell lines. **B.** Cell viability of MOLAS and ISOHAS upon treatment with chemotherapeutic drugs paclitaxel and doxorubicin. *, $p < 0.05$.

3.6 Survival Analysis

In keeping with the abovementioned data, *SPP1* gene expression was correlated with poorer overall survival (HR 1.84, 95% CI 1.07–3.18, $p = 0.0288$) and poorer progression-free survival (HR 1.74, 95% CI 0.98–3.06, $p = 0.0566$) (Fig. 8).

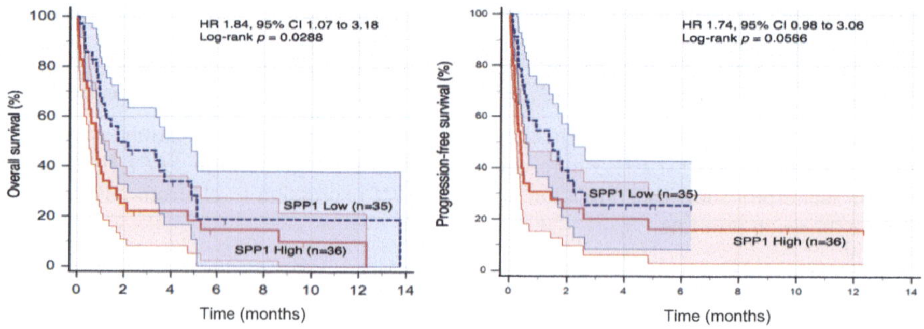

Fig. 8. Kaplan–Meier curves showing survival probability in patients with *SPP1*-high versus *SPP1*-low AS.

4 Discussion

This study demonstrated that *SPP1* overexpression is significantly associated with chemoresistance and poorer survival outcomes in human angiosarcoma (AS). Next-generation molecular diagnostics have only recently begun to characterize the pathobiology of AS, which is an otherwise enigmatic disease with an aggressive clinical phenotype [5, 11]. Moreover, despite recent studies into combination therapies in the management of AS, response to first-line systemic therapy is poor because of the absence of predictive biomarkers and significant intra-tumor heterogeneity [12]. Thus far, apart from one study that has employed a morphological approach to identify a potential marker of chemoresistance in AS, that is $CD31^{low}$ cells, which relies on enhanced YAP signaling to improve redox status and is doxorubicin-resistant in AS [13], to my knowledge, this is the first study to identify potential biomarkers of chemoresistance in AS via an omics approach.

From gene expression profiling, a gene expression signature associated with chemoresistance is identified in AS. The most significantly upregulated gene in the gene expression signature was secreted phosphoprotein 1 (*SPP1*), which is well known to be involved in the regulation of many tumor-associated biological processes, including tumorigenesis, tumor progression, and the tumor immune microenvironment. *SPP1* was suggested as a potential target for evaluating prognosis and immunotherapy response in multiple cancers [13, 14].

Notably, *SPP1* overexpression has been associated with resistance to chemoradiotherapy in various other cancers. For example, in lung adenocarcinoma, *SPP1* expression on TAMs was found to correlate with poor prognosis and chemoresistance [15]. In ovarian cancer, the *SPP1-CD44* axis facilitated cancer cell chemoresistance via PI3K/AKT signaling and ATP-binding cassette drug efflux transporter activity [16]. Similarly, fibroblast-derived *SPP1* was found to contribute to resistance of hepatocellular carcinoma to sorafenib and lenvatinib treatment [17], and in oral squamous cell carcinoma, the *SPP1*-integrin $\alpha_v\beta_3$ axis was crucial for 5-fluorouracil resistance [18]. Most recently, *Helicobacter pylori* infection-induced *SPP1* activation was found to promote chemoresistance and T cell inactivation in gastric cancer cells [19]. The role of *SPP1* in cancer thus appears to vary according to cancer subtype, suggesting the need for further research tailored to specific cancer types to explore its clinical implications.

In this study, a correlation was observed between *SPP1* overexpression and the upregulation of the myeloid compartment, cytokine and chemokine signaling, and metastasis/matrix remodeling pathways in chemoresistant tumors, which is in line with the cell-type analysis that suggested the enrichment of myeloid cells (neutrophils and macrophages), as well as NK cells. Moreover, several genes in the Nanostring analysis were differentially expressed between *SPP1*-high and *SPP1*-low AS tumors. A gene-specific analysis characterized the top *SPP1*-mediated pathways in AS, including the myeloid compartment pathway with the significant overexpression of *SLC11A1*, *CXCL3*, *CXCL2*, *TREM1*, *IL1β*, and *TLR2* genes, as well as the cytokine and chemokine signaling pathway, with significant overexpression of the *CXCL3*, *IL6*, *CXCL2*, *IL1β* genes.

Another significant finding is the potential crosstalk between *SPP1*, interleukin(*IL)1β*, and myeloid cells such as TAMs which could contribute to the development of a signaling pathway causing chemoresistance in the AS TME. *IL1β* is a pro-inflammatory cytokine produced by myeloid cells including TAMs. *IL1β*-induced *IL6* and *sIL6R* have been found to trigger *IL6* trans-signaling, contributing to the upregulation of *SPP1* in THP-1 macrophages [20]. Moreover, a pan-cancer single-cell transcriptional analysis of tumor infiltrating myeloid cells suggests that *SPP1* and *IL1β* were both highly produced by TAMs [21].

Taken together with the DEGs associated with *SPP1*-high tumors (Fig. 3A & Table 3, Appx), a potential mechanism regulating chemoresistance in angiosarcoma involves the over-secretion of *IL1β* by TAMs in the AS TME. This induces *IL6* and *sIL6R* expression, triggering *IL6* trans-signaling that contributes to the overexpression of *SPP1* within the TAMs. Subsequently, *SPP1* is secreted into the TME and binds to its major integrins, $\alpha_v\beta_3$ and $\alpha_v\beta_5$. This then activates the *TREM1* pathway, which amplifies pro-inflammatory cytokines including *IL1β*, *IL6*, *CXCL2*, and *CXCL3*. This results in a positive feedback loop that not only exacerbates *IL1β* over-secretion by TAMs, but also leads to recruitment of more TAMs and neutrophils to the TME via *CXCL2* and *CXCL3* signaling. Concurrently, the binding of *SPP1* to $\alpha_v\beta_3$ and $\alpha_v\beta_5$ could trigger *PLOD2*-mediated extracellular matrix remodeling, such as cross-linking and stiffening, thus reducing drug penetration. This then counters paclitaxel or other chemotherapy-induced apoptosis.

Curiously, the initial trigger of myeloid cell (such as TAM) recruitment to AS tumors, as well as the over-secretion of *IL1β* by these myeloid cells remains unknown, but from my clinicopathological analysis, oncogenic mutations are likely not the initial trigger due to the lack of any association between high *SPP1* expression and TMB. Instead, a possible initial trigger could be the release of damage-associated molecular patterns (DAMPs) due to AS tumors' characteristic extensive vascular instability and tissue damage, which could activate pattern recognition receptors such as *TLR2*, which was the seventeenth highest DEG in *SPP1*-high tumors, on TAMs and other myeloid cells. This could then trigger *IL1β* over-secretion via the toll-like receptor signaling pathway.

Separately, *SPP1* has also been extensively studied in the context of rheumatoid arthritis and osteoarthritis [22], where it promotes osteoclastogenesis and inflammatory responses via integrin and *CD44*-mediated pathways. Currently, drugs like methotrexate have been able to reduce serum concentrations of *SPP1* in patients with juvenile idiopathic arthritis [23], while anti-*SPP1* monoclonal antibodies like mAb 23C3 have been capable of alleviating symptoms of rheumatoid arthritis in a murine collagen-induced arthritis model [24], thus suggesting possible repurposing of such therapeutics for angiosarcoma patients.

Furthermore, repurposing various integrin inhibitors in the context of angiosarcoma as inhibitors of *SPP1*-mediated tumor progression is another possibility. One promising option could be AOM1 (Pfizer Inc.), a fully human IgG2 which inhibits *SPP1* binding to $\alpha_v\beta_3$, that significantly inhibited growth of large metastatic non-small cell lung adenocarcinoma tumors in genetically engineered mouse models both as a single agent or in combination with carboplatin [25]. Finally, other potential inhibitors include brefelamide, a novel inhibitor of *SPP1* that suppressed the invasive ability of A549 lung carcinoma cells [26], entrectinib, a tropomyosin receptor kinase inhibitor that showed excellent binding affinity against *SPP1* protein *in silico* and thus has the potential to be a *SPP1* inhibitor to curtail cervical cancer progression [27], as well as cilengitide, a cyclicised arginine-glycine-aspartic acid-containing pentapeptide that selectively and potently inhibits the major *SPP1* integrin receptors $\alpha_v\beta_3$ and $\alpha_v\beta_5$ [28].

5 Future Directions

Future work includes functional validation studies, such as *SPP1* knockdown or overexpression in an AS cell line via RNA interference (RNAi) or CRISPR-mediated transcriptional activation, to confirm the causative role of *SPP1* in regulating chemoresistance. Additionally, *in vivo* studies can be conducted to validate the results in murine models. Upon proof-of-concept, the efficacy and viability of chemotherapy with anti-*SPP1* therapeutics can then be assessed in clinical trials as a novel targeted combination therapy for chemoresistant angiosarcoma patients. Finally, another avenue for future research is to decipher the trigger and mechanism behind myeloid cell recruitment by *SPP1*-high AS tumors, which could unveil the underlying cause of *SPP1* overexpression in AS, thereby identifying additional upstream targets to enhance *SPP1* inhibition as a precision therapy.

6 Limitations

The limitations of our study include its retrospective approach and relatively small sample size of 72 due to the rarity of angiosarcoma and thus difficulty in obtaining patient samples. Additionally, tumor heterogeneity and potential sampling issues could influence the reliability of the findings, since this will affect gene expression signatures and immune profiles. Thus, the contribution of *SPP1* overexpression to chemotherapy resistance will need to be validated in a larger independent prospective study.

7 Conclusion

This study describes the successful transcriptomic profiling of angiosarcoma for identification of a gene expression signature indicative of chemoresistance and survival outcomes in angiosarcoma. The overexpression of *SPP1*, which was the top candidate gene in angiosarcoma tumors, was demonstrated. Through investigating the clinical significance of *SPP1*, its viability as both a novel prognostic biomarker as well as predictive biomarker of chemotherapy response for angiosarcoma was also validated. This deepens our understanding of the potential *SPP1*-regulated pathway underlying chemoresistance and facilitates the development of novel targeted combination therapies for enhanced chemosensitivity in angiosarcoma patients via precision medicine.

Appendix

Table 1. Angiosarcoma patient characteristics stratified by high and low *SPP1* levels.

		SPP1 Levels			Total
		High	Low	*p*	
n (%)					72 (100%)
Sex					
	Male	21 (58.3%)	26 (72.2%)	0.219	47 (65.3%)
	Female	15 (41.7%)	10 (27.8%)		25 (34.7%)
Age at diagnosis (years)					
	< 65	21 (58.3%)	15 (41.7%)	0.16	36 (50.0%)
	≥ 65	15 (41.7%)	21 (58.3%)		36 (50.0%)
Ethnicity					
	Chinese	27 (75.0%)	31 (86.1%)	0.237	58 (80.6%)
	Other	9 (25.0%)	5 (13.9%)		14 (19.4%)
FNCLCC tumor grade					
	3	13 (36.1%)	6 (16.7%)	0.005	19 (26.4%)
	2	19 (52.8%)	14 (38.9%)		33 (45.8%)
	1	4 (11.1%)	16 (44.4%)		20 (27.8%)
Site of primary tumor					
	Head and neck	17 (47.2%)	25 (69.4%)	0.058	42 (58.3%)
	Others [a]	19 (52.8%)	11 (30.6%)		30 (41.7%)
Human herpesvirus-7					
	Positive	25 (69.4%)	22 (61.1%)	0.461	25 (34.7%)
	Negative	11 (30.6%)	14 (38.9%)		47 (65.3%)
UV signature					
	Present	11 (30.6%)	10 (27.8%)	0.797	51 (70.8%)
	Absent	25 (69.4%)	26 (72.2%)		21 (29.2%)
Epithelioid histology					
	Present	23 (63.9%)	13 (36.1%)	0.019	36 (50.0%)
	Absent	13 (36.1%)	23 (63.9%)		36 (50.0%)
Disease state at diagnosis [b]					
	Metastatic	14 (38.9%)	8 (22.9%)	0.147	22 (31.0%)
	Non-metastatic	22 (61.1%)	27 (77.1%)		49 (69.0%)

[a] Others: breast (n = 7), liver (n = 4), abdominal wall (n = 2), arm (n = 2), forearm (n = 2), peritoneum (n = 2), small bowel (n = 2), brachial plexus (n = 1), chest wall (n = 1), lower limb (n = 1), pleura (n = 1), prostate (n = 1), retroperitoneum (n = 1), spleen (n = 1), thigh muscle (n = 1), and vagina (n = 1)

[b] Unknown disease state: n = 1

Table 2. Top 10 candidate genes overexpressed (greatest log2fold change) in chemoresistant tumors

	log2 fold change	standard error (log2)	lower confidence limit (log2)	upper confidence limit (log2)	*p*-value	BY *p*-value	gene sets
SPP1	−3.49	0.632	−4.72	−2.25	3.99E−06	0.0112	Costimulatory Signaling, Matrix Remodeling and Metastasis, PI3K-Akt
CXCL13	−2.57	0.468	−3.49	−1.65	4.66E−06	0.0112	Cytokine and Chemokine Signaling, Lymphoid Compartment
CD48	−2.97	0.572	−4.09	−1.85	1.04E−05	0.0166	Costimulatory Signaling, Lymphoid Compartment
TAF3	2.19	0.443	1.32	3.06	2.36E−05	0.0247	
CLEC5A	−2.55	0.526	−3.58	−1.52	3.11E−05	0.0247	Immune Cell Adhesion and Migration, Myeloid Compartment
GBP4	−2.44	0.513	−3.44	−1.43	3.82E−05	0.0247	Interferon Signaling
GMIP	−2.25	0.475	−3.18	−1.32	4.15E−05	0.0247	
GBP2	−1.81	0.385	−2.57	−1.06	4.38E−05	0.0247	Interferon Signaling
GBP1	−2.23	0.476	−3.16	−1.3	4.65E−05	0.0247	Interferon Signaling
CD7	−2.53	0.547	−3.61	−1.46	5.76E−05	0.027	Lymphoid Compartment

Table 3. Top 10 candidate genes overexpressed (greatest log2fold change) in *SPP1*-high tumors

	log2 fold change	standard error (log2)	lower confidence limit (log2)	upper confidence limit (log2)	*p*-value	BY *p*-value	gene sets
SPP1	3.81	0.313	3.2	4.43	6.22E-19	2.87E-15	Costimulatory Signaling, Matrix Remodeling and Metastasis, PI3K-Akt
SLC11A1	2.97	0.418	2.15	3.79	7.96E-10	0.00000184	Myeloid Compartment
PLOD2	2.14	0.329	1.49	2.78	1.02E-08	0.0000134	Matrix Remodeling and Metastasis
CXCL3	2.88	0.446	2.01	3.76	1.16E-08	0.0000134	Cytokine and Chemokine Signaling, Myeloid Compartment
IL6	2.88	0.474	1.95	3.81	5.58E-08	0.0000511	Cytokine and Chemokine Signaling, Hypoxia, JAK-STAT Signaling, Metabolic Stress, PI3K-Akt
FSTL3	2.37	0.393	1.6	3.14	6.64E-08	0.0000511	Angiogenesis
CXCL2	1.98	0.357	1.28	2.68	4.82E-07	0.000318	Cytokine and Chemokine Signaling, Myeloid Compartment
TREM1	2.91	0.551	1.83	3.99	1.41E-06	0.000815	Myeloid Compartment

(*continued*)

Table 3. (*continued*)

	log2 fold change	standard error (log2)	lower confidence limit (log2)	upper confidence limit (log2)	*p*-value	BY *p*-value	gene sets
IL1B	2.53	0.482	1.58	3.47	1.62E-06	0.00083	Cytokine and Chemokine Signaling, MAPK, Myeloid Compartment
CCL3/L1	2.06	0.401	1.28	2.85	2.28E-06	0.00105	Cytokine and Chemokine Signaling

References

1. Young, R.J., Brown, N.J., Reed, M.W., Hughes, D., Woll, P.J.: Angiosarcoma. Lancet Oncol. **11**, 983–991 (2010)
2. Koh, Y.S., Chan, J.Y., Looi, W.S., et al.: Outcomes of head and neck angiosarcoma with different treatment modalities: a 20-year single institutional experience. Precis. Cancer Med. **5** (2022)
3. Cao, J., Wang, J., He, C., Fang, M.: Angiosarcoma: a review of diagnosis and current treatment. Am. J. Cancer Res. **9**(11), 2303 (2019)
4. Painter, C.A., et al.: The Angiosarcoma project: enabling genomic and clinical discoveries in a rare cancer through patient partnered research. Nat. Med. **26**, 181–187 (2020)
5. Chan, J.Y., et al.: Multiomic analysis and immunoprofiling reveal distinct subtypes of human angiosarcoma. J. Clin. Investig. **130**, 5833–5846 (2020)
6. Khor, G.M.S., et al.: High-throughput transcriptomics identifies chemoresistance-associated gene expression signatures in human angiosarcoma. Int. J. Mol. Sci. **25**(19), 10863 (2024)
7. Lin, E.Y.H., Xi, W., Aggarwal, N., Shinohara, M.L.: Osteopontin (Opn)/SPP1: from its biochemistry to biological functions in the innate immune system and the central nervous system (Cns). Int. Immunol. **35**, 171–180 (2023)
8. Tu, Y., Chen, C., Fan, G.: Association between the expression of secreted phosphoprotein—related genes and prognosis of human cancer. BMC Cancer **19**, 1230 (2019)
9. Zhang, Y., Du, W., Chen, Z., Xiang, C.: Upregulation of PD-L1 by SPP1 mediates macrophage polarization and facilitates immune escape in lung adenocarcinoma. Exp. Cell Res. **359**, 449–457 (2017)
10. Klement, J.D., et al.: An osteopontin/CD44 immune checkpoint controls CD8+ T cell activation and tumor immune evasion. J. Clin. Investig. **128**, 5549–5560 (2018)
11. Loh, J.W., et al.: Spatial transcriptomics reveal topological immune landscapes of Asian head and neck angiosarcoma. Commun. Biol. **6**, 461 (2023)
12. D'Angelo, S.P., et al.: Angiosarcoma: outcomes of systemic therapy for patients with metastatic disease. Oncology **89**, 205–214 (2015)
13. Venkataramani, V., et al.: CD31 expression determines redox status and chemoresistance in human angiosarcomas. Clin. Cancer Res. **24**, 460–473 (2018)

14. Zeng, P., Zhang, X., Xiang, T., Ling, Z., Lin, C., Diao, H.: Secreted phosphoprotein 1 as a potential prognostic and immunotherapy biomarker in multiple human cancers. Bioengineered **13**, 3221–3239 (2022)
15. Wei, R., Wong, J.P.C., Kwok, H.F.: Osteopontin—a promising biomarker for cancer therapy. J. Cancer **8**, 2173–2183 (2017)
16. Matsubara, E., et al.: SPP1 derived from macrophages is associated with a worse clinical course and chemo-resistance in lung adenocarcinoma. Cancers **14**, 4374 (2022)
17. Qian, J., et al.: Cancer-associated mesothelial cells promote ovarian cancer chemoresistance through paracrine osteopontin signaling. J. Clin. Investig. **131**, e146186 (2021)
18. Eun, J.W., et al.: Cancer-associated fibroblast-derived secreted phosphoprotein 1 contributes to resistance of hepatocellular carcinoma to sorafenib and lenvatinib. Cancer Commun. **43**, 455–479 (2023)
19. Nakamura, T., et al.: Osteopontin-integrin α(V)β(3) axis is crucial for 5-fluorouracil resistance in oral squamous cell carcinoma. FEBS Lett. **589**, 231–239 (2015)
20. Song, H., Yao, X., Zheng, Y., Zhou, L.: Helicobacter pylori infection induces POU5F1 upregulation and SPP1 activation to promote chemoresistance and T cell inactivation in gastric cancer cells. Biochem. Pharmacol. **225**, 116253 (2024)
21. Uchibori, T., Matsuda, K., Shimodaira, T., Sugano, M., Uehara, T., Honda, T.: IL-6 trans-signaling is another pathway to upregulate Osteopontin. Cytokine **90**, 88–95 (2017)
22. Martín-Márquez, B.T., Sandoval-García, F., Corona-Meraz, F.I., et al.: Osteopontin: a bone-derived protein involved in rheumatoid arthritis and osteoarthritis immunopathology. Biomolecules **13**(3), 502 (2023)
23. Masi, L., Ricci, L., Zulian, F., et al.: Serum osteopontin as a predictive marker of responsiveness to methotrexate in juvenile idiopathic arthritis. J. Rheumatol. **36**(10), 2308–2313 (2009)
24. Du, J., Hou, S., Zhong, C., et al.: Molecular basis of recognition of human osteopontin by 23c3, a potential therapeutic antibody for treatment of rheumatoid arthritis. J. Mol. Biol. **382**(4), 835–842 (2008)
25. Farrokhi, V., Chabot, J.R., Neubert, H., Yang, Z.: Assessing the feasibility of neutralizing osteopontin with various therapeutic antibody modalities. Sci. Rep. **8**, 7781 (2018)
26. Zhang, J., Yamada, O., Kida, S., et al.: Identification of brefelamide as a novel inhibitor of osteopontin that suppresses invasion of A549 lung cancer cells. Oncol. Rep. **36**(4), 2357–2364 (2016)
27. Poleboyina, P.K., Alagumuthu, M., Pasha, A., Ravinder, D., Pasumarthi, D., Pawar, S.C.: Entrectinib a plausible inhibitor for osteopontin (SPP1) in cervical cancer-integrated bioinformatic approach. Appl. Biochem. Biotechnol. **195**(12), 7766–7795 (2023)
28. Guo, D., Chen, Z., Fu, Y., et al.: Cilengitide inhibits osteoclast adhesion through blocking the αvβ3-mediated FAK/Src signaling pathway. Heliyon. **9**(7), e17841 (2023)

Harnessing miR-138 Modulating Agents to Combat Glioblastoma Multiforme (GBM)

Tanvi Nanni, Chrislyn Charles, and Natalie Su-Yin Ng(✉)

NUS High School of Mathematics and Science, Singapore 129957, Singapore
h2110105@nushigh.edu.sg

Abstract. Glioblastoma Multiforme (GBM) is a highly aggressive brain cancer with less than 5% 5-year survival rate. Due to the heterogeneous permeability of the brain-tumour barrier, current multimodal treatment methods are limited in effectiveness. MicroRNAs are short, non-coding RNAs implicated in cancer. MicroRNA-138 (miR-138) appears upregulated in treatment-resistant glioma stem cells. miR-138 modulating compounds consists of Antisense Oligonucleotide-138 (ASO-138) and cell-penetrating peptide (CPP) conjugated together. To investigate the efficacy of miR-138 modulating compounds at reducing miR-138 levels, GBM cells were cultured then treated with ASO-138/CPP at different concentrations and durations. Cell viability and miR-138 expression were then measured via MTS assay and RT-qPCR respectively. At high concentrations, ASO-138/CPP exhibits low cytotoxicity, demonstrating its biocompatibility with GBM cells. A shorter treatment duration of 24 h is also more effective at downregulating expression of miR-138. The 4 CPPs tested had different properties and exhibited varying efficacies at downregulating expression of miR-138, ranging from no to extremely high statistical significance ($p < 0.0001$). Thus, to maximise therapeutic outcomes, the right CPP must be chosen and patients should be treated every 24 h for greater reduction in miR-138 levels. For further research, in vivo mouse studies can be utilised to determine systemic toxicity and potential off-target effects. Delivery of ASO-138/CPP through the brain-tumour barrier must also be optimised to translate our in vitro findings into a feasible and effective treatment method. By addressing critical challenges in GBM treatment, this project hopes to demonstrate the potential of miRNA modulating agents in providing effective, novel alternatives for hard-to-treat cancers.

Keywords: Glioblastoma Multiforme · microRNA · cell-penetrating peptide

1 Introduction

Glioblastoma Multiforme (GBM) is a highly aggressive and malignant brain cancer that exhibits a heterogeneous cellular composition, making it challenging to treat. Originating from astrocytes, Glioblastoma accounts for 50.1% [1] of all malignant brain tumours globally. In Singapore, about 100 cases of glioblastoma are diagnosed every year [2].

Classified as a grade IV glioma, Glioblastoma's progression is inevitable, with 15–18 months of median overall survival and a 5-year survival rate of less than 5% [3].

H. Guo et al. (Eds.): IRC-SET 2025, PTT, pp. 21–29, 2026.
https://doi.org/10.1007/978-981-95-9630-0_2

Currently, treatment involves surgical resection followed by chemoradiotherapy [4, 5]. Due to its proximity to crucial brain regions, surgical resection aims to remove as much of it as safely possible, often with the implantation of carmustine wafers [6] to release chemotherapy locally. To target residual cancer cells, radiation therapy and temozolomide, an oral chemotherapy drug, are used concurrently as maintenance therapy to slow tumour progression [7].

However, the effectiveness of these methods are limited due to the presence of the blood-brain as well as the brain-tumour barrier [8]. The heterogeneous permeability of the brain-tumour barrier creates an immunosuppressive tumour microenvironment that reduces the efficacy of immune-based treatments [9]. Despite the current multimodal treatment methods utilised [10], frequent recurrence of GBM tumours has been associated with chemo and radio-resistant glioma stem cells (GSCs) [11], making it challenging to treat.

MicroRNAs (miRNAs) are short, non-coding RNAs that have a post-transcriptional regulatory function in gene expression and can be linked to the pathogenesis of various human diseases. In cancers, they can act as tumour suppressor genes or oncogenes [12, 13]. Thus, variation in miRNA expression may potentially be a key factor in cancer initiation, progression and metastasis [14, 15]. MicroRNA-138 (miR-138) appears to generally be upregulated in GSCs and inhibition with antimiR-138 decreases GSC proliferation. miR-138 also indirectly modulates the expression of several genes involved in tumorigenesis to increase cell growth, highlighting its crucial role in maintenance and survival of GSCs [16].

In this project, the miR-138 modulating compound consists of 2 components - Antisense oligonucleotide-138 (ASO-138) and cell-penetrating peptide (CPP). ASO-138, a single-stranded RNA fragment, anneals to miR-138 via complementary base pairing to form a double-stranded RNA complex. This prevents miR-138 from functioning or marks it for degradation by RNAses [17]. This aims to bring the level of expression of miR-138 down to normal. However, as both ASO-138 and lipid bilayer of the plasma membrane bear a negative charge, it is difficult for ASO-138 to enter the cell via diffusion. While various chemical modifications involving ASO-138 are possible, a CPP can be conjugated with it to allow the whole compound to enter the cell via endocytosis or direct penetration. Through careful peptide design, these versatile peptides can be customised to deliver specific molecules that range from small molecules drugs to large plasmid DNA to specific target cells and tissues [18].

2 Hypothesis

Thus, this project aims to investigate the efficacy of miR-138 modulating compounds, consisting of ASO-138 and a CPP conjugated together (ASO-138/CPP), at suppressing tumour growth via their impact on the level of expression of miR-138 of GBM cells.

We expect that the expression of miR-138 will be lower in cells treated with ASO-138/CPP than in the scrambled control (SCR-138/CPP). We also predict that the longer the treatment duration, the lower the miR-138 transcript abundance. Additionally, we experimented with 4 different CPPs synthesized in our lab to investigate their differences in efficacy.

3 Methodology

3.1 Cell Culture

U87-GBM cells were seeded on a petri dish at an initial concentration of $2x10^4$ cells per mL and incubated at 37 °C and 5% CO_2. After 72 h, the cells were passaged into 2 sub-cultures and cultured for another 72 h at the same conditions in the incubator.

3.2 Treatment

Some cells were passaged into 96 wells on a 96-well plate. Then, cells were treated with either ASO-138, SCR-138 (control) or CPP at serially diluted concentrations (factor of 0.5x) starting from 1000nM for 48 h. A second 96-well plate was used for replication purposes. Another portion of cells were plated on a 12-well plate and treated with 1μM CPP conjugated with either ASO-138 or SCR-138 for either 24 or 48 h, each using 3 wells for replication purposes.

3.3 MTS Assay

After treatment, 20μL of CellTiter 96® AQueous One Solution was added to each well. Following this, the cells were incubated in the dark at 37 °C and 5% CO_2 for 1 h. After removal from the incubator, absorbance at 490 nm was recorded using a 96-well plate reader. The absorbance of CellTiter 96® AQueous One Solution sans any cells was also measured by the same machine to determine the baseline absorbance at 490nm.

3.4 RNA Isolation

After the treatment duration, cells on the 12-well plate were lysed. To extract the RNA, the sample solutions (ASO-138 (24 h), ASO-138 (48 h), SCR-138 (24 h), SCR-138 (48 h)) were transferred to separate centrifuge tubes. 400μL of ethanol was added to each centrifuge tube then the samples were centrifuged for 30 s with the supernatant discarded. The RNA precipitated by ethanol was then washed 3 times using a washing solution via a similar centrifugation process. 15μL of DNase and 100μL of enzyme incubation buffer was added to each centrifuge tube. Thereafter, the solutions were incubated for 45 min at room temperature before being washed 2 more times via a similar centrifugation process. 10μL of DI water was added to each centrifuge tube then a nanophotometer was used to determine the quantity and purity of RNA present in each centrifuge tube to ensure there was sufficiently pure RNA in large enough quantities for further experimentation. The eluted RNA was stored at -20 °C until the RT-qPCR step commenced.

3.5 RT-qPCR

Each of the samples were diluted with DI water to form 2 solutions at concentrations of 100ng of RNA/μL and 10g of RNA/μL. 5μL of each of the resultant solutions were added to separate PCR tubes. Then, 5.66μL nucleus-free water, 1.5μL 10x RT buffer, 0.15μL 100μM dNTP, 0.19μL RNase inhibitor, 1.5μL 20x miR-138 RT primer and 1μL multiscribe reverse transcriptase were added to each PCR tube. The same procedure was performed using a second set of PCR tubes for the same set of sample solutions but replaced the miR-138 RT primer with the U6 gene primer. Both sets of PCR tubes were centrifuged then placed in a thermal cycler (BioRad) for 1.5 h to produce cDNA. 1.5μL of each of the resultant sample solutions were transferred into a 96-well plate to determine the miR-138 transcript abundance using the Quantstudio machine.

4 Results

4.1 miR-138 Modulating Compound Exhibits Low Cytotoxicity Towards GBM Cells at High Concentrations

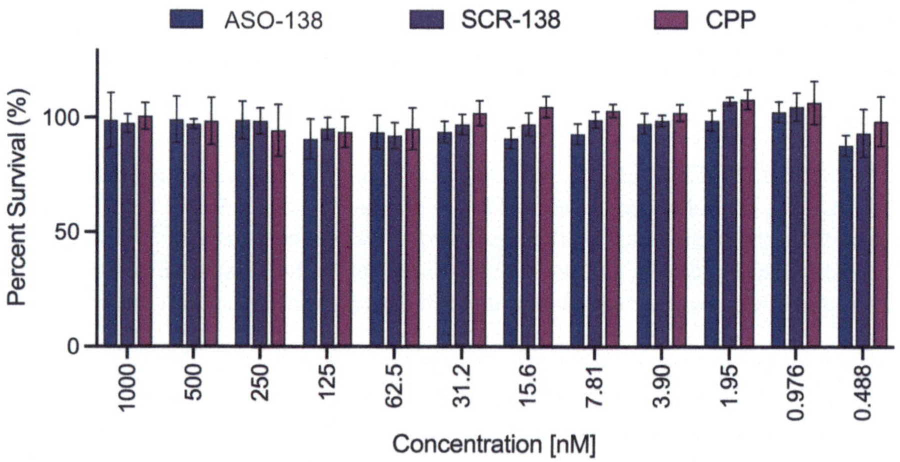

Fig. 1. Cell viability (%) of U87-MG cells treated with ASO-138, SCR-138 or CPP for 48 h at varying concentrations

To determine the cytotoxicity of the 2 components of the miR-138 modulating compound, an MTS assay was conducted after 48 h of incubation. The results show that cell viability remained consistently above 85% for all tested concentrations from 1000nM to 0.488nM when treated with ASO-138, SCR-138 and CPP. There was no significant differences in cell viability when treated with ASO-138 or with SCR-138 and CPP, indicating that ASO-138 has minimal cytotoxic effects at the tested concentrations. Hence, the data supports the biocompatibility of ASO-138 with the U87-MG. The low cytotoxicity supports ASO-138 as a potential therapeutic agent in glioblastoma treatment without adversely affecting cell viability (Fig. 1).

4.2 24 h of Treatment is More Effective at Reducing miR-138 Levels Than 48 h of Treatment

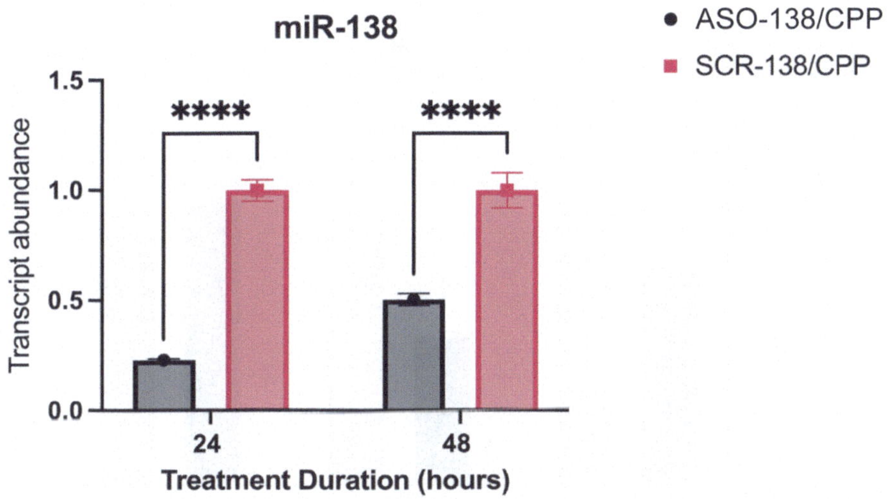

Fig. 2. miR-138 Transcript Abundance in cells treated with ASO-138/CPP for 24 h and 48 h.

Based on the analysed RT-qPCR results, we first note that treatment with ASO-138/CPP is indeed effective at downregulating the expression of miR-138 in GBM cells for both 24- and 48-h duration due to the high level of statistical significance (p-value < 0.0001) for both durations (Fig. 2).

However, we also observed that the relative miR-138 levels were higher in the cells treated for 48 h than the cells treated for the shorter duration of 24 h. This runs counter to our initial hypothesis that the longer the treatment duration with ASO-138/CPP, the lower the miR-138 transcript abundance in the cell. Hence, we conclude that a shorter treatment duration of 24 h is more effective at downregulating expression of miR-138. Thus, patients should be treated every 24 h with ASO-138/CPP for greater effect on reducing miR-138 levels.

4.3 Choice of CPP Determines the Effectiveness of the miR-138 Modulating Compound at Downregulating miR-138 Expression

From a chemistry point of view, CPPs are usually 5–35 amino acids long and are either cationic or amphipathic in character, with few reported examples of anionic CPPs in literature. While amphipathic CPPs can interact with both the hydrophilic and hydrophobic parts of the plasma membrane, many CPPs are cationic due to the presence of many cationic lysine and arginine residues in their primary structure. While multiple mechanisms of internalisation have been proposed, the exact mechanisms utilised are still not fully understood. Besides endocytosis, various models such as the "carpet" or "membrane-thinning" effect have been proposed to describe the direct translocation

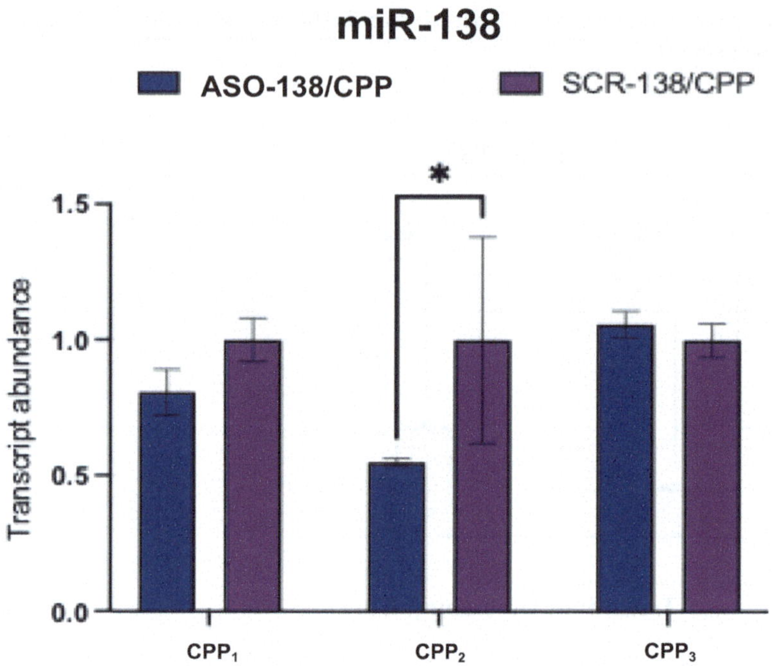

Fig. 3. miR-138 Transcript Abundance in cells treated with ASO-138 conjugated with CPP_1, CPP_2, or CPP_3 for 24 h.

of positively charged CPPs through the plasma membrane. These mechanisms involve interaction between the positively charged CPP and the negatively charged components of the plasma membrane like heparan sulphate or the phospholipid heads to disrupt or rearrange the plasma membrane to allow the CPP to pass through it. Pore formation via the barrel-stave or the toroidal model are also proposed mechanisms for how amphipathic CPPs enter cells [19, 20] (Fig. 3).

CPPs can come in a wide variety of structures to transport different cargos into the cell. Thus, the choice of CPP is also important at determining the effectiveness of its cargo. Hence, in the 3rd part of our research, we conjugated 3 other CPPs: CPP_1, CPP_2 and CPP_3, with ASO-138 and SCR-138 for control at the same concentration as our earlier experiments. These 6 combinations (ASO-138/CPP_1, ASO-138/CPP_2, ASO-138/CPP_3, SCR-138/CPP_1, SCR-138/CPP_2, SCR-138/CPP_3) were then added to separate wells with GBM cells for 24 h before RT-qPCR was conducted on the resulting samples to determine the miR-138 transcript abundance.

Compared to the CPP used earlier, these 3 CPPs were all less effective at reducing miR-138 levels. The cells treated with ASO-138/CPP_3 had a higher mean transcript abundance compared to its control. While CPP_1 and CPP_2 samples both showed down-regulation of miR-138 compared to their respective controls, only CPP_2 showed statistical significance (p-value < 0.05). Even so, the relative decrease in transcript abundance is smaller as compared to the 1st CPP used for the same duration. Overall, this demonstrates the importance of choosing the right CPP to transport ASO-138 into the cancer cell to maximise efficacy of the miR-138 modulating compound.

5 Discussion

Through this study, we explored the application of ASO-138 conjugated with CPPs as a potential therapeutic approach for glioblastoma (GBM). Our experiments demonstrated the ability of these conjugates to modulate miR-138 levels in tumour cells, highlighting their potential role in addressing challenges associated with GBM, such as treatment resistance and limited therapeutic options. The findings provide a strong foundation for further research into miRNA-modulating agents as a novel strategy for glioblastoma therapy. We experimented using CPPs conjugated with ASOs to reduce miR-138 levels tumour cells, varying the time after introduction to the cells as well as the type of CPP as well.

With the relatively recent discovery of miRNAs, the finding that downregulating miR-138 in this brain tumour could potentially be a part of devising a greater variety of treatment options for patients suffering from glioblastoma.

The data from our experiment shows that 24-h treatment is more effective than 48-h treatment at downregulating miR-138 expression. This finding indicates that over long durations, there are cellular mechanisms such as endosomal entrapment, exocytosis or enzyme degradation that leads to the availability of the modulating compound reducing. Hence, this underscores the importance of optimising treatment schedule for maximum efficacy. Our results show that daily administration of ASO-138/CPP is optimal and should be validated through further studies.

Another important consideration is the method of administration for ASO-138/CPP. CPP is likely to be destroyed by digestive enzymes such as pepsin, it would not be ideal to orally administer it. The most feasible option based on past research would be to administer it via injections. Since it enters the bloodstream directly, it would not encounter digestive enzymes and degrade, allowing CPP to perform its function. It is crucial that future studies optimise injection formulation for optimal cell penetration during clinical translations.

Another challenge would be the penetration of ASO-138/CPP through the heterogenous blood-tumour barrier which typically only allows limited penetration of various drugs from the bloodstream. Either the CPP attached must be able to penetrate this barrier or something else must help it do so. One possible method is MRI-guided focused ultrasound (MRI-g-FUS), which can temporarily increase the permeance of the barrier to various drugs at precise locations. Thus, future studies can consider the usage of MRI-g-FUS to aid ASO-138/CPP to cross this barrier while not permanently disrupting them at indiscriminate locations [21].

Future work should also investigate the effects of downregulating miR-138 on other types of cells and tissues, to ensure that it will not yield harmful side effects to other key functions of the body. In GSCs and chemotherapy resistant GBM cells, miR-138 is upregulated due to its oncogenic function [16]. However, in many other human cancers, miR-138 acts as a tumour suppressor and is therefore downregulated in cancer cells [22].

While glioblastoma only metastasizes in about 0.4–2% of cases [23], it is still crucial to determine if a patient requires miR-138 inhibitors or miR-138 upregulators in order to treat the condition effectively. Hence, when clinical trials are involved, patients chosen should be screened for GBM metastasis or any other types of cancer. This also demonstrates the need for targeted delivery of ASO-138/CPP to the GBM tumour specifically, to inhibit cancer progression and improve patient outcome.

However, this study only investigated the cytotoxicity factor of the cell and whether the introduction of the CPPs affected cell viability. In further investigation, we could explore other long-term effects that this treatment could pose to the patient's body, through in vivo studies using mouse models, clinical trials focusing on systemic toxicity, and comprehensive monitoring of immune responses or potential off-target effects.

6 Conclusion

This study opens a new avenue of exploration into the immense potential of ASO-138/CPP as miR-138 modulating agents, with implications spanning translational medicine, therapeutic innovation, and precision oncology. By addressing critical challenges in glioblastoma treatment, such as targeted delivery and molecular specificity, this approach has the potential to revolutionize the way aggressive brain tumours are treated. We hope that this project will serve as a foundational building block, inspiring future research and fostering the development of novel, effective treatment options for glioblastoma and other hard-to-treat cancers.

References

1. About Glioblastoma - National Brain Tumor Society. National Brain Tumor Society (2024). https://braintumor.org/events/glioblastoma-awareness-day/about-glioblastoma/
2. Begum, S. New therapy shows promise in treating deadly brain cancer, advanced abdominal cancer. The Straits Times (2023). https://www.straitstimes.com/singapore/new-therapy-shows-promise-in-treating-deadly-brain-cancer-advanced-abdominal-cancer
3. Oronsky, B., Reid, T.R., Oronsky, A., Sandhu, N., Knox, S.J.: A review of newly diagnosed glioblastoma. Front. Oncol. (2021). https://doi.org/10.3389/fonc.2020.574012
4. Holland, E.C.: Glioblastoma multiforme: the terminator. Proc. Natl. Acad. Sci. **97**(12), 6242–6244 (2000). https://doi.org/10.1073/pnas.97.12.6242
5. Weingart, J.: Glioblastoma Multiforme (GBM): Advancing Treatment for a Dangerous Brain Tumor. John Hopkins Medicine. https://www.hopkinsmedicine.org/health/conditions-and-diseases/glioblastoma-multiforme-gbm-advancing-treatment-for-a-dangerous-brain-tumor. Accessed 11 Mar 2025
6. Hart, M.G., Grant, R., Garside, R., Rogers, G., Somerville, M., Stein, K.: Chemotherapy wafers for high grade glioma. Cochrane Database Syst Rev. **2011**(3), CD007294 (2011). https://doi.org/10.1002/14651858.CD007294.pub2
7. Stupp, R., et al.: Radiotherapy plus concomitant and adjuvant temozolomide for glioblastoma. New Engl. J. Med. **352**(10), 987–996 (2005). https://www.nejm.org/doi/pdf/10.1056/NEJMoa043330?articleTools=true
8. Rayati, M., Mansouri, V., Ahmadbeigi, N.: Gene therapy in glioblastoma multiforme: can it be a role changer? Heliyon (2024). https://doi.org/10.1016/j.heliyon.2024.e27087

9. Rocha, P.S.L., et al.: Immunotherapy in glioblastoma treatment: Current state and future prospects. World J. Clin. Oncol. **14**(4), 138–159 (2023). https://doi.org/10.5306/wjco.v14.i4.138
10. Garden, A.S., et al.: Outcome and patterns of failure following limited-volume irradiation for malignant astrocytomas. Radiother. Oncol. **20**(2), 99–110 (1991)
11. Bao, S., et al.: Glioma stem cells promote radioresistance by preferential activation of the DNA damage response. Nature **444**, 756–760 (2006)
12. Esquela-Kerscher, A., Slack, F.J.: Oncomirs—microRNAs with a role in cancer. Nat. Rev. Cancer **6**(4), 259–269 (2006)
13. Slack, F.J., Weidhaas, J.B.: MicroRNAs as a potential magic bullet in cancer. Future Oncol. **2**(1), 73–82 (2006)
14. Calin, G.A., Croce, C.M.: MicroRNA signatures in human cancers. Nat. Rev. Cancer **6**(11), 857–866 (2006)
15. Ma, L., Teruya-Feldstein, J., Weinberg, R.A.: Tumour invasion and metastasis initiated by microRNA-10b in breast cancer. Nature **449**, 682–688 (2007)
16. Chan, X.H.D., et al.: Targeting glioma stem cells by functional inhibition of a Prosurvival OncomiR-138 in malignant gliomas. Cell Rep. **2**(3), 591–602 (2012). https://doi.org/10.1016/j.celrep.2012.07.012
17. Lima, J.F., Cerqueira, L., Figueiredo, C., Oliveira, C., Azevedo, N.F. Anti-miRNA oligonucleotides: a comprehensive guide for design. RNA Biol. **15**(3), 338–352 (2018). https://doi.org/10.1080/15476286.2018.1445959
18. Copolovici, D.M, Langel, K., Eriste, E., Langel, Ü.: Cell-penetrating peptides: design, synthesis, and applications. ACS Nano (2014). https://doi.org/10.1021/nn4057269
19. Gori, G., Lodigiani, S.G., Colombarolli, G., Bergamaschi, A.V.: Cell penetrating peptides: classification, mechanisms, methods of study, and applications. ChemMedChem (2023). https://doi.org/10.1002/cmdc.202300236
20. Shin, M.C., et al.: Cell-penetrating peptides: achievements and challenges in application for cancer treatment. J. Biomed. Mater. Res. A. **102**(2), 575–587 (2015). https://doi.org/10.1002/jbm.a.34859
21. Bunevicius, A., McDannold, N.J., Golby, A.J.: Focused ultrasound strategies for brain tumor therapy. Oper. Neurosurg. (Hagerstown) **19**(1), 9–18 (2020). https://doi.org/10.1093/ons/opz374. PMID: 31853548; PMCID: PMC7293897
22. Yeh, M., Oh, C.S., Yoo, J.Y., Kaur, B., Lee, T.J.: Pivotal role of microRNA-138 in human cancers. Am. J. Cancer Res. **9**(6), 1118–1126 (2019). PMID: 31285946; PMCID: PMC6610051
23. Yuen, C.A., Pekmezci, M., Bao, S., Kong, X.T.: Metastatic glioblastoma to the lungs: a case report and literature review. CNS Oncol. **13**(1), 2351789 (2024). https://doi.org/10.1080/20450907.2024.2351789

Explainable Diagnosis of Migraine via Deep Learning Through the Use of EEG Data

Wee Zen Neo[1], Yew Kee Peh[1], Kai Jun Elgin Koh[1](✉), Wan Yu Grace Ang[2], and Hai Leong Chieu[2]

[1] NUS High School of Math and Science, 20 Clementi Avenue 1, Singapore 129957, Singapore
dso.neuro.arp@gmail.com

[2] DSO National Laboratories, 12 Science Park Dr, Singapore 118225, Singapore

Abstract. Migraines are a highly prevalent and costly disorder which is challenging to diagnose and typically requires a specialist reviewing a patient's history. As a result, migraines remain underdiagnosed and hence undertreated. Electroencephalography (EEG) data has previously been used to diagnose various neurological disorders such as epilepsy, motivating the use of this data to develop a model for the automated diagnosis of migraines. In this paper, we propose a straightforward approach to automated migraine diagnosis via the fine-tuning of the ResNet-50 architecture on spectrograms of EEG data. We demonstrate that our proposed model has comparable performance to recent methods of automated migraine diagnosis at 96.3% accuracy. Furthermore, we show that we can apply methods in model explainability to highlight aspects of EEG data which our model places more importance on, making it more suitable for clinical use where the explainability of model predictions play an important factor in clinical adoption.

Keywords: Explainable artificial intelligence · xAI · Deep Learning · Migraine · Migraine diagnosis · Electroencephalogram · Short-time Fourier transform · Transfer Learning

1 Introduction

Migraine is a complex neurological disorderthat affects more than a billion individuals every year. Globally, migraine ranked as the second leading cause of disability-adjusted life-years (DALYs) lost to neurological disorders in 2016, contributing 16.3% [95% uncertainty interval (UI): 11.7–20.8] of the total attributable DALYs [1]. The International Classification of Headache Disorders (ICHD-3) defines migraine as a recurrent primary headache disorder that lasts for 4–72 h. The headache is usually single, pulsatile, moderate or severe, aggravated by physical activity and accompanied by nausea, photophobia and phonophobia [2].

The diagnosis of migraine is usually carried out by a specialist through analysing the patient's history, physical examination and fulfilment of the criteria given by the ICHD-3 [2, 3]. However, migraine remains underdiagnosed and undertreated [4], and

H. Guo et al. (Eds.): IRC-SET 2025, PTT, pp. 30–38, 2026.
https://doi.org/10.1007/978-981-95-9630-0_3

may sometimes be misdiagnosed, even by healthcare professionals [5]. As such, methods for more accurate diagnosis of migraine are necessary.

Electroencephalography (EEG) is a non-invasive method for recording the brain's spontaneous and rhythmic electrical activity, from the placement of electrodes on the scalp. EEG has been used to diagnose various neurological disorders such as epilepsy [6] and sleep disorders [7]. Several studies have reported frequency-specific differences in power spectra between interictal migraine sufferers and healthy patients. O'Hare et al. [8] analysed resting-state alpha band oscillations in the visual areas of the brain. They found an increase in lower alpha band (8 – 10 Hz) power in migraineurs compared to control. Bjørk et al.[9] reported that migraineurs had globally increased theta activity during rest.

Cao et al. [10] found that in interictal migraineurs, EEG power was lower in fronto-central and parietal regions in all frequency bands excluding gamma. Chamanzar et al. [11] analysed EEG signals of interictal migraine sufferers and healthy controls exposed to visual and auditory stimuli and at rest. They found abnormal brain connectivity in migraineurs with or without stimulus. Many more reported correlates, sometimes contradictory, exist.

Because these feature-based correlates are contradictory and disputible in the literature, some researchers have turned to machine and deep learning diagnosis methods which can also reveal EEG correlates that are indicative of migraines. Aslan et al. [12] used Tunable Q-Factor wavelet transform and ensemble learning techniques to distinguish between the EEG signals of migraine patients in interictal state and healthy controls, achieving a 89.6% accuracy. Subasi et al. [13] used random forest and discrete wavelet transform, alongside photic stimulation of migraineurs during the experiment, achieving an accuracy of 85.95%. They also found that EEG signals acquired during photic stimulation increased classification accuracy. Lastly, Göker used a bidirectional long-short term memory deep learning model, achieving a performance of 95.99% [14].

In this paper, we introduce a deep learning approach to diagnosing migraines. Our proposed model applies a short-time Fourier transform on EEG data, before passing the data into a fine-tuned version of ResNet-50. Our model performs on par, if not better than existing models, while requiring fewer channels than most models. In addition, our model provides visual explainability with HiResCAM, allowing us to identify the frequency bands and electrodes it prioritizes. This not only provides quantifiable insights into the outputs of the deep learning model but also allows for targeted insights into migraine correlates which could be used in future research.

2 Methodology

(See Fig. 1).

2.1 Task Definition

We structure our task as a binary classification problem, where the aim of our model is to correctly classify given multi-channel EEG samples of duration L as either from migraineurs or healthy control samples. In this paper, L is of duration 4 s.

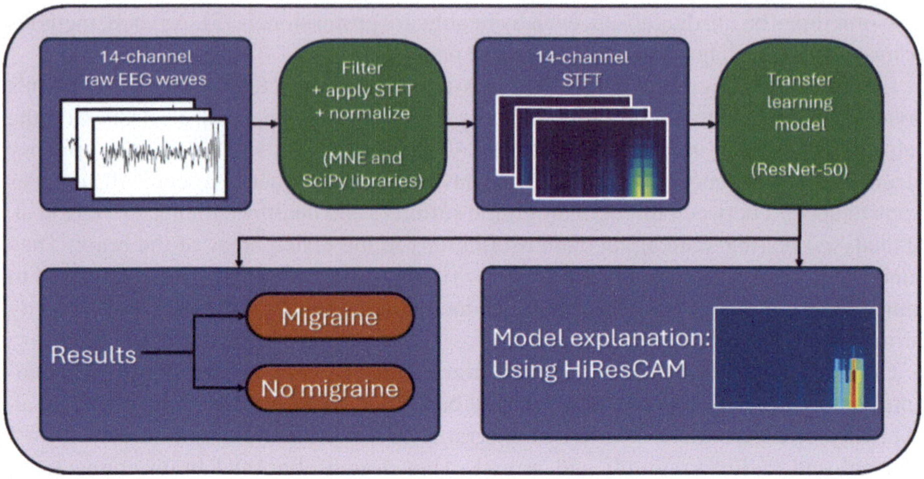

Fig. 1. Overview of our approach to model development and explainability. Given 14 channel raw EEG waves segmented into 4 s intervals, we apply a short-time Fourier transform (STFT) and normalize afterwards to the interval of [0, 1] for better model performance. We then train ResNet-50 on this data to produce our final model with the capability to diagnose, based on given 14-channel EEG waves, whether the source of the EEG waves was a migraineur or non-migraineur. Furthermore, we apply HiResCAM, a class activation mapping method, to determine which parts of the EEG the model places more importance on.

2.2 Dataset

For our dataset, we used a publicly available dataset of high-density (128 electrodes) EEG recordings, of 17 individuals with migraine in interictal periods, and 18 control subjects [15]. These recordings were acquired during resting state along with auditory tones and visual checkboard stimuli. All recordings were used for training, evaluation, and testing, with a train-eval-test split of 64:16:20. Out of all 128 channels available in the dataset, we selected channels Fpz, Oz, F7, F8, C3, C4, P3, P4, O1, O2, C5, C6, T7, and T8 channels, based on Ullah et al., 2024 [16] (Fig. 2).

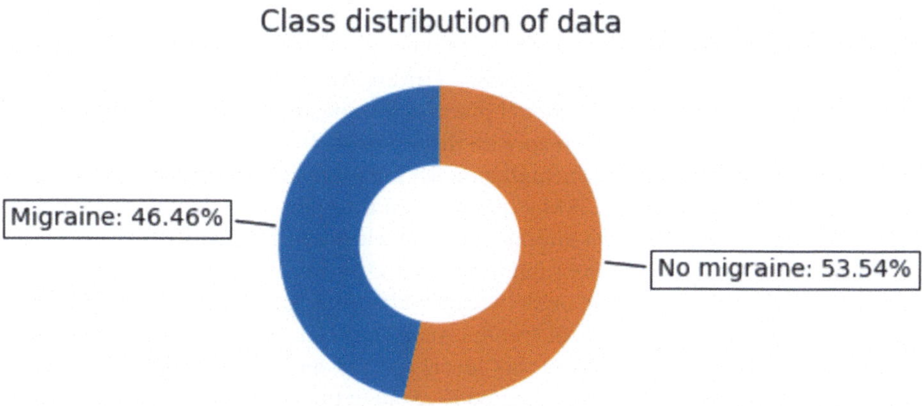

Fig. 2. Donut chart showing the class distribution of migraine/no migraine episodes in our data. The episodes are of roughly even distribution.

2.3 Data Preprocessing

For our data preprocessing, we made use of an online GitHub repository as a template for our code [20]. We filtered these signals using a Butterworth filter to remove frequencies below 0.5 Hz and above 32 Hz to isolate beta, alpha, theta and delta frequency bands, and a notch filter at 60 Hz to remove oscillations from mains electricity.

After segmenting EEG recordings into 4-s episodes, we applied the short-time Fourier transform (STFT) onto each episode with a sampling frequency of 512 Hz to match the sampling frequency of the dataset, with the window being shifted by 20 samples in each step. This is to generate spectrogram images which are suitable as image data for the model we are training. Utilising STFT as opposed to other approaches to data preprocessing such as continuous wavelet transform also allows for clearer explanations as each bin in our spectrogram directly corresponds to a specific frequency band at a clear time interval. For the STFT windowing function, we utilised a symmetrical Gaussian window with a size of 128 samples, and a standard deviation of 8. The magnitude from the STFT was normalised by dividing every value by the largest magnitude in the STFT for each episode. The result of our data preprocessing is 14 normalised spectrograms of the EEG data, 1 for each channel.

2.4 Model Description

After preprocessing of our data, we fine-tuned ResNet-50 to obtain our model; the first layer was replaced with a 2D convolutional neural network (CNN) layer (kernel size 7 $\times$ 7, stride 2 $\times$ 2, padding 3 $\times$ 3, no bias), with the required input of 14 normalised spectrograms of filtered EEG data. The final layer was replaced with fully connected network, with output as a singular probability of whether the source of the EEG data is a migraineur or non-migraineur. Only the first and last layer are trained, with the rest of the ResNet-50 layers frozen. For training, we used the Adam optimizer with learning rate of 0.005, and as this is a binary classification task, we used binary cross entropy loss.

2.5 Model Explanation

To obtain our model explanation, we applied HiResCAM onto our fine-tuned ResNet-50, to explain the importances the model places upon specific frequencies. HiResCAM element-wise multiplies a feature map with the gradients in a selected layer of a CNN. Higher values in the output indicate higher model importance for a particular area of the feature map. Our use of HiResCAM expands upon the approach taken upon in Aslan, 2023 [18], where Grad-CAM, a similar model explanation technique, was also used for explainability, but provides less faithful visual explanations than HiResCAM [17]. Grad-CAM visualisations of our model are included in Appendix B. HiResCAM was applied onto the first CNN layer of the fine-tuned ResNet-50 model. An explanation for why HiResCAM was applied to this layer, and not any of the ResNet-50 blocks, is include in Appendix C. We then proceed to average the HiResCAM values for each channel over all 4-s episodes.

2.6 Statistical Analysis

We then perform a statistical analysis for frequency bands with the highest average HiResCAM values. After squaring the magnitude of the STFT to obtain the power spectral density, we perform Welch's t-test to see whether the difference in band power between migraineurs and non-migraineurs is statistically significant.

3 Results

(See Table 1).

Table 1. Comparison of our method to previously reported methods for migraine diagnosis. Our method is comparable to current methods while also having the crucial benefit of explainability for practical usage. All methods use the same dataset by Chamanzar et al., 2020, except for Subasi et al., 2019.

	Classification method	Number of channels used	Accuracy (%)
Aslan, 2021 [12]	Tunable Q-Factor wavelet transform, ensemble learning	128	89.6%
Subasi et al., 2019 [13]	Discrete wavelet transform, random forest	18	86.0%
Göker, 2023 [14]	Welch's method, Bidirectional long-short term memory	128	96.0%
Ullah et al., 2024 [16]	Logistic regression	14	99.7%

(continued)

Table 1. (*continued*)

	Classification method	Number of channels used	Accuracy (%)
Aslan, 2023 [18]	Continuous wavelet transform, CNN	128	100%
Orhanbulucu et al., 2023 [19]	Continuous wavelet transform, transfer learning via AlexNet	64	99.7%
Proposed method	Short-time Fourier transform, fine-tuning via Resnet-50	14	96.3%

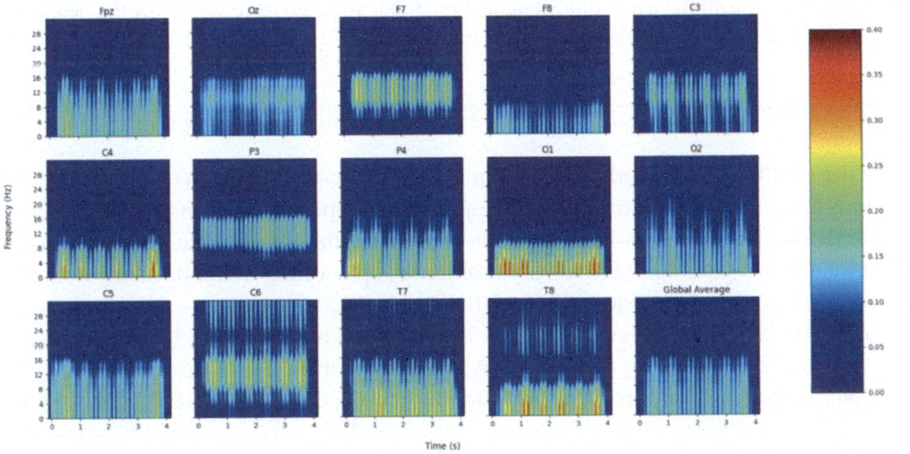

Fig. 3. HiResCAM values for each channel, averaged over all 4-s episodes. Bands with higher values from HiResCAM in the heatmap represent regions (corresponding to frequency bands in time) of higher saliency used in model predictions. In Appendix A, we show that such visualisations are accurate in place of visualisations for only migraines or only non-migraines, as model importances align for both migraines and non-migraines. The bottom right box refers to the average HiResCAM values of all channels, across all episodes. Grad-CAM visualisations of our model are included in Appendix B.

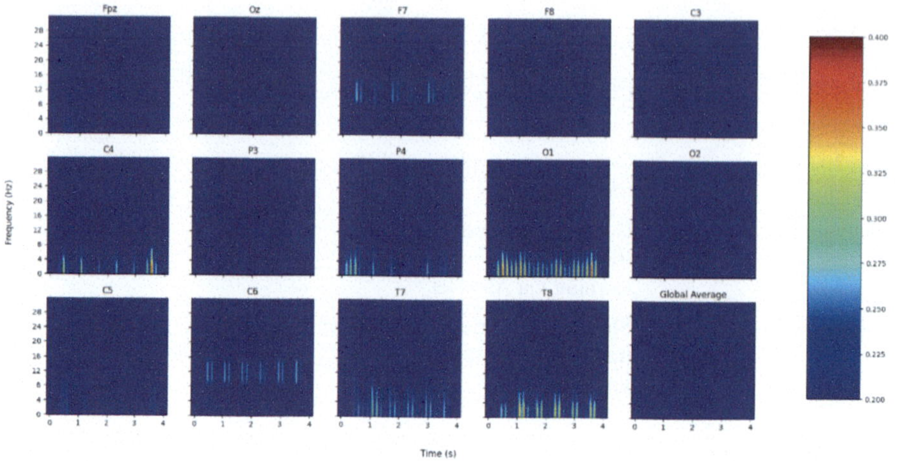

Fig. 4. Produced with the same procedure as Fig. 3, but with the lower cutoff of the heatmap at 0.2 instead of 0, for better visibility of more significant model importances.

4 Discussion

From the results, we conclude that our fine-tuning approach performs comparably with previous research, while outperforming approaches proposed in Aslan, 2021[12], Subasi et al., 2019[13], and Göker, 2023[14]. We also show that we are able to achieve model performance comparable to previously reported deep learning methods, using data from just a subset of electrodes identified in Ullah et al., 2024.

Importantly, our analyses of features learnt by the model reveal EEG correlates of migraine activity. Based on the HiResCAM values, our model places emphasis on the electrode-frequency combinations (in order of their appearance in Fig. 4), C4: 0–8, O1: 0–8, T8: 0–8. Further statistical analysis was carried out on these electrode-frequency combinations, to confirm that significant differences exist between migraineurs and control.

For C4, band power from 4–8 Hz and 0–4 Hz were significantly less for migraineurs compared to control ($p < 0.01$ and $p < 0.05$ respectively). For O1, band power from 0–8 Hz was significantly less for migraineurs compared to control ($p < 0.01$). For T8, band power from 0–8 Hz was significantly less for migraineurs compared to control ($p < 0.01$).

The correlates found from C4 have some consistency with prior research by Cao et al., 2016, which found that in interictal migraineurs, power was lower in fronto-central and parietal regions in all frequency bands except gamma [10].

Other electrode-frequency combinations that the model placed emphasis on were, F7: 8–16, P4: 0–8, C6: 8–16, T7: 0–8.

With high accuracy and low electrode use count, our model can provide a cheaper and less time-consuming method of migraine diagnoses, with only slight decreases to accuracy. Moreover, the explainability of our model allows for clinicians to view our model with a higher level of trust, as they are able to understand the reasoning behind the outputs of our model.

In our research, we were limited in our research into other pre-trained models other than ResNet-50, due to our limited computational resources; a possible avenue for further investigation would be replacing ResNet-50 with different architectures. Additional future work could include investigating whether further removing certain electrodes without compromising on accuracy is possible, because certain electrodes like F8, P3, P4 and O1 are not of particular importance to the model according to Fig. 3

5 Conclusion

In this paper, we have shown a 14-channel explainable fine-tuning approach to migraine diagnosis, with comparable if not better accuracy than other models. Using HiResCAM for explainability, we have shown how our model corroborates with some pre-existing neural correlates, while finding new ones in the process.

References

1. Stovner, L.J., et al.: Global, regional, and national burden of migraine and tension-type headache, 1990–2016: a systematic analysis for the Global Burden of Disease Study 2016. Lancet Neurol. **17**(11), 954–976 (2018)
2. Arnold, M.: Headache classification committee of the international headache society (IHS) the international classification of headache disorders. Cephalalgia **38**(1), 1–211 (2018)
3. Pescador Ruschel, M.A., De Jesus, O.: Migraine Headache. StatPearls, StatPearls Publishing (2024)
4. Miller, S., Matharu, M.S.: Migraine is underdiagnosed and undertreated. Practitioner **258**(1774), 19–24 (2014)
5. Viana, Michele, et al. "Poor patient awareness and frequent misdiagnosis of migraine: findings from a large transcontinental cohort." *European journal of neurology* 27.3 (2020): 536–541
6. Smith, S.J.: EEG in the diagnosis, classification, and management of patients with epilepsy. J. Neurol. Neurosurg. Psychiatry **76**(suppl 2), ii2–ii7 (2005)
7. Siddiqui, M.M., et al.: EEG signals play major role to diagnose sleep disorder. Int. J. Electron. Comput. Sci. Eng. (IJECSE) **2**(2), 503–505 (2013)
8. O'Hare, L., Menchinelli, F., Durrant, S.J.: Resting-state alpha-band oscillations in migraine. Perception **47**(4), 379–396 (2018)
9. Bjørk, M.H., et al.: Interictal quantitative EEG in migraine: a blinded controlled study. J. Headache Pain **10**, 331–339 (2009)
10. Cao, Z., et al.: Resting-state EEG power and coherence vary between migraine phases. J. Headache Pain **17**, 1–9 (2016)
11. Chamanzar, A., et al.: Abnormalities in cortical pattern of coherence in migraine detected using ultra high-density EEG. Brain Commun. 3(2), fcab061 (2021)
12. Aslan, Z.: Migraine detection from EEG signals using tunable Q-factor wavelet transform and ensemble learning techniques. Phys. Eng. Sci. Med. **44**(4), 1201–1212 (2021)
13. Subasi, A., et al.: "Effect of photic stimulation for migraine detection using random forest and discrete wavelet transform. Biomed. Signal Process. Control **49**, 231–239 (2019)
14. Göker, H.: Automatic detection of migraine disease from EEG signals using bidirectional long-short term memory deep learning model. SIViP **17**(4), 1255–1263 (2023)
15. Chamanzar, A., Behrmann, M., Grover, P.: Ultra high-density EEG recording of interictal migraine and controls: sensory and rest. R1 (2020)

16. Ullah, M.E., et al.: A comparative analysis of stimuli response among people with migraine classification: a machine learning approach. In: 2024 3rd International Conference on Advancement in Electrical and Electronic Engineering (ICAEEE). IEEE (2024)
17. Draelos, R.L., Carin, L.: Use HiResCAM instead of Grad-CAM for faithful explanations of convolutional neural networks. arXiv preprint arXiv:2011.08891 (2020)
18. Aslan, Z.: Deep convolutional neural network-based framework in the automatic diagnosis of migraine. Circ. Syst. Signal Process. **42**(5), 3054–3071 (2023)
19. Orhanbulucu, F., Latifoğlu, F., Baydemir, R.: A new hybrid approach based on time frequency images and deep learning methods for diagnosis of migraine disease and investigation of stimulus effect. Diagnostics **13**(11), 1887 (2023)
20. Boonjong, K.: A Convolutional Neural Network for Classification of EEG Signals from SSVEP Responses of Migraine. figshare (2022). https://doi.org/10.6084/m9.figshare.28238639.v1

Alien: Covenant - Cytoneme Biology of Polyaneuploid Cancer Cells to Confer Chemotherapy Resistance

K. Lim[1,2](✉)

[1] Raffles Institution, 1 Raffles Institution Lane, Singapore 575954, Singapore
25YKIER171G@student.ri.edu.sg
[2] Department of Pathology, NUS, Singapore, Singapore

Abstract. Cytonemes, specialized actin-based filopodial structures, facilitate the intercellular transfer of signaling proteins and organelles. While well-documented in developmental biology, their role in cancer progression remains underexplored. In this study, we investigate the involvement of cytonemes in mediating chemoresistance in High-Grade Serous Ovarian Carcinoma (HGSOC), an aggressive cancer subtype. Our research focuses on poly-aneuploid giant cancer cells (PACCs), which arise under stress conditions and exhibit unique survival mechanisms. We demonstrate that PACCs act as mitochondrial reservoirs, transferring functional mitochondria to neighboring smaller aneuploid cancer cells (ACCs) via cytonemes. This mitochondrial transfer enables ACCs to withstand the cytotoxic effects of chemotherapy, contributing to the development of chemoresistance. Using advanced time-lapse microscopy, we visualized these dynamic intercellular interactions and metabolic adaptations. We further examined the effects of cytoneme inhibitors, such as Vismodegib and ARI 1, under chemotherapeutic stress. Our findings reveal that PACCs form significantly more cytonemes than ACCs and that the inhibition of cytoneme formation disrupts mitochondrial transfer, reducing ACC survival. This study highlights a novel mechanism of chemoresistance mediated by mitochondrial transfer through cytonemes and underscores the therapeutic potential of cytoneme inhibitors in overcoming resistance in HGSOC.

Keywords: Cancer · Tumours Microenvironment · Polyaneuploid Cancer Cells (PACCs) · Mitochondria · Cytonemes · Chemoresistance · Cytoneme Inhibitors

1 Background

1.1 Cancer Biology and Tumour Microenvironment

Cancer arises when normal cellular regulatory mechanisms fail, leading to unchecked proliferation and evasion of apoptosis [1]. It is fundamentally a disease of the genome, characterized by mutations in driver genes—genes essential for regulating cell growth and division. These mutations, which may accumulate due to environmental factors

H. Guo et al. (Eds.): IRC-SET 2025, PTT, pp. 39–50, 2026.
https://doi.org/10.1007/978-981-95-9630-0_4

or replication errors, result in oncogenesis, a multi-step process that eventually leads to tumor formation [2, 3]. Contrary to the traditional view of tumors as homogeneous masses of cancer cells, research now shows that tumors are complex ecosystems. They consist of cancerous cells and various non-cancerous stromal cells, including immune cells, fibroblasts, and endothelial cells [1, 9]. This diverse cellular composition contributes to a dynamic tumor microenvironment where heterotypic interactions promote tumor growth, metastasis, and resistance to therapy. The cancer cells may also form subpopulations that may interact with each other to facilitate growth and survival, and one such subpopulation that is of interest to our current study is a type of polyploid giant cells known as the polyaneuploid cancer cells (PACCs).

1.2 Polyanueploid Giant Cancer Cells

One emerging area of interest in cancer biology is the role of PACCs in chemotherapy resistance. PACCs are characterized by their large size, abnormal nuclei with more than 2n chromosome content, and ability to withstand environmental stresses. These cells can form via processes like endoreplication, mitotic slippage, cytokinesis failure, or cell fusion, all of which lead to polyploidization [6–8]. PACCs occur naturally at very low frequencies in cell lines and in tumours, and appear to increase in frequency with drug treatment [6]. PACCs exhibit unique survival strategies, including the ability to exit the cell cycle, avoiding DNA damage and apoptosis. They serve as reservoirs of genetic diversity, giving rise to tumour heterogeneity, and thus highly adaptable progeny that drive tumor evolution and therapy resistance [6, 9].

1.3 Cytonemes and Intercellular Communication

Cytonemes are specialized, actin-based cellular extensions that facilitate the direct transfer of signaling molecules between cells. First observed in Drosophila melanogaster, cytonemes have since been implicated in various biological processes, including tissue development, regeneration, and cancer progression [19–23]. Structurally, they consist of actin filaments encased in a lipid bilayer, and their role in signal transduction is critical for establishing signaling gradients and promoting cellular differentiation [19–23]. Recent evidence highlights the involvement of cytonemes in cancer, where they mediate intercellular communication and metabolic exchange [14–16]. Cytonemes enable the transfer of key survival factors or even organelles, such as mitochondria, between cancer cells [14–16]. This ability is particularly significant under chemotherapy-induced stress, where the exchange of mitochondria via cytonemes may enhance the survival and chemoresistance of recipient cells.

1.4 Our Hypothesis: Mitochondrial Transfer by PACCs and Chemoresistance

Our study investigates the hypothesis that PACCs play a pivotal role in mitochondrial transfer through cytonemes, conferring chemoresistance to smaller aneuploid cancer cells (ACCs) in ovarian cancer. HGSOC is an aggressive subtype with limited treatment options, often associated with poor prognosis due to its high resistance to chemotherapy.

Using advanced imaging techniques, we observed that PACCs form a greater number of cytonemes compared to ACCs, actively transferring mitochondria under chemotherapeutic stress. This transfer likely provides recipient cells with the bioenergetic resources needed to endure cytotoxic conditions. Importantly, our findings demonstrate that the inhibition of cytoneme formation using agents such as Vismodegib (an FDA-approved drug) and ARI-1 significantly reduces cytoneme connectivity between PACC and the surrounding ACCs. By elucidating the mechanisms underlying cytoneme-mediated mitochondrial transfer, this study advances our understanding of cancer cell cooperation and offers potential therapeutic strategies for overcoming chemoresistance in ovarian cancer.

2 Methodology

2.1 Observation of Cytonemes in PACCs

Induction of PACCs: The human ovarian carcinoma cell line, HEY, will be the one we use for our investigation. GFP-labelled HEY cancer cells (500,000 cells on a 10-cm plate) were seeded in complete RPMI medium (10 mL) supplemented with 10% fetal bovine serum (FBS) and antibiotics. To induce PACC formation, cells were treated with 100 nM Barasertib (an Aurora kinase inhibitor that inhibits cell division while promoting genome replication). Another set of HEY cancer cells were treated with a control - dimethyl sulfoxide (DMSO) – and are denoted as ACC. The resulting PACCs and ACCs are GFP-positive. Cultures were incubated at 37 °C in a humidified incubator with 5% CO_2 for 5 days.

Cell Preparation for Co-Culture: Cells were harvested using trypsin to detach adherent cells, followed by neutralization with RPMI medium supplemented with FBS. Cells were centrifuged at 300 g for 5 min, and the pellet was resuspended in fresh medium. Viability and cell count were determined using a hemocytometer and trypan blue exclusion. Labeling of the mitochondria was performed using MitoTracker Deep Red (100nM, 30 min at 37 °C). Labeled cells were washed three times with phosphate-buffered saline (PBS) and resuspended in a staining medium.

Co-Culture Setup: Wild-type HEY cells were co-cultured with 1% labeled PACC or ACC cells in 24-well plates for a total of 200,000 cells per well. Co-cultures were maintained at 37 °C in a humidified incubator for 24 h. Cells were then harvested for flow cytometry (Fig. 1).

2.2 Exploring Mitochondrial Transfer Dynamics

Confocal Microscopy: For the imaging of cytonemes, cells were fixed with 4% paraformaldehyde for 15 min, permeabilized with 0.1% Triton X-100 for 15 min and stained with Alexa Fluor 488 phalloidin for 30 min. Cells were washed twice with PBS and imaged using the Olympus FV3000 confocal microscope. A series of images were acquired along the Z-axis. Z-projection images were then formed using the Olympus FV31S-SW. For time-lapse confocal microscopy, the imaging was performed on a Yokogawa CQ1 spinning disc confocal microscope.

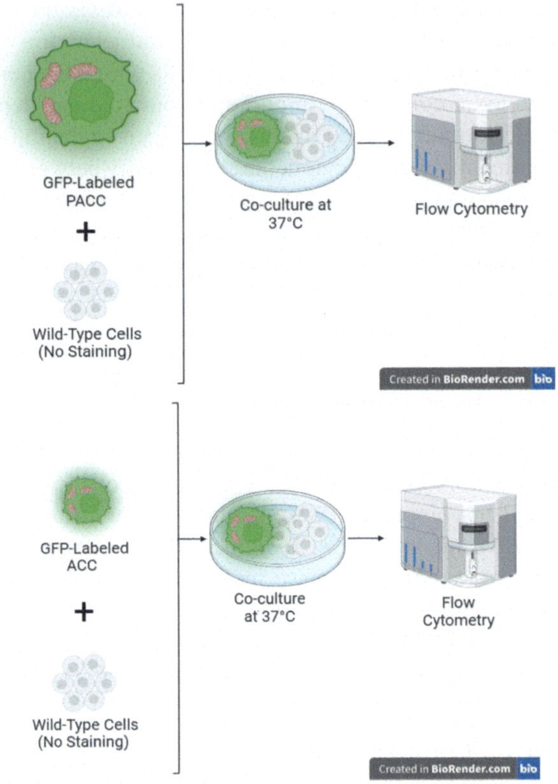

Fig. 1. Generalised Experimental Design of Co-Culture Setups for PACCs and ACCs

2.3 Cytonemes Under Chemotherapeutic Stress

Flow Cytometry: Co-cultured cells were trypsinized and analysed via the CytoFlex LX flow cytometer. The non-GFP cells were identified as the co-cultured wild type cells. These cells are analysed for any indication of mitochondrial transfer i.e. positivity for Mitotracker Deep Red.

Viability Assay: Co-cultured cells (at 10% PACC), pure PACCs and pure control DMSO-treated cells were treated with increasing doses of carboplatin (a commonly used drug in chemotherapy, to stimulate chemotherapeutic stress) for 3 days in 96 well plates (4,000 cells per well). Viability was assessed using the CellTiter-Glo assay. Hypothetical or expected results for co-culture were calculated based on the viability data of pure PACC and pure control cells, and then compared to the experimental data of the co-cultured cells.

2.4 Testing Efficiency of Cytoneme Inhibitors

Inhibitor Treatment: PACCs were co-cultured with wild-type HEY cells in the presence of cytoneme inhibitors targeting Hedgehog (Vismodegib, Sonidegib) and ROR1

(ARI-1, Box5, Zilovertamab) signaling pathways. Experimental conditions included varying concentrations of PACCs (1%, 3%, 100%) in combination with DMSO. Cells were fixed, permeabilized, and stained with phalloidin-FITC.

Microscopy and Analysis: Confocal microscopy was used to assess the effects of inhibitors on cytoneme formation. Images were analyzed to compare cytoneme prevalence and structural integrity across treatment groups.

3 Results and Analysis

3.1 PACCs Form More Cytonemes than ACCs

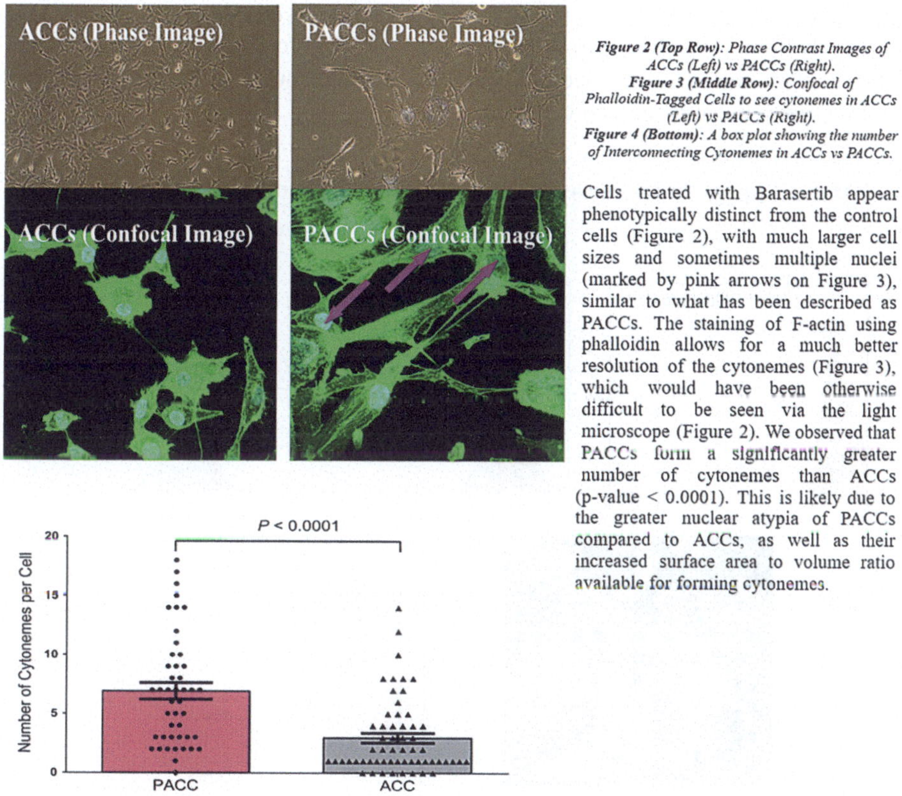

Fig. 2. (Top Row): Phase Contrast Images of ACCs (Left) vs PACCs (Right). (Middle Row): Confocal of Phalloidin-Tagged Cells to see cytonemes in ACCs (Left) vs PACCs (Right). (Bottom): A box plot showing the number of Interconnecting Cytonemes in ACCs vs PACCs.

Cells treated with Barasertib appear phenotypically distinct from the control cells (Fig. 2), with much larger cell sizes and sometimes multiple nuclei (marked by pink arrows on Fig. 2), similar to what has been described as PACCs. The staining of F-actin using phalloidin allows for a much better resolution of the cytonemes (Fig. 2), which would have been otherwise difficult to be seen via the light microscope (Fig. 2). We observed that PACCs form a significantly greater number of cytonemes than ACCs (p-value < 0.0001). This is likely due to the greater nuclear atypia of PACCs compared to ACCs, as well as their increased surface area to volume ratio available for forming cytonemes.

3.2 Mitochondria are Transferred from PACCs to ACCs

We observed that wild-type cells appear to gain MitoTracker Deep Red staining in the presence of PACCs under time-lapse confocal microscopy (Fig. 3) and were able to capture mitochondria in the cytonemes formed between a PACC and an adjacent cell, implying that mitochondria transfer was mediated via the cytonemes (Fig. 3). We demonstrated quantitatively via flow cytometry that in a co-culture comprising 1% PACC, a significantly higher percentage of the co-cultured wild-type cells received MitoTracker-labelled mitochondria. Taken together, these data strongly suggest that PACCs transfer mitochondria to the surrounding cells via cytonemes.

3.3 PACCs Appear to Increase Population-Wide Chemoresistance

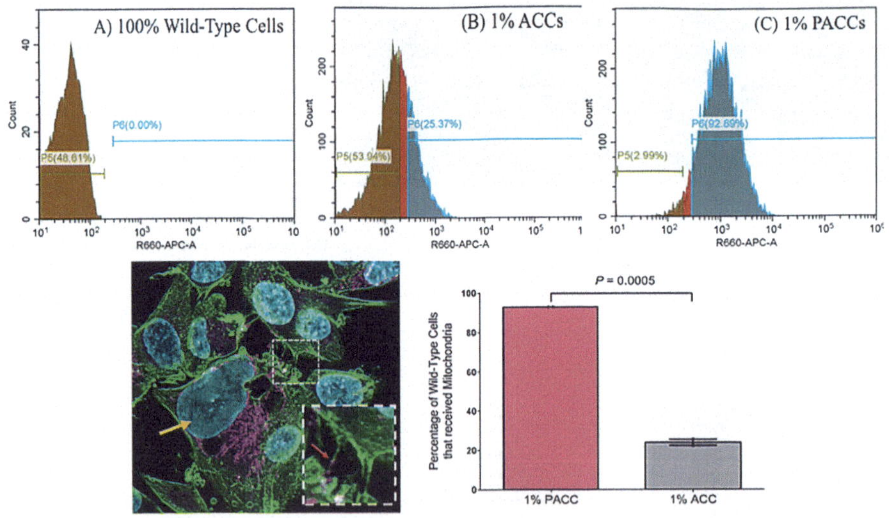

Fig. 3. (Bottom Left): Confocal microscopy image showing mitochondria (red arrow, inset) stained by MitoTracker Deep Red (pink) inside of a cytoneme that was formed between a PACC (brightly stained with MitoTracker Deep Red and having an abnormally large nucleus) (orange arrow) and a neighbouring cell. (Top): Histogram Graphs, Flow Cytometry Analysis of (A) 100% Wild-Type cells, (B) 1% ACC and (C) 1% PACC. (Bottom Right): Percentage of Wild-Type Cells that received Mitochondria from Stained Cells in 1% PACC co-culture vs 1% ACC

3.4 Vismodegib and ARI-1 are the Best Cytoneme Inhibitors

To interrogate the link between mitochondrial transfer from PACCs (via cytonemes) and the enhanced overall chemoresistance, we explored ways to regulate cytoneme formation. We judged the effectiveness of cytoneme inhibition based on the mean number of cytonemes formed per cell, recording them in Table 1, and the phenotypic change in the cytonemes as shown in Fig. 4. We treated the PACC with different cytoneme inhibitors, each at their optimal concentration. The inhibitors we used are as follows:

- CONTROL SET-UP: No Cytoneme Inhibitors Added
- **SONIDEGIB:** Sonidegib is an inhibitor of the Hedgehog (Hh) signaling pathway, binding to and inhibiting the smoothened (SMO) transmembrane protein, which suppresses GLI and by extension downstream signaling and thus reduces tumour proliferation [24]. It is primarily used to treat locally advanced basal cell carcinoma for patients who are not eligible for, or present cancer recurrence, after surgery and/or radiotherapy [25]. From our experiment, with the addition of Sonidegib the cytonemes are of regular shape and thickness and the cytonemes that were able to be formed, were still noticeably able to connect adjacent cells. **Sonidegib is an ineffective inhibitor of cytonemes.**
- **VISMODEGIB:** Vismodegib, similar to Sonidegib, is an inhibitor of the Hedgehog (Hh) signaling pathway, binding to and inhibiting the smoothened (SMO) transmembrane protein, which suppresses GLI and by extension downstream signaling and thus reduces tumour proliferation. However, unlike Sonidegib, it is primarily used to treat metastatic basal cell carcinoma for patients who are not eligible for, or present cancer recurrence, after surgery and/or radiotherapy [26]. From our experiment, with the addition of Vismodegib the cytonemes are much shorter in length, and crumpled and feathery in shape. Much fewer cytonemes formed in the first place, and those that did were much less likely able to connect with an adjacent cell. Thus, **Vismodegib is an effective inhibitor of cytonemes.**
- **ARI-1:** ARI-1 is an inhibitor of the receptor tyrosine kinase-like orphan receptor 1 (ROR1) signaling pathway. ARI-1 specifically targets the extracellular Frizzled domain of ROR1, suppressing non-small cell lung cancer (NSCLC) cell proliferation and migration by regulating PI3K/AKT/mTOR signaling in a ROR1-dependent manner [27]. Other than this, little is known about ARI-1. From our experiment, with the addition of ARI-1 the cytonemes formed by the PACCs cultured with ARI-1 were much shorter and thinner, and more wispy and feathery in texture. Most importantly, practically all the cytonemes formed were noticeably too short to even connect to an adjacent cell. Thus, **ARI-1 is an effective inhibitor of cytonemes.**
- **BOX5:** Box5 is an inhibitor of the Wnt5a signaling pathway, which is prevalent in the aggressive skin cancer, melanoma. Box5 directly inhibits Wnt5a signaling, abolishing cell adhesion, migration and invasion - all of which are crucial to tumour proliferation and inhibits the basal migration and invasion of Wnt5a-expressing HTB63 melanoma cells [28]. From our experiment, with the addition of Box5 the cytonemes are of regular shape, and if anything are even thicker than in the control. Most importantly, the cytonemes are still able to connect to adjacent cells. This was the most prominent occurrence of this phenomenon from all inhibitor co-cultures. This indicates that **Box5 may not even be an inhibitor of cytonemes.**

- **ZILOVERTAMAB:** Zilovertamab is an anti-ROR1 antibody that inhibits the ROR1 signaling pathway. In chronic lymphocytic leukaemia (CLL) cells, Zilovertamab can reduce expression of p62 and target genes induced by activation of NF-κB, mTORC1, and NRF2, enhancing the cytotoxicity of anti-cancer drugs [29]. From our experiment, with the addition of Zilovertamab the cytonemes are of regular shape and thickness. Most importantly, the cytonemes that were able to be formed, are still able to connect to adjacent cells. **This indicates that Zilovertamab may not be an inhibitor of cytonemes.**

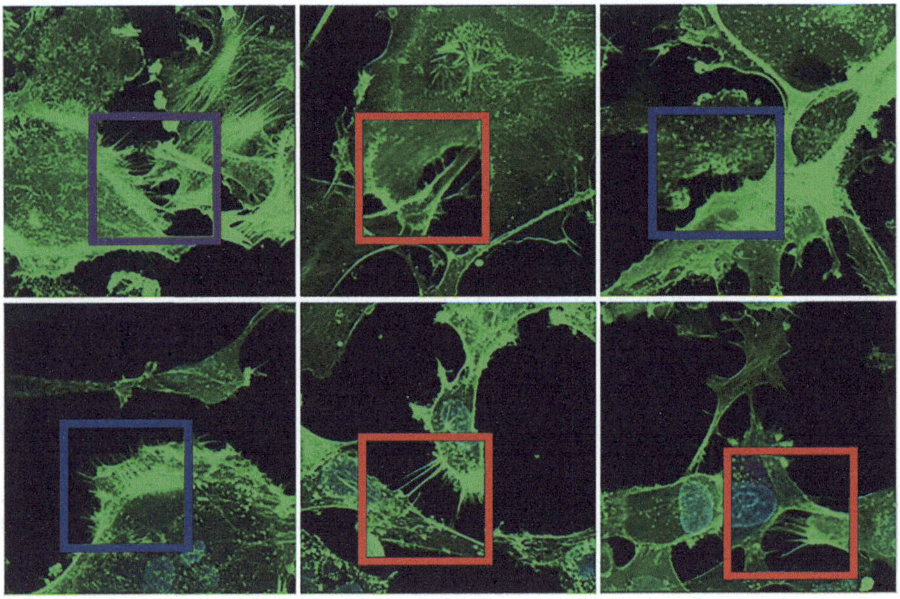

Fig. 4. Representative confocal microscopy photos found in PACCs that were treated with various inhibitors - (Clockwise from Left) Control, Sonidegib, Vismodegib, Zilovertamab, Box5, ARI-1

Table 1. The number of (countable and visible) cytonemes in each confocal image of PACCs under various inhibitors, as compared to the number and percentage of disconnected cytonemes

Inhibitor	Mean Total Number of Cytonemes per cell	Mean Number of Disconnected Cytonemes per cell	Mean Percentage of Disconnected Cytonemes
Sonidegib	28	21	75.00
Vismodegib	12	11	91.67
ARI-1	54	54	100.00
Box5	10	6	60.00
Zilovertamab	29	13	44.83

4 Discussion

The focal point of our research was examining, in greater detail, the bioenergetic dynamics between poly-aneuploid giant cancer cells (PACCs) and their surrounding smaller wild-type cells in the tumour microenvironment, how this dynamic contributes to chemoresistance in cancer cells, and how we can tailor future chemotherapy treatment to better combat this form of chemoresistance. The study of cytonemes in the specific context of PACCs is a niche one, and we believe that our research sheds new light on the subject:

- One, studies on mitochondria transfer have largely been focused on heterocellular transfer between two different cell types. An example would be between immune cells and mesenchymal stem cells [21], as well as cancer cells and stromal cells, and between cancer cells and immune cells [1]. However, this research focused on mitochondrial transfer between subpopulations of cancer cells within the same population.
- Two, while previous research papers focused on the signalling proteins (e.g. Wnt) being transferred between cytonemes [20, 21], this study on the bioenergetic transfer of mitochondria from PACCs to other subpopulations of cancer cells.
- Three, unlike other studies, we explore drugs that specifically target and inhibit the signalling pathways required for cytoneme formation (i.e. Vismodegib and ARI-1) as a novel way to reduce drug resistance towards standard-of-care (SOC) treatment, including carboplatin.

Four key findings of this research are presented in this paper below, each corresponding to their respective stage.

1. We demonstrated that PACCs form more cytonemes than ACCs. This is possibly due to their greater surface area to volume ratio available for forming cytonemes, as well as their nuclear atypia due to their higher concentration of mutant DNA, and their greater volume of cytoplasm [8]. It was this finding in the preliminary stage of our research that led us on to hypothesisc that there is a greater transfer of organelles such as mitochondria from PACCs to other cells.
2. We showed that mitochondria flows from PACCs to neighbouring cells. In this way, it confirms our theory that PACCs act as mitochondria reservoirs for smaller cancer cells, and transfer these mitochondria to them via the use of cytonemes.
3. We demonstrated that in the presence of PACCs, the overall chemoresistance of the population was enhanced. We thus propose that the mitochondria transferred by the PACCs to the smaller cells enables smaller cells to continue undergoing aerobic respiration under hypoxic conditions and therefore survive, as well as resist the cytotoxic effects of chemotherapy, thus contributing to the overall chemoresistance of the cancer cell.
4. We validated two effective inhibitors of cytoneme formation for ovarian cancer - Vismodegib and ARI-1. Since we have already established the essential role cytonemes play in tumours resisting chemotherapy. Where Vismodegib shines in Hedgehog-positive tumours, ARI-1 may be effective for ROR1-positive tumours. We propose

that the delivery of either Vismodegib and ARI-1, *in tandem* with existing chemotherapeutic treatment like carboplatin, could considerably improve the effectiveness of the chemotherapy and overcome chemoresistance.

5 Limitations and Challenges

- **Manual Image Analysis**: In Stage 1, confocal microscopy images were manually scanned to count cytonemes, introducing potential for human error. Differentiating between cytonemes, merged structures, and filopodia was challenging, particularly due to variable image quality. Switching from GFP to Phalloidin dye improved image clarity, but the manual process left room for inaccuracies. Despite this, observer consistency validated results showing a statistically significant increase in cytoneme formation in Barasertib-induced PACCs compared to wild-type cells ($P < 0.0001$).
- **Population Sampling**: High-resolution microscopy revealed clear cytoneme changes under inhibitors, such as reduced density and structural abnormalities, but these observations were limited to sampled cells. Without analyzing the entire population, the possibility of unobserved variations remains. However, qualitative data was sufficient for Stage 4, and findings were consistent across samples.
- **Quantitative Methodologies**: Quantitative data in Stages 2 and 3 were obtained through robust methods. Fluorescence intensity thresholds accurately quantified mitochondrial transfer in Stage 2, while flow cytometry provided reliable mitochondrial uptake data. These approaches complemented qualitative observations, ensuring the reliability of key findings.

6 Conclusion

In conclusion, this study proposes a novel model to chemoresistance, akin to the interconnected nature of an alien hive mind, as portrayed in Ridley Scott's Alien: Covenant (2017), from which this paper draws its title. Poly-aneuploid giant cancer cells (PACCs) function as a central, alien hivemind for smaller wild-type cancer cells to ensure their survival by their tentacled appendages - the cytonemes. Cytonemes, known up until now to transport signalling proteins, are also capable of transferring mitochondria, and PACCs serve as a mitochondrial reservoir to transport mitochondria to other smaller cancer cells via these cytonemes, thereby improving their resistance to chemotherapy, hypoxia and cytotoxic stresses. The drugs Vismodegib and ARI-1 are excellent inhibitors of these cytonemes, and while Vismodegib shines in quashing Hedgehog-positive cells, its less-famous counterpart ARI-1 is potentially excellent for treating ROR1-positive cells. Co-administration of these inhibitors with existing treatments like carboplatin could revolutionise treatment outcomes.

References

1. Hanahan, D., Weinberg, R.A.: Hallmarks of cancer: the next generation. Cell **144**(5), 646–674 (2011). https://doi.org/10.1016/j.cell.2011.02.013
2. https://frontlinegenomics.com/an-overview-of-cancer-biology/

3. Campbell, N.A., Reece J.B.: Biology *8th Ed, Pearson Education, Inc. (2008)
4. Brooker, R.J.: Genetics: Analysis and Principles, 2nd edn. McGraw-Hill (2005)
5. Front Line Genomics. (n.d.). An overview of cancer biology. Front Line Genomics. https://frontlinegenomics.com/an-overview-of-cancer-biology/
6. Bukkuri, A., Pienta, K.J., Austin, R.H., et al.: A life history model of the ecological and evolutionary dynamics of Poly-Aneuploid cancer cells. Sci. Rep. **12**, 13713 (2022). https://doi.org/10.1038/s41598-022-18137-4
7. Was, H., et al.: Polyploidy formation in cancer cells: how a trojan horse is born. Semin. Cancer Biol. **81**, 24–36 (2022)
8. Liu, P., Wang, L., Yu, H.: Polyploid giant cancer cells: origin, possible pathways of formation, characteristics, and mechanisms of regulation. Front. Cell Dev. Biol. **12**, 1410637 (2024). https://doi.org/10.3389/fcell.2024.1410637
9. Coward, J., Harding, A.: Size does matter: why polyploid tumor cells are critical drug targets in the war on cancer. Front. Oncol. **4**, 123 (2014). https://doi.org/10.3389/fonc.2014.00123
10. Casotti, M.C., et al.: Computational biology helps understand how polyploid giant cancer cells drive tumor success. Genes, 14(4), 801 (2023). https://doi.org/10.3390/genes14040801
11. Gong, Y., Peng, X., Kang, H., Zhao, M., Ge, G.: Polyploid giant cancer cells (PGCCs): The evil roots of cancer. Front. Oncol. **11**, 724781 (2021). https://doi.org/10.3389/fonc.2021.724781
12. Saini, G., et al.: Polyploid giant cancer cell characterization: new frontiers in predicting response to chemotherapy in breast cancer. Semin. Cancer Biol. **81**, 220–231 (2022). https://doi.org/10.1016/j.semcancer.2021.03.017
13. Schoentgen, F.: Insight on the role of RKIP in cancer through key protein partners and cellular protrusions. Prognostic and Therapeutic Applications of RKIP in Cancer, Elsevier, pp.3–35, 2020, ff10.1016/B978-0–12-819612-0.00001-8ff. Ffhal-02870563f
14. Kretschmer, A., et al.: Stress-induced tunneling nanotubes support treatment adaptation in prostate cancer. Sci. Rep. **9**, 7826 (2019)
15. Desir, S., et al.: Chemotherapy-induced tunneling nanotubes mediate intercellular drug efflux in pancreatic cancer. Sci. Rep. **8**, 9484 (2018)
16. Hanna, S.J., McCoy-Simandle, K., Leung, E., Genna, A., Condeelis, J., Cox, D.: Tunneling nanotubes, a novel mode of tumor cell–macrophage communication in tumor cell invasion. J. Cell Sci. **132**(3), jcs223321 (2019)
17. Ramírez-Weber, F.A., Kornberg, T.B.: Cytonemes: cellular processes that project to the principal signaling center in Drosophila imaginal discs. Cell **97**, 599–607 (1999)
18. Hsiung, F., Ramirez-Weber, F.A., Iwaki, D.D., Kornberg, T.B.: Dependence of Drosophila wing imaginal disc cytonemes on Decapentaplegic. Nature **437**, 560–563 (2005)
19. Fereres, S., Hatori, R., Hatori, M., Kornberg, T.B.: Cytoneme-mediated signaling essential for tumorigenesis. PLoS Genet. **15**, e1008415 (2019)
20. Casas-Tintó, S., Portela, M.: Cytonemes, their formation, regulation, and roles in signaling and communication in tumorigenesis. Int. J. Mol. Sci. **20**(22), 5641 (2019). https://doi.org/10.3390/ijms20225641
21. Portela, M.: Tumorigenesis: using cytonemes to engage mesenchymal cells in epithelial tumours. Current biology: CB **30**(10), R441–R443 (2020). https://doi.org/10.1016/j.cub.2020.03.017
22. Portela, M.: Cytonemes in Tumourigenesis. In Encyclopedia (2019). https://encyclopedia.pub/entry/161
23. Routledge, D., et al.: The scaffolding protein flot2 promotes cytoneme-based transport of wnt3 in gastric cancer eLife, **11**, e77376 (2022)
24. Jain, S., Song, R., Xie, J.: Sonidegib: mechanism of action, pharmacology, and clinical utility for advanced basal cell carcinomas. Onco. Targets. Ther. **10**, 1645–1653 (2017). https://doi.org/10.2147/OTT.S130910

25. Villani, A., Fabbrocini, G., Costa, C., et al.: Sonidegib: safety and efficacy in treatment of advanced basal cell carcinoma. Dermatol Ther (Heidelb) **10**, 401–412 (2020). https://doi.org/10.1007/s13555-020-00378-8
26. Aditya, S., Rattan, A.: Vismodegib: a smoothened inhibitor for the treatment of advanced basal cell carcinoma. Indian Dermatol. Online J. **4**(4), 365–368 (2013). https://doi.org/10.4103/2229-5178.120685
27. Liu, X., et al.: Novel ROR1 inhibitor ARI-1 suppresses the development of non-small cell lung cancer. Cancer Lett. **458**, 76–85 (2019). https://doi.org/10.1016/j.canlet.2019.05.016
28. Jenei, V., et al.: A t-butyloxycarbonyl-modified Wnt5a-derived hexapeptide functions as a potent antagonist of Wnt5a-dependent melanoma cell invasion. Proc. Natl. Acad. Sci. U.S.A. **106**(46), 19473–19478 (2009). https://doi.org/10.1073/pnas.0909409106
29. Burger, J.A., Kipps, T.J.: ROR1: An orphan becomes apparent. Blood **140**(14), 1583–1592 (2022). https://doi.org/10.1182/blood.2022016152
30. All diagrams are created in https://BioRender.com

IoT

Detection Tools for Outbound C2 Traffic

Charisee Zhi Ling Yip(✉), Clarabel Jinghui Teo, Tee Kiat Chua, Jing Rui Goh, Brandon Jia Le Loo, Huaqun Guo, Liming Lu, and Kan Chen

Infocomm Technology Cluster, Singapore Institute of Technology, Singapore, Singapore
{2203215,2201877,2201304,2202074,2201940}@sit.singaporetech.edu,
{huaqun.guo,liming.lu,kan.chen}@singaporetech.edu.sg

Abstract. Ransomware attacks have rapidly evolved in sophistication and pose severe threats to organizations by encrypting critical data and demanding ransom payments. A key component in ransomware campaigns is the Command and Control (C2) communication where compromised systems establish outbound connections with malicious servers to receive instructions and exfiltrate data. As such, detecting C2 outbound traffic is crucial for early detection and mitigation of ransomware attacks. This paper provides a comprehensive analysis and comparison of existing detection techniques, and their respective tools specifically aimed at identifying ransomware C2 traffic.

Keywords: security posture · ransomware · command and control · detection technique · tools

1 Introduction

Ransomware has become an increasingly prominent threat to organizations worldwide. As such, significant focus has been brought to the detection of Command and Control (C2) outbound traffic related to ransomware. Adversaries commonly leverage C2 channels to establish constant and persistent communication with compromised systems in order to send instructions for further malicious activities such as exfiltrating and encrypting data. There has also been an increased use of ransomware-as-a-service (RaaS) models, which allows adversaries to outsource their tools, increasing the already present challenges of detecting such traffic, such as evasion techniques, encryption and obfuscation of data. In response, various detection tools that implement multiple detection techniques have emerged, including those that leverage deep packet inspection (DPI) and threat intelligence feeds to detect suspicious traffic patterns that indicate ransomware activity. However, rapid evolution of attack techniques requires continuous adaptation and refinement of these tools such as hybrid approaches to provide more comprehensive detection against sophisticated ransomware campaigns. These insights underscore the importance of a robust, multi-layered defense strategy that incorporates the latest advances in analysis and detection to stay ahead of evolving ransomware threats [1]. Hence, to address this issue, this survey paper mainly provides analysis and comparison of the different C2 ransomware traffic detection techniques and tools and offers recommendations tailored to various scenarios.

H. Guo et al. (Eds.): IRC-SET 2025, PTT, pp. 53–63, 2026.
https://doi.org/10.1007/978-981-95-9630-0_5

2 Literature Review

2.1 Ransomware and Its Prevalence in the Cyber World

Ransomware is a type of malware that infiltrates systems and encrypts files, disrupting normal operations to coerce victims into paying a ransom. The rapid adoption of digital technologies during the COVID-19 pandemic, including widespread remote work has exposed new vulnerabilities that ransomware operators have exploited. Many organizations were unprepared for the surge in security risks associated with remote work, often lacking strong cybersecurity measures and clear guidelines for safe online practices. The economic downturn and increased unemployment rates during the pandemic have led to a rise in cybercriminal activity [2], as financial hardship can drive more individuals toward cybercrime [3]. Given the threat landscape, companies, especially those in critical sectors, must strengthen their defenses against ransomware attacks.

2.2 Overview of C2 Traffic and Threat Landscape

The Cyber Kill Chain framework includes the C2 phase as a critical and vital role in a cyber attack. C2 Traffic is extremely important and plays an essential role in malware operations, mainly acting as the communication link between compromised hosts and the adversaries' control server. Detecting and disrupting C2 communications is a key defensive measure that can halt attacks before reaching the final stage and thus mitigating potential damage. The evolving threat landscape demonstrates increasing sophistication and complexity in C2 communication strategies as adversaries turn to encryption and obfuscation techniques to hide their presence. For instance, they often utilize domain generation algorithms (DGA) and domain fronting which generates dynamic rendezvous points to evade detection as these leverage legitimate services to mask C2 traffic [4, 5].

2.3 Diversity of Threats and Techniques

The diversity and continuous evolution of C2 techniques employed by adversaries to evade detection necessitates agile and adaptive detection tools are required in this evolving threat landscape. Given the use of novel attack patterns such as obfuscated and encrypted traffic, it is crucial to understand the strengths and limitations of different detection techniques to determine the effectiveness of these tools in identifying evasive C2 traffic [6].

2.4 Alignment with Organizational Security Posture

Each organization has their own unique security needs and hence it is essential to align detection tools that can meet the security needs of the organization. Comparing across various detection techniques and frameworks can help organizations determine which solution best integrates into their existing architecture and risk management strategies. By evaluating various metrics such as detection accuracy, false positive rates and other evaluation criteria, organizations can better determine which tools are best suited for their security needs and posture [7, 8].

3 Detection Techniques and Tools

There are five detection techniques below, namely signature, anomaly, behavioral, reputation and machine learning that have been identified as the key detection mechanisms for detecting outbound C2 traffic.

3.1 Signature-Based Detection

Signature-based detection identifies network traffic patterns that match predefined signatures representing known malicious behavior or communication patterns. In the context of C2 traffic, this approach entails the development and deployment of signatures that are tailored to the unique characteristics of C2 communication, such as recognizable packet structures, specific command sequences, or URL paths. Security tools can detect attempts by C2 servers to communicate with compromised systems by comparing network traffic against these signatures [9]. Some of the signature-based tools include Snort, Suricata and Zeek.

3.2 Anomaly-Based Detection

Anomaly-based detection identifies unusual patterns or deviations from a model of "normal" behavior within a network or system, making it effective for discovering new or evolving threats that do not match known signatures. For example, anomaly-based methods can detect advanced persistent threats (APTs) and zero-day attacks by flagging irregularities in user behavior or system activities, such as unusual login times, increased data transfers, or network traffic anomalies. The following tools - Suricata, ELK Stack and Splunk, implement anomaly-based detection.

3.3 Behavioral-Based Detection

Behavioral-based detection identifies patterns in C2 traffic indicative of malicious activity, especially in network behavior rather than relying solely on static signatures and is versatile for identifying evolving C2 techniques. Combining the dynamic features and behavioral indicators will enhance accuracy and enable models to further refine the detection process and achieve even higher accuracy of distinguishing ransomware activities from benign traffic [10]. Furthermore, it has been proven that analyzing pre-encrypted network traffic is effective against malicious activities as they alert administrators through identification of early C2 communications, allowing for them to obtain a critical window of intervention before the ransomware completes its activities [11]. The tools discussed for behavioral-based detection are Cortex XDR, FireEye Network Security and Cisco Stealthwatch.

3.4 Reputation-Based Detection

Reputation-based C2 traffic detection is a method that leverages threat intelligence databases and entity reputation scores to block malicious communications associated with ransomware attacks. It leverages on the history of malicious activity from known Indicators of Compromise (IoCs) such as IP addresses, domain names, URLs and SSL/TLS certificates in order to evaluate outbound network connections and detect potential malicious traffic. By utilizing up-to-date reputation lists, various security tools can promptly block and/or alert suspicious traffic, preventing ransomware on compromised hosts from communicating with its C2 servers. Reputation-based detection can quickly recognize known threats due to an adversary's reputation score even if the specific threat pattern is unknown. The tools discussed under reputation-based detection target 3 different areas of reputation - IP and Domain, URL and Certificate.

3.5 Machine Learning-Based Detection

Machine learning (ML) based detection involves the usage of algorithms to study data from various sources to be able to identify malicious or suspicious behavior in networks and files. It leverages on pattern recognition and behavior-based characteristics to identify and detect new and evolving threats that are often undiscovered by traditional means of malware detection, resulting in the vulnerability of zero-day attacks. ML-based detection uses machine learning as its core engine to autonomously learn from large-scale data and detect both known and unknown C2 traffic without relying on predefined rules. Tools such as WildFire, DarkTrace, Microsoft Defender and CrowdStrike are some of the most used advanced ML-based malware threat detection solutions in the industry.

4 Result and Insight

This section details the result and insights gained from the review and analysis of the various tools mentioned in Section 3.

4.1 Comparison Metrics Used

To compare the tools and techniques, a list of comparison metrics were used and the rationale for its usage are detailed in the following subsections.

Detection Accuracy. Detection accuracy is fundamental for assessing effectiveness of detection tools in identifying ransomware C2 traffic. It measures the proportion of correctly detected attacks (true positives) relative to the total attempts [12].

False Positive Rates. False positive rate is a critical metric that reflects the number of benign traffic instances that have been incorrectly flagged as malicious. High false positive rates will impact security teams negatively as they may be overwhelmed with unnecessary alerts, causing potential actual threats to be missed. Excessive false positive rates can also hinder the practical deployment of C2 traffic detection systems making it difficult for organizations to handle and manage security alerts [13].

Machine Learning Capabilities. Given the complexity of modern ransomware that employ advanced evasion techniques, the ability of a tool to leverage on ML capabilities is important. The implementation of ML algorithms within these tools are able to improve detection rates without relying on static signatures [14].

Real-Time Monitoring. Real-time monitoring capabilities assess the tool's ability to identify malicious activities instantly and without delays. With real time monitoring, early detection and rapid response can be enabled, allowing security teams to intervene before significant damage such as data encryption by ransomware occurs.

Threat Intelligence. Threat intelligence capabilities include the integration of updated information about known threats, attack patterns and adversary tactics into the tool. With this metric, evaluation of how well the tool leverages external threat data and insights to enhance detection capabilities for new attack techniques and emerging threats can be done.

Cost of Implementation. As some of these tools have different licenses and rate of subscriptions for different infrastructures, it is important to measure the cost across different scales to ensure fair comparison. This also takes into account the resource consumption when implementing the tool.

Ease of Integration. Integration into current networks and systems are important and critical when implementing a new tool to be used, especially for security monitoring as they involve multiple components. Comparing the ease of integration between the tools can help to evaluate the time and effort it takes to ensure smooth operations.

Zero-Day Detection. This metric measures the tool's ability to identify previously unknown threats which may not be available from threat intelligence data sources. The ability to detect zero-day attacks can be beneficial as they allow for effective response before the threats evolve into something more sinister [15].

Customizability. Customizability is key to tailoring detection tools for specific organizational needs and threat profiles. Allowing the flexibility to modify detection rules and integrate other threat intelligence sources will enhance the tools effectiveness against diverse environments.

4.2 How Scores Are Derived

The performance of each tool will be rated upon the score of 5 for each comparison metric. The scores are then totaled and compared against other tools within the same detection technique in order to determine the best tool within the technique. Among the best tools for each detection technique, they will then be compared against each other to determine the best tool across all the 5 different techniques discussed.

4.3 Comparison of Signature-Based Detection Tools

Zeek is the top performer with a score of 37.5 in the comparison of tools. It is exceptional in several critical areas, such as real-time monitoring, threat intelligence, cost-effectiveness, and simplicity of integration. While Snort and Suricata provide similar functionalities, Zeek's overall efficacy is further enhanced by its customizability and zero-day detection capabilities, which render it the most reliable and balanced tool among the 3.

4.4 Comparison of Anomaly-Based Detection Tools

Splunk is the most suitable tool for anomaly-based detection with a score of 35.5 out of 45. It excels in areas such as detection accuracy, threat intelligence, and real-time monitoring. While it may be slightly more costly compared to the other tools, its capabilities justify the investment, especially for larger organizations. While ELK Stack and Suricata offer competitive features and benefits, Splunk's overall performance and versatility make it the preferred choice for robust behavioral-based detection.

4.5 Comparison of Behavioral-Based Detection Tools

FireEye Network Security is the best tool, with a score of 37.5 out of 45. It is relatively more affordable compared to the other two tools and is dependent on the scale of the organization, which is great for smaller infrastructures. It has high detection accuracy rates and extremely low false positive rates with proven statistics, and is also very capable for real-time threat monitoring, threat intelligence and zero-day detection. Although it is slightly weaker in its ML capabilities compared to the others, this has the potential to be improved, which may increase its accuracy even more. FireEye also offers comparable customization options and ease of integration among all 3 tools.

4.6 Comparison of Reputation-Based Detection Tools

The comparison was done among the tools for each type of reputation-based detection, namely IP and Domain, URL and Certificate, before a final comparison was done among the best tools for each type of reputation-based detection to determine the overall best reputation-based detection tool. Figures 1, 2 and 3 show the best IP and Domain, URL and Certificate reputation-based detection tools respectively.

Tool	Detection Accuracy	False Positive Rate	ML capabilities	Real time monitoring	Threat intelligence	Cost	Ease of integration	Zero day detection	Customizability	Total
DAbR	4	3	5	5	5	3	5	2	4	**36**
DNS Graph Mining	4	5	5	5	5	3	5	5	4	**41**

Fig. 1. Comparison results of IP and domain reputation-based detection tools

Tool	Detection Accuracy	False Positive Rate	ML capabilities	Real time monitoring	Threat intelligence	Cost	Ease of integration	Zero day detection	Customizability	Total
DAbR	4	3	5	5	5	3	5	2	4	**36**
DNS Graph Mining	4	5	5	5	5	3	5	5	4	**41**

Fig. 2. Comparison results of URL reputation-based detection tools

Tool	Detection Accuracy	False Positive Rate	ML capabilities	Real time monitoring	Threat intelligence	Cost	Ease of integration	Zero day detection	Customizability	Total
DAbR	4	3	5	5	5	3	5	2	4	**36**
DNS Graph Mining	4	5	5	5	5	3	5	5	4	**41**

Fig. 3. Comparison results of certificate reputation-based detection tools

Best Overall Reputation-Based Detection Tool. Among PAN-OS, DNS Graph Mining and Two-Stage Detection Mechanism, PAN-OS is ranked to be best followed by DNS Graph Mining and finally Two-stage Detection Mechanism. PAN-OS provides deep SSL/TLS inspection across multiple layers, including certificates, IP addresses, domains, and URLs. This comprehensive approach enables it to detect threats in all areas of network traffic, not just DNS queries or URLs. Moreover, PAN-OS excels with its deep inspection capabilities, especially in decrypting SSL/TLS traffic, which is a crucial advantage over DNS Graph Mining and URL Filtering. Finally, PAN-OS offers detailed customization options for SSL/TLS inspection policies, allowing security teams to fine-tune settings according to their specific organizational needs. This granular control helps optimize detection, reduce false positives, and align with compliance requirements. Therefore, PAN-OS is ranked highest due to its versatility, deep inspection capabilities, real-time threat intelligence integration, and the ability to handle multiple layers of traffic, including SSL/TLS decryption, IP, domain, and certificate filtering.

4.7 Comparison of Machine Learning-Based Detection Tools

CrowdStrike Falcon is the tool with the highest rating score of 38 out of 45. This is due to CrowdStrike Falcon having one of the highest scores for detection accuracy, false positive rate, real-time monitoring, threat intelligence and zero-day detection. It has a detection accuracy of 99.7% with a 100% protection rate against a wide variety of malwares. Its false positive rate is 0% which further proves that its ML capabilities are the best among the other tools. Its outstanding performance and advanced features make it a worthwhile investment for organizations looking for a robust and effective anti-malware detection system.

4.8 Overall Comparison Between All the Tools Across the 5 Techniques

Fig. 4 shows the 5 best tools from the above 5 detection techniques. It was determined that Crowdstrike Falcon was the best tool for C2 traffic detection with the highest score based on the following reasons.

Advanced Use of Machine Learning. Crowdstrike Falcon uses advanced machine learning algorithms specifically designed to detect anomalies and malicious activities based on behavior, which is crucial for identifying sophisticated C2 techniques.

Zero-Day Threat Detection. Crowdstrike Falcon has the ability to detect new and emerging threats using AI, significantly enhancing capability to intercept C2 traffic that may not be known or listed in threat databases yet.

Tool	Detection Accuracy	False Positive Rate	ML capabilities	Real time monitoring	Threat intelligence	Cost	Ease of integration	Zero day detection	Customizability	Total (/45)
Zeek	4	4	3	4.5	4.5	5	3.5	4	5	37.5
Splunk	4.5	4	4	4	4.5	2	4.5	4	4	35.5
FireEye	5	4	3.5	5	5	3	3.5	5	3.5	37.5
PAN-OS	5	3	4	5	5	2	3	5	5	37
Crowdstrike Falcon	4.5	4.5	4	5	5	2	4	5	4	38

Fig. 4. Comparison ratings for the best tools across the 5 techniques

Integration and Adaptability. Crowdstrike Falcon is able to integrate extensive threat intelligence and adapt detection mechanisms in real-time, aligning with the dynamic nature of C2 traffic patterns.

5 Recommendations

This section provides tool recommendations that organizations can adopt to improve their detection of ransomware C2 traffic based on existing research, findings from above and analysis.

5.1 Tool Recommendations based on Score

Based on the above analysis, Crowdstrike Falcon with the highest score is recommended, due to its robust ML algorithms, threat intelligence integration and advanced behavioral analytics. Although it is costly, its capabilities enable it to detect and mitigate sophisticated C2 activities with high accuracy and minimize false positives. Furthermore, its user-friendly interface and scalability options for different infrastructures make it an adaptable choice for diverse environments. While other tools have commendable features, Crowdstrike Falcon had an edge in C2 traffic detection, particularly in scenarios involving complex and stealthy adversarial techniques.

5.2 Tool Recommendations based on Budget and Scale

Commercial tools which offer high detection accuracy on a large scale may incur high operational and installation costs, deterring many organizations from significantly investing in their security posture, leading to potential security incidents. With that, recommendations catered to different budgets will be made for the tools compared.

Limited Budget. The Suricata and Zeek tools recommended here are open source with no cost and offer a cost-effective solution compared to the rest. Both Suricata and Zeek are network analysis tools and offer strong performance in signature-based and anomaly-based detection. As they focus on real-time traffic analysis and pattern recognition, they are suitable for organizations with budget constraints. They are also comparable against other tools with the same detection in metrics such as usage of external threat intelligence, real-time monitoring capabilities and zero-day detection capabilities [13].

Larger Budget. Wildfire by Palo Alto Networks is known for its advanced machine learning capabilities and integrated threat intelligence. Wildfire is very effective in detecting sophisticated C2 traffic with extremely high accuracy that is close to perfection. Its deep-learning model with a well-trained neural network of millions of data allows it to be able to detect ransomware C2 traffic easily [16].

6 Future Research Direction

6.1 Analysis of Hybrid Detection Models

Analysis on hybrid detection models can be conducted to enhance the research into C2 detection techniques and tools. This includes models that integrate various detection techniques, for example combining signature and machine learning approaches. Although some tools currently support such capabilities, deeper analysis and understanding of such hybrid models need to be conducted in order to determine their effectiveness against C2 attacks.

6.2 Enhancing Zero-Day C2 Detection Approaches

Another research direction would be into the enhancement of zero-day C2 traffic detection, by delving into approaches to improve the identification of novel and suspicious traffic patterns across diverse network environments. If the organization is able to identify suspicious traffic patterns without prior knowledge, this will allow them to focus on improving sensitivity to zero-day C2 behaviors across diverse network environments.

7 Conclusion

In this research survey paper, a variety of detection tools for C2 outbound traffic were explored, and examination of their capabilities across different techniques and evaluation metrics was conducted. By analyzing tools like Snort, Suricata, Cortex XDR, PAN-OS and Crowdstrike Falcon, their strengths and limitations were assessed based on their features, with focus on various aspects such as detection accuracy, false positive rates, ML capabilities, zero-day detection, among others. Among all the tools being compared, Crowdstrike Falcon emerged as the best solution with the highest rating score due to its outstanding performance and advanced features. The findings of this research highlight the importance of leveraging advanced detection tools to combat evolving cyber threats effectively. By adopting a proactive approach to tool selection and integration, alongside the findings of this research paper, it is hoped that organizations can enhance their defenses against C2 outbound traffic, thereby strengthening their security posture and resilience against cyber attacks.

Acknowledgement. The authors deeply acknowledge the professors from the Singapore Institute of Technology (SIT) for their help and support for this project research. The authors would like to declare that there is no conflict of interest.

References

1. "Understanding Ransomware Threat Actors: LockBit | CISA," Cybersecurity and Infrastructure Security Agency CISA (2023). https://www.cisa.gov/news-events/cybersecurity-advisories/aa23-165a
2. Lang, M., Connolly, L., Taylor, P., Corner, P.J.: The evolving menace of ransomware: a comparative analysis of pre-pandemic and mid-pandemic attacks. Digital Threats: Research and Practice, vol. 4, no. 4. Association for Computing Machinery (ACM), pp. 1–22 (2023). https://doi.org/10.1145/3558006
3. Beaman, C., Barkworth, A., Akande, T.D., Hakak, S., Khan, M.K.: Ransomware: Recent advances, analysis, challenges and future research directions. Computers & Security, vol. 111. Elsevier BV, p. 102490 (2021). https://doi.org/10.1016/j.cose.2021.102490
4. Li, Z., et al.: Identification domain fronting traffic for revealing obfuscated C2 communications. 2021 IEEE Sixth International Conference on Data Science in Cyberspace (DSC, pp. 91–98). IEEE (2021). https://doi.org/10.1109/dsc53577.2021.00020
5. Ren, F., Jiang, Z., Wang, X., iu, J.: A DGA domain names detection modeling method based on integrating an attention mechanism and deep neural network. Cybersecurity **3**(1) (2020). https://doi.org/10.1186/s42400-020-00046-6
6. Novo, C., Morla, R.: Flow-based detection and proxy-based evasion of encrypted malware C2 traffic. Proceedings of the 13th ACM Workshop on Artificial Intelligence and Security, pp. 83–91 (2020). https://doi.org/10.1145/3411508.3421379
7. Divya, P.J., George, R.S., Madhusudhan, G., Padmasree, S.: Organization-wide IOC monitoring and security compliance in endpoints using open source tools. 2022 IEEE 3rd Global Conference for Advancement in Technology (GCAT), pp. 1–6 (2022). https://doi.org/10.1109/gcat55367.2022.9971999
8. Demertzis, K., Tsiknas, K., Takezis, D., Skianis, C., Iliadis, L.: Darknet traffic Big-Data analysis and network management for Real-Time automating of the malicious intent detection process by a weight agnostic neural networks framework. Electronics **10**(7), 781 (2021). https://doi.org/10.3390/electronics10070781
9. Macedo, E.: Signature-based IDS for Encrypted C2 Traffic Detection. thesis (2022). https://repositorio-aberto.up.pt/bitstream/10216/142718/2/572020.pdf
10. Hwang, J., Kim, J., Lee, S., Kim, K.: Two-Stage ransomware detection using dynamic analysis and machine learning techniques. Wireless Pers. Commun. **112**(4), 2597–2609 (2020). https://doi.org/10.1007/s11277-020-07166-9
11. Moussaileb, R., Cuppens, N., Lanet, J.-L., Bouder, H.L.: Ransomware network traffic analysis for pre-encryption alert. In: Lecture Notes in Computer Science, pp. 20–38 (2020). https://doi.org/10.1007/978-3-030-45371-8_2
12. Salehi, S., Shahriari, H., Ahmadian, M.M., Tazik, L.: A Novel Approach for Detecting DGA-based Ransomwares. 2018 15th International ISC (Iranian Society of Cryptology) Conference on Information Security and Cryptology (ISCISC), pp. 1–7. IEEE (2018). https://doi.org/10.1109/iscisc.2018.8546941
13. Gardiner, J., Nagaraja, S.: On the reliability of network measurement techniques used for malware traffic analysis. Lecture Notes in Computer Science, pp. 321–333. Springer International Publishing (2014). https://doi.org/10.1007/978-3-319-12400-1_31
14. Cusack, G., Michel, O., Keller, E.: Machine learning-based detection of ransomware using SDN. Proceedings of the 2018 ACM International Workshop on Security in Software Defined Networks & Network Function Virtualization, pp. 1–6 (2018). https://doi.org/10.1145/3180465.3180467

15. Ganame, K., Allaire, M.A., Zagdene, G., Boudar, O.: Network behavioral analysis for zero-day malware detection – a case study. Lecture Notes in Computer Science, pp. 169–181. Springer International Publishing (2017). https://doi.org/10.1007/978-3-319-69155-8_13
16. Almousa, M., Basavaraju, S., Anwar, M.: API-based ransomware detection using machine learning-based threat detection models. 2021 18th International Conference on Privacy, Security and Trust (PST), pp. 1–7. IEEE (2021). https://doi.org/10.1109/pst52912.2021.9647816

Computer Science

Improving Semantic Segmentation Performance on Occlusions Using Attention Mechanisms and Synthetic Datasets

Christina Pan[1(✉)] and Xing Zian[2]

[1] Raffles Institution, 1 Raffles Institution Ln, Singapore 575954, Singapore
25ypanw345a@student.ri.edu.sg
[2] River Valley High School, 6 Boon Lay Ave, Singapore 649961, Singapore

Abstract. Dealing with occlusions remains a significant challenge to solve in semantic segmentation tasks, especially in the context of pedestrian tracking; if not performed properly, the lives of pedestrians and drivers will be endangered. In this study, we aim to use a hybrid approach to utilise both attention mechanisms and synthetic datasets to enhance the performance of the model against occlusions. Specifically, the Convolutional Block Attention Module (CBAM) is integrated into the DeepLabV3 + architecture to enhance its focus on important features through channel attention and spatial attention submodules. Additionally, synthetically occluded images were created to expose the model to different occlusion scenarios to enhance the model's generalisation's ability when training. We validated our model on the Cityscapes dataset that had been augmented with the synthetically occluded images. Mean Intersection over Union (mIoU) was used to evaluate the performance of the model. Our results show that our new model demonstrated a noticeable improvement in occlusion performance when segmenting occluded objects like pedestrians and vehicles as compared to the baseline model. These discoveries demonstrate the potential of a hybrid approach involving both attention mechanisms and synthetic datasets as a practical way to approach segmentation tasks in pedestrian tracking contexts when dealing with occlusions.

Keywords: semantic segmentation · computer vision · occlusions

1 Introduction

The evolution of autonomous driving has enabled vehicles to perceive their environment without any human interference. The systems behind autonomous driving aim to perform intelligent operations like object detection in road scenes, adaptive cruise control, lane keep assistance and collision avoidance. [1] A critical component of autonomous driving is the perception system, [2] which relies on heavily robust computer vision techniques for real-time perception. Its role is to identify, classify and localise various objects in images or videos [3].

However, occlusions—caused by factors such as vehicles, pedestrians, or natural elements like trees—can significantly hinder the accuracy of these perception systems

H. Guo et al. (Eds.): IRC-SET 2025, PTT, pp. 67–76, 2026.
https://doi.org/10.1007/978-981-95-9630-0_6

[4]. By obscuring critical information, occlusions compromise a system's ability to detect its surroundings, placing road users into potential danger. To mitigate this, improving the effectiveness of such models against occlusions is essential. Currently, research into occlusion-robust semantic segmentation is ongoing and growing.

Zhang et al. [5] proposed an occlusion face detection algorithm based on attention mechanisms—specifically using Selective Enhancement, self-attention and spatial attention modules—to improve the visibility of key image features. However, the network model proposed is computationally intense, and miniaturising the model for lightweight deployment remains challenging. Zeng et al. [6] refined this idea by using CBAM-RepVit as the backbone network, where the squeeze-and-excitation blocks integrated in the RepVit backbone network are replaced by CBAM modules to enhance spatial and channel feature representations. Yet, the limited diversity of data constrained its overall generalisability, underscoring the need for more varied datasets. As for data augmentation techniques, Dinsley et. al [7] artificially occluded the dataset with common features such as trees, bushes and graffiti. The image was randomly scaled and rotated to simulate real-world occlusions. However, the simplicity of the scenes and unrealistic occlusion patterns limited their effectiveness.

Though DeepLabV3 + [8] is highly regarded for semantic segmentation, it notably struggles to exploit multi-scale information during feature extraction, which often leads to detail loss and lower accuracy [9]. To address these shortcomings, we integrated CBAM modules into DeepLabV3 + to enhance feature representation under occlusion. This integration refines feature maps via spatial and channel attention. Additionally, we introduced a synthetic occlusion dataset to expose the model to varied scenarios during training. Together, these strategies produce relative improvements occlusion robustness, as reflected in our higher mIoU scores.

2 Aims and Hypothesis

Our aims are summarised as follows: (1) To evaluate the performance of the proposed approach on the Cityscapes [10] dataset under various occlusion conditions. (2) To analyse the effectiveness of combining data augmentation with attention mechanisms in mitigating the impact of occlusions. We hypothesise that this strategy will give the greatest improvement over synthetic datasets or attention modules alone.

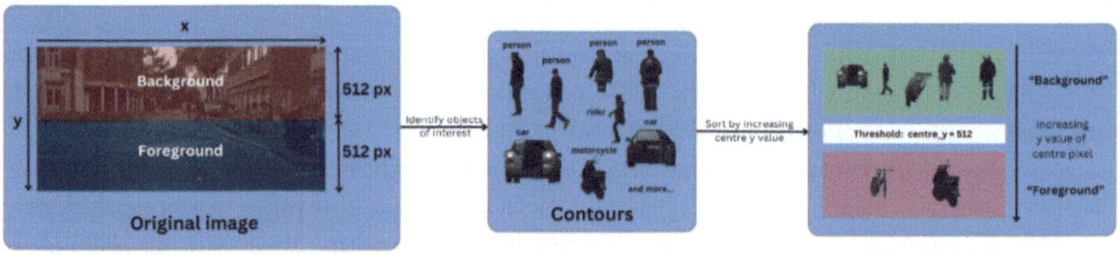

Fig. 1. The process of categorising occludees belonging to the relative background and foreground of the image.

3 Methodology

3.1 Generating Original Synthetic Occlusion Dataset

To generate the synthetic occlusion dataset, we begin by extracting occluder images from the Cityscapes dataset. Objects belonging to the 'vehicle' category were shortlisted as occluders due to their relevance in urban scenes. Each occluder was manually reviewed to remove those with irregular edges or unrealistic features.

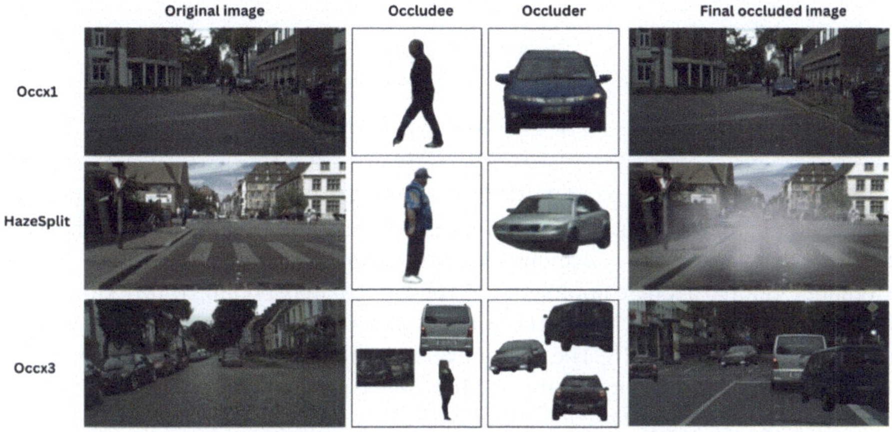

Fig. 2. Types of images produced for the dataset.

The generation process incorporates various enhancements to make images more relevant:

1. Occluder images were strategically pasted over contours of objects most hazardous if obscured in real-world scenarios e.g. pedestrians or vehicles, categorised under 'human' and 'vehicle' respectively.
2. Spatial consistency was maintained by classifying images as 'foreground' or 'background'.

A summary of the procedure is shown in Fig. 1, while examples of generated images are shown in Fig. 2.

Three datasets were generated: Cityscapes-Occx1 with one occlude, Cityscapes-Occx3 with three occluders, and Cityscapes-Hazesplit with a 50/50 blend of original and haze-covered Occx1 images. These simulate varying degrees of occlusion, from partial to severe, including weather-based occlusions. Occx1 challenges the model to detect partially obscured features, while Occx3 and HazeSplit require inferring objects from minimal visual data and using global context.

3.2 CBAM Modules

CBAM, [11] proposed in 2018, is a lightweight attention mechanism that can be used to enhance its focus on target features and improve its localisation ability for smaller objects [12]. We will briefly explain the relevant two sequential attention sub-modules (Fig. 3).

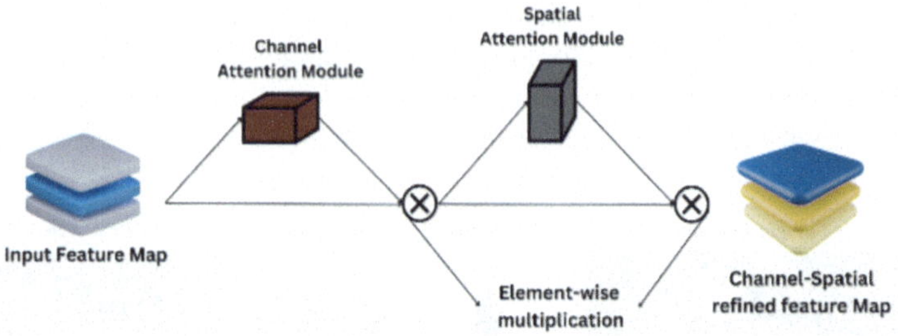

Fig. 3. Structure of CBAM

The Channel Attention Module (CAM) highlights the most informative channels in the feature map by assigning higher weights to more informative edge channels. Its structure is shown in Fig. 4. Given input feature maps of the dimensions H × W × C, CAM applies global average pooling (GAP) and global max pooling (GMP) to generate two C x 1 x 1 descriptors [13]. These two vectors are forwarded through a shared Multilayer Perceptron (MLP). The output feature vectors are merged element-wise, [14] and activated through a sigmoid activation function to produce channel attention weights [15]. It is computed using these equations below:

$$M_c(F) = \sigma(MLP(AvgPool(F)) + MLP(MaxPool(F))) \quad (1)$$

$$M_c(F) = \sigma(FC_2(FC_1(AvgPool(F))) + FC_2(FC_1(MaxPool(F)))) \quad (2)$$

σ is a nonlinear sigmoid function which produces outputs between 0 and 1. F is the input feature map. M_c is the channel attention feature map. Finally, an element-wise multiplication is performed between the channel attention weights and the input feature map to produce the refined feature map, as shown by the equation below. F_r represents the refined feature map after CAM is applied to the input feature map F.

$$F_r = M_c \otimes F + F \quad (3)$$

The Spatial Attention Module (SAM) assigns higher weights to highlight the more informative spatial regions [16]. Its structure is shown in Fig. 5. It reduces the channel-refined features to feature maps of sizes 1 × H × W by applying max pooling and average pooling across the channel axis. The pooled maps are concatenated to form a 2 × H × W feature map. A 7 x 7 convolution layer followed by a sigmoid activation layer is applied to generate the spatial attention weights. These weights are multiplied element-wise with the refined feature maps [17] to generate the final refined output. The spatial attention is computed as:

$$M_s(F) = \sigma\left(f_7 \times_7 \left(\left[MaxPool(F); AvgPool(F)\right]\right)\right) \quad (4)$$

$$F_s = M_s \otimes F + F \quad (5)$$

σ denotes the sigmoid function, M_s represents the spatial attention feature map. ';' represents a concatenation of the feature maps along the channel axis after their respective pooling operations. $f_{7\times7}$ denotes a convolutional operation. F_s is the refined feature map after SAM is applied to the input feature map F.

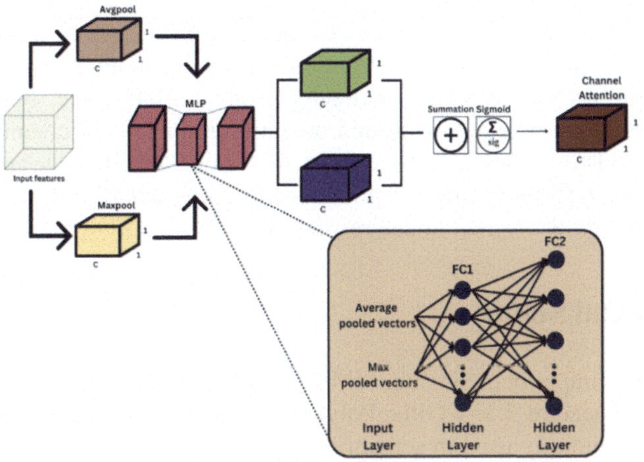

Fig. 4. Structure of CAM.

Fig. 5.Structure of SAM.

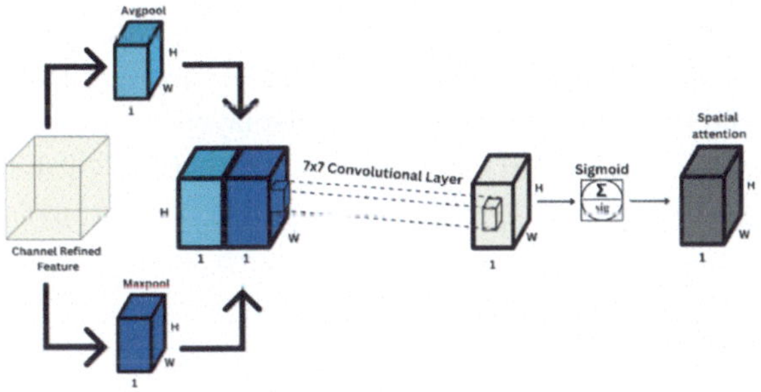

Fig. 5. Structure of SAM.

3.3 Modified DeepLabv3 + Architecture

The backbone architecture uses Residual Neural Network (ResNet), characterised by the stacking of residual blocks. ResNet utilises the skip connection technique to solve the vanishing and exploding gradients problem [18]. In order to further enhance the model's focus on relevant features and spatial information, CBAM was integrated into the encoder and decoder paths.

In the encoder, the CBAM module was immediately employed after ResNet's output. The CAM computes spatial information using average pooling and max pooling which is then forwarded to a shared network consisting of multi-layer perceptrons applied to each descriptor to produce the channel attention map. Subsequently, the refined features were fed to the Atrous Spatial Pyramid Pooling (ASPP) module to better extract the multi-scale features using atrous convolutions.

In the decoder, the high-level features extracted from the ASPP module were first upsampled by 4 using bilinear interpolation. Subsequently, another CBAM module was used to refine the low-level features extracted from the backbone. Next, these low-level features undergo a 1×1 convolution operation to have their channel dimensions reduced and minimise computational overhead. After concatenation, we applied multiple 3 x 3 convolutions to further refine the combined features. Finally, bilinear upsampling restores the output to its original image resolution.

4 Experimental Setup

DeepLabv3 + was implemented using MMSegmentation [19]. LR was standardized at 0.005 with a batch size of 2×1 GPU. PolyLR learning rate scheduling technique was used since it is commonly used in semantic segmentation tasks for better convergence of segmentation tasks over longer training schedules [20]. Binary cross entropy loss is used as the loss function. All code was run in the Google Colab Pro environment using the PyTorch DL software framework with built-in Python version 3.11.0.

We tested:

1. Original DeepLabV3 + on original Cityscapes
2. DeepLabV3 + on Cityscapes-Occx1 and Cityscapes-Occx3
3. DeepLabV3 + with CBAM at various layers. Integration of CBAM was optimal at layers 3 and 4, and thus only these results are displayed
4. Hybrid model containing CBAM and trained on synthetic occlusions

We employed mean Intersection over Union (mIoU).

5 Results and Discussion

5.1 Effect of Increased Occlusion on the Original Model

From the training results of the original DeepLabv3 + model, performance consistently degrades when occlusions are introduced. For instance, mIoU decreased by 2.25% for Cityscapes-Occx1, and 1.13% for Cityscapes-Occx3 in Table 1 compared to the original DeepLabv3 + model trained on the unedited Cityscapes dataset. Visually, mislabeled regions appear in the image as the occlusion severity increases, as evidenced by the images in the first column of Fig. 6. As occlusions obscure critical features in the training image, the model without any attention mechanisms is unable to recognise them as well.

5.2 Effect of Attention Mechanisms

The addition of CBAM modules to the original model in all layers also consistently improves performance across all three datasets. When comparing mIoU, vehicle IoU and human IoU of the three models on the same dataset, most exhibit at least slight increases when CBAM is present compared to when it is not. The largest improvement is found in the models trained on Cityscapes-Occx1, as seen in Table 1: adding CBAM after every bottleneck increased mIoU by 3.83%, while selectively adding CBAM to layers 3 and 4 increased mIoU by 5.47%. Visually, in this case, the addition of CBAM also removes inaccuracies, such as a mislabeled sign slightly occluded by a car and the mislabeled reflection of a sign in a window (in the red and green boxes). The enhanced focus on relevant spatial and channel features provided by CBAM likely contributes to this improvement.

Table 1. Results.

Model	Dataset	$mIoU_1$ (%)	$mIoU_2$ (%)	Vehicle IoU_2 (%)	Human IoU_2 (%)
DeepLabv3 +	Cityscapes	66.58	65.92	92.20	81.02
	Cityscapes-Occx1	64.33	63.81	91.46	78.09
	Cityscapes-Occx3	65.45	64.52	91.81	79.45
	Cityscapes-HazeSplit	65.16	64.39	91.24	80.38
DeepLab-CBAM	Cityscapes	66.76	66.11	92.01	80.97
	Cityscapes-Occx1	68.16	67.48	92.39	81.65
	Cityscapes-Occx3	65.64	66.01	91.84	81.02
	Cityscapes-HazeSplit	68.88	66.17	92.40	81.83
DeepLab-CBAM (after layers 3 & 4)	Cityscapes	69.27	66.35	**91.89**	83.60
	Cityscapes-Occx1	**69.80**	68.12	93.11	81.73
	Cityscapes-Occx3	61.61	67.13	92.34	82.48
	Cityscapes-HazeSplit	69.75	**69.58**	93.19	82.08

Note: $mIoU_1$ refers to training results, validated on the **original** Cityscapes dataset. $mIoU_2$ refers to testing results, tested on the **synthetic** occlusion dataset

However, this improvement is not always present when occlusions are more severe. In the case of the Cityscapes-Occx3 dataset, the addition of CBAM to all layers improves the training mIoU by only 0.19% compared to the baseline, while selective addition to deeper layers (layers 3 and 4) worsens model performance, as evidenced by a decrease of 3.84%. Accounting for the fact that Cityscapes-Occx3 images have significantly fewer unobstructed objects of interest i.e. vehicles and cars, this finding could indicate that the models struggle to compensate for the substantial information loss caused by high occlusion levels.

Further analysis of per-class IoU reveals that the overall improvement in mIoU is accompanied by corresponding gains in the 'vehicle' and 'human' IoU metrics. Since these two categories were specifically selected to be occluded during training, the results confirm that the increase in overall mIoU is not merely due to improvements in unrelated classes and demonstrate the model's enhanced ability to accurately segment the occluded categories.

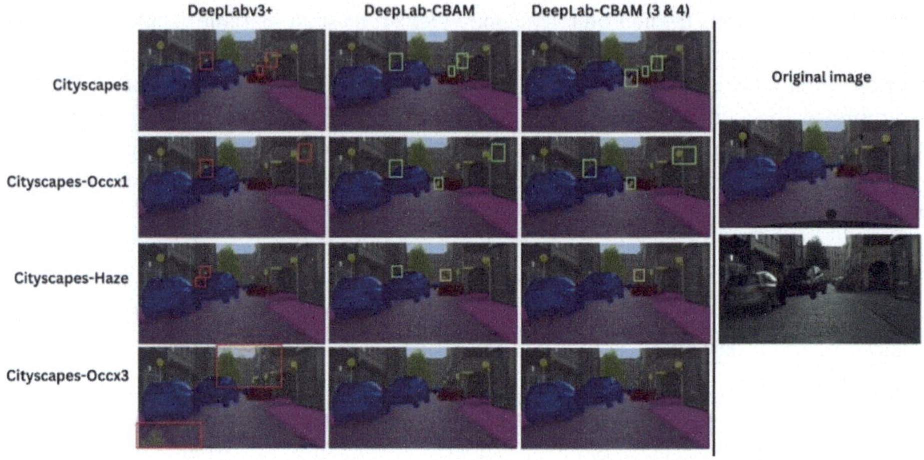

Fig. 6. Visual comparison of model performance through inference.

5.3 Layer-Specific CBAM Integration

Beyond simply comparing the presence and absence of attention mechanisms, we also investigated the effect of purposeful CBAM placement in the model. Through our ablation studies, we identified that CBAM integration at layers 3 and 4 consistently yielded the best balance between model complexity and performance for moderate occlusions amongst all of the combinations of layers possible e.g. layers 1 and 2, layers 2 and 3 etc. As seen from both tables of results, applying CBAM to layers 3 and 4 achieves the highest training and testing mIoU on all datasets except Cityscapes-Occx3.

Interestingly, this layer-specific integration comes with a trade-off. On the Cityscapes-Occx3 dataset, training performance decreases by 3.84% compared to the full CBAM configuration. On the testing set, however, it outperforms the standard model by 2.61%. This discrepancy can likely be attributed to the presence of synthetic occlusions in the test set used in Table 1, which are absent in the images used in Table 1. As a result, the model's layer-specific integration seems to better adapt to occlusion scenarios during inference, despite a slight compromise in overall training performance. Moreover, it is also possible that the model had overfitted to recognise heavily occluded features from less information and lost the ability to recognise whole occluders in the process. With objects of interest sometimes being occluded by two or more objects, the model is unable to recognise as many features for successful segmentation during training.

6 Conclusion

This study investigated the impact of integrating CBAM into DeepLabv3 + for handling occlusions in semantic segmentation. The results demonstrate that while CBAM enhances segmentation performance, its placement within the network varies with occlusion severity. A uniform application of CBAM across all layers provides a balanced improvement for both moderate and severe occlusions. In contrast, layer-specific integration amplifies performance for moderate occlusions but struggles with higher levels of occlusion where objects might be obscured beyond recognition. From this study, the ideal approach for a realistic urban scene would be a combination of attention mechanisms in the later layers and a synthetic dataset, such that both occlusion instances and fully visible objects provide a well-rounded source of information.

One limitation of this study would be resource constraints. The experiments were conducted on a single GPU with a batch size of two, whereas optimal results for DeepLabv3 + are obtained with multi-GPU setups. This limited the number of epochs and thereby the model's ability to converge fully. Another limitation would be the lack of hyperparameter tuning.

The study promises further avenues of exploration, such as investigating how attention mechanisms can be dynamically adjusted during inference—an idea inspired by the inconsistent improvements observed in severe occlusion cases. In this case, CBAM is applied uniformly across all layers or selectively to specific layers. Dynamic adjustment, if possible, would allow the model to identify varying levels of occlusion in the image and modify the behaviour of attention mechanisms based on the image's characteristics. This would allow the model to better target moderate or severe occlusions within the same image during inference.

References

1. Muhammad, K., Ullah, A., Lloret, J., Del Ser, J., de Albuquerque, V.H.C.: Deep learning for safe autonomous driving: current challenges and future directions. IEEE Trans. Intell. Transp. Syst. **22**(7), 4316–4336 (2020)
2. Hao, X.: MSC-Bench: benchmarking and analysing multi-sensor corruption for driving perception. arXiv preprint arXiv:2501.01037 (2025). https://doi.org/10.48550/arXiv.2501.01037
3. Brunetti, A., Buongiorno, D., Trotta, G., Bevilacqua, V.: Computer vision and deep learning techniques for pedestrian detection and tracking: a survey. Neurocomputing **300**, 17–33 (2018)
4. Al Jaberi, S.M., Patel, A., Al-Masri, A.N.: Object tracking and detection techniques under GAN threats: a systemic review. Appl. Soft Comput. **139**, 110224 (2023). https://doi.org/10.1016/j.asoc.2023.110224
5. Zhang, K., Zhu, Q., Li, W.: Experimental research on occlusion face detection method based on attention mechanism. J. Phys.: Conf. Ser. **2258**(1), 012078 (2022). https://doi.org/10.1088/1742-6596/2258/1/012078
6. Zeng, W., He, M.: Rice disease segmentation method based on CBAM-Carafe-DeepLabv3+. Crop Prot. **180**, 106665 (2024). https://doi.org/10.1016/j.cropro.2024.106665
7. Dineley, A., Natalia, F., Sudirman, S.: Data augmentation for occlusion-robust traffic sign recognition using deep learning. ICIC Express Lett. Part B Appl. **15**(4), 381–388 (2024). https://doi.org/10.24507/icicelb.15.04.381

8. Chen, LC., Zhu, Y., Papandreou, G., Schroff, F., Adam, H.: Encoder-Decoder with Atrous Separable Convolution for Semantic Image Segmentation. In: Ferrari, V., Hebert, M., Sminchisescu, C., Weiss, Y. (eds.) Computer Vision – ECCV 2018. ECCV 2018. Lecture Notes in Computer Science(), vol 11211. Springer, Cham (2018). https://doi.org/10.1007/978-3-030-01234-2_49
9. Hou, X., Chen, P., Gu, H.: LM-Deeplabv3+: a lightweight image segmentation algorithm based on multi-scale feature interaction (2024). Preprint at https://doi.org/10.20944/preprints 202401.1052.v1
10. Cordts, M., et al.: The Cityscapes dataset for semantic urban scene understanding. arXiv preprint arXiv:1604.01685 (2016). https://doi.org/10.48550/arXiv.1604.01685
11. Woo, S., Park, J., Lee, J.-Y., Kweon, I.S.: CBAM: convolutional block attention module. In: Ferrari, V., Hebert, M., Sminchisescu, C., Weiss, Y. (eds.) ECCV 2018. LNCS, vol. 11211, pp. 3–19. Springer, Cham (2018). https://doi.org/10.1007/978-3-030-01234-2_1
12. Chen, L., Yin, T., Zhou, S., Yi, G., Fan, D., Zhao, J.: Underwater target detection algorithm based on feature fusion enhancement. Electronics **12**(13), 2756 (2023). https://doi.org/10.3390/electronics12132756
13. Shafique, A., Seydi, S.T., Alipour-Fard, T., Cao, G., Yang, D.: SSVIT-HCD: a spatial–spectral convolutional vision transformer for hyperspectral change detection. IEEE J. Sel. Top. Appl. Earth Obs. Remote Sens. **16**, 6487–6504 (2023). https://doi.org/10.1109/jstars.2023.3251646
14. Hao, J., et al.: Sheep face detection based on an improved RetinaFace algorithm. Animals **13**(15), 2458 (2023). https://doi.org/10.3390/ani13152458
15. Eftekhari, A., Samadzadegan, F., Dadrass Javan, F.: Building change detection using the parallel spatial-channel attention block and edge-guided deep network. Int. J. Appl. Earth Obs. Geoinf. **117**, 103180 (2023). https://doi.org/10.1016/j.jag.2023.103180
16. Zagoruyko, S., Komodakis, N.: Paying more attention to attention: Improving the performance of convolutional neural networks via attention transfer. arXiv:1612.03928 [cs.CV] (2016). https://doi.org/10.48550/arXiv.1612.03928
17. Liu, Q., et al.: An improved Deeplab v3+ network based coconut CT image segmentation method. Front. Plant Sci. **14**, 1139666 (2023). https://doi.org/10.3389/fpls.2023.1139666
18. Xu, G., Wang, X., Wu, X., Leng, X., Xu, Y.: Development of skip connection in deep neural networks for computer vision and medical image analysis: a survey. ResearchGate preprint (2024). https://doi.org/10.13140/RG.2.2.19870.75847
19. MMSegmentation Contributors: MMSegmentation: OpenMMLab semantic segmentation toolbox and benchmark. https://github.com/open-mmlab/mmsegmentation. Accessed 13 Jul 2025
20. Mishra, P., Sarawadekar, K.: Polynomial learning rate policy with warm restart for deep neural network. In: IEEE Region 10 Conf. (TENCON 2019), pp. 2087–2092. IEEE (2019). https://doi.org/10.1109/tencon.2019.8929465

Investigation on the Convergence Properties of a Markov Chain

Aden Chong(✉) and Nguyen Tien Minh

NUS High School of Mathematics and Science,
20 Clementi Ave 1, Singapore, Singapore
adencyt@gmail.com

Abstract. This study investigates the convergence properties of a recursive sequence defined by $s_0 = 2$ and $s_{i+1} = 2R(1, s_i)$, where $R(a, b)$ is a random integer uniformly distributed between a and b. We aim to determine the expected number of steps for the sequence to exceed a threshold n, denoted $E(n)$. Via forming a system of equations from probabilistic analysis and the use of matrices, the research is able to form equations to predict $E(n)$ for large n. Computational experimenting reveals a linear trend as n tends to infinity, with an asymptotic approximation of $E(n) \approx 2.51778n$. We also explored an alternative approach involving binary partitions, a well-known function studied by Euler. Through this research, we gained insight on the hidden patterns lying behind this sequence and its connections to Markov chains, with potential applications in randomized algorithms and recursive systems.

Keywords: Markov Chains · Recursive Sequence · Convergence Properties · Asymptotic Analysis

1 Introduction

A recursive sequence often poses numerous interesting properties and has many potential applications in analysis. By combining it with probability, it has the potential of connecting many fields together. By observing the behaviour of the sequence, usually as the sequence goes to infinity, we can discover valuable information about the sequence itself. In this research, we want to investigate a recursive sequence defined by a random function, which possesses several useful properties that might be beneficial when it comes to recursion analysis and chaotic algorithms.

Denote $R(a, b)$ as a random (independent) integer uniformly distributed between a and b, inclusive.

We formulate our question statement. Set $s_0 = 2$, and define

$$s_{i+1} = 2R(1, s_i).$$

Let Q_n be the minimal integer that satisfies $s_{Q_n} \geq n$. Then let $E(n)$ denote $\mathbb{E}(Q_n)$ for convenience. With note of $E(2k + 1) = E(2k)$, we shall investigate properties of $E(n)$ with even n.

H. Guo et al. (Eds.): IRC-SET 2025, PTT, pp. 77–86, 2026.
https://doi.org/10.1007/978-981-95-9630-0_7

Simply put, we apply "2Randint(1, Ans)" to a scientific calculator and pushing the equals button repetitively. This function returns twice a random integer between 1 and the previous number. By doing this repetitively, we expected that the number we obtain would surpass a threshold n after some point: We aim to find the average number of times we must apply this function.

2 Simple Results

First, we try to obtain some simple results for small n using synthetic methods.

Lemma 1. $E(4) = 2$.

Proof. Recall that Q_4 is the random variable for the minimal number of steps to reach 4, in a sequence. Recall that $s_0 = 2$, so $s_1 = 2$ or 4, each with chance $\frac{1}{2}$.

$$\begin{aligned}\mathbb{E}[Q_4] &= \mathbb{E}[Q_4 \mid s_1 = 2]\mathbb{P}(s_1 = 2) + \mathbb{E}[Q_4 \mid s_1 = 4]\mathbb{P}(s_1 = 4)\\ \mathbb{E}[Q_4] &= (1 + \mathbb{E}[Q_4]) \cdot \frac{1}{2} + (1) \cdot \frac{1}{2}\\ \mathbb{E}[Q_4] &= 2\end{aligned}$$

where we used $\mathbb{E}[Q_4 \mid s_1 = 2] = 1 + \mathbb{E}[Q_4]$ because one step is consumed and we are back to the original state. Thus, we conclude that $E(4) = 2$. □

Lemma 2. $E(6) = 5$.

Proof. First, notice if i is the smallest index such that $s_i \geq 6$, then $s_{i-1} = 4$ is the only possible value of s_{i-1}.
Also denote $T_{a,b}$ as the number of steps to reach b from a. (Hence, $Q_b = T_{2,b}$.)
Let j be any number satisfying $s_j = 4$. Then,

$$\begin{aligned}\mathbb{E}(T_{4,6}) &= \mathbb{E}(T_{4,6} \mid s_{j+1} = 2)\mathbb{P}(s_{j+1} = 2) + \mathbb{E}(T_{4,6} \mid s_{j+1} = 4)\mathbb{P}(s_{j+1} = 4)\\ &\quad + \mathbb{E}(T_{4,6} \mid s_{j+1} = 4)\mathbb{P}(s_{j+1} \geq 6)\\ &= \frac{1}{4}(\mathbb{E}(T_{4,6}) + 1 + \mathbb{E}(Q_4)) + \frac{1}{4}(\mathbb{E}(T_{4,6}) + 1) + \frac{2}{4}\\ \mathbb{E}(T_{4,6}) &= 3\end{aligned}$$

Now, we notice

$$\mathbb{E}(Q_6) = \mathbb{E}(Q_4) + \mathbb{E}(T_{4,6}) = 2 + 3 = 5$$

where we frequently used $\mathbb{E}(Q_4) = E(4) = 2$. Therefore, $E(6) = 5$. □

To summarize: In order to have the sequence reach the number 6 for the first time, one will expect (on average) to click the calculator, 5 times.

3 General Results

Clearly $E(2k-1) = E(2k)$ so finding E of odd numbers has no point. So, let's generalise to all even n. We now let $a_{i,n}$ be n variables with $1 \leq i \leq n$. This variable is dependent on n. We abuse notation to now let $a_{i,n} = a_i$ when the value n is clear from context.

Lemma 3. *Consider a set of* n *linear equations, such that*

$$2ka_k = 2k + \sum_{i=1}^{\min(n,2k)} a_i$$

for $1 \leq k \leq n$. *We solve for* a_1. *Then* $E(2n+2) = a_1$ *for all* n.

Proof. Let $a_k = \mathbb{E}[Q_{2n+2}|s_0 = 2k]$. Our goal is to find $a_1 = E(2n+2)$.

The expected value a_k can be defined recursively. One step after state $2k$ is the new state $2i$, where i is chosen uniformly from $\{1, 2, \ldots, 2k\}$. The expected steps is thus 1 plus the average of the expected steps from all possible subsequent states:

$$a_k = 1 + \frac{1}{2k}\sum_{i=1}^{2k} a_i^*$$

where a_i^* is the expected number of *additional* steps from state $2i$. If $2i < 2n+2$, then $a_i^* = a_i$. If $2i \geq 2n+2$, the process terminates and $a_i^* = 0$.

First, suppose $k < (n+1)/2$. Then the maximum possible next state is $4k < 2n+2$. Thus, none of the next states terminate the process, so $a_i^* = a_i$ for all $i \in \{1, \ldots, 2k\}$.

$$2ka_k = 2k + \sum_{i=1}^{2k} a_i$$

Next, for $k \geq (n+1)/2$, the process terminates if $2i \geq 2n+2$, which means $i \geq n+1$. For these states, the additional expected steps $a_i^* = 0$.

$$a_k = 1 + \frac{1}{2k}\left(\sum_{i=1}^{n} a_i + \sum_{i=n+1}^{2k} 0\right)$$

$$2ka_k = 2k + \sum_{i=1}^{n} a_i$$

Combining these two cases proves the lemma.

No neat solutions result from evaluating n x n determinants through Cramer's rule. Figure 1 shows the value of $E(2n+2)/2n$ for $1 \leq n \leq 150$.

In fact, by looking at data produced by our code, it appears that $\lim_{n\to\infty} E(2n+2)/n = (4\ln 2 - 2)/(1 - \ln 2) \approx 2.51778$. We will try to prove this interesting result that $E(n)$ is linear in n, by solving the system of equations.

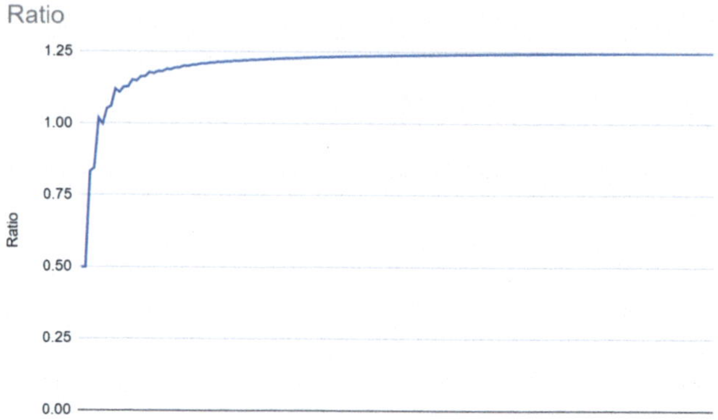

Fig. 1. The value of $E(2n+2)/(2n+2)$ for $1 \leq n \leq 150$.

Define $a_{i,n}$ with $1 \leq i \leq n$ as the n solutions to the system of n linear equations, similarly defined as the lemma above. Let $n \cdot j' = j$ and $n \cdot b_{j',n} = a_{n \cdot j',n}$. Denote $b_{j'} = \lim_{n\to\infty} b_{j',n}$. (This definition is scaling down a_i.)

We now provide a formula for $b_{j'}$, in terms of b_1.

Theorem 1. *Let $k \in \mathbb{Z}_0^+$, then for $\frac{1}{2^{k+1}} \leq j' < \frac{1}{2^k}$, we have*

$$b_{j'} = \frac{b_1}{j'} \sum_{i=0}^{k} \frac{\ln^i(2^i j')}{2^i i!}.$$

Proof. Our system of linear equations have augmented matrix:

$$\left[\begin{array}{cccccccccccc|c}
-1 & 1 & 0 & 0 & 0 & 0 & \cdots & 0 & 0 & 0 & \cdots & 0 & -2 \\
1 & -3 & 1 & 1 & 0 & 0 & \cdots & 0 & 0 & 0 & \cdots & 0 & -4 \\
1 & 1 & -5 & 1 & 1 & 1 & \cdots & 0 & 0 & 0 & \cdots & 0 & -6 \\
\vdots & \vdots & \vdots & \vdots & \vdots & \vdots & \ddots & \vdots & \vdots & \vdots & \cdots & \vdots & \vdots \\
1 & 1 & 1 & 1 & 1 & 1 & \cdots & 1-n & 1 & 1 & \cdots & 0 & 2-n \\
1 & 1 & 1 & 1 & 1 & 1 & \cdots & 1 & -1-n & 1 & \cdots & 1 & -n \\
\vdots & \vdots & \vdots & \vdots & \vdots & \vdots & \ddots & \vdots & \vdots & \vdots & \cdots & \vdots & \vdots \\
1 & 1 & 1 & 1 & 1 & 1 & \cdots & 1 & 1 & 1 & \cdots & -(2n-1) & -2n
\end{array}\right]$$

Base case $k = 0$: We wish to show, for $\frac{1}{2} \leq j' < 1$, we have $b_{j'} = \frac{b_1}{j'}$. We have $\frac{n}{2} \leq j < n$. Subtract row j from row n to get

$$\begin{aligned}
2ja_{j,n} &= 2na_{n,n} - 2(n-j) \\
ja_{j,n} &= na_{n,n} + j - n \\
n^2 j' b_{j',n} &= n^2 b_{1,n} + nj' - n \\
b_{j',n} &= \frac{b_{1,n}}{j'} + \frac{1}{n} - \frac{1}{nj'} \\
b_{j'} &= \frac{b_1}{j'} \quad \text{by taking } n \to \infty, \text{finishing.}
\end{aligned}$$

For the inductive step, assume the statement is true for $k-1$. $(k \geq 2)$ Let $\frac{n}{2^{k+1}} \leq j < \frac{n}{2^k}$. Subtract row j from row n to get

$$2ja_{j,n} + \sum_{i=2j+1}^{n} a_{i,n} = 2na_{n,n} - 2n + 2j$$

$$2n^2 j' b_{j',n} + n^2 \sum_{i=2j+1}^{n} b_{i/n,n} \cdot \frac{1}{n} = 2n^2 b_{1,n} - 2n + 2nj'$$

Divide by n^2 and take $n \to \infty$. The Riemann Sum turns into an integral:

$$2j' b_{j'} + \int_{2j'}^{1} b_x dx = 2b_1$$

To simplify future work, we solve the integral in a lemma. □

Lemma 4. *If* $\frac{1}{2^{k+1}} \leq j' < \frac{1}{2^k}$, *then*

$$\int_{2j'}^{1} b_x dx = -b_1 \sum_{l=0}^{k-1} \frac{\ln^{l+1}(2^{l+1} j')}{2^l (l+1)!}.$$

Proof. Evaluating the integral, we get:

$$\int_{2j'}^{1} b_x dx = \sum_{i=0}^{k-2} \int_{1/2^{i+1}}^{1/2^i} b_x dx + \int_{2j'}^{1/2^{k-1}} b_x dx$$
$$- b_1 \left(\sum_{i=0}^{k-2} \int_{1/2^{i+1}}^{1/2^i} \frac{1}{x} \sum_{l=0}^{i} \frac{\ln^l (2^l x)}{2^l l!} dx + \int_{2j'}^{1/2^{k-1}} \frac{1}{x} \sum_{l=0}^{k-1} \frac{\ln^l (2^l x)}{2^l l!} dx \right)$$

Make a u-substitution $u = \ln(2^l x)$:

$$\int_{2j'}^{1} b_x dx = b_1 \left(\sum_{i=0}^{k-2} \left[\sum_{l=0}^{i} \frac{\ln^{l+1}(2^l x)}{2^l (l+1)!} \right]_{1/2^{i+1}}^{1/2^i} + \left[\sum_{l=0}^{k-1} \frac{\ln^{l+1}(2^l x)}{2^l (l+1)!} \right]_{2j'}^{1/2^{k-1}} \right)$$
$$= b_1 \left(\sum_{i=0}^{k-2} \sum_{l=0}^{i} \frac{\ln^{l+1}(2^{l-i}) - \ln^{l+1}(2^{l-i-1})}{2^l (l+1)!} \right.$$
$$\left. + \sum_{l=0}^{k-1} \frac{\ln^{l+1}(2^{l-k+1})}{2^l (l+1)!} - \sum_{l=0}^{k-1} \frac{\ln^{l+1}(2^{l+1} j')}{2^l (l+1)!} \right)$$

The telescoping sum cancels with the third term, and we obtain

$$\int_{2j'}^{1} b_x dx = -b_1 \sum_{l=0}^{k-1} \frac{\ln^{l+1}(2^{l+1} j')}{2^l (l+1)!}.$$

□

From the lemma,

$$2j'b_{j'} = 2b_1 + b_1 \sum_{l=0}^{k-1} \frac{\ln^{l+1}(2^{l+1}j')}{2^l(l+1)!} = 2b_1 + 2b_1 \sum_{l=1}^{k} \frac{\ln^l(2^l j')}{2^l l!}$$
$$= 2b_1 + 2b_1 \sum_{l=0}^{k} \frac{\ln^l(2^l j')}{2^l l!} - 2b_1$$
$$b_{j'} = \frac{b_1}{j'} \sum_{l=0}^{k} \frac{\ln^l(2^l j')}{2^l l!}$$

□

We can now express any solution (with $n \to \infty$), $b_{j'}$, in terms of b_1. We are now interested in representing b_0 as a constant multiple of b_1.

Proposition 1. $b_0 = b_1/(1 - \ln 2)$.

Proof. We begin with a lemma that will help us prove the result rigorously.

Lemma 5.

$$\lim_{k\to\infty} 2^{k+1} \sum_{i=0}^{k} \frac{(i-k-1)^i}{i!} \left(\frac{\ln 2}{2}\right)^i = \lim_{k\to\infty} 2^{k+1} \sum_{i=0}^{\infty} \frac{(i-k-1)^i}{i!} \left(\frac{\ln 2}{2}\right)^i$$

Proof. This proof, is tedious and long but yet holds no significant meaningful value. We outline the sketch of the proof, leaving the details out.
We use *Stirling's approximation* to show

$$\lim_{k\to\infty} 2^{k+1} \sum_{i=k+1}^{\infty} \frac{(i-k-1)^i}{i!} \left(\frac{\ln 2}{2}\right)^i \le \lim_{k\to\infty} 2^k \sum_{i=0}^{\infty} \exp\left((i+k)\ln\left(\frac{i}{i+k} \cdot \frac{e\ln 2}{2}\right)\right)$$

Let $f(i) = (i+k)\ln\left(\frac{i}{i+k} \cdot \frac{e\ln 2}{2}\right)$, treating k as a constant for now. We showed $f(i) < -0.752k$ for all positive i via differentiation and elementary bounds. Then

$$\lim_{k\to\infty} \sum_{i=0}^{\infty} 2^k e^{f(i)} \le \lim_{k\to\infty} \left(\sum_{i=0}^{10^3 k} 2^k e^{-0.752k} + \sum_{i=10^3k+1}^{\infty} 2^k \exp\left((i+k)\ln\left(\frac{ie\ln 2}{2(i+k)}\right)\right)\right)$$
$$\le \lim_{k\to\infty} (10^3 k + 1)(e^{k\ln 2 - 0.752k}) + \lim_{k\to\infty} \sum_{i=10^3k+1}^{\infty} 2^k e^{-0.059(i+k)} = 0$$

□

Theorem 1 with $x = j' = \frac{1}{2^{k+1}}$ and $k \to \infty$ gives us

$$b_0 = b_1 \lim_{k\to\infty} 2^{k+1} \sum_{i=0}^{k} \frac{\ln^i(2^{i-k-1})}{2^i i!}$$

The *Lagrange Inversion Formula* applied on the rth power of the *Lambert W Function* $W_0(x)$ rewritten as a Taylor expansion gives the well known identity

$$e^{-rW_0(x)} = \sum_{i=0}^{\infty} \frac{r(i+r)^{i-1}}{i!}(-x)^i. \tag{1}$$

and holds for $x \in (-\frac{1}{e}, \frac{1}{e})$ and $r \in \mathbb{C}$. Differentiating both sides with respect to x then dividing throughout by r, we have

$$W_0'(x)e^{-rW_0(x)} = \sum_{i=1}^{\infty} \frac{(i+r)^{i-1}}{(i-1)!}(-x)^{n-1} = \sum_{i=0}^{\infty} \frac{(i+1+r)^i}{i!}(-x)^i$$

Substitute $r = -k-2$ and $x = -\frac{\ln 2}{2} \in (-\frac{1}{e}, \frac{1}{e})$, we have:

$$W_0'\left(-\frac{\ln 2}{2}\right) e^{(k+2)W_0(-\frac{\ln 2}{2})} = \sum_{i=0}^{\infty} \frac{(i-k-1)^i}{i!}\left(\frac{\ln 2}{2}\right)^i$$

Use the identities $W_0(\frac{\ln x}{x}) = \ln x$ and $1/W_0'(x) = x + e^{W_0(x)}$ to get

$$W_0\left(-\frac{\ln 2}{2}\right) = -\ln 2, \qquad W_0'\left(-\frac{\ln 2}{2}\right) = \frac{2}{-\ln 2 + 2e^{W_0(-\frac{\ln 2}{2})}} = \frac{2}{1-\ln 2}.$$

Substitute the values into the above equation to get

$$\frac{2e^{-(k+2)\ln 2}}{1-\ln 2} = \sum_{i=0}^{\infty} \frac{(i-k-1)^i}{i!}\left(\frac{\ln 2}{2}\right)^i$$

Our original sum now gives

$$\begin{aligned} b_0 &= b_1 \lim_{k\to\infty} 2^{k+1} \sum_{i=0}^{k} \frac{\ln^i(2^{i-k-1})}{2^i i!} = b_1 \lim_{k\to\infty} 2^{k+1} \sum_{i=0}^{k} \frac{(i-k-1)^i}{i!}\left(\frac{\ln 2}{2}\right)^i \\ &= b_1 \lim_{k\to\infty} 2^{k+1} \sum_{i=0}^{\infty} \frac{(i-k-1)^i}{i!}\left(\frac{\ln 2}{2}\right)^i = b_1 \lim_{k\to\infty} 2^{k+1} \frac{2e^{-(k+2)\ln 2}}{1-\ln 2} \\ &= \frac{1}{1-\ln 2} b_1 \end{aligned}$$

Where the third equality follows from the lemma above. □

Now, we let $k_1 = 2$ and $k_m = 2m/\left(2m - \prod_{i=\lceil m/2\rceil}^{m-1} k_i\right)$ for $n \geq 2$. Then

Lemma 6.

$$b_1 = \lim_{n\to\infty} \frac{\sum_{k=1}^{n} \prod_{j=k}^{n} k_j}{n^2}.$$

Proof. By performing Gaussian elimination on our system of linear equations in the proof of 1, we convert it to an upper triangular matrix, from which we can find b_1. We perform the following operations:

$$R_n - R_{n-1} \to R_{n-1} - R_{n-2} \to \cdots \to R_2 - R_1$$

$$\left[\begin{array}{cccccccccccc|c} -1 & 1 & 0 & 0 & 0 & 0 & 0 & 0 & \cdots & 0 & 0 & -2 \\ 2 & -4 & 1 & 1 & 0 & 0 & 0 & 0 & \cdots & 0 & 0 & -2 \\ 0 & 4 & -6 & 0 & 1 & 1 & 0 & 0 & \cdots & 0 & 0 & -2 \\ 0 & 0 & 6 & -8 & 0 & 0 & 1 & 1 & \cdots & 0 & 0 & -2 \\ \vdots & \vdots & \vdots & \vdots & \vdots & \vdots & \vdots & \vdots & \cdots & \vdots & \vdots & \vdots \\ 0 & 0 & 0 & 0 & 0 & 0 & 0 & 0 & \cdots & 2(n-1) & -2n & -2 \end{array}\right]$$

Let position (i, j) denote the element in the ith row and jth column.

We follow the elimination process: Multiply Row R_i by k_i, and subtract it from the row R_{i+1}, for $i \geq 1$. Clearly $k_1 = 2$. Then we add R_1 to R_2 to get $[0 \quad -2 \quad 1 \quad 1 \quad \ldots \quad 0]$. Then we need $k_2 = 2$. Before the elimination process, for row R_m, $m \geq 2$, there is always a $-2m$ element at (m, m). Observe column m. The only non-zero elements are 1 at $(\lceil m/2 \rceil, m)$ and $-2m$ at (m, m). After the process of:

$$k_1 R_1 \to R_2 + R_1 \to k_2 R_2 \to R_3 + R_2 \to k_3 R_3 \to \cdots \to R_m + R_{m-1}$$

The element at (m, m) becomes $\prod_{i=\lceil m/2 \rceil}^{m-1} k_i - 2m$. In order to cancel out with the element $2m$ at position $(m+1, m)$, we have $k_m = \frac{2m}{2m - \prod_{i=\lceil m/2 \rceil}^{m-1} k_i}$. This continues until we find the value k_n. Then the element at (n, n) shall be $k_n(\prod_{i=\lceil n/2 \rceil}^{n-1} k_i - 2n) = -2n$. Now consider the augmented column. The bottommost element after all the row operations becomes:

$$k_n(k_{n-1}(\ldots(k_2(k_1(-2) - 2) - 2)\ldots) - 2) = -2 \sum_{k=1}^{n} \prod_{j=k}^{n} k_j$$

Then the desired solution is:

$$a_{n,n} = \frac{-2 \sum_{k=1}^{n} \prod_{j=k}^{n} k_j}{-2n} = \frac{\sum_{k=1}^{n} \prod_{j=k}^{n} k_j}{n}$$

$$b_1 = \lim_{n \to \infty} \frac{\sum_{k=1}^{n} \prod_{j=k}^{n} k_j}{n^2}$$

We are done. □

Conjecture 1.

$$b_1 = \lim_{n \to \infty} \frac{\sum_{k=1}^{n} \prod_{j=k}^{n} k_j}{n^2} = 4 \ln 2 - 2.$$

Remark 1. So far, we have not proved this conjecture. However, there is a accurate approximation of $k_n \approx 1 + \frac{2}{n} + (-1)^n \frac{2}{n^2} - \cos\left(\frac{\pi}{4}(2n+1)\right) \frac{1}{2n^3}$ provided by user Gary in [1].

Assuming the above conjecture is true, Proposition 1 and Lemma 3 gives

$$\lim_{n \to \infty} \frac{E(2n+2)}{n} = \frac{4 \ln 2 - 2}{1 - \ln 2}$$

which verifies our hypothesis that $E(n)$ scales linearly as $\frac{2 \ln 2 - 1}{1 - \ln 2} n \approx 1.25889n$.

4 Conclusion

Our main aim was to find properties of $E(n)$, and we propose that $E(n)$ approaches $\frac{2\ln 2-1}{1-\ln 2}n \approx 1.25889n$ as $n \to \infty$, in which we left the final piece of the puzzle as a conjecture. Further research into the problem might be able to construct a complete, rigorous proof for $E(n)$ behaviour.

5 Future Work

Our work sheds light on many fields: The problem statement is a combination of Probability and Recursion, specifically Markov chains and hitting times, while our analysis brought us to Linear Algebra, Calculus and infinite sums. To further this research, asymptotic analysis of k_i may allow us to prove the conjecture. We could also generalise the sequence into various forms and analyse their behaviour as well (see Tables 1 and 2).

Acknowledgements. We would like to thank our mentor, Dr. Lee Chan Lye, for his guidance throughout this journey.

A Appendix

Table 1. Table of $E(n)$ values for even n from 2 to 50.

n	E(n)	n	E(n)	n	E(n)	n	E(n)
2	1	16	16.96367521	30	34.92744209	44	52.34625659
4	2	18	20.16873278	32	37.27699089	46	55.01160801
6	5	20	22.17631975	34	40.10616216	48	57.4065546
8	6.75	22	24.81237607	36	42.33808791	50	60.11762834
10	10.2	24	27.06975665	38	44.95366391		
12	11.98684211	26	29.95333071	40	47.29704678		
14	14.75	28	32.1857053	42	50.01285974		

Table 2. Relative error of the asymptotic approximation for $E(n)$ for $n = 2^k$. Error is calculated as $\frac{A-E(n)}{E(n)}$ where $A = \frac{4\ln 2-2}{1-\ln 2}n$.

n	$E(n)$	$\frac{4\ln 2-2}{1-\ln 2}n$	Relative Error (%)
2	1	2.5177	151.7783%
4	2	5.0355	151.7783%
8	6.75	10.0711	49.2019%
16	16.9636	20.1422	18.7376%
32	37.2769	40.2845	8.0681%
64	77.5929	80.5690	3.8356%
128	157.9427	161.1380	2.0231%
256	318.3976	322.2761	1.2181%
512	638.9959	644.5523	0.8696%
1024	1279.9198	1289.1047	0.7176%
2048	2561.5432	2578.2094	0.6506%
4096	5124.5204	5156.4189	0.6225%
8192	10250.2227	10312.8379	0.6109%

References

1. Math Stack Exchange: Limit of a recurrence relation. (2024). https://math.stackexchange.com/questions/5020478/limit-of-a-recurrence-relation-lim-n-to-infty-frac1n2-sum-k-1n. Accessed 21 May 2024
2. Weisstein, E.W.: Lambert W-Function. from MathWorld–a wolfram web resource. https://mathworld.wolfram.com/LambertW-Function.html. Accessed 21 May 2024

Sensor-Assisted Gamified Assessment of Cognitive, Socio-Emotional, and Motor Domains in Older Adults: Design and Feasibility Study

Chng Yu Ye Kelly[1], Wang Xinyue[2], and Aung Aung Phyo Wai[2(✉)]

[1] Hwa Chong Institution (College), Singapore, Singapore
211706x@student.hci.edu.sg
[2] Nanyang Technological University, Singapore, Singapore
apwaung@ntu.edu.sg

Abstract. Age-related neurocognitive disorders often erode cognition, socio-emotional engagement, and motor coordination in parallel, yet most digital tools assess these domains in isolation and rely on self-report or game scores alone. We introduce TriADL game suit, a sensor-assisted, gamified assessment suite using multimodal wearable sensors to address cognition, motor and socio-emotional needs of the elderly. The suite consists of different levels of mini-game modules to assess and train (i) spatial working memory, (ii) bimanual rhythmic fluency, and (iii) autobiographical recall and affect. Each module subscribes to a shared analytics service that streams θ/α cognitive-load indices, scan-path entropy, and gesturelatency symmetry every 250 ms, allowing real-time difficulty scaling and comprehensive log export. This paper details the design rationale, sensor integration workflow, calibration procedures, and data schema, and then reports initial feasbility evaluation with different sensors. Next step will be planing multiple subjects evaluation with longitudinal deployments aimed for personalised training across the cognitive–emotional–motor spectrum in older adults.

Keywords: Hand-Eye Coordination · Reminiscence Therapy · Spatial Attention · Multimodal sensors

1 Background

Older adults living with mild cognitive impairment (MCI) and early-stage dementia frequently experience a triple decline—diminishing executive function and memory, reduced social–emotional engagement, and progressive motor slowing—which undermines independence in activities of daily living [22,?]. Singapore longitudinal ageing study demonstrates the motor slowing component, how functional mobility decline (measured by Timed-Up-and-Go test) precedes incident MCI and early dementia [15]. Study by Subramaniam et al. highlights social-emotional engagement decline by individuals with MCI in Singapore experienced marked disability, clinical levels of depression and

H. Guo et al. (Eds.): IRC-SET 2025, PTT, pp. 87–98, 2026.
https://doi.org/10.1007/978-981-95-9630-0_8

anxiety [27]. Digital interventions have begun to tackle each domain in isolation with the significant gains in cognition and behaviours [4, 20]. Recent randomized controlled trials have demonstrated the efficacy of gamified interventions for cognitive enhancement in older adults with MCI, with 8-week game training showing significant improvements in cognition and reductions in depression symptoms compared to control groups [34]. The gamified cognitive platforms such as BrainHQ and mobile n-back applications reported the modest gains in attention and working memory after 6- to 12-week programmes [20]. Large-scale citizen-science titles like Sea Hero Quest demonstrate the diagnostic value of virtual way-finding for early spatial disorientation [11]; Leap-Motion–based exergames improve upperlimb dexterity and quality of life in randomized trials with community-dwelling elders [23]; and immersive reminiscence experiences delivered in virtual reality reduce depressive symptoms while preserving mood and autobiographical recall in longitudinal studies of nursing-home residents [18].

However, three major gaps remain in such gamified solutions. Firstly, most serious games remain single domain such as cognitive, motor or socio-emotional—despite evidences that integrated, multi-domain stimulation yields larger functional effects in MCI meta-analyses [16]. Secondly, outcome metrics rely heavily on in-game scores or questionnaires; real-time physiological indices such as EEG-derived engagement and gaze-based workload are only now being explored in small laboratory samples [1]. Lastly, existing feasibility studies seldom report sensor reliability, latency, or home-use practicality—factors identified as barriers in recent scoping reviews of serious digital games for dementia care [21]. To address these gaps, we propose TriADL-Game suite that combines low-cost multimodal sensors integrated with data-driven assessment across three thematically linked game modules. The present paper outlines the system design rationale, sensor-integration workflow, and preliminary feasibility evaluations that pave the system development and performance evaluation towards patient-centric holistic dementia care.

2 Proposed Solution: TriADL Game Suite

The TriADL-Games suite, Fig. 1, introduces two core innovations differentiating from existing games for older adults besides triple domains coverage. First, the platform features real-time adaptive game engine powered by sensory datadriven besides performance-driven customization. By continuously analysing EEG, eye-tracking, and motion data, the system dynamically adjusts game difficulty, pacing, and support cues to maintain optimal challenge and engagement throughout each session. This closed-loop adaptation ensures that interventions are both responsive and personalised, which is rarely achieved in current gamebased assessment tools. Second, TriADL-Games incorporates multimodal performance profiling. Data-driven inference models fuse cognitive, motor, and socioemotional signals from multiple sensors to construct comprehensive user profiles. These profiles enable more precise identification of individual strengths and challenges, supporting a holistic approach to assessment and targeted intervention.

By using Lab Streaming Layer, incoming data and event streams are timestamped and synchronized on a shared clock and performed basic quality checks ensuring data integrity and reliability. The adaptive engine fuses sensor features with live performance (errors, pauses, completions) to infer moment-to-moment state (e.g., workload, inattention, fatigue) and sends closed-loop adjustments back to the game modules, dynamically tuning difficulty, pacing, and cue modality. A profiling and analytics layer combines synchronized signals and events to compute per-trial indices (cognitive load, gaze efficiency, etc.), visualize progress, and summarize feasibility metrics (uptime, latency, dropouts). Personalisation provides user baselines, accessibility preferences, target goals, and safety limits that constrain the engine's adaptations. Any personally identifiable context is stored in an encrypted Privacy Vault, with only de-identified tags used for analytics and exports. So TriADL combines these components to form a safe, individualized, and quantifiable closed-loop personalise game-based training and assessment system.

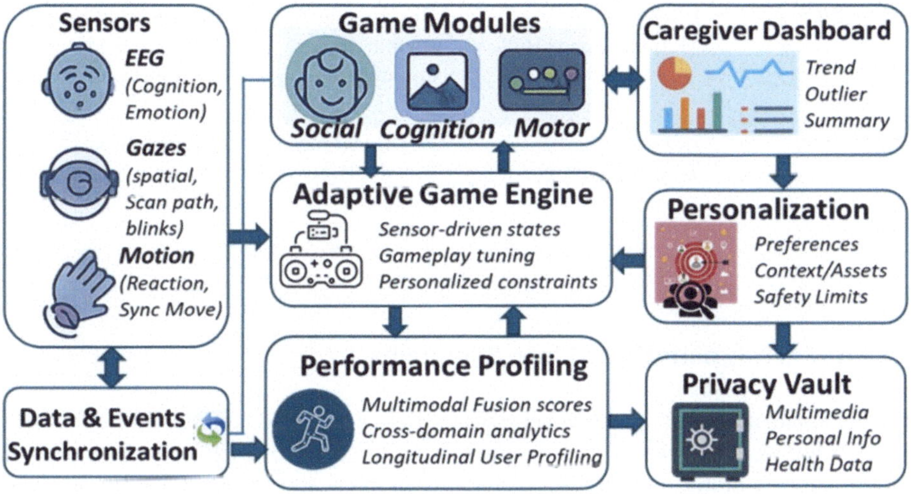

Fig. 1. Tri-ADL Game Suite System Architecture and Design

As shown in Table 1, the first game module Memory Market + aims to address Spatial disorientation & route-finding unmet needs by simulating familiar-route way-finding with adaptive EEG/eye hints. The second game, Motion Memoirs, use rhythm-guided, gesture-tracked routines that restore coordination to resolve common issues of motor slowing and bimanual decline. The third game, Life Threads + provides interactive life-story reconstruction with voice, photos, emotion mapping to issues like identity erosion & social withdrawal.

To systematically capture user requirements and preferences for the TriADLGames suit, we conducted an online survey using Google Forms structured across five sections with total forty-five questions. These sections are designed to gather 1) Screening and Demographics 2) Functional Assessment and Training Priorities 3) Technology Acceptance and Privacy Attitudes 4) Usability and Accessibility Preferences 5) Game-Specific Feature Evaluation respectively. The survey employed 5-point Likert scales for quantitative measures, supplemented by targeted free-text response fields to capture nuanced user perspectives. This multi-method approach was structured to inform design decisions including: requirement specification, feature prioritisation, safety establishment, and potential adoption barriers related to setup, user comfort, and privacy concerns. Response data underwent both descriptive and thematic analysis to generate actionable design specifications for the feasibility evaluation and prototype development.

Table 1. Game specifications and measurement focus

Game Title	Sensor(s) Used	Measurement Focus	ADL Simulations
Memory Market +	EEG, Eye Tracker	Working memory, spatial attention	Grocery shopping, Navigating shops, Spatial Orienting
Motion Memoirs	EEG, Leap Motion, Arcade Button	Bimanual fluency, rhythm, motor coordination	Cooking routines, Bilateral hand coordination, sequential following
Life Threads +	EEG, Eye Tracker, Microphone	Emotional recall, social interaction, facial Emotions	Conversation, Storytelling, Memory reflection

3 Game Design and Use Case Scenarios

We will explain details of the three thematically linked gamified modules of the TriADL-Games Suite as shown in Table 1 using synchronised EEG–gaze–motion sensing to enable cognitive, motor, and autobiographical assessment and training for early dementia and MCI.

3.1 Memory Market+

The goal is to train spatial memory, sequencing, cue recognition, and decisionmaking in personally familiar environments (e.g., home-to-market routes) [10]. A grocery-store way-finding game that trains and assesses spatial memory, planning, and selective attention as shown in Fig. 2 (a). Players follow familiar routes to locate list items while EEG features (e.g., θ/α) and gaze features guide real-time hints (ghost paths, aisle arrows). This module targets early disorientation—an established marker in prodromal dementia—and builds on evidence that virtual navigation tasks sensitively index real-world navigation ability. e.g., Sea Hero Quest findings [11]. The targeted skills include Spatial orientation, visual scanning of landmarks, memory encoding/route recall, and inhibition/planning to play around 6 to 8 min including tutorial and 3–4 challenge levels.

Personalised maps (Google Street View stitches, family photos) and local landmark cues (e.g., MRT sign, kopitiam) were used to define each route with cueing of ambient street sounds at early levels but quieter soundscape at higher difficultly levels. Primarly, EEG is used for assistance triggers, eye tracking for hesitation/confidence, optional hand motion sensors for gesture-based navigation [6]. Upon reaching destination, user will be asked to do daily chores like making popular snacks or foods of preferences as shown in Fig. 2 (b). Hints and feedbacks are adaptive to real-time sensor-driven assessment such as prolonged eye dwell elicits a confirmation prompt, elevated EEG stress slows pacing or adds in-situ visual/audio cues, uncertain gestures evoke a gentle retry nudge—mechanics [2]. The novel elements include personalised ecological validity, sensor-adaptive gameplay, memory-driven decision points, emotionally anchored navigation, and familiarity-based difficulty scaling [26]. Both in-game metrics such as route efficiency, landmark-guided decision accuracy, and reversalnavigation success and sensor-driven metrics such has EEG-based load trends, gaze scanning/hesitation, and gesture smoothness—derived using the same adaptive criteria [5].

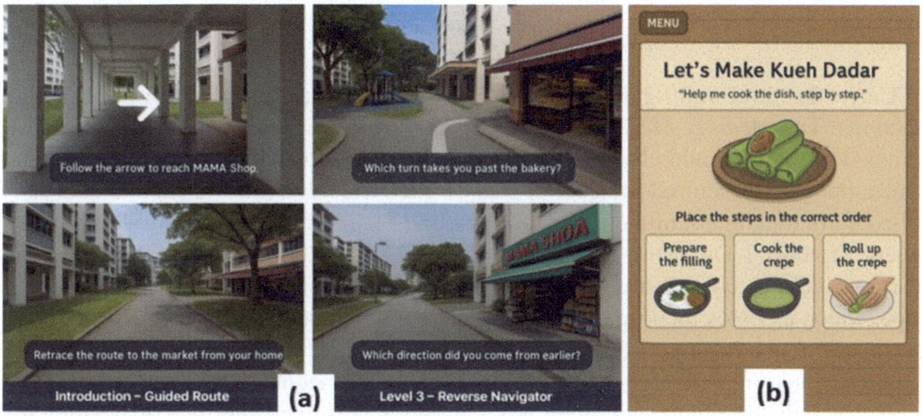

Fig. 2. Sample screenshots of first game module (a) Wayfinder (b) Snack Maker (Game scene generated by GenAI tool based on initial game description text)

3.2 Motion Memoirs

This game module is designed for short, low-fatigue sessions (6 – 10 min) motor coordination tasks with gentle, non-punitive feedback based on recognized neural and physical events to encourage confidence and adherence [35]. A rhythmguided, gesture-tracked games can elicit bilateral reaches and sequencing akin to kitchen routines [19]. Different sensors such as webcam/motion sensor-based kinematics quantify the initiation latency, peak-velocity symmetry, and movement smoothness as well as adaptive pacing [24]. Prior work shows Leap-Motion exergames are feasible for older adults and can improve upper-limb dexterity and engagement, motivating this design choice [14]. Our goal is to train and assess bimanual fluency, rhythm entrainment, hand–eye response timing, inhibition, and expressive movement using emotionally meaningful content. For better user engagement, game user interfaces will be personalised by seeding caregiversupplied photos, object icons, and nostalgic music (language/tempo) pieces [29, 30].

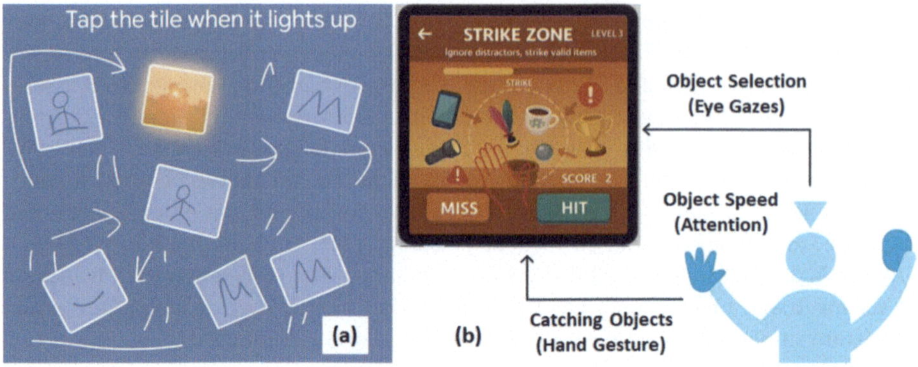

Fig. 3. Sample screenshots of second game module (a) Memory Mosaic (b) Memory Drift (both game scenes generated using GenAI tool based on initial game concept description)

Currently, three mini-games are designed to target specific cognitive and motor functions. The *Memory Mosaic* as shown in Fig. 3 (a) is a rhythm-based tapping task designed to measure and challenge attentional control and spatial orientation [8]. The game employs a structured protocol of four scripted trial blocks designed to create controlled variations in cognitive load: 1) a baseline Static, 2) a Slow Float with moving targets, 3) a Fast + Distractors phase, and 4) a final Static Recovery phase. The outcome measures include reaction time (RT) distribution, hit rate, false alarms, and the load-RT slope, which assesses the impact of increasing cognitive demand on performance. The *Memory Drift* as shown in Fig. 3 (b) engages users in a three-phase loop designed to train executive functions and bimanual motor skills [3]. The sequence requires participants to first catch drifting items with their left and right hands, then choose a disposition for each item (e.g., Keep, Discard, Return), and finally return the item to a contextually appropriate location, such as a shelf or laundry line. Performance is quantified using metrics such as the left/right hand use balance, hand-path smoothness, the frequency of specific decisions, and the tracking of error-correct cycles. The *Fun Moves* is an expressive dance activity intended to reinforce motor control and emotional expression. The game guides participants through a series of symbolic gestures (e.g., "wave to a friend," "pick flowers") that gradually progress into short, memorable sequences. An optional freestyle dance segment is also included to encourage creativity and spontaneous movement. The therapeutic assessments combines enhancing coordination, rhythm, posture, and emotional expression with cognitive training in an enjoyable context [17].

3.3 Life Threads+

This module is designed to strengthen identity, foster social engagement, and stimulate emotional well-being while enabling longitudinal monitoring of socio–emotional changes [13]. An interactive reminiscence module where players narrate life episodes using family photos, places, and music supplied by caregivers [32]. Eye/EEG features and simple affect cues titrate prompt depth and timing, while conversational and recall metrics capture socio-emotional engagement. The approach aligns with evidence that reminiscence—especially in immersive, personalised formats—supports mood and autobiographical memory in older adults. Assessing and training autobiographical memory, narrative fluency, and social communication is vital for preserving identity, sustaining relationships, and supporting emotional well-being in older adults [9]. Deficits in these domains are linked to social withdrawal and faster cognitive decline [13]. Interactive storytelling anchored in one's life journey offers ecologically valid, emotionally salient stimuli that enhance engagement and promote long-term retention [12, 33] (Fig. 4).

Fig. 4. Sample screenshots of third game module (Left) Memory Matching (b) Life Reel Narration (both game scenes generated using GenAI tool based on initial game concept description)

The targeted skills include autobiographical (episodic and semantic) recall, verbal fluency, executive sequencing, empathy, perspective-taking, and reality monitoring. Per-stage metrics include gaze patterns, EEG workload/engagement indices, gesture dynamics, postural expressiveness, vocal tone/latency, sequence accuracy, and narrative richness [28]. The game summaries and sensor-based evaluations are fed into clinician dashboards and caregiver reports, supporting longitudinal analysis over multiple weeks. The Life Threads + module comprises five interconnected mini-games designed to stimulate autobiographical memory, emotional awareness, and narrative fluency. In Memory Match – Moments Recalled, players link personal photos with the correct captions or voice descriptions and are encouraged to share related stories, fostering recognition accuracy and emotional engagement. Emotion Mapping – The Heart Remembers builds emotional labelling skills by having players associate life moments with emotion icons and explain their choices. Sequence Builder – Life Timeline challenges players to arrange shuffled personal photos into chronological order, enhancing temporal sequencing and executive planning. In Memory Challenge – What Doesn't Belong?, players identify and identify false elements that were deliberately inserted in their timeline, strengthening reality monitoring and reasoning. Finally, Life Reel Narration – Your Story, Your Words invites players to narrate their life story using their arranged timeline, producing a recorded highlight reel that supports identity expression, autobiographical fluency, and socio-emotional engagement [25].

4 Feasibility Study and Evaluation

Home-based gamified cognitive training has shown acceptable feasibility in clinical populations, with 63% of participants completing more than 50% of recommended sessions and 81% voluntarily continuing training beyond the prescribed intervention period [31]. From initial analysis of our survey results as shown in Fig. 5 strengthen the fesability of TriADL game suit by demonstrating strong motor domain games needs (90.9% acceptance to hand-eye motor gameplay), preference on playing groups or family/caregivers requiring social contexts and 77.2% current and interested non-users with cognitive training and assessment.

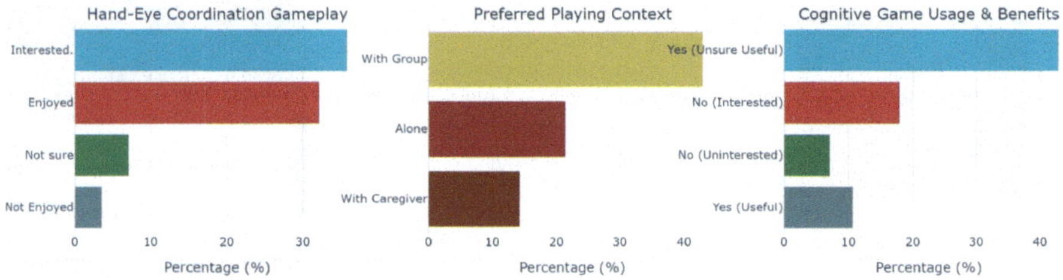

Fig. 5. Comparison of user responses to three domains of TriADL game suite

As part of future development, we aim to further enhance TriADL game suit with additional AI-driven functionalities. Planned improvements include generative AI for personalised content creation, such as custom memory prompts and story elements; explainable AI dashboards to provide clinicians with transparent insights into how sensor data influence in-game adaptations and outcomes; and AI-based session recommendation algorithms that tailor training schedules and game sequences to each user's unique profile and progress. These features will further advance the platform's capacity for individualized, effective, and engaging support for older adults experiencing cognitive, emotional, and motor decline.

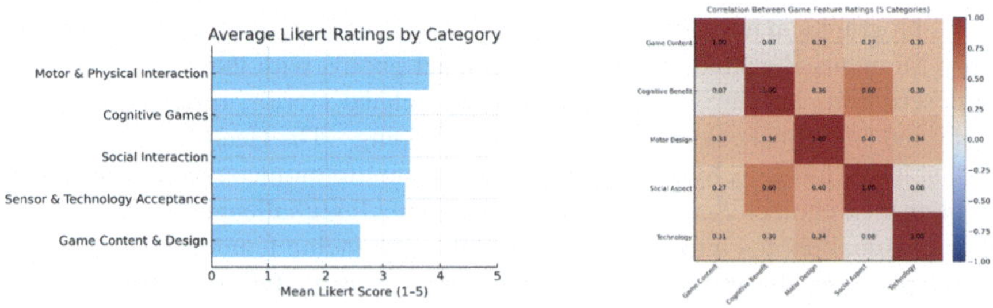

Fig. 6. Average Likert Ratings and Inter-feature Correlation of Game Design Categories (Left) Mean Likert scores 1 (Least Preferred) to 5 (Most Preferred) across five thematic categories (Right) Pearson correlation heatmap between average ratings of the same five categories

The left panel in Fig. 6 shows the average Likert ratings for each category. Motor & Physical Interaction achieved the highest mean score ($M = 3.9$), indicating strong participant interest in physically engaging activities integrated into the game. Sensor & Technology Acceptance ($M = 3.6$) and Social Interaction ($M = 3.5$) received similarly positive evaluations, suggesting openness toward sensor integration and the value of socially oriented gameplay. Cognitive Games also rated comparably ($M = 3.4$), highlighting appreciation for mental stimulation features. In contrast, Game Content & Design ($M = 2.9$) was rated lowest, suggesting that aesthetic, narrative, or interface design aspects could benefit from further refinement.

The right panel in Fig. 6 shows the correlation matrix between the five categories. Strong positive correlations emerged between Game Content & Design and Cognitive Games ($r = 0.87$), as well as between Cognitive Games and Social Interaction ($r = 0.62$). These relationships indicate that perceptions of well-designed content are closely tied to perceived cognitive benefits, and that mentally stimulating games are also associated with enhanced social engagement. Moderate correlations were observed between Motor & Physical Interaction and both Game Content & Design ($r = 0.33$) and Sensor & Technology Acceptance ($r = 0.31$), suggesting that physical interactivity is moderately linked to both design appeal and technological openness. Lower correlations (e.g., Social Interaction and Sensor & Technology Acceptance, $r \approx 0.08$) indicate dimensions perceived more independently.

5 Discuss and Conclusion

Recent findings from gamified dual-task training paradigms demonstrated the potential of integrated, multi-domain interventions for addressing the needs of individuals with cognitive impairment [7]. By integrating real-time adaptive gameplay and AI-driven multimodal performance profiling across three engaging mini-games, TriADL game suite offers a holistic and responsive approach to early dementia and MCI assessment and training. Our survey findings indicate that physical engagement and sensor-based interactivity are highly valued by older adults, supporting the inclusion of motion-based tasks, balance exercises, and real-time feedback in game design.

The relatively high ratings for Sensor & Technology Acceptance suggest that barriers to adopting sensor-assisted systems may be lower than expected, provided usability is addressed. However, the lower mean rating for Game Content & Design reveals a design gap. While functional and interactive components are rated positively, the overall appeal of the game may be limited by narrative depth, visual polish, or thematic relevance. Improving these elements could yield cross-domain gains due to their strong association with cognitive and social perceptions. Finally, the relatively low correlations between certain dimensions (e.g., Social Interaction and Sensor & Technology Acceptance) suggest that design improvements may need to be targeted separately. For example, enhancing social engagement features (e.g., cooperative tasks, multiplayer modes) may not automatically improve perceptions of technology, and vice versa—requiring parallel but distinct design strategies.

In summary, Memory Market + combines personalised, sensor-adaptive wayfinding, object-memory, and rule-switching mini-games to holistically train and assess spatial orientation, working memory, and executive flexibility in ecologically meaningful contexts.

Motion Memoirs unifies personalised nostalgia with bimanual, rhythm-guided interaction across three complementary sub-games, producing analysable motor and attention signals aligned to everyday actions. Life Threads + uniquely blends personalised, multi-sensor, multi-domain cognitive training with emotionally resonant storytelling, false-memory detection, and adaptive difficulty. By integrating caregiver co-creation, AI-assisted content generation, and social sharing, it addresses gaps in personalisation, ecological validity, and longitudinal tracking while enhancing cultural sensitivity and clinical utility. So the proposed thematic mini-games with synchronised neurophysiological signals with in-game events enable adaptive support that is both individualized and evidence-based addressing the current user needs and applications gaps. Looking forward, future enhancements including AI-generated personalised content, clinician-facing explainable analytics, and session recommendation algorithms will further strengthen the platform's impact. Collectively, these advances position TriADL as a framework for scalable, home-based cognitive, emotional, and motor health support in age-related neurocognitive disorders.

References

1. Ahmed, Y., Ferguson-Pell, M., Adams, K., Ríos Rincón, A.: Eegbased engagement monitoring in cognitive games. Sensors **25**(7) (2025). https://doi.org/10.3390/s25072072
2. Aricò, P., et al.: Adaptive automation triggered by eeg-based mental workload index: A passive brain-computer interface application in realistic air traffic control environment. Front. Hum. Neurosci. **10**, 539 (2016). https://doi.org/10.3389/fnhum.2016.00539
3. Balasubramanian, S.P., Melendez-Calderon, A., Roby-Brami, A., Burdet, E.: Onthe analysis of movement smoothness. J. Neuroeng. Rehabil. **12**(1), 112 (2015). https://doi.org/10.1186/s12984-015-0090-9
4. Bonnechère, B., Langley, C., Sahakian, B.J.: Cognitive training using tablet-basedserious games for mild cognitive impairment: a randomized controlled trial. J. Med. Internet Res. **23**(10), e26498 (2021). https://doi.org/10.2196/26498
5. Boone, A.P., Maghen, B., Hegarty, M.: Instructions matter: Individual differencesin navigation strategy and ability. Mem. Cognit. **47**(7), 1401–1414 (2019). https://doi.org/10.3758/s13421-019-00941-5
6. Bécu, M., et al.: Landmark-based spatial navigation across the human lifespan. eLife **12**, e81318 (2023). https://doi.org/10.7554/eLife.81318
7. Chua, L.K., et al.: Gamified dual-task training for individuals with parkinson disease: An exploratory study on feasibility, safety, and efficacy. Int. J. Environm. Res. Pub. Health **18**(23) (2021). https://doi.org/10.3390/ijerph182312384
8. Cisek, P., Kalaska, J.F.: Neural mechanisms for interacting with a worldfull of action choices. Annu. Rev. Neurosci. **33**, 269–298 (2010). https://doi.org/10.1146/annurev.neuro.051508.135409
9. Conway, M.A., Pleydell-Pearce, C.W.: The construction of autobiographical memories in the self-memory system. Psychol. Rev. **107**, 261–288 (2000). https://doi.org/10.1037/0033-295X.107.2.261
10. Coughlan, G., Laczó, J., Hort, J., Minihane, A.M., Hornberger, M.: Spatial navigation deficits—overlooked cognitive marker for preclinical alzheimer disease? Nat. Rev. Neurol. **14**(8), 496–506 (2018). https://doi.org/10.1038/s41582018-0031-x
11. Coutrot, A., et al.: Global determinants of navigation ability. Curr. Biol. **28**(17), 2861-2866.e4 (2018)

12. Ekman, P., Rosenberg, E.L.: What the face reveals: Basic and applied studies ofspontaneous expression using the facial action coding system (FACS). Oxford University Press (2005). https://doi.org/10.1093/acprof:oso/9780195179644.001.0001
13. El Haj, M., Antoine, P., Nandrino, J.L., Kapogiannis, D.: Autobiographical memory decline in alzheimer's disease, a theoretical and clinical overview. Ageing Res. Rev. **24**, 15–26 (2015). https://doi.org/10.1016/j.arr.2015.07.001
14. Fernández-González, P., et al.: Leap motion controlled video game-based therapy for upper limb rehabilitation in patients with parkinson's disease: a feasibility study. J. Neuroeng. Rehabil. **16**(1), 1–11 (2019). https://doi.org/10.1186/s12984-019-0593-x
15. Gao, Q., et al.: Functional mobility decline and incident mild cognitive impairment and early dementia in community-dwelling older adults: the singapore longitudinal ageing study. Age and Ageing **51**(9), afac182 (2022). https://doi.org/10.1093/ageing/afac182
16. Gavelin, H.M., et al.: Combined physical and cognitive training for older adults with and without cognitive impairment: A systematic review and network meta-analysis of randomized controlled trials. Ageing Res. Rev. **66**, 101232 (2021). https://doi.org/10.1016/j.arr.2020.101232
17. Gavelin, H.M., et al.: Combined physical and cognitive training for older adults with and without cognitive impairment: A systematic review and network meta-analysis of randomized controlled trials. Ageing Res. Rev. **66**, 101232 (2021). https://doi.org/10.1016/j.arr.2020.101232
18. Khirallah Abd El Fatah, N., Abdelwahab Khedr, M., Alshammari, M., Mabrouk Abdelaziz Elgarhy, S.: Effect of immersive virtual reality reminiscence versus traditional reminiscence therapy on cognitive function and psychological well-being among older adults in assisted living facilities: A randomized controlled trial. Geriatric Nursing **55**, 191–203 (2024). https://doi.org/10.1016/j.gerinurse.2023.11.010
19. Lagarrigue, V., Puyjarinet, F., Dalla Bella, S.: At-home training with a rhythmicvideo game for improving motor and cognitive processing of rhythm in parkinson's disease: a pilot study. J. Neurol. **269**(6), 3104–3115 (2022). https://doi.org/10.1007/s00415-021-10903-5
20. Lampit, A., Hallock, H., Valenzuela, M.: Computerized cognitive training in cognitively healthy older adults: a systematic review and metaanalysis of effect modifiers. PLoS Med. **16**(11), e1002987 (2019). https://doi.org/10.1371/journal.pmed.1002987
21. Longley, V., Wilkey, J., , C.O.: Outcome measurement of cognitive impairment and dementia in serious digital games: a scoping review. Disability and Rehabilitation: Assistive Technology **20**(3), 499–509 (2025). https://doi.org/10.1080/17483107.2024.2405894, pMID: 39311520
22. Morris, J.C., et al.: Mild cognitive impairment represents earlystage alzheimer disease. Arch. Neurol. **58**(3), 397–405 (2001). https://doi.org/10.1001/archneur.58.3.397
23. Pirbabaei, E., Amiri, Z., Sekhavat, Y.A., Goljaryan, S.: Exergames for hand rehabilitation in elders using leap motion controller: A feasibility pilot study. Int. J. Human-Comp. Stud. **178**, 103099 (2023). https://doi.org/10.1016/j.ijhcs.2023.103099
24. Proffitt, R., Lange, B.: A novel methodology for extracting and evaluatingtherapeutic movements from unconstrained, unmonitored game-based rehabilitation. J. Neuroeng. Rehabil. **15**(1), 1–15 (2018). https://doi.org/10.1186/s12984-018-0453-0
25. Sabat, S.R., Harré, R.: The construction and deconstruction of self in alzheimer'sdisease. Ageing Soc. **12**(4), 443–461 (1992)
26. Sailer, M., Hense, J.U., Mayr, S.K., Mandl, H.: How gamification motivates: An experimental study of the effects of specific game design elements on psychological need satisfaction. Comput. Hum. Behav. **69**, 371–380 (2017). https://doi.org/10.1016/j.chb.2016.12.033
27. Subramaniam, M., et al.: Prevalence, lifestyle correlates, and psychosocial functioning among multi-ethnic older adults with mild cognitive impairment in singapore: Preliminary findings from a 10/66 population study. Int. J. Environ. Res. Public Health **18**(5), 2702 (2021). https://doi.org/10.3390/ijerph18052702

28. Subramaniam, P., Woods, B.: Digital life storybooks for people with dementialiving in care homes: an evaluation. Clin. Interv. Aging **11**, 1263–1276 (2016). https://doi.org/10.2147/CIA.S111097
29. Swinnen, S.P.: Intermanual coordination: from behavioural principles to neuralnetwork interactions. Nat. Rev. Neurosci. **3**(5), 348–359 (2002). https://doi.org/10.1038/nrn807
30. Thaut, M.H., Abiru, M.: Rhythmic auditory stimulation in rehabilitation of movement disorders: A review of current research. Music Perception: An Interdisciplinary Journal **27**(4), 263–269 (2010). https://doi.org/10.1525/mp.2010.27.4.263
31. van de Weijer, S.C., et al.: Feasibility of a cognitive training game in parkinson's disease: The randomized parkin'play study. European Neurology **83**(4), 426–432 (2020). https://doi.org/10.1159/000509685
32. Woods, B., O'Philbin, L., Farrell, E., Spector, A., Orrell, M.: Reminiscencetherapy for dementia. Cochrane Database of Systematic Reviews (3) (2018). https://doi.org/10.1002/14651858.CD001120.pub3
33. Woods, B., O'Philbin, L., Farrell, E.M., Spector, A.E., Orrell, M.: Reminiscencetherapy for dementia. Cochrane Database of Systematic Reviews, CD001120 (2018). https://doi.org/10.1002/14651858.CD001120.pub
34. Xue, B., Xiao, A., Luo, X., Li, R.: The effect of a game training intervention oncognitive functioning and depression symptoms in the elderly with mild cognitive impairment: a randomized controlled trial. Int. J. Methods Psychiatr. Res. **30**(4), e1887 (2021)
35. Zander, T.O., Kothe, C.: Towards passive brain-computer interfaces: applyingbrain-computer interface technology to human-machine systems in general. J. Neural Eng. **8**(2), 025005 (2011). https://doi.org/10.1088/17412560/8/2/025005

Plant Science

In-House Production and Modification of Biocellulose

Ming Cong Jerrell Chng(✉) and Hanyu Jia

Hwa Chong Institution, Singapore, Singapore
{202213n,202678r}@student.hci.edu.sg

Abstract. The use of cellulose has increased in recent years, resulting in widespread damage to the environment, most critically deforestation, where forests are cleared for wood or the cultivation of crops that produce cellulose. Biocellulose produced by *Komagataeibacter xylinus* offers a promising alternative, with a unique nanostructure and mechanical strength. This project aims to investigate the effectiveness of modifying biocellulose, using esterification with carboxylic acids and metal ion cross-linking, in increasing the tensile strength of biocellulose and its effect on hydrophobicity. Results showed that carboxylic acids of longer carbon chain length corresponded with a greater increase in biocellulose tensile strength. Esterification of biocellulose led to a reduction in swelling ratio, indicating a rise in hydrophobicity. Metal ion cross-linking of biocellulose increased the tensile strength, with trivalent metal ions having a greater impact on wet biocellulose, and divalent metal ions for dry biocellulose. The properties of increased tensile strength and hydrophobicity of biocellulose may prove to be useful in the medical, manufacturing and food industries, providing a more durable material while simultaneously avoiding the issues that come with the high swelling ratio and low hydrophobicity of unmodified biocellulose.

Keywords: Biocellulose · Esterification · Metal Ion Cross-linking · Tensile Strength · Hydrophobicity

1 Background and Purpose of Research Area

In the wake of our world's rapid advancement and development, the hunt for sustainable materials has intensified, and of these, biocellulose offers itself as a promising alternative to conventional plant cellulose.

Biocellulose is a strong polymer consisting of nanofibrillar structures, forming a large specific surface area and microporous structure. This creates many opportunities for modification and the production of various composite materials with better characteristics [11]. Furthermore, biocellulose produced by bacteria, such as *Komagataeibacter xylinus*, displays properties far better than plant cellulose, including a unique nanostructure, degree of polymerisation and mechanical strength [9]. Moreover, it is free from lignin, pectin and hemicellulose, conferring high purity and crystallinity.

H. Guo et al. (Eds.): IRC-SET 2025, PTT, pp. 101–109, 2026.
https://doi.org/10.1007/978-981-95-9630-0_9

Of these properties, the tensile strength of biocellulose lends it many uses in the medical, cosmetic, manufacturing and food industries, as temporary skin substitutes, skincare facial masks and packaging. Biocellulose needs to be strong and durable to fulfil these functions. Its high tensile strength arises from thin fibrils that are interconnected through inter and intra-molecular hydrogen bonds that stabilise the reticulate structure [14]. However, improvements can still be made to increase the tensile strength of biocellulose and increase its resistance to breakage.

Research has shown that the esterification of biocellulose, using carboxylic acids, results in the formation of cellulose esters with the carboxylic acids. Esterification is commercially used to increase the mechanical strength of biocellulose through the formation of ester bonds between cellulose fibres.

Multivalent metal ion cross-linking of wet bacterial cellulose has shown further improved inter-fibril interactions, through electrostatic forces between the metal ions and the adjacent hydroxyl group oxygen atoms that stabilise the cellulose, with some transition metal ions being able to form coordinate bonds with the adjacent hydroxyl group oxygen atoms [8], further contributing to the high tensile strength of biocellulose.

2 Objectives and Hypotheses of the Research

This study aims to investigate the effectiveness of esterification, using carboxylic acids with differing carbon chain lengths, and the number of hydroxyl (OH) groups, and of metal-ion cross-linking, using divalent and trivalent metal ions, in increasing the tensile strength of biocellulose and its effect on the hydrophobicity of the biocellulose.

We hypothesise that carboxylic acids with a longer carbon chain and with a larger number of OH groups, and that trivalent metal ions will be more effective in increasing the tensile strength of biocellulose.

3 Materials and Methods

3.1 Cultivation of *Komagataeibacter Xylinus*

Falcon tubes were filled with 25 ml of Hestrin–Schramm (HS) medium, followed by inoculation with 2 ml of stock *Komagataeibacter xylinus*. The culture was incubated at 26.0°C for 7 days to obtain a seed culture.

3.2 Synthesis of Biocellulose

Seed culture of *Komagataeibacter xylinus* was added to HS medium in a ratio of 1:10, in a sterile plant tissue culture container. The culture was then incubated at 25°C to obtain a biocellulose membrane with a wet thickness of 0.4 cm. The biocellulose membranes were harvested and washed with ethanol, then boiled with 0.01 mol dm^{-3} sodium hydroxide (NaOH) at 80°C for 15 min, before being soaked with deionised water overnight [16].

3.3 Esterification of Biocellulose

Solutions of 0.5 mol dm^{-3} of oxalic acid, malonic acid, succinic acid, malic acid and tartaric acid were prepared. To catalyse the esterification reaction, 0.05 ml of concentrated H_2SO_4 was added to each of the acid solutions. To each of the solutions, a biocellulose sheet was added at 80°C for 5 h. Biocellulose sheets were then washed thoroughly with deionised water after removal.

3.4 Modification of Biocellulose with Metal Ions

Biocellulose sheets were pre-stretched to 20% of their maximum tensile strength. Solutions of 0.25 mol dm^{-3} of manganese (II) nitrate [$Mn(NO_3)_2 \cdot 4H_2O$], zinc nitrate [$Zn(NO_3)_2 \cdot 6H_2O$], iron (III) nitrate [$Fe(NO_3)_3 \cdot 9H_2O$] and chromium (III) nitrate [$Cr(NO_3)_3 \cdot 9H_2O$] were prepared. A biocellulose sheet was immersed in 20 ml of each solution at 25°C for 96 h. Biocellulose sheets were then washed thoroughly with deionised water after removal.

3.5 Fourier Transform Infrared (FTIR) Spectroscopy of Biocellulose

FTIR Spectroscopy was performed on the biocellulose sheets, before and after esterification with carboxylic acids. Samples of biocellulose sheets were crushed with solid potassium bromide (KBr) with a mortar and pestle, before being pressed into discs with a pellet press. Pressed discs were analysed with FTIR Spectroscopy and OPUS Spectroscopy software.

3.6 Tensile Strength of Biocellulose

To measure the tensile force that the biocellulose can withstand, the dried and wet biocellulose sheets were fitted into the Kelvin material tester 841812. The biocellulose was then stretched until breakage. The maximum tensile force of the biocellulose was recorded.

3.7 Hydrophobicity of Biocellulose

A piece of unmodified biocellulose was used to compare contact angles. 0.05 cm^3 of water was applied to the strip. The spreading and angle of the drop were examined using a Dino digital microscope.

The initial mass of a dry strip was first measured using an electronic balance. Next, the strip was immersed in deionised water. Each strip was placed in a vial containing 20 cm^3 of deionised water and incubated at room temperature. The swollen membranes were weighed after removing excess surface water at 10 min intervals, from 0 min to 60 min. The swelling ratio was determined using the equation as follows:

$$SR(\%) = (m_{wet} - m_{dry}/m_{dry}) \times 100$$

m_{wet} = mass of swollen biocellulose, m_{dry} = mass of biocellulose after oven drying.

4 Results and Discussion

4.1 Fourier Transform Infrared (FTIR) Spectroscopy of Esterified Biocellulose

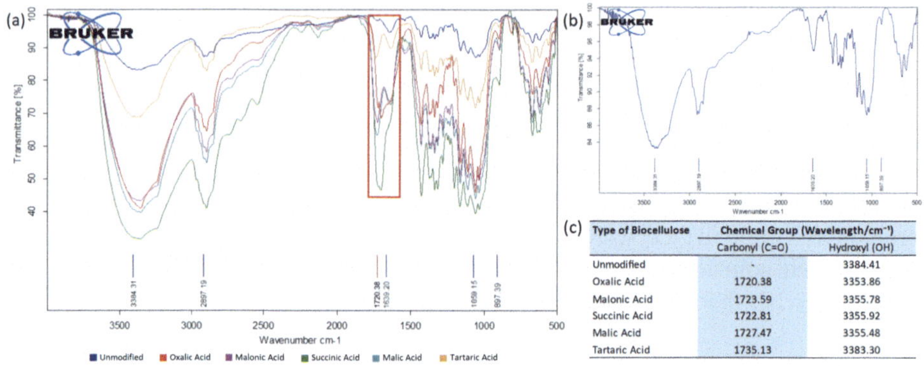

Type of Biocellulose	Chemical Group (Wavelength/cm⁻¹)	
	Carbonyl (C=O)	Hydroxyl (OH)
Unmodified	-	3384.41
Oxalic Acid	1720.38	3353.86
Malonic Acid	1723.59	3355.78
Succinic Acid	1722.81	3355.92
Malic Acid	1727.47	3355.48
Tartaric Acid	1735.13	3383.30

Fig. 1. FTIR spectra of biocellulose modified with (a) oxalic, malonic, succinic, malic, tartaric acids and (b) unmodified. (c) IR of hydroxyl and carbonyl groups in biocellulose.

Figure 1(b) shows the FTIR spectra of unmodified biocellulose, with wavelength peaks at 3384.31 cm^{-1}, 2897.19 cm^{-1}, and 897.39 cm^{-1}, characteristic of the stretching vibration of hydroxyl groups, the symmetric C–H stretching vibration, and the β-glycosidic linkages of the glucose ring, respectively, present in biocellulose [1]. The peak at 1059.15 cm^{-1} correlates to the C–O–C stretching of biocellulose [6]. From Fig. 1(a) and (c), the respective modified samples exhibit peaks in the range of 1720–1735 cm^{-1}, characteristic of the C=O ester vibrational stretching band. A shift in the peak corresponding to hydroxyl groups from 3384 cm^{-1} to 3355 cm^{-1} is observed in biocellulose esterified with oxalic, malonic, succinic, and malic acids (Fig. 2).

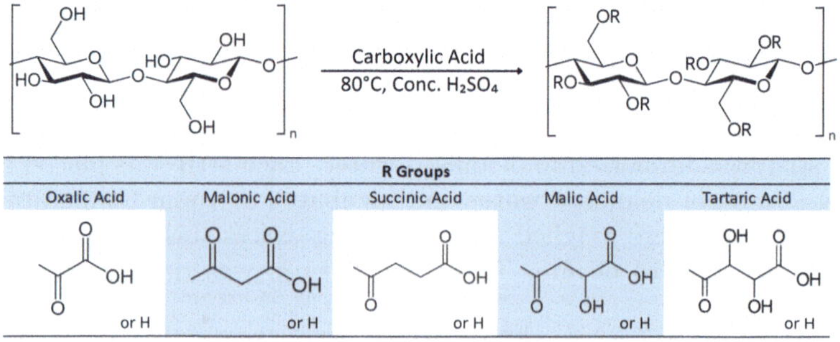

Fig. 2. Schematic diagram of esterification of biocellulose with carboxylic acids.

This can be attributed to the reaction between the hydroxyl group on the cellulose chains and the carboxyl group of the carboxylic acids through esterification. The shift of the hydroxyl group peak, along with a lower transmittance in the regions 3300–2500 cm^{-1} (O–H stretch), 1760–1690 cm^{-1} (C=O stretch), 1320–1210 cm^{-1} (C–O stretch), and 1440–1395 cm^{-1} and 950–910 cm^{-1} (O–H bend), indicates an increase in carboxyl groups attached to the biocellulose structure due to the attachment of carboxylic acids [4], confirming the successful esterification of biocellulose.

4.2 Tensile Strength of Esterified Biocellulose

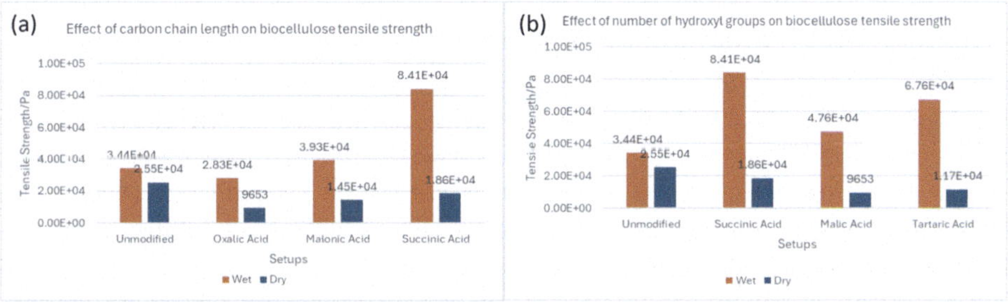

Fig. 3. Effect of (a) carbon chain length and (b) number of hydroxyl (OH) groups on the tensile strength of biocellulose.

Figure 3(a) shows an increase in the tensile strength of wet biocellulose as the carbon chain length of the carboxylic acid esterified with biocellulose increases. Carboxylic acids with longer carbon chains can form crosslinks between biocellulose fibres [2]. Carboxylic acids with shorter carbon chains, such as oxalic acid, are unable to form crosslinks, as most oxalic acid molecules are bonded to the fibres through only one of their two carboxyl groups, leaving the other free [7]. The disruption of hydrogen bonding during esterification and the lack of crosslinking results in biocellulose esterified with oxalic acid being weakened.

Figure 3(b) shows no significant correlation between the number of hydroxyl (OH) groups and the tensile strength of wet biocellulose. Almost all hydroxyl groups in biocellulose are occupied with hydrogen bonds [5], so the addition of more OH groups does not lead to the formation of additional hydrogen bonds. Moreover, introducing more OH groups may disrupt existing hydrogen bonding between biocellulose fibres, weakening the biocellulose structure.

Figure 3(a) and (b) show a significant decrease in the tensile strength of biocellulose after oven drying, due to the collapse of polymer chains that form the structural network of biocellulose [15]. Hence, carboxylic acids with longer carbon chains are more effective in increasing the tensile strength of biocellulose.

4.3 Hydrophobicity of Esterified Biocellulose

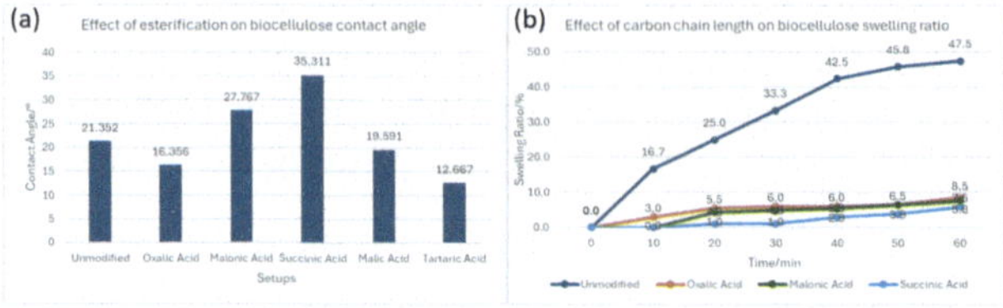

Fig. 4. Effect of esterification on (a) contact angle and (b), (c) swelling ratio of biocellulose.

Figure 4 shows an increase in hydrophobicity, as indicated by an increase in contact angle and a decrease in swelling ratio for esterified biocellulose, with increasing carbon chain length and decreasing number of hydroxyl groups. An increase in carbon chain length enhances the degree of crosslinking, forming a more rigid and less porous structure, resulting in a significant increase in the hydrophobicity of the biocellulose [3]. The reduction in hydroxyl groups also contributes to increased hydrophobicity, due to the hydrophilic nature of hydroxyl groups, which form hydrogen bonds with water molecules and increase water adsorption. Hence, the esterification of biocellulose with carboxylic acids possessing longer carbon chains and fewer hydroxyl groups is more effective in increasing the hydrophobicity of biocellulose.

4.4 Tensile Strength of Biocellulose Modified with Metal Ions

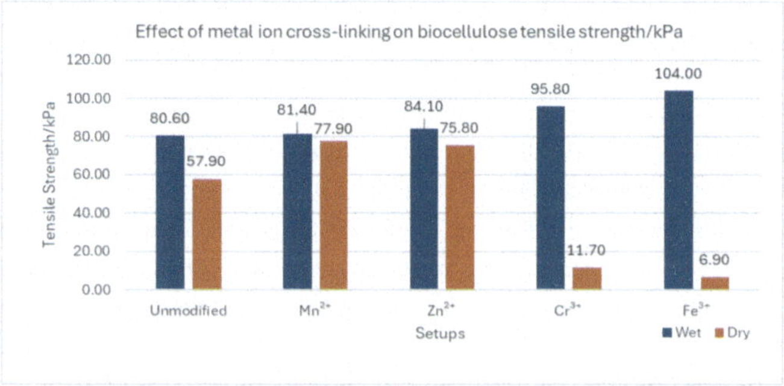

Fig. 5. Effect of metal ion cross-linking on biocellulose tensile strength.

Figure 5 shows a greater increase in the tensile strength of wet biocellulose modified with trivalent metal ions than with divalent metal ions. Trivalent metal ions are more effective in increasing the tensile strength of biocellulose due to (a) a higher charge density, leading to stronger electrostatic forces between the metal ions and adjacent hydroxyl group oxygen atoms, and (b) a higher nuclear charge and electron affinity, resulting in the formation of stronger and more coordination bonds with adjacent hydroxyl group oxygen atoms.

Figure 5 also shows an increase in the tensile strength of dry biocellulose modified with divalent metal ions, but a decrease when modified with trivalent metal ions. Trivalent metal ions in wet biocellulose possess two full hydration shells and a portion of a third [12], which leads to fewer coordination bonds due to the high concentration of water molecules. However, when dried, the trivalent metal ions lose their hydration shells, forming more coordination bonds and creating a rigid and brittle biocellulose structure.

4.5 Hydrophobicity of Biocellulose Modified with Metal Ions

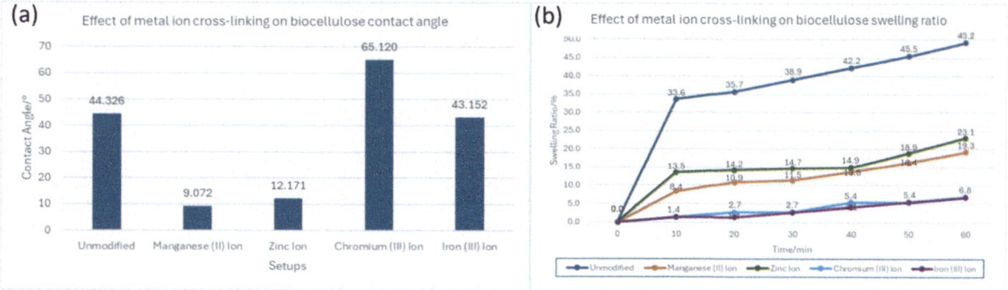

Fig. 6. Effect of metal ion cross-linking on (a) contact angle and (b) swelling ratio of biocellulose.

Figure 6(a) shows a greater contact angle for biocellulose modified with trivalent metal ions compared to divalent metal ions, indicating a more hydrophobic biocellulose surface. Trivalent metal ions form coordination bonds with the polar groups (OH, C=O, COOH) of the biocellulose, creating a more compact structure and limiting the exposure of hydrophilic groups [10]. The greater contact angle for Cr^{3+} can be attributed to its higher ligand field stabilisation energy compared to Fe^{3+} [13].

Figure 6(b) shows a lower swelling ratio for biocellulose modified with trivalent metal ions than with divalent metal ions, indicating a more hydrophobic biocellulose due to lower porosity resulting from more extensive crosslinking in biocellulose modified with trivalent metal ions. Hence, trivalent metal ions are more effective in increasing the hydrophobicity of biocellulose.

5 Conclusions and Recommendations for Future Work

Esterification of biocellulose with carboxylic acids was successful, resulting in the formation of C=O ester bonds, with carboxylic acids of longer carbon chain length forming crosslinks that effectively increase the tensile strength of biocellulose. Esterification led to a general reduction in swelling ratio, indicating a rise in hydrophobicity. Among the carboxylic acids used, succinic acid esterification displayed ideal qualities, combining high tensile strength, high contact angle, and low swelling ratio, signifying a significant improvement in mechanical strength and hydrophobicity.

Metal ion cross-linking of biocellulose effectively increased tensile strength, with trivalent metal ions having a more significant impact on wet biocellulose, while divalent metal ions were more effective for dry biocellulose. Cross-linking with trivalent metal ions led to an increase in contact angle, and both divalent and trivalent metal ions resulted in a decreased swelling ratio. Among the metal ions tested, trivalent ions exhibited ideal qualities, including high wet tensile strength and comparatively greater hydrophobicity.

The combined properties of increased tensile strength and hydrophobicity of biocellulose suggest promising applications in the medical, cosmetic, manufacturing, and food industries, providing a more durable material while simultaneously addressing the high swelling ratio and low hydrophobicity issues of unmodified biocellulose. This supports its potential use in improving wound dressings, edible packaging films, and water-resistant coatings. Furthermore, the antibacterial properties of metal ions may further enhance the medical applications of biocellulose.

References

1. Bastida, G.A., et al.: Effect of oxalic acid concentration and different mechanical pre-treatments on the production of cellulose micro/nanofibers. Nanomaterials **12**(17), 2908 (2022). https://www.mdpi.com/2079-4991/12/17/2908
2. Belay, M., Tyeb, S., Rathore, K., Kumar, M., Verma, V.: Synergistic effect of bacterial cellulose reinforcement and succinic acid crosslinking on the properties of agar. Int. J. Biol. Macromol. (2018). https://doi.org/10.1016/j.ijbiomac.2020.10.144
3. Bonetti, L., De Nardo, L., Farè, S.: Chemically crosslinked methylcellulose substrates for cell sheet engineering. Gels **7**(3), 141 (2021). https://www.mdpi.com/2310-2861/7/3/141
4. CUBoulder: IR Spectroscopy Tutorial: Carboxylic Acids. OrgChemBoulder (2011). https://orgchemboulder.com/Spectroscopy/irtutor/tutorial.shtml
5. Gorgieva, S., Trček, J.: Bacterial cellulose: production, modification and perspectives in biomedical applications. Nanomaterials **9**(10), 1352 (2019). https://doi.org/10.3390/nano9101352
6. Kara, H.T., Anshebo, S.T., Sabir, F.K.: Adsorptive removal of Cd(II) ions from wastewater using maleic anhydride nanocellulose. J. Nanotechnol. (2021). https://www.hindawi.com/journals/jnt/2021/9966811/
7. Li, D., Henschen, J., Ek, M.: Esterification and hydrolysis of cellulose using oxalic acid dihydrate in a solvent-free reaction suitable for preparation of surface-functionalised cellulose nanocrystals with high yield. Green Chem. (2017). https://doi.org/10.1039/C7GC02489D

8. Liu, X., Jiang, Y., Wei, Y., Wei, X.: Strengthening and toughening mechanisms induced by metal ion cross-linking in wet-drawn bacterial cellulose films. Mater. Des. (2022). https://doi.org/10.1016/j.matdes.2022.111431
9. Mohammad, S.M., Rahman, N.A., Khalil, M.S., Abdullah, S.R.S.: An overview of Biocellulose production using acetobacter xylinum culture. ResearchGate (2014). https://www.researchgate.net/publication/270742325_An_Overview_of_Biocellulose_Production_Using_Acetobacter_xylinum_Culture
10. Nayanathara, R.M.O.N.O., et al.: A general Metal-Ion-Modification route for preparing hydrophobic paper and tableware from lignocellulose fibers. Chem. Eng. J. (2023). https://doi.org/10.1016/j.cej.2023.141596
11. Niyazbekova, Z., Kurmanbayev, A., Nagmetova, G.: An overview of bacterial cellulose applications. ResearchGate (2018). https://www.researchgate.net/publication/328239710_AN_OVERVIEW_OF_BACTERIAL_CELLULOSE_APPLICATIONS
12. Persson, I.: Structure and size of complete hydration shells of metal ions and inorganic anions in aqueous solution. Dalton Trans. (2024). https://doi.org/10.1039/D4DT01449A
13. Schaller, C.: 10.3.3: Ligand Field Stabilization Energy. Chemistry LibreTexts (2024). https://chem.libretexts.org/Bookshelves/Inorganic_Chemistry/Inorganic_Chemistry_(LibreTexts)/10%3A_Coordination_Chemistry_II_-_Bonding/10.03%3A_Ligand_Field_Theory/10.3.03%3A_Ligand_Field_Stabilization_Energy
14. Ul-Islam, M., Jung, H.H., Khan, T., Joong, K.P.: Effects of glucuronic acid oligomers on the production, structure and properties of bacterial cellulose. Carbohydr. Polym. (2012). https://www.sciencedirect.com/science/article/abs/pii/S0144861712009782
15. Vasconcellos, V.M., Farinas, C.S.: The effect of the drying process on the properties of bacterial cellulose films from *Gluconacetobacter hansenii*. Chem. Eng. Trans. (2018). https://www.aidic.it/cet/18/64/025.pdf
16. Zakaria, J., Mohamad, S., Siew, K.W.: Enhancement of Biocellulose production in mixed medium culture (2012). http://umpir.ump.edu.my/2862/1/ENHANCEMENT_OF_BIOCELLULOSE_PRODUCTION_IN_MIXED.pdf

Life Sciences

Forever Chemicals then and Now: Comparative Toxicity Assessment of Emerging Versus Legacy PFAS on Synechocystis Growth

Elijah Chew(✉), Karina Gin, and Song Qixuan

National University of Singapore, 21 Lower Kent Ridge Road, Singapore 119077, Singapore
ecsf3688@gmail.com

Abstract. Per- and polyfluoroalkylated substances (PFAS), the "forever chemicals", are a vast family of pollutant substances that can be divided into legacy and emerging PFAS. The former have been progressively phased out over environmental and health dangers and replaced with shorter-chain emerging PFAS, which are highly under-researched, comparatively unregulated and yet are found in many novel applications such as lithium-ion batteries. PFAS can cause growth inhibition, membrane peroxidation, cell signalling disruption and protein damage through production of excess reactive oxygen species and inducing oxidative stress. In this study, we compare the impact of high and low concentrations of legacy and emerging PFAS on model cyanobacterium Synechocystis after 21 days of exposure. We evaluate cell density as well as chlorophyll, carotenoid and allophycocyanin concentration. All growth parameters experienced toxicity from emerging PFAS, and there is no significant difference in the growth inhibition and photosystem oxidative damage caused by different PFAS types at the same concentration. Most parameters show significant inhibition occurs at high PFAS concentrations and stimulatory effects occur at low PFAS concentration for all types. This suggests both legacy and emerging PFAS cause environmental toxicity or at low concentrations trigger harmful algal blooms. We therefore bridge the gap in emerging PFAS toxicity research, concluding that emerging PFAS are not an improvement over legacy PFAS and stricter regulation is required for the former. Further research can differentiate the mechanisms of uptake and cellular damage of longer and shorter chains between legacy and emerging PFAS, allowing clearer understanding of their environmental impact.

Keywords: PFAS · environmental microbiology · water pollution · emerging PFAS · legacy PFAS · Synechocystis · PFOA · PFOS · toxicity · cytotoxicity · reactive oxygen species

1 Introduction

Per- and polyfluoroalkylated substances (PFAS) are a vast family of pollutant substances that can be divided into legacy and emerging PFAS. Older or "legacy" PFAS tend to possess longer-chain backbones and have been progressively phased out over environmental and health concerns and replaced with shorter-chain "emerging" PFAS [1]. The

H. Guo et al. (Eds.): IRC-SET 2025, PTT, pp. 113–123, 2026.
https://doi.org/10.1007/978-981-95-9630-0_10

shorter-chain variant degrades more rapidly in the environment while the longer-chain variants have extremely long half-lives and tend to bioaccumulate in living tissue, hence being referred to as "forever chemicals" [2].

PFAS have been detected in recent years on a global scale within food, drinking water, air and living tissues, including those of humans [9]. Outfall discharge from fluorochemical manufacturing plants is one key source of PFAS pollution, emitting it into rivers and thus into global waterways [4]. PFAS has also been located far from major sources, further suggesting atmospheric means of distribution and deposition from diffuse non-point sources without a specific source in addition to point sources such as wastewater treatment plant discharge [9].

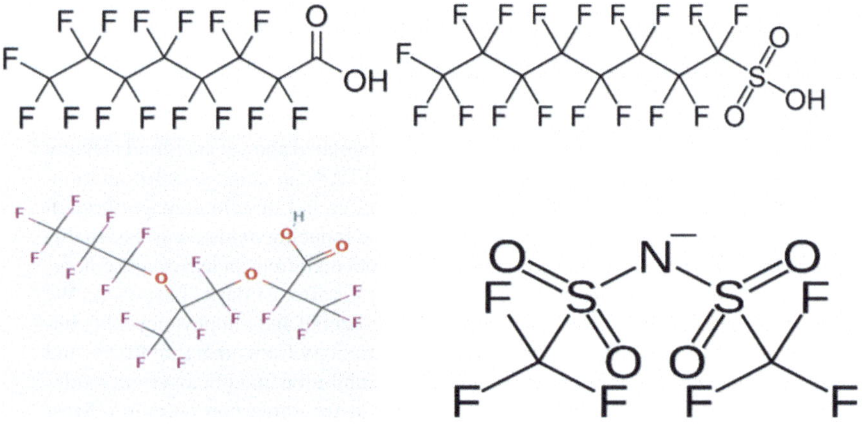

Fig. 1. Molecular structures (clockwise) of PFOA, PFOS, TFSI- and HFPO-TrA respectively.

PFAS are typically classified as legacy PFAS and emerging PFAS based on their historical use, structural characteristics, and environmental and health impacts. Legacy PFAS have been found in a wide variety of industries such as cosmetics and textiles, and for production of non-stick surface linings [8]. Legacy PFAS have demonstrated the ability to cause oxidative stress and related growth inhibition to microalgae such as chlorella species [10], causing elevated levels of reactive oxygen species (ROS), triggering oxidative stress which can lead to growth inhibition and photosystem damage. Loss of chlorophyll and other membrane pigments alongside drops in levels of soluble proteins such as phycocyanin or allophycocyanin were recorded after long exposure [11]. At low concentrations, PFAS can cause small increases in ROS levels which can instead stimulate cellular proliferation by signalling in metabolic pathways [5].

Furthermore, emerging PFAS are used in many novel technologies. For instance, one major source of emerging PFAS is lithium-ion batteries, a new sustainable energy source that nonetheless utilise Bis-FASIs as solvents [4].

Legacy PFAS have had their production, use and disposal increasingly strictly regulated. Notably, the Environmental Protection Agency of the United States has recently finalised new regulations to push public drinking water systems to curb the amount of PFOA and PFOS, amongst other legacy PFAS, in their water supply, with further funding from the presidential administration of $1 billion in order to help states test and eliminate PFAS contamination from their water [3].

As previously noted, emerging PFAS are widely used in several contemporary applications, such as anionic conduction in lithium-ion batteries making use of LiTFSI. Yet, disturbingly, there is little to no regulation of it, and the amount of information collected on the sources, use and disposal of emerging PFAS remains scarce [4]. However, the fact that emerging PFAS can be detected already in international water bodies from Europe to the South China Sea suggests that it is being released without comprehensive oversight to prevent it from polluting natural water systems [6, 13].

Emerging PFAS are under-regulated and under-studied in terms of their toxic properties. Nonetheless, it has been shown that they are widely distributed in the environment and thus present a hidden danger to the ecosystems contaminated with them [4]. Furthermore, due to the lack of legislation on such PFAS, their harms will continue to compound as they continue to be released without restrictions. PFAS pollution can bring major damage. For instance, oxidative stress from PFAS such as PFOA or PFOS can damage aquatic ecosystems within which the balance of phytoplankton such as green and blue microalgae plays a vital role in maintaining food chains that depend on them while at the same time preventing harmful algal blooms if their growth is overly stimulated, which can result in strongly negative impacts on the ecosystem due to release of cyanotoxins [12].

A direct comparison of the biological toxicity from legacy and emerging PFAS remains elusive, and thus this study's aim is to determine if their harms are similar at varying concentrations or if emerging PFAS may be an improvement over legacy PFAS. This would serve to shed some light on the danger of emerging PFAS, and provide direction for future changes in regulation that may be necessary to protect aquatic life.

This study uses Synechocystis sp. PCC 6803 as a biological model to compare the effects of perfluorooctanoic acids (PFOA), perfluorooctanesulfonic acids (PFOS), bis((trifluoromethyl)sulfonyl)amide ($TFSI^-$) and Hexafluoropropylene oxide tetramer acids (HFPO-TrA) (Fig. 1) on algal activity and hence the toxicity of legacy and emerging PFASs. We assessed the toxicity of varying concentrations of legacy and emerging PFAS by measuring cell density, insoluble photosystem proteins chlorophyll a and carotenoids, as well as soluble light-harvesting protein allophycocyanin (APC). We had the following research questions: 1. How do legacy and emerging PFAS differ in toxicity at the same concentration? 2. How do legacy and emerging PFAS respectively differ in toxicity based on their concentration? 3. Based on the earlier questions, are emerging PFAS safer for the environment than legacy PFAS?

2 Hypothesis

In this study, we hypothesise that shorter-chain emerging PFAS do significantly less oxidative damage to Synehocystis than legacy PFAS at the same concentration, that both legacy and emerging PFAS have significantly more toxicity at higher concentrations than at lower concentrations and that therefore emerging PFAS are safer than legacy PFAS.

3 Methods

3.1 Calibration Curve

A species of blue algae, Synechocystis sp. PCC 6803, was cultivated for 168h under 12h/12h day-night cycles in an orbital shaker incubator in BG11 medium. A calibration curve was prepared of absorbance against cell density using samples in duplicate which were diluted by 20, 30, 40, 50, 100, 200 and 500x. Cell density counting was performed with NanoEntek C-chip haemocytometer, and absorbance of all concentration samples were then measured with the Thermo Scientific™ Orion™ AquaMate 8000 UV-Vis Spectrophotometer using wavelength of 680 nm. The following calibration curve was then constructed as: Cell density (Cells/mL) = 22200000*10^7 * OD680 – 148957 (Fig. 2).

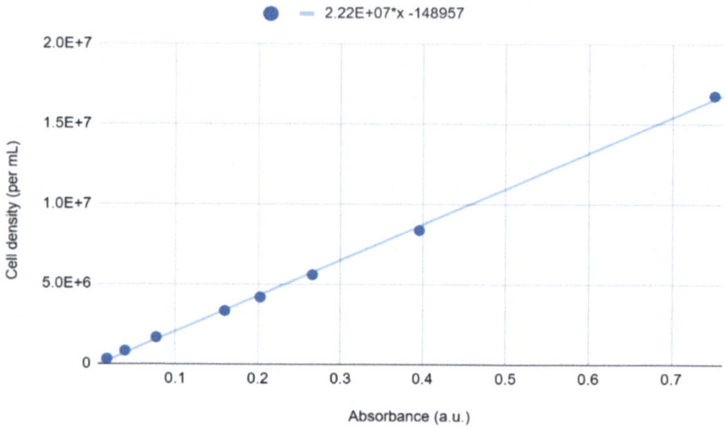

Fig. 2. Absorbance against cell density of Synechocystis

3.2 Experimental Setup

Eight solutions of PFAS were prepared in methanol, two each of PFOS, PFOA, TFSI- and HFPO-TrA with concentrations of 5 mg/L and 500 mg/L respectively. Following the construction of the calibration curve, Synechocystis was added to BG11 medium and thoroughly shaken, and 27 conical flasks were prepared each receiving 100 mL of this mixture. They were then divided into 9 groups in triplicate. The eight experimental setups received 0.5 mL each of one of the eight PFAS solutions listed above, while the control received 0.5 mL of methanol with no PFAS added, creating 25μg/L and 2.5 mg/L solutions respectively. Following this, the samples were cultivated in BG11 medium under 12h/12h day-night cycles in an orbital shaker incubator. On days 0, 3, 7, 10, 14, 17 and 21, samples were taken for testing of various growth parameters.

3.3 Growth Parameter Analysis

All spectrophotometer readings are done using the Thermo Scientific™ Orion™ AquaMate 8000 UV-Vis Spectrophotometer.

To measure cell density (CD), 1.5 mL of Synechocystis from each setup was sampled and their absorbance at 680 nm was measured with a spectrophotometer zeroed with deionised water. The CD was then calculated according to the algae density formula in 3.1. (Fig. 3).

To measure chlorophyll a and carotenoid concentration [14, 15], 1.5 mL of Synechocystis was centrifuged at 8000 rpm for 10 min. The pellet was then dissolved in 1.5 mL of 95% ethanol, and left to stand at 4 °C overnight. It was then centrifuged again at 8000 rpm for 10 min, before its absorbances at 470, 649 and 665 nm were measured with a spectrophotometer zeroed using 95% ethanol. The following equations were used to calculate raw concentrations: Chl a (mg/L) = 13.95*OD665–6.88*OD649; Carot (mg/L) = (1000*OD470–2.05*Chl a)/245. (Fig. 4).

Furthermore, on the first and last days of sampling, APC levels were analysed [7]. To measure APC concentration, 1.5 mL of Synechocystis was centrifuged at 10000 rpm for 10 min. The pellet was then dissolved in 1.5 mL of 95% ethanol, crushed in an ultrasonic cell crusher for 10 min at 4 °C, and then centrifuged at 7000 rpm for 15 min at 4 °C. Its absorbance at 615 and 652 nm was measured with a spectrophotometer zeroed using 95% ethanol. The following equations were used to calculate the raw concentration: APC (mg/L) = (OD652 - 0.208*OD615)/5.09. (Fig. 5).

4 Results

We set the null hypotheses that there is no significant difference in toxicity between emerging and legacy PFAS cell densities at each concentration and that there is no significant difference in toxicity between high and low concentrations of each of the four PFAS, and that there is no significant difference between the four at the same concentration. ANOVA was then conducted to determine if there were significant differences ($p < 0.05$) caused by changing the concentration or PFAS type.

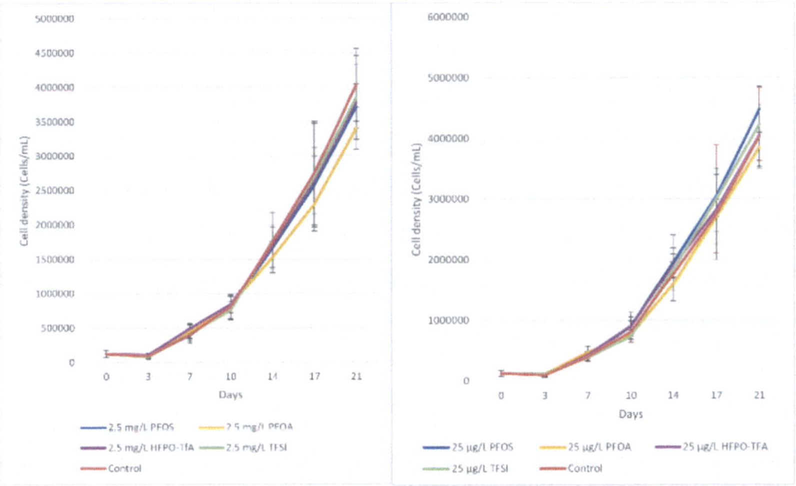

Fig. 3. Change in cell density by concentrated and diluted PFAS types respectively

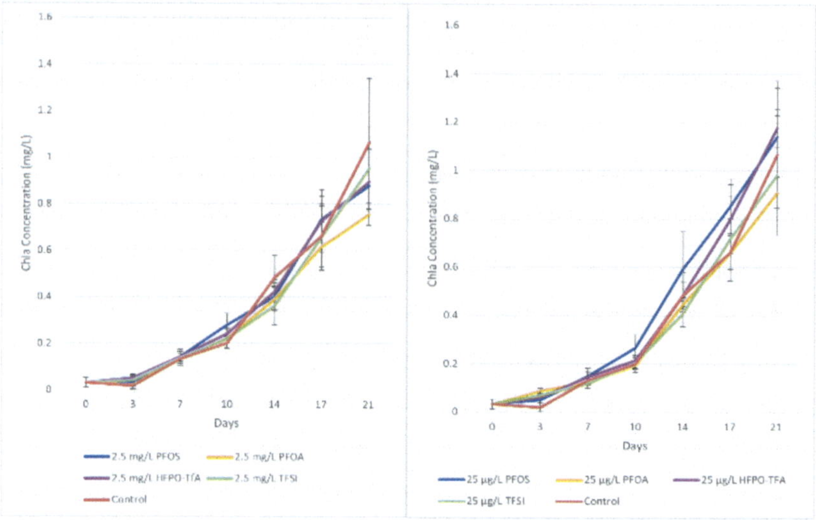

Fig. 4. Change in chlorophyll a by concentrated and diluted PFAS types respectively

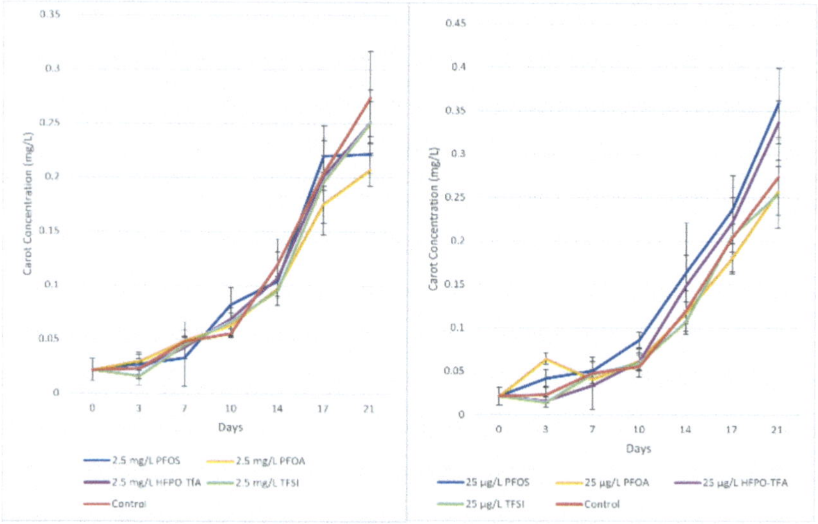

Fig. 5. Change in carotenoid by concentrated and diluted PFAS types respectively

4.1 Growth Inhibition

Growth inhibition was calculated with the following formula: Inhibition (%) = (CD of control group – CD of experimental group)*100/CD of control group.

After 21 days of exposure, cell growth was markedly inhibited in the 2.5 mg/L PFAS groups, with toxicity decreasing in the order PFOA > PFOS > HFPO-TrA > TFSI-, with maximum inhibition of 15.6% ± 7.57% recorded for 2.5 mg/L PFOA. Furthermore, apart from PFOA, the remaining PFAS types all had a stimulatory effect on Synechocystis cell proliferation at 25 μg/L, with stimulatory effect decreasing in the order PFOS > TFSI-- > HFPO-TrA, with only PFOA showing positive growth inhibition of 4.40% ± 5.83% and PFOS showing the highest stimulation of 11.0% ± 9.37% at 25 μg/L. No significant difference ($p < 0.05$) existed between all groups at the same concentration of 2.5 mg/L or 25 μg/L respectively, but a significant difference ($p < 0.05$) existed between the 2.5 mg/L and 25 μg/L groups for PFOS. (Fig. 6).

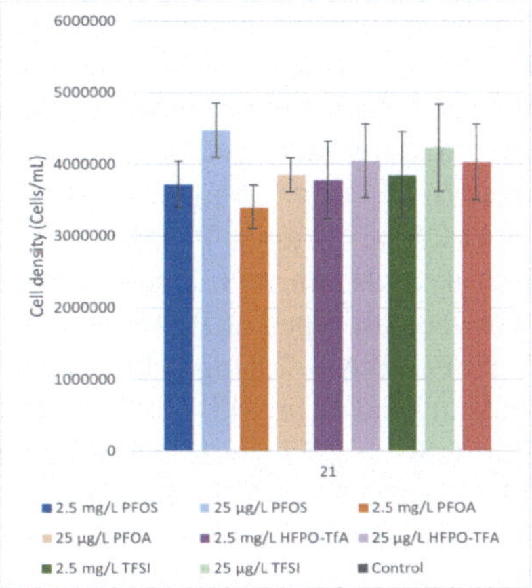

Fig. 6. Cell density by PFAS type and concentration after 21 days.

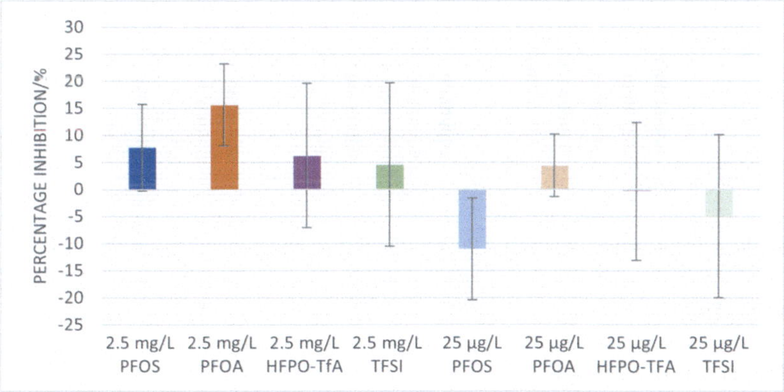

Fig. 7. Growth inhibition by PFAS type and concentration after 21 days

4.2 Photosystem Damage

Chla level showed a similar trend to cell density, with no significant differences recorded between all PFAS types at the same concentration and a significant difference ($p < 0.05$) between the 2.5 mg/L and 25μg/L groups for PFOS and HFPO-TrA. HFPO-TrA and PFOS showed stimulatory effect at 25 μg/L, with HFPO-TrA showing the highest concentration of 1.18 ± 0.0805 mg/L, compared with control group concentration of 1.07 ± 0.276 mg/L. Overall, cell density increased in the order 25 μg/L HFPO-TrA > 25 μg/L PFOS > Control > 25 μg/L TFSI- > 2.5 mg/L TFSI- > 25 μg/L PFOA > 2.5 mg/L HFPO-TrA > 2.5 mg/L PFOS > 2.5 mg/L PFOA. (Fig. 7).

Contrary to the above trend, there was a significant difference ($p < 0.05$) between carotenoid levels of PFOS and PFOA compared to those of HFPO-TrA and TFSI- at 2.5 mg/L, with the former showing stronger inhibitory effect. The lowest concentration was recorded with 2.5 mg/L PFOA at 0.207 ± 0.0151 mg/L compared with the control at 0.274 ± 0.0427 mg/L. The highest concentration was recorded with 25 μg/L PFOS at 0.359 ± 0.0396 mg/L. Overall, toxicity increased in the order 25 μg/L PFOS > 25 μg/L HFPO-TrA > Control > 25 μg/L PFOA ≈ 25 μg/L TFSI-- > 2.5 mg/L HFPO-TrA ≈ 2.5 mg/L TFSI- > 2.5 mg/L PFOS > 2.5 mg/L PFOA. (Fig. 8).

In a similar vein to cell density and Chla, APC levels showed no significant difference at the same concentration between all types of PFAS, toxicity decreasing in the order 2.5 mg/L PFOA > 2.5 mg/L HFPO-TrA > 2.5 mg/L PFOS > 2.5 mg/L TFSI- ≈ 25 μg/L PFOA, with the lowest concentration being 4.04 ± 2.14 ug/L with 2.5 mg/L PFOA compared with control group 6.86 ± 2.52 mg/L. Stimulating effect decreased in the order PFOS > TFSI- > HFPO-TrA at 25 μg/L with the highest concentration recorded with 25 μg/L PFOS, 9.95 ± 1.99 μg/L. Significant differences ($p < 0.05$) were noted between 2.5 mg/L and 25 μg/L samples for PFOS, HFPO-TrA and TFSI-. (Fig. 4.8).

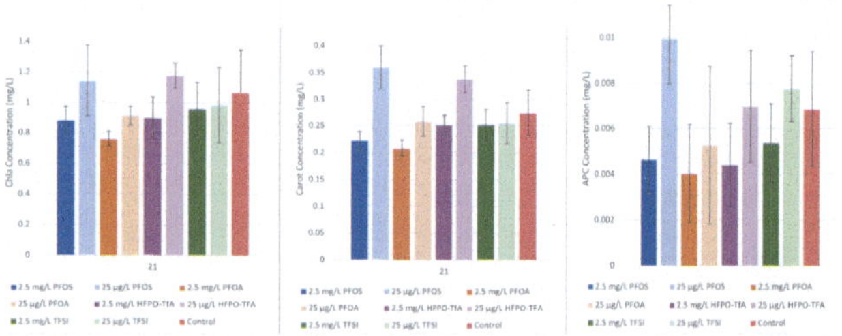

Fig. 8. Chla, Carotenoids and APC by PFAS type and concentration respectively.

5 Discussion

5.1 Interpretation of Results

From the above results, it can be clearly seen that all PFAS show toxicity at 2.5 mg/L in order HFPO-TrA > PFOA > PFOS > TFSI-. Similarly, apart from PFOA, the PFAS show stimulatory effects on Synechocystis growth at low concentrations. However, more notably, there is little significant difference between PFAS types at the same concentration for most growth parameters. Furthermore, it seems that while there are toxicity differences between PFAS types, this does not correlate positively or negatively with their classification as legacy or emerging.

The results with respect to legacy PFAS seem to follow the trends shown in the literature, with a large degree of growth inhibition at high concentrations and some stimulation at low concentrations [5, 10]. While the literature on emerging PFAS toxicity is sparse, the results show little deviation from the observed pattern for legacy PFAS.

5.2 Areas for Further Research

A vital area for further research would be to investigate the differences in molecular mechanisms of cell damage between various PFAS to determine what factor or combination of factors contributes the most to toxicity and oxidative damage.

For most growth parameters except for APC, the difference between 2.5 mg/L and 25 μg/L groups were significant ($p < 0.05$) for legacy PFAS and not emerging PFAS, likely due to the strongly stimulating effect of PFOS at 25 μg/L compared with the strongly inhibiting effect of 2.5 mg/L PFOA. However, this trend was reversed with APC having a more significant difference in toxicity between concentrations for emerging PFAS than legacy PFAS, since both HFPO-TrA and TFSI- had significant differences but only PFOS and not PFOA.

This difference may be an aberration due entirely due to the uniquely high toxicity of PFOA which caused inhibitory effects for both low and high concentrations of it in APC, suggesting that it may be able to bind directly to the soluble protein and damage it, compared with damaging other cell components mostly through production of ROS. However, this cannot be confirmed through this study.

Another area for further research would therefore be to identify which parts of the cell legacy and emerging PFAS respectively are more likely to damage and why, which could then further aid in identification and remediation of PFAS waste in natural water bodies.

In addition, one limitation of this study is that there are widely differing responses amongst setups with the same PFAS type and concentration, resulting in several unusual results (notably for growth inhibition and APC concentration.) A potential area for further research would thus be to clarify the rate and mechanism of uptake of different types of PFAS, which could help in investigating other factors that could impact toxicity, as well as possibly conducting replicates using a larger sample size to confirm the results with greater reliability.

5.3 Policy Implications

The implications of our results are far-reaching. This study gives evidence that emerging PFAS poses a significant toxicity threat to the environment and there is no significant difference between emerging and legacy PFAS in terms of environmental impact. Both result in damage to photosynthetic pigments and growth inhibition at high concentrations and stimulate cyanobacterial proliferation at low concentrations, which can result in harmful algal blooms. In both scenarios, there is notable disruption to the ecosystem and long-term pollution of water bodies with emerging PFAS will result in similar consequences to those of legacy PFAS. This also further serves to show that emerging PFAS are not a very useful improvement over emerging PFAS, given that they have comparable toxicity.

Therefore, this study suggests that despite the many useful applications of emerging PFAS, there should be comparable strict regulations on them as with legacy PFAS. The process would depend upon first clearly defining emerging PFAS similar to how the USA's EPA has defined legacy PFAS, with a clear glossary of chemical products considered as PFAS and their precursors [3]. After this list is created, methods could be developed to accurately and quickly identify emerging PFAS in drinking water and water bodies, as well as to remove it, and this would then be progressively applied around the world to assess PFAS levels and target water treatment to those bodies affected. Legislative decisions also show strong possibility of effectively controlling emerging PFAS pollution. Because many of these emerging PFAS are being used in increasingly common electronic devices such as lithium-ion batteries, there should be relevant regulations introduced globally to advise corporations to avoid using such substances as well as on the national scale to ensure that citizens dispose of these devices in a way that minimises environmental damage.

While there will be short-term economic consequences, there would be sufficient environmental benefits to justify a more stringent framework for controlling the release and production of emerging PFAS just as there are for legacy PFAS.

6 Conclusion

This study investigates multiple growth parameters for Synechocystis cyanobacteria to compare toxicity of differing concentrations of legacy and emerging PFAS. It was determined that emerging PFAS pose serious environmental dangers and cause similar oxidative damage to photosystems and growth inhibition to cyanobacteria compared with legacy PFAS at both high and low concentrations, thus suggesting that there is no significant difference in their polluting effects on the environment. However, different PFAS do have different levels of toxicity, for instance with HFPO-TrA being more toxic than PFOS in general. Therefore, the toxicity of the emerging PFAS is not as weak as expected compared to legacy PFAS, therefore, the use of emerging PFAS also needs to be strictly monitored.

References

1. Brase, R. A., Mullin, E. J., & Spink, D. C. (2021, January 20). Legacy and emerging per- and polyfluoroalkyl substances: Analytical techniques, environmental fate, and health effects. International journal of molecular sciences. https://pmc.ncbi.nlm.nih.gov/articles/PMC7863963/
2. Buck, R. C., Franklin, J., Berger, U., Conder, J. M., Cousins, I. T., de Voogt, P., Jensen, A. A., Kannan, K., Mabury, S. A., & van Leeuwen, S. P. J. (2011, October). Perfluoroalkyl and polyfluoroalkyl substances in the environment: Terminology, classification, and origins. Integrated environmental assessment and management. https://pmc.ncbi.nlm.nih.gov/articles/PMC3214619/

3. EPA. (2024, December 2). Final PFAS National Primary Drinking Water Regulation. https://www.epa.gov/sdwa/and-polyfluoroalkyl-substances-pfas
4. Guelfo, J.L., Ferguson, P.L., Beck, J., et al.: Lithium-ion battery components are at the nexus of sustainable energy and environmental release of per- and polyfluoroalkyl substances. Nat. Commun. **15**, 5548 (2024). https://doi.org/10.1038/s41467-024-49753-5
5. Hong, Y., Boiti, A., Vallone, D., & Foulkes, N. S. (2024, March 1). Reactive oxygen species signaling and oxidative stress: Transcriptional regulation and evolution. MDPI. https://www.mdpi.com/2076-3921/13/3/312
6. Neuwald, I.J., et al.: Hans Peter H. Arp, and Daniel Zahn, Environmental Science & Technology **56**(10), 6380–6390 (2022). https://doi.org/10.1021/acs.est.1c07949
7. Jayakumar, S., Bhuyar, P., Pugazhendhi, A., Rahim, M. H. A., Maniam, G. P. & Govindan, N. (2021) Effects of light intensity and nutrients on the lipid content of marine microalga (diatom) Amphiprora sp. for promising biodiesel production.Science of the Total Environment 768, 145471
8. Kotthoff M, Müller J, Jürling H, Schlummer M, Fiedler D. Perfluoroalkyl and polyfluoroalkyl substances in consumer products. Environ Sci Pollut Res Int. 2015 Oct;22(19):14546–59. https://doi.org/10.1007/s11356-015-4202-7. Epub 2015 Feb 19. PMID: 25854201; PMCID: PMC4592498
9. Kurwadkar, S., Dane, J., Nadagouda, M., Cawdrey, R. W., Ambade, B., Struckhoff, G., & Wilkin, R. (2022, February 25). Per- and polyfluoroalkyl substances in water and wastewater: A critical review of their global occurrence and distribution. Science of The Total Environment. https://www.sciencedirect.com/science/article/pii/S0048969721060812
10. Mao , W., Li , M., Xue, X., Cao, W., Wang, X., Xu, F., & Jiang, W. (2023, January 31). Bioaccumulation and toxicity of perfluorooctanoic acid and perfluorooctane sulfonate in marine algae chlorella sp.. Science of The Total Environment. https://www.sciencedirect.com/science/article/abs/pii/S0048969723004977
11. Mojiri, A., Nazari Vishkaei, M., Ansari, H. K., Vakili, M., Farraji, H., & Kasmuri, N. (2023, January 26). Toxicity effects of perfluorooctanoic acid (PFOA) and perfluorooctane sulfonate (PFOS) on two green microalgae species. International journal of molecular sciences. https://pmc.ncbi.nlm.nih.gov/articles/PMC9916455/#:~:text=Liu%20et%20al.,and%20bisphenols
12. NOAA. (2024, August 7). NOAA National Ocean Service Education: Coastal pollution tutorial. NOAA's National Ocean Service. https://oceanservice.noaa.gov/education/tutorial-coastal/harmful-algal-blooms/habs02.html#:~:text=HABs%20harm%20human%20and%20animal,and%20people%20in%20different%20ways
13. Qi Wang, Yuefei Ruan, Linjie Jin, Lily S. R. Tao, Han Lai, Guifeng Li, Leo W. Y. Yeung, Kenneth M. Y. Leung, and Paul K. S. Lam Environmental Science & Technology (2023) 57 (22), 8355–8364 https://doi.org/10.1021/acs.est.3c00374
14. Zhang, D., Zhang, S., Wang, Z., Li, P.: Effects of nitrogen recovery on the growth of nitrogen starved Microcystis aeruginosa. Journal of Southern Agriculture **50**(11), 2592–2598 (2019)
15. Zhang, X.W., Kai, Z.P., Song, W.G., Chen, S.S.: Ecotoxicological effects of 50 kinds of fragrance materials on Microcystis aeruginosa. China Environ. Sci. **41**(3), 1429–1435 (2021)

Study of Enhancing Moringa Oleifera Lam Seeds as Coagulants Using Turmeric

Sathana Shankar(✉), Yueyao Qin, Lashirah Ebansha Aaisha, and Chow Siew Foong

Victoria Junior College, 20 Marine Vista, Singapore 449035, Singapore
sathana.research@gmail.com

Abstract. Heavy metal contamination from sources like Copper(II) solutions is a significant environmental concern. Natural coagulants like Moringa oleifera offer a sustainable alternative to chemical treatments, but their efficiency can be limited. This study investigates a novel, fully natural system to enhance Cu(II) removal by exploring the synergy between Moringa oleifera seeds and turmeric extract. We direct this development towards addressing copper disposal in school labs, which is one of the sources of copper pollution, besides copper pipe erosion and industrial wastes. A series of batch coagulation-flocculation experiments were conducted to determine the optimal dosages for removal. Moringa oleifera seeds alone achieved a maximum removal of 36.5% at 96 g/L. The addition of turmeric extract significantly enhanced this process, with the combined system achieving a peak removal of 45.0% at an optimal turmeric dose of 4 mg/L. However, excess turmeric was found to inhibit the process, likely due to the formation of neutral complexes. The findings validate a synergistic, dual mechanism where curcumin first chelates Cu(II) ions into larger complexes that are then more effectively coagulated by moringa. This work presents a successful proof-of-concept for a low-cost, entirely plant-based system for water purification, contributing to the development of sustainable environmental technologies.

Keywords: heavy metal pollution · copper · Moringa Oleifera · natural coagulant aid · turmeric · synergy · sustainability

1 Introduction

Copper(II) solution is a common reagent often utilised in school laboratories by students for practical sessions. It is often poured down the sink into the drainage system after the usage of the reagent. As Cu(II) is a heavy metal, this may lead to water pollution if not properly removed. [11]. Copper is an essential nutrient that helps with tissue and bone development as well as enzyme synthesis. However, a concentration of more than 3 mg/l in water causes gastrointestinal symptoms [14] and accumulation can affect vital organs [19] Thus, a method of treating the solution, to remove a certain amount of copper(II) ions from the solution, is required before its disposal.

The processes of coagulation and flocculation can be utilised in the purification of wastewater. There are several methods of water purification, such as utilizing activated carbon, membrane filtration, reverse osmosis, precipitation and UV processes [10].

H. Guo et al. (Eds.): IRC-SET 2025, PTT, pp. 124–133, 2026.
https://doi.org/10.1007/978-981-95-9630-0_11

However, coagulation is an inexpensive method [15]. It occurs when the coagulant disrupts and neutralizes the charges of particles in the mixture [18]. These form micro flocs that are not visible to naked eye, which then form visible flocs through flocculation [18]. These can then be removed from the mixture by filtration. Moringa Oleifera Lam seeds act as an effective natural coagulant, which has been proven as a viable and safer alternative to chemical coagulants such as aluminum sulfate [10], as such chemical coagulants may also act as pollutants and may worsen the quality of the water, with metal residues even being shown to cause conditions like Alzheimer's while natural coagulants do not, conversely possessing medicinal properties [5]. Studies have shown that Moringa seeds are effective in removing Cu(II) ions from water as they contain water-soluble cationic proteins, which are utilised as a primary coagulant which can be used to treating wastewater by reducing turbidity [2].

Turmeric is known for its various medicinal properties and has been used to treat diseases. [7]. It has also been utilised as a coagulant aid for a chemical coagulant, alum, showing that it is able to improve the effectiveness of alum in removing boron ions from wastewater due to its ability to form complexes with boric acid [7]. Curcumin (diferuloylmethane) is an active principle of the perennial Curcuma longa herb and is the major component in turmeric. Curcumin also can be classified as a chelating agent due to the present of 1,3-diketone structure [7], thus being able to bind to boron ions to form flocs, while aiding alum in the coagulation process by enhancing the agglomeration of the particle to form larger and heavier flocs which would be easier to remove via filtration [7].

While moringa seeds are recognised as efficient adsorbents and turmeric as an efficient coagulant aid, the synergy of the two in removing heavy metal ions is yet to be explored. We want to test a fully natural removal system of high removal efficiency using products that are cheap, widely available and effective. We hypothesised that the active agent of turmeric, curcumin, will form complexes with Cu(II) ions, thereby forming larger flocs for easier removal, with inference from curcumin's interaction with boron. This will improve the efficiency of moringa seeds on adsorbing Cu (II) ions. Therefore, this study investigates the enhancement of coagulation of Cu (II) ions by moringa seeds when combined with turmeric extract and evaluates the overall performance of this novel, natural system.

2 Materials and Method

Moringa concentration, turmeric concentration, pH and contact time of adsorbent and adsorbate are some factors that affect the efficiency of Cu(II) ions removal [13]. The pH was not monitored as Moringa seeds act as natural buffer solutions [18]. Jar tests were conducted to determine the optimum moringa and turmeric concentrations separately. A long duration of 2h was followed for every jar test to ensure the maximum required contact is achieved. This is because after optimal contact time is reached, removal efficiency becomes unchanged for subsequent timings [8].

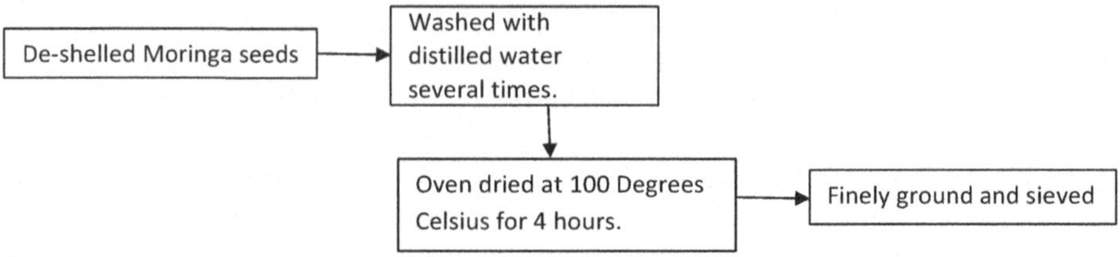

Fig. 1. Preparation of moringa seeds as biosorbents

2.1 Preparation of Moringa Oleifera Seed Powder

Moringa Oleifera Lam Seeds were bought from Shopee. The dry seeds were peeled by hand and then washed several times with distilled water, until the water they were submerged in was clear. The seeds were then dried in the oven at 100 degrees Celsius for 4 h. They were then ground into powder and sieved using a kitchen sieve available in the school biology laboratory. The powder was stored in airtight bottles and labeled clearly as "Dry moringa seed powder". This methodology follows [8], as shown in Fig. 1.

2.2 Preparation of Turmeric Extract

Turmeric powder was bought online from Shopee. 0.04 g of powder was mixed with 200 ml of 95% concentrated ethanol and stirred until most of the turmeric dissolved. The solution was then filtered using filter paper and funnel, to get rid of undissolved turmeric, and obtain a clear 200 mg/l turmeric extract stock solution, that was stored in a dark tincture bottle.

2.3 Preparation of Wastewater

A Cu(II) stock solution of 0.2 mol/dm^3 was prepared to simulate wastewater contaminated with Cu(II) ions released during practical lessons and experiments conducted by students in school laboratories. This concentration was confirmed with our lab technician. It was prepared by weighing 6.4 g of copper sulfate powder and dissolved in 200 ml of distilled water.

2.4 To Determine the Optimum Moringa Seed Extract Concentration

A series of jar tests were conducted to determine the optimum moringa seed extract concentration to remove Cu(II) ions from 50 ml of simulated wastewater. 200 ml beakers were filled with 50 ml of 32 g/l of Cu(II) solution. This concentration was chosen as schools use 0.2 mol/dm^3 of Cu(II) solution. 50 ml of moringa seed extract of varying concentrations, from 64 g/l to 320 g/l were added to the solution, keeping the volume ratio of Cu(II) solution to moringa extract solution 1:1. The range of concentration for moringa extract solution was chosen to determine the optimum ratio of concentration of Cu(II) solution to moringa extract solution that yields maximum removal of Cu(II) ions. Hence, for each test, with copper concentration kept constant at 32g/l (which gets diluted to 16 g/l in 100ml reaction mixture to mimic dilution of Cu(II) solutions poured down the sink), the concentration ratio of Cu(II) : Moringa seed solution were varied from 1:2 (1 part Cu(II) solution to 2 parts Moringa seed solution), 1:3,1:4 to 1:10. Each test was carried out at constant room temperature and pH of 5. They were placed on a magnetic stirrer and stirred at speed 8 for 1 min, then at speed 3 for 2 h each [7]. The initial high speed ensures homogenous distribution of coagulants, to allow greater contact of Cu(II) ions with the binding sites and the subsequent slower speed allows floc size to grow [8].

2.5 Analysis of Cu(II) Concentration After Jar Test with Moringa Extract

A UV Vis spectrophotometer was used to measure the concentration of Cu(II) ions. It is first blanked with sample of distilled water. A standardised curve was plotted using known Cu(II) ion concentration, from 10 g/l to 32 g/l. After completion of each jar test, the solution was filtered twice, using filter paper and funnel, and a sample of the filtrate was collected in a plastic cuvette to measure the absorbance at the wavelength most absorbed during standardisation. Standardisation was conducted each day and the concentration of Cu(II) ions in filtrates from jar tests were based on the standard graphs obtained each day, ensuring more accurate final Cu(II) concentrations were obtained from the jar tests.

2.6 To Determine the Optimum Turmeric Extract Concentration

A series of jar tests were again performed to determine optimum turmeric extract concentration. For each jar test, 50 ml of 32 g/l of Cu(II) solution, 50 ml of moringa seed extract of optimum concentration and 5 ml of turmeric extract of varying concentration, from 20 mg/l to 200 mg/l, were added to 200 ml beaker. The solutions were mixed at speed 8 on the magnetic stirrer for 1 min to agitate the sample and its contents and then at speed 3 for 2 h to allow for flocculation. The range of concentration of turmeric was chosen based on a 2012 science journal on using turmeric extract as coagulant aid with alum to remove boron from wastewater [7].

2.7 Analysis of Cu(II) Concentration After Jar Test with Turmeric Extract

Cu(II) concentrations after completion of jar tests were analysed using the same procedure as Sect. 2.4.

3 Results and Discussions

The percentage of Cu(II) ions removed by Moringa Oleifera seeds, with and without aid of turmeric have been analysed in this study.

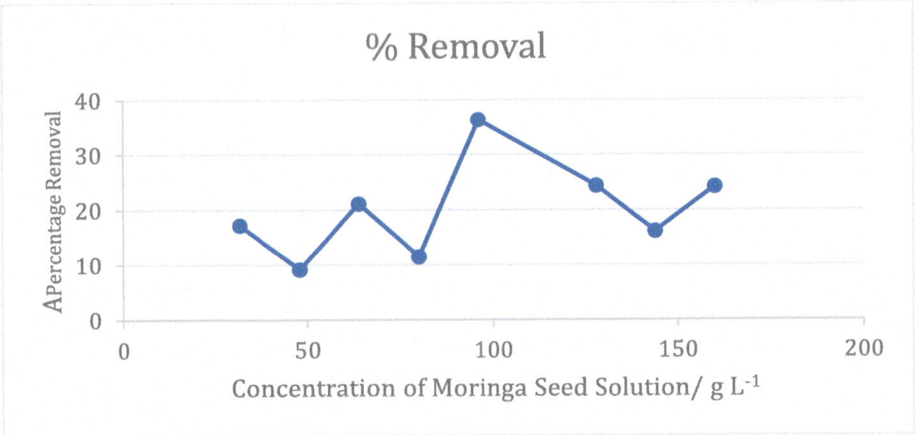

Fig. 2. Graph showing the percentage removal of Cu(II) ions by varying concentrations of moringa seed solution concentration, showing the effectiveness of the moringa solutions in removing Cu(II)

3.1 The Effect of Dose of Moringa Solution

Figure 2 shows a general increase in percentage removal of Cu(II) ions until 96 g/l moringa seed solution concentration, after which there's a general decrease, with some fluctuations on each side of the peak. The optimum dose of moringa seeds solution during the coagulation process was 96 g/l, which yielded the highest removal percentage of 36.5%, as shown in Fig. 1. Since the initial Cu(II) solution concentration is fixed at 16 g/l in the reaction mixture and such a sample requires 96 g/l for optimum removal, optimum ratio of initial concentration of Cu(II): moringa seed solution is 1:6. However, it is to be noted that the optimum here is only for this specific range of moringa seed solution concentrations and using a wider range can reveal a new optimum than can have a potentially higher removal rate than 36.5%.

The trend observed in the removal efficiency can be explained as follows. The increase in removal efficiency upon increasing moringa seed solution concentration from 32 g/l to 96 g/l is due to increasing number of available active sites present on the surface of the moringa seed, where positively charged Cu ions can bind to [6]. These active sites contain negatively charged amino acid groups which attract positively charged Cu(II) ions [3]. This electrostatic attraction enabling binding via the ion exchange mechanism, through which Cu(II) is known to be coagulated by moringa seeds [4]. The general decrease in percentage removal beyond 96 g/l is likely due to accumulation of the moringa biosorbent, which decreases the exposed surface area, decreasing the efficiency of absorption process [16]. A larger range of moringa concentrations than the ones currently utilised could be used to show a clearer trend, as the current range of concentrations were limited, enabling the further confirmation of the optimum moringa concentration.

However, it can be seen that moringa seeds are able to remove Cu(II) ions by a considerable amount, when used at optimum concentration, as the initial concentration of Cu(II) is much higher than afterwards [17].

3.2 The Effect of Dose of Turmeric Extract

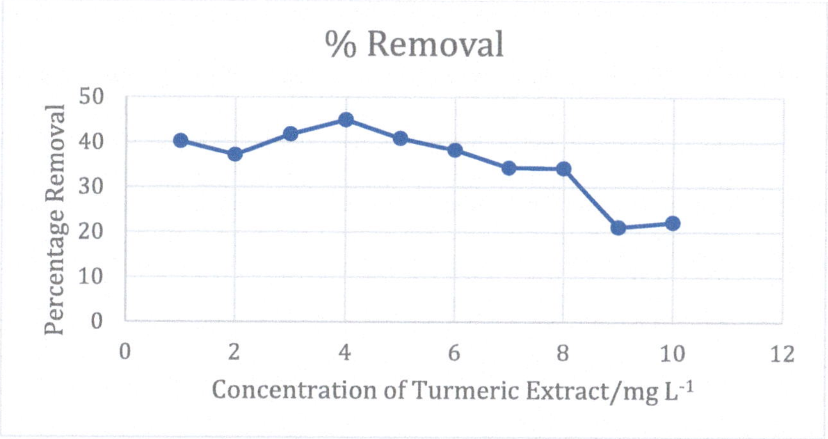

Fig. 3. Graph showing the effect of turmeric extract concentration on removal of Cu(II) ions by varying concentrations of turmeric extract, keeping moringa seed solution constant at optimum, showing the effectiveness of their synergy in removing Cu(II) ions.

Figure 3 shows a generally increasing trend of removal (up to 4 mg/L) and then a decreasing trend in percentage removal of Cu(II) ions. The optimum dose of turmeric extract using coagulation process was determined to be 4 mg/l, where the percentage removal of Cu(II) ions is the highest at 45.0%.

From Fig. 3, we observe that the effectiveness of removal of Cu(II) ions has increased for the turmeric concentration range of 1 mg/L to 6 mg/L, while from 7 mg/L to 10 mg/L, the percentage removal is smaller than what 96 g/l of moringa seed solution was able to achieve by itself, which was 36.5%. This suggests that turmeric concentrations beyond 6 mg/L impede the moringa seed solution of 96 g/l from achieving its optimal removal capacity of 36.5%. This presents a notable contrast to the findings of [7], where turmeric consistently enhanced the performance of alum for boron removal. We propose that this critical difference is explained by the charge of the metal-curcumin complex formed, which is a key affecting factor in the coagulation mechanism.

Alum works as a coagulant by neutralising negatively charged contaminants, thereby destablising them [7], the formation of larger and negatively charged boron-curcumin complexes enable more effective adsorption due to greater molecular size of the contaminants and the ion exchange mechanism takes place undisrupted regardless of the size of the contaminants. This enables turmeric to increase the effectiveness of alum in removing boron to be high at all concentrations of alum.

However, from our investigation, turmeric has been shown to form a negatively charged tetrahedral complex with curcumin [12]. It has also been found to form complexes with Cu(II) ions [1]. This is because turmeric roots consist of curcuminoids, which is an organic molecule with a symmetrical structure. As [9], reported, β-diketo moiety on a curcumin molecule undergoes keto-enol tautomerism, forming a hook structure that has high chelating ability, making curcumin molecules tend to bind with positively charged species. This enables Cu(II) ions to bind to the curcumin structure, forming Cu-Curcumin complex. [1].

In contrast to [7], these Cu-curcumin complexes formed can be positively charged or neutral [1]. The positively charged complex formed allows for larger and heavier flocs to be formed, allowing more Cu to be removed, with its peak at 45% with 4 mg/l of turmeric. This is because moringa seeds show better adsorption of ions with larger radii due to the surface of moringa containing pores that favour the binding of larger molecules, hence the larger complexes get easily adsorbed [4] and the ion exchange mechanism between the moringa seed particles and the positive complex is undisrupted. Since the percentage removal of Cu(II) ions in higher than 36.5% from 1 mg/L to 6 mg/L, it is likely that positively charged complexes were formed at this range.

However, as turmeric concentration increases, it is likely that more neutral Cu-Curcumin complexes were formed as curcumin can help to neutralise copper (II) ions. The zero net charge on the neutral Cu-Curcumin complex cannot be disrupted and flocs from neutral Cu-Curcumin complex cannot be formed. Thus, as turmeric concentration increases from 7 mg/l to 10 mg/l, the percentage of Cu removed decreases below 36.5% due to the inability of moringa seed particles to adsorb the neutral complexes. This shows that turmeric started to decrease the effectiveness of moringa seeds as more of the Cu(II) ions form neutral complexes that are unable to bind with moringa seeds via the ion exchange mechanism, impeding the adsorption capacity of moringa seeds.

Therefore, the charge of the intermediate complex is the pivotal factor determining whether turmeric acts as an effective coagulant aid, explaining both our results and the difference observed in the literature.

3.3 Comparison Between the Coagulation Process with and Without the Aid of Turmeric Extract

By applying the optimum parameters that had been obtained in this study (96 g/l of moringa seed solution and 4 mg/l of turmeric extract), the percentage of Cu(II) ion removal has been compared with and without the addition of turmeric extract. The results have been demonstrated as stated in Table 1.

Table 1. The comparison between results from the coagulation flocculation process with and without the assistance of turmeric extract

	Without turmeric extract	With turmeric extract
Mean percentage of Cu(II) Ions removed	36.5	45.0
Dose of moringa solution in g/l	96.0	96.0
Dose of turmeric extract in mg/l	nil	4.0

These results confirm the effectiveness of the synergy between optimal concentrations of 96 g/l moringa seed solution and 4 mg/l of turmeric extract, as they show an improvement in removal efficiency from 36.5% to 45.0%.

This effectiveness can be attributed to the dual-system mechanism, where the distinct properties of the two compounds come into play.

(I) The chelating property of curcumin molecules enables the formation of larger complexes that are positively charged.
(II) The coagulation by moringa seeds through the ion exchange mechanism is enhanced due to larger radii of the complexes.

These findings validate the proposed dual-mechanism and confirm that turmeric acts as an effective coagulant aid for moringa. In this study, we utilised filtration as the separation technique to remove the flocs from wastewater sample. However, this may have caused some of the smaller floc particles to go uncaptured by the filter paper, causing a potential underestimation of the removal rate. Hence, future studies could employ a centrifuge to remove the flocs to a greater extent.

4 Conclusion

Moringa seeds have good potential to remove Cu(II) ions, by removing about 36.5%, when its concentration is 96 g/l, showing that out of the range of concentration ratios of Cu(II) to moringa seeds chosen, 1:6 was the most efficient. Its effectiveness is enhanced by turmeric, as it is a coagulant aid by itself [7], albeit, not all concentrations of turmeric enhance the performance of moringa seeds, as at higher concentrations, they are likely to form more neutral complexes, thereby disrupting coagulation of Cu(II) by moringa seeds.

Further research using the synergy confirmed can be done in several directions. Our research chose to focus on using dry moringa powder as it would be the easier to implement on a large scale as it is more cost and time-efficient to produce, but, the system could be further optimized by applying the synergistic mechanism to a greater range of moringa concentration for a better optimum concentration of unmodified moringa seeds that have a higher removal percentage of 70% or to more potent biosorbents, such as de-oiled moringa seeds which have shown baseline removal rates of around 70% [13]. However, oil extraction from seeds is expensive due to the specific apparatus needed, such as Soxhlet extraction. Furthermore, the practical application of this system can be tested with a small-scale apparatus in labs, by designing and testing an apparatus that filters out the copper solution before the lab wastewater mixes with wastewater from other sources, as our aim was to eventually implement this in schools. Lastly, while we chose to focus on copper (II) ion removal as it is the most commonly used ion in school lab experiments, the combination of moringa and turmeric can also be tested on various other metal solutions used in school labs, like zinc, aluminum and manganese solutions.

In conclusion, this study validates a low-cost and entirely natural system for enhancing heavy metal removal, contributing to the development of more sustainable water purification technologies.

References

1. Addicoat, M.A., Metha, G.F., Kee, T.W.: Density functional theory investigation of Cu(I)- and Cu(II)-curcumin complexes. J. Comput. Chem. **32**(3), 429–438 (2011). https://doi.org/10.1002/jcc.21631
2. Al-Jadabi, N., Laaouan, M., El Hajjaji, S., Mabrouki, J., Benbouzid, M., Dhiba, D.: The dual performance of *Moringa Oleifera Seeds* as eco-friendly natural coagulant and as an antimicrobial for wastewater treatment: a review. Sustainability **15**(5), 4280 (2023). https://doi.org/10.3390/su15054280
3. Amagloh, F., Benang, A.: Effectiveness of Moringa oleifera seed as coagulant for water purification. Afr. J. Agricu. Res. **4**(1), 119–123 (2009). Retrieved from: http://www.academicjournals.org/AJAR
4. Apori, S.O., Atiah, K., Hanyabui, E., Byalebeka, J.: Moringa oleifera seeds as a low-cost biosorbent for removing heavy metals from wastewater. STED Journal. **2**(1), 45–52 (2020). https://doi.org/10.7251/STED2002045O
5. Bahrodin, M.B., Zaidi, N.S., Hussein, N., Sillanpää, M., Prasetyo, D.D., Syafiuddin, A.: Recent advances on coagulation-based treatment of wastewater: Transition from chemical to natural coagulant-current pollution reports. SpringerLink (2021). https://doi.org/10.1007/s40726-021-00191-7
6. Babel, S., Kurniawan, T.A.: Cr(VI) removal from synthetic wastewater using coconut shell charcoal and commercial activated carbon modified with oxidizing agents and/or chitosan. Chemosphere **54**(7), 951–967 (2004). https://doi.org/10.1016/j.chemosphere.2003.10.001
7. Halim, A.A., Thaldiri, N.H., Awang, N., Latif, M.T.: Removing boron from an aqueous solution using turmeric extract-aided coagulation-flocculation. Am. J. Environ Sci. **8**(3), 322–327 (2012). https://doi.org/10.3844/ajessp.2012.322.327
8. Jayadipraja, E., et al.: The effectiveness of moringa oleifera seed powder and mixing time (fast-low Speed) using jartest in reducing nitrate levels in river water. Public Health Indonesia **10**(3), 361–372 (2024). https://doi.org/10.36685/phi.v10i3.826
9. Kanhathaisong, S., Rattanaphani, S., Rattanaphani, V., Manyum, T.: A spectroscopic investigation of the complex of turmeric dye with copper(II) in aqueous solution. Suranaree J. Sci. Technol. **18**(2), 159–165 (2011). Retrieved from: https://ird.sut.ac.th/journal/sjst/#/home
10. Koul, B., Bhat, N., Abubakar, M., Mishra, M., Arukha, A.P., Yadav, D.: Application of natural coagulants in water treatment: a sustainable alternative to chemicals. Water **14**(22), 3751 (2022). https://doi.org/10.3390/w14223751
11. Liu, Y., Wang, H., Cui, Y., Chen, N.: Removal of copper ions from wastewater: a review. Int. J. Environ. Res. Public Health **20**(5), 3885 (2023). https://doi.org/10.3390/ijerph20053885
12. Mohammed, F., Rashid-Doubell, F., Cassidy, S., Henari, F.: A comparative study of the spectral, fluorometric properties and photostability of natural curcumin, iron-and boron-complexed curcumin. Spectrochim. Acta Part A Mol. Biomol. Spectrosc. **183**, 439–450 (2017). https://doi.org/10.1016/j.saa.2017.04.027
13. Nwagbara, V.U., Chigayo, K., Iyama, W.A., Kwaambwa, H.M.: Removal of lead, cadmium, and copper from water using *Moringa oleifera* seed biomass. J. Water Clim. Change **13**(7), 2747–2760 (2022). https://doi.org/10.2166/wcc.2022.091
14. Potera, C.: Copper in drinking water: using symptoms of exposure to define safety. Environ. Health Perspect. **112**(10), A568–A569 (2004). https://doi.org/10.1289/ehp.112-a568b
15. Rana, V., et al.: Effective and affordable water purification technologies for rural development. Water Resour. Manage. Rural Dev. 91–106 (2024). https://doi.org/10.1016/b978-0-443-18778-0.00016-7
16. Rao, M.M., Ramana, D.K., Seshaiah, K., Wang, M.C., Chien, S.W.C.: Removal of some metal ions by activated carbon prepared from Phaseolus aureus hulls. J. Hazard. Mater. **166**(2), 1006–1013 (2009). https://doi.org/10.1016/j.jhazmat.2008.12.002

17. Stanikina, N.: Metal ions removal from wastewaters using moringa oleifera seeds. BIO Web Conf. (2023). https://doi.org/10.1051/bioconf/20235803005
18. Sulaiman, M., Zhigila, D.A., Mohammed, K., Umar, D.M., Babale, A., Manan, F.A.: Moringa oleifera seed as alternative natural coagulant for potential application in water treatment: a review. **30**(1), 1–11 (2017). Retrieved from https://www.akademiabaru.com/submit/?journal=arms
19. Tchounwou, P.B., Yedjou, C.G., Patlolla, A.K., Sutton, D.J.: Heavy metal toxicity and the environment. Experientia Suppl. **101**, 133–164 (2012). https://doi.org/10.1007/978-3-7643-8340-4_6

Multicilin Redundancy in the Zebrafish MCC Programme as a Result of Evolutionary Divergence

Ian Ee Eleos Lo(✉)

NUS High School of Mathematics and Science, 20 Clementi Avenue 1, Singapore 129957, Singapore
h2210101@nushigh.edu.sg

Abstract. Cilia play a crucial role in left-right symmetry and fetus development. Cells with dozens of cilia are termed as multiciliated cells (MCCs). Improper MCC differentiation may result in impeded fluid flow, a cause of airway infections, primary ciliary dyskinesia, situs inversus, and retinal degradation, classified under a broad term RGMC (reduced generation of motile cilia). In humans, multicilin (MCI) is inhibited by geminin (GMNN), and overexpression resulted in ectopic MCC differentiation. We sought to explore whether mci functions similarly in zebrafish. By cloning and transforming mci into *E. coli*, then extracting it and transcribing it, we injected them into 1-cell stage zebrafish gmnn knockout embryos. We then stained these embryos for cilia, basal bodies, and nuclei, before checking for the presence of ectopic cilia under confocal microscopy. Overexpression of mci did not result in complete ectopic MCC differentiation, with isolated ectopic cilia in various tissues. This means that gmnn may not be completely inhibiting mci in zebrafish, rather it is somewhat dispensable in the zebrafish MCC differentiation programme. This may be due to is evolutionary history, as its sequence diverges significantly from other mci homologues. These findings altogether may indicate that the zebrafish MCC programme diverges from humans and may lay the foundation for understanding causes of RGMC and how it may be treated.

Keywords: mci · gmnn · zebrafish · cilia · ectopic differentiation · MCC

1 Background

Cilia are membrane bound projections from cells supported by a microtubule cytoskeleton, [16]. The structure of a cilium is segregated into three parts, the axoneme, basal body, and transition zone [15]. The sub-ciliary compartment is anchored to the plasma membrane by a basal body, a barrel-like compartment of microtubules [8]. Such cells possessing more than two, typically dozens of cilia each are thus referred to as multiciliated cells (MCCs), beating in a coordinated and polarised manner to drive fluid flow across a surface [5, 10]. Such MCCs are often differentiated from progenitor cells with inhibited notch (N) signalling and the activation of geminin coiled-coil domain-containing protein 1 (gmnc), and multicilin (mci) [18]. MCCs can be found in the lungs,

H. Guo et al. (Eds.): IRC-SET 2025, PTT, pp. 134–142, 2026.
https://doi.org/10.1007/978-981-95-9630-0_12

brain, and epididymis, rhythmically beating to ensure constant fluid flow. They aid in mucus clearance and cerebrospinal fluid flow among others [4, 12]. Diseases related to cilia motility, such as primary ciliary dyskinesia, airway infections, reduced fertility, situs inversus and retinal degradation [2, 4, 6] are typically genetic and tend to be life-long [1].

The master regulator of cilia formation is GMNC, being proven to be sufficient and essential in inducing MCC differentiation [22]. GMNC, proven to act downstream of N signalling [22], complexes with transcription factors E2F4, E2F5, and DP1, acting as the earliest known transcriptional activator [20]. Another gene, FOXJ1, plays an essential role in ciliogenesis [21]. Loss of FOXJ1 activity resulted in compromised formation of motile cilia, and foxj1 was proven to induce and transcriptionally activate key genes involved in motile ciliogenesis [21]. In addition to N signalling, geminin (GMNN) also downregulates the GMNC/E2F4/E2F5/DP1 complex (EDG complex) formation [20] which downplays MCI and its complexing with E2F4, E2F5, and Dp1 (EDM complex) [11], whose main function is to initiate centriole duplication and activate motile ciliogenesis genes and transcription factors such as c-Myb, FOXJ1, and Rfx. The EDG complex first forms and activates mci and other downstream ciliary transcription factors. MCI then forms the EDM complex to further enhance expression of ciliary transcription factors [9].

MCI interacts with GMNN via the CC domain, inhibiting it. Conversely, GMNN also downplays the ability of MCI to induce MCC differentiation [19]. GMNN is hypothesised to sequester GMNC or MCI away from complexing [20], repressing differentiation when GMNN levels are high. As ciliogenesis progresses, GMNN and N signalling levels eventually diminish, allowing GMNC and MCI to increase.

Mutations in MCI have been proven to result in mucociliary clearance disorders and reduced generation of multiple motile cilia (RGMC) in humans [3]. It is characterised by tissue, organ, and gamete dysfunction, hindering mucus clearance and resulting in chronic respiratory diseases that further affect fertility, brain development, and laterality [14].

Recently, a similar MCI-like protein has been reported in zebrafish, highlighting possible similarities in motile ciliogenesis programs in vertebrates [23]. Furthermore, mouse Mci has been also proven to induce ectopic ciliation in wild-type zebrafish embryos, raising questions about the similarities of zebrafish mci and gmnn.

Ectopic ciliation is the phenomenon when MCCs are formed in other types of tissue not typically possessing them. In humans, ectopic overexpression of human MCI is sufficient to induce ectopic ciliation. This forces cells to exit the cell cycle and become post mitotic, a hallmark characteristic of MCCs. They produce large amounts of basal bodies, and also experience subsequent activation of other ciliary transcription factors and basal ciliary components [5]. However, overexpression of zebrafish mci does not induce ectopic ciliation in embryos, hinting at possible disruption or inhibition by the presence of gmnn.

The mechanism by which gmnn acts in the zebrafish MCC pathway is relatively unknown. Hence, we were curious about the function of gmnn and whether it continues to inhibit MCC differentiation. We aimed to explore whether ectopic ciliation could still occur in gmnn knockouts and mci overexpression. We suspected that with gmnn knockouts, mci overexpression should result in ectopic ciliation. We sought to explore this by conducting gmnn knockouts in zebrafish, then injecting them with synthetic zebrafish mci mRNA and checking for the presence of ectopic ciliation.

Our findings indicate the function of zebrafish mci and gmnn, showing that mci is insufficient alone to induce ectopic ciliation in the absence of gmnn. Rather, an EDM complex is more likely to induce the same effect. These results lay the groundwork for further understanding of the zebrafish MCC differentiation pathway, perhaps allowing further advancements into why RGMC occurs and how it can be treated.

2 Materials and Methods

2.1 Zebrafish Genotyping

Fish were fin clipped and genotyped for a homozygous 5 base pair (bp) gmnn knockout mutation targeting exon 3 (Fig. 1). First, 40 µl of lysis buffer (50% 50 mM NaOH; 40% 0.5 mM EDTA; 10% nuclease free water (dH_2O), Invitrogen) was added to polymerase chain reaction (PCR) strip tubes. Fish were individually dipped in tricane until unresponsive. The fish was transferred to a petri dish and a small part of its tail fin was cut off. The piece of tail cut off was added to a PCR tube. The procedure was repeated for all six fish.

The PCR strip tube was then incubated at 95 °C for 30 min to lyse the tailfin. PCR amplification was then carried out. 5 µl of Myfi Polymerase (Fisher Scientific), 0.5 µl of forward and reverse primers for mci (Integrated DNA Technologies), 0.5 µl of template DNA, and 4 µl of dH_2O (Invitrogen) were added to 6 PCR strip tubes. Denaturation was at 95 °C for 25 s, annealing at 59 °C for 30 s, and extension at 72 °C for 1 min over 35 cycles.

2 µl of the product for each tube was aliquoted and mixed with 1 µl of loading dye (New England Biolabs) and ran in a 3% agarose small gel for 17 min at 160 V with 5 µl of 100 bp DNA ladder (New England Biolabs). The product (8 µl), ExoI exonuclease (0.5 µl, Thermo Fisher Scientific), Fast Alkaline Phosphatase (1 µl, Thermo Fisher Scientific), and dH_2O (7 µl, Invitrogen) were added together in a PCR strip tube for a 16.5 µl reaction. It was incubated at 37 °C for 20 min, then 85 °C for 15 min and sent for sequencing by Biobasic (Fig. 2).

2.2 Bacterial Transformation

1 µl of cloned pRD67 plasmid with a HA tag, sp6 promoter and zebrafish mci was added to 50 µl dH2O (Invitrogen). 3 µl was then added to an Eppendorf tube with competent *E. coli*. The mixture was incubated in ice for 20 min, then heated at 42 °C for 1 min and cooled for 15 min. 100 µl of SOC medium (Invitrogen) was then added and the bacteria were incubated at 37 °C for 45 min in a rotary incubator. Bacteria were spread on an ampicillin plate and grown overnight at 37 °C.

Two colonies were picked a day after and grown in 4 ml of ampicillin LB broth for 4 h. pRD67-mci-HA was isolated using the Qiagen Miniprep kit, following manufacturer protocol. Concentration was measured by nanodrop spectrometry.

The plasmid was then digested to check for an insert. 0.5 µl of XhoI and BamHI (New England Biolabs) were added to a PCR tube. 6.7 µl and 5.1 µl of plasmid were added, followed by 2.3 µl and 3.9 µl of rCutsmart buffer (New England Biolabs) for a 10 µl reaction. It was incubated at 37 °C for 1 h and ran in a 1% 50 ml agarose small gel for 15 min at 150 V, with 5 µl of a 1 kb DNA ladder.

2.3 mRNA Synthesis

mRNA was synthesised for microinjection into gmnn double mutant zebrafish. pRD67-mci-HA (10 µl), rCutsmart buffer (2 µl, New England Biolabs), XbaI restriction enzyme (2 µl, New England Biolabs) and dH_2O (6 µl, Invitrogen) were added to a 1.5 ml Eppendorf tube, mixed, and incubated at 37 °C for 2 h. It was then run in a 1% 50 ml small agarose gel with 1000 bp DNA ladder (5 µl) for 15 min at 150 V (Fig.3).

The band corresponding to the linearised plasmid was gel extracted and purified using the Qiagen Miniprep kit, following manufacturer protocol. The concentration was measured by nanodrop spectroscopy. The linearised plasmid was concentrated via D-AQ for 8 min in an Eppendorf concentrator plus.

Transcription was carried out using the Invitrogen mmessage mmachine SP6 kit following manufacturer protocol. Contents were then concentrated under D-AL for 5 min followed by D-AQ for another 5 min. The pellet was resuspended in 20 µl dH_2O (Invitrogen) and the concentration was measured by nanodrop spectrometry.

2.4 Microinjection and Staining

Male and female gmnn homozygous mutant fishes were set up and crossed. All eggs were collected and used for microinjection at 1-cell stage. Mci mRNA was further diluted to of 250 ng/µl by addition of dH_2O. Microinjection was carried out using a Leica MZ125 micromanipulator and a World Precision Instruments PV850 injector. Embryos were injected with 0.5 nl of mRNA (0.125 pg) and 0.75 nl of mRNA (0.1875 pg). Injected embryos were then transferred to petri dishes with methylene blue infused egg water and incubated at 28 °C for 1 day.

15 injected and 15 noninjected embryos at 20–30% epiboly stage were collected 6 h after microinjection washed with 1 ml of phosphate buffered saline (PBS). PBS was removed and 1 ml of fish fix (1 g PFA, 1 g sucrose, 25 ml 1x PBS, 2 drops 6 M NaOH) was added. It was stored overnight at 4 °C and stained with anti-HA antibody.

The remaining embryos were also collected 24 h after fertilisation and fixed in fish fix before being stained with rabbit anti-acetylated α-tubulin antibody and mouse anti-γ-tubulin antibody for the presence of cilia and basal bodies respectively.

Embryos were incubated for 3 h at room temperature on a nutating rotor. The fish fix was removed and embryos were transferred to a cavity dish and flooded with 1 ml PBS. They were incubated on an orbital shaker at 45 rpm for 30 min. They were removed, washed with PBS, and incubated another 3 times. 250 µl of blocking solution (2% sheep serum in PBDT) was added and the dishes were incubated on an orbital shaker for 1 h. Blocking solution was replaced with primary mouse anti-HA antibody in a 1:500 dilution with dH_2O (Invitrogen). 250 µl of primary antibody solution was added to each dish and incubated overnight at 4 °C on an orbital shaker.

The next day, the embryos were brought back to room temperature and the primary antibody solution was removed. It was washed with 1 ml of PBS and incubated for 30 min on an orbital shaker. This washing was repeated 3 more times. PBS was then removed and 250 µl of secondary antibody solution (Alexa Fluor 555 donkey anti-mouse, Thermo Fisher Scientific) in a 1:500 dilution with PBDT buffer was added. It was incubated in a box on an orbital shaker for 4 h at room temperature. 0.3 µl of DAPI was then added to each dish and incubated for 30 min.

The secondary antibody solution was then removed and replaced with 1 ml of PBDT buffer. The dishes were incubated for 20 min on an orbital shaker. Washing was repeated twice more. PBS was removed and replaced with 1 ml of 70% glycerol in DI water and stored at −20 °C.

24 hpf embryos were stained similarly to the 20% epiboly embryos. Embryos were first treated with 1 ml of chilled acetone and incubated at −20 °C for 7 min to increase permeability. Acetone was then removed, and 1 ml of PBS was added. It was incubated for 5 min. PBS washing was repeated another three times. Embryos were deyolked and mounted onto microscope slides in glycerol and imaged under an upright confocal microscope (Olympus FluoView 3000) (Fig. 4).

3 Results

3.1 gmnn Knockout Fish Possess a 5 bp Deletion in Exon 3

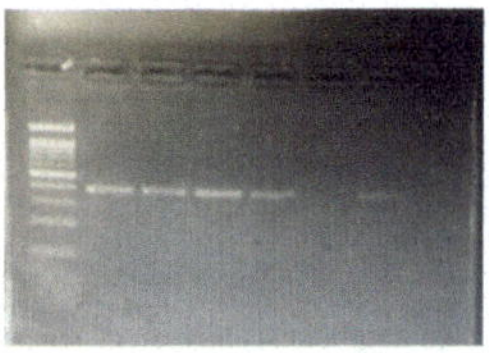

Fig. 1. 3% gel run with fin clipped PCR products. All fish were homozygous for the gmnn 5 bp deletion mutation. Leftmost column: 100 bp DNA ladder

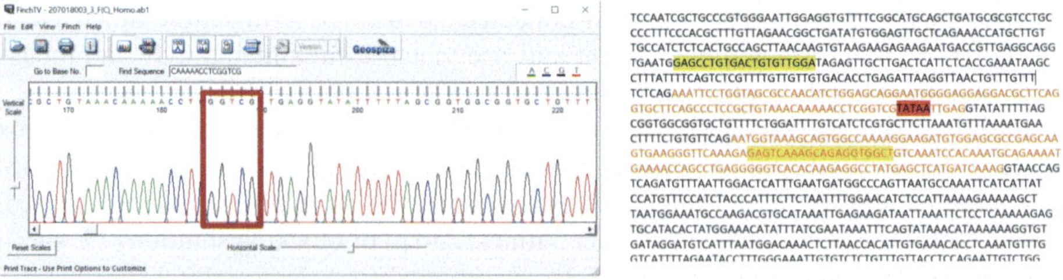

Fig. 2. Left: Sequencing results of the PCR product. The red box highlights sequence before the deletion (right, in red). Right: sequence of mci with deletion (red). Primers for the PCR are yellow.

The sequencing result of the PCR product also confirmed the deletion mutation. Analysis of the DNA sequence before the deletion (fig. 2 right, in red) showed that it was deleted in the sequenced result. A heterozygous fish would have had overlapping reads after the GGTCG sequence in the red box as both the wild-type and frameshift mutant allele would be read together, leading to an inconclusive sequencing result.

3.2 Transformed, Linearized PRD67-HA Served as a Transcription Template

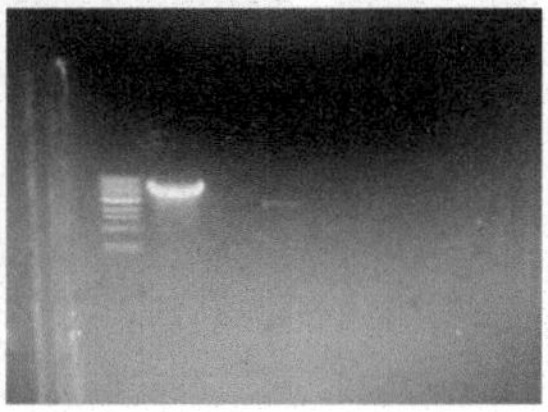

Fig. 3. Gel run with linearised pRD67 mci HA plasmid. Left: 1 kb DNA ladder. Centre: linearised plasmid. Right: circular plasmid.

Two colonies were picked and isolated plasmids had a concentration of 151.4 and 197.8 ng/μl. The linearised plasmid had a concentration of 111.2 ng/μl. mRNA was synthesized by following the in vitro transcription protocol. Its was 404.1 ng/μl. 2ul of mRNA was loaded in 1% agarose gel and a quick gel run (150V, 5 min) was done. A single clean band was seen that indicated the good quality of the synthesized mRNA.

3.3 Overexpression of Zebrafish Mci in Gmnn Knockout Embryos May Induce Ectopic Ciliation

Only two out of 16 embryos scanned showed a positive case of ectopic ciliation (Fig. 4: right -top, bottom), when staining with acetyl-tubulin for cilia, gamma-tubulin for basal bodies, and DAPI for nuclei. We cannot for certain assert that overexpression can induce ectopic ciliation in gmnn knockout embryos. More embryos should be screened and observed to see if a significant number show a similar effect. Nevertheless, the results may indicate that ectopic ciliation could be possible. The dosage of mRNA injected may be too little to see ectopic ciliation in a larger number of injected embryos. Therefore, a dosage increase might show proper effect.

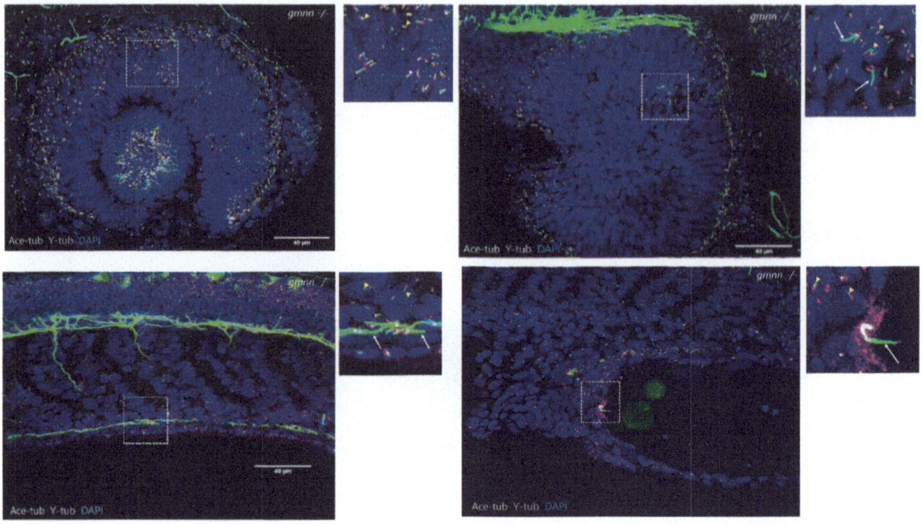

Fig. 4. 25 hpf embryos were stained with acetyl-tubulin (cilia), gamma-tubulin (basal bodies), and DAPI (nuclei). The results of the staining are shown above in four separate embryos, with ectopic cilia outlined in dotted white boxes, further magnified in the inset. Embryos viewed under laser confocal microscopy with three channels set at 488, 495, and 555 nm. Top left and right: eye section. Bottom left: pronephric duct. Bottom right: nose. All ectopic cilia are outlined in white boxes and shown in the insets.

4 Discussion

In this study, we aimed to further investigate the function of the zebrafish mci paralogue. In the absence of gmnn, MCC differentiation is still able to proceed normally in the notochord and muscle fibres. However, overexpression of zebrafish mci in gmnn knockout embryos is unable to induce ectopic ciliogenesis, as compared to mouse mci or gmnc [22]. There are a few possible reasons for such a divergence in function.

First, mci function is not typically needed for MCC differentiation in zebrafish as compared to other tetrapods such as *Xenopus*, mice, and humans [9, 11]. Mci is not only essential for MCC differentiation but also forms the EDM complex which binds and activates other genes required for centriole duplication. This may be because the recently characterised zebrafish mci paralogue shows little amino acid sequence homology to human, mouse, and *xenopus* mci proteins. The CC domain, for example, shares only 13% sequence conservation between humans and zebrafish [23]. In contrast, the CC domains between human, mouse, and *xenopus* mci show much higher rates of conservation. This divergence in amino acid sequence over time may have also contributed to divergence in function and essentiality of zebrafish mci within the zebrafish MCC differentiation program. Mci function may be enhanced by binding to E2f4, E2f5, and Dp1 to induce ectopic ciliogenesis.

Secondly, mci may have been recruited into the MCC differentiation program later than gmnn and gmnc. Gmnn homologues were detected in placozoans, an early evolutionary organism, predating the development of a nervous system [17, 23]. Similarly, gmnc was found to also be relatively primitive in comparison to mci. The protein was found in jawless vertebrates and the elephant shark, a member of the oldest lineage of jawed vertebrates, and other bony vertebrates such as zebrafish, *Xenopus*, and humans. However, mci is absent from earlier primitive organisms and homologues were only found in more recent bony vertebrate genomes. Given that all eukaryotic species possess some form of cilia, the origin of MCCs thus predates gmnc and mci [7, 13, 18]. This may hint that gmnc and mci functions were acquired by organisms along the path of evolution. As mci was recruited into the MCC differentiation pathway later than gmnc, it served as an additional factor to induce MCC differentiation rather than play a central role in ciliogenesis. Gmnc thus had sufficient time to adapt and take the place of a critical regulator of MCC differentiation before the introduction of mci.

5 Conclusion

Cilia play a major role in embryo development and are closely related to several major diseases which severely impact normal body function. Gmnn eventually inhibits MCC differentiation in humans. In zebrafish, the role of gmnn on ectopic MCC differentiation is unclear. We aimed to explore MCC differentiation in zebrafish, by synthetically overexpressing mci in gmnn knockout fish. By transcribing and injecting mci mRNA into embryos, and then staining them, we observed the effect on ectopic MCC differentiation. Ectopic cilia formation was observed occasionally, hinting at redundance in zebrafish. This may be because mci was recruited later than gmnc into the MCC differentiation programme, and its sequence diverges from human, xenopus, or mouse mci, which may cause its function to differ significantly. Lastly, zebrafish possess comparatively fewer basal bodies than humans, which may also allude to a simpler MCC differentiation programme. In conclusion, zebrafish mci is not crucial for ectopic MCC differentiation, which may be due to its evolutionary history.

Acknowledgements. The author would like to thank the following for mentorship, guidance, and provision of laboratory resources: Dr Sudipto Roy and Dr Dheeraj Rayamajhi (Agency for Science, Technology, and Research/Institute of Molecular and Cell Biology; Singapore). The author would also like to thank the following for advice and mentorship over the course of the project: Ms Fong Kit Ching, Dr Chiam Sher-Yi, Dr Maury Julien Jean Pierre (NUS High School of Mathematics and Science; Singapore).

References

1. Afzelius, B.: Cilia-related diseases. J. Pathol. **204**(4), 470–477 (2004). https://doi.org/10.1002/path.1652
2. Bisgrove, B.W., Yost, H.J.: The roles of cilia in developmental disorders and disease. Development **133**(21), 4131–4143 (2006). https://doi.org/10.1242/dev.02595
3. Boon, M., et al.: MCIDAS mutations result in a mucociliary clearance disorder with reduced generation of multiple motile cilia. Nat. Commun. **5**(1), 4418 (2014). https://doi.org/10.1038/ncomms5418

4. Boutin, C., Kodjabachian, L.: Biology of multiciliated cells. Curr. Opin. Genet. Dev. **56**, 1–7 (2019). https://doi.org/10.1016/j.gde.2019.04.006
5. Brooks, E.R., Wallingford, J.B.: Multiciliated Cells. Curr. Biol. **24**(19), R973–R982 (2014). https://doi.org/10.1016/j.cub.2014.08.047
6. Brown, J.M., Witman, G.B.: Cilia and Diseases. Bioscience **64**(12), 1126–1137 (2014). https://doi.org/10.1093/biosci/biu174
7. Jékely, G., Arendt, D.: Evolution of intraflagellar transport from coated vesicles and autogenous origin of the eukaryotic cilium. BioEssays **28**(2), 191–198 (2006). https://doi.org/10.1002/bies.20369
8. Kilburn, C., Winey, M.: Basal bodies. Curr. Biol. **18**(2), R56–R57 (2008). https://doi.org/10.1016/j.cub.2007.10.057
9. Lu, H., et al.: *Mcidas* mutant mice reveal a two-step process for the specification and differentiation of multiciliated cells in mammals. Development **146**(6), dev172643 (2019). https://doi.org/10.1242/dev.172643
10. Lyu, Q., Li, Q., Zhou, J., Zhao, H.: Formation and function of multiciliated cells. J. Cell Biol. **223**(1), e202307150 (2024). https://doi.org/10.1083/jcb.202307150
11. Ma, L., Quigley, I., Omran, H., Kintner, C.: Multicilin drives centriole biogenesis via E2f proteins. Genes Dev. **28**(13), 1461–1471 (2014). https://doi.org/10.1101/gad.243832.114
12. Meunier, A., Azimzadeh, J.: Multiciliated cells in animals. Cold Spring Harb. Perspect. Biol. **8**(12), a028233 (2016). https://doi.org/10.1101/cshperspect.a028233
13. Mitchell, D.R.: The Evolution of Eukaryotic Cilia and Flagella as Motile and Sensory Organelles. In: Eukaryotic Membranes and Cytoskeleton (Vol. 607, pp. 130–140). Springer New York (2007). https://doi.org/10.1007/978-0-387-74021-8_11
14. Mitchison, H.M., Valente, E.M.: Motile and non-motile cilia in human pathology: from function to phenotypes. J. Pathol. **241**(2), 294–309 (2017). https://doi.org/10.1002/path.4843
15. Reiter, J.F., Blacque, O.E., Leroux, M.R.: The base of the cilium: roles for transition fibres and the transition zone in ciliary formation, maintenance and compartmentalization. EMBO Rep. **13**(7), 608–618 (2012). https://doi.org/10.1038/embor.2012.73
16. Satir, P., Christensen, S.T.: Overview of structure and function of Mammalian Cilia. Annu. Rev. Physiol. **69**(1), 377–400 (2007). https://doi.org/10.1146/annurev.physiol.69.040705.141236
17. Schierwater, B., et al.: The enigmatic Placozoa part 1: exploring evolutionary controversies and poor ecological knowledge. BioEssays **43**(10), 2100080 (2021). https://doi.org/10.1002/bies.202100080
18. Spassky, N., Meunier, A.: The development and functions of multiciliated epithelia. Nat. Rev. Mol. Cell Biol. **18**(7), 423–436 (2017). https://doi.org/10.1038/nrm.2017.21
19. Terré, B., et al.: GEMC 1 is a critical regulator of multiciliated cell differentiation. EMBO J. **35**(9), 942–960 (2016). https://doi.org/10.15252/embj.201592821
20. Vladar, E.K., Mitchell, B.J.: It's a family act: The geminin triplets take center stage in motile ciliogenesis. EMBO J. **35**(9), 904–906 (2016). https://doi.org/10.15252/embj.201694206
21. Yu, X., Ng, C.P., Habacher, H., Roy, S.: Foxj1 transcription factors are master regulators of the motile ciliogenic program. Nat. Genet. **40**(12), 1445–1453 (2008). https://doi.org/10.1038/ng.263
22. Zhou, F., Narasimhan, V., Shboul, M., Chong, Y.L., Reversade, B., Roy, S.: Gmnc Is a master regulator of the multiciliated cell differentiation program. Curr. Biol. **25**(24), 3267–3273 (2015). https://doi.org/10.1016/j.cub.2015.10.062
23. Zhou, F., et al.: Conservation as well as divergence in Mcidas function underlies the differentiation of multiciliated cells in vertebrates. Dev. Biol. **465**(2), 168–177 (2020). https://doi.org/10.1016/j.ydbio.2020.07.005

Green Enzyme Engineering: Bioremediation of Methylene Blue-Contaminated Wastewater Using Hydrogel-Immobilised Enzyme Extract Synthesised from Spent Mushroom Compost

Cherelle Wong, Joyce Wang, Suchita Manikandan, Abigayle Ng, and Grace Lim(✉)

Raffles OpenLab, Raffles Institution, Singapore 575954, Singapore
grace.lim@ri.edu.sg

Abstract. This study aims to synthesise reusable, cost-effective and sustainable spent mushroom extract-alginate beads (SME-AB), investigate its bioremediation abilities of textile dye pollutant, methylene blue (MB), and enhance its removal efficiency ABTS mediation. MB is a toxic, non-biodegradable textile dye. Spent mushroom compost (SMC) waste, has proven its effectiveness in removing a variety of pollutants, including MB, due to its adsorption properties and ligninolytic enzymes like laccase. Recent studies show *Pleurotus ostreatus* and *Hericium erinaceus* SME effectively remove dyes and pharmaceuticals, respectively. However, the effectiveness of an SME-mediator hydrogel system in dye removal has not been studied. Enzymes were extracted from SMC by sectioning mycelium, homogenising, centrifuging (10000rpm, 4°C) and vacuum filtering (pore size 0.2 μm). *P. ostreatus* SME was 2.25 and 1.32 times more effective than *H. erinaceus* in decolourising MB at 5 mg/L and 10 mg/L initial MB concentrations, respectively. *P. ostreatus* SME was immobilised in alginate to improve recovery and enzyme stability and had a swelling degree of 6009.2%, indicating strong mechanical stability. 10.0g SME-AB was added to 200 ml 20 mgL^{-1} MB and placed in a shaking incubator (150 rpm, 35°C). SME-AB decolourised 26.0% of MB in 24 h, through enzymatic degradation and adsorption, with only an 8.0% drop in removal between the second and third uses. Adding laccase-mediator ABTS to the SME-AB composition further enhanced efficiency by 28.3%. This demonstrates that dried ABTS-SME-AB, repurposed from SMC waste, is a promising low-cost, sustainable technology for textile wastewater treatment.

Keywords: Mycoremediation · Spent Mushroom Compost · Ligninolytic Enzymes · Methylene Blue · Hydrogel Immobilisation · ABTS

1 Introduction

In recent years, the booming textile industry has led to extensive use and discharge of synthetic dyes into water bodies, generating 300,000 tons of dye-contaminated wastewater annually [1]. Methylene blue (MB) is a widely used azo dye that is toxic, carcinogenic and non-biodegradable, posing significant threat to the environment and human health [2].

H. Guo et al. (Eds.): IRC-SET 2025, PTT, pp. 143–154, 2026.
https://doi.org/10.1007/978-981-95-9630-0_13

Spent mushroom compost (SMC) has garnered attention as a low-cost and eco-friendly alternative to conventional bioremediation methods using algae and bacteria [3]. Its effectiveness arises from a combination of adsorption due to its affinity for pollutants including heavy metals [4][5], and textile dyes [6], and presence of ligninolytic enzymes [7] which are capable of degrading them. However, SMC is often discarded after harvest; for instance, Fogo Fungi, a local gourmet mushroom farm, disposes of about 4 tonnes monthly.

Spent mushroom extract (SME) from *Pleurotus ostreatus*, commonly known as oyster mushroom, has been well-documented in its bioremediation of textile dyes, and abundance of laccase, a ligninolytic enzyme [8]. Laccase is known for its unique ability to oxidise a wide range of chemical compounds without requiring H_2O_2 [9][10]. More recently, SME of *H. erinaceus*, commonly known as lion's mane mushroom, has shown promise in the removal of acetaminophen and sulfanomides [11]. However, despite their potential, real-life industrial application of free enzymes is limited [12] due to difficulty in recovery from wastewater and sensitivity to environmental factors including temperature and pH. Enzyme immobilisation addresses these issues, with alginate gel entrapment being a simple and low-cost method [13]. Immobilisation enhances enzyme stability and extends shelf life, enabling longer storage [14]. Adding the laccase-mediator, 2,2-azino-bis-(3-ethylbenzthiazoline-6-sulfonic acid) (ABTS), to immobilised laccase further improves degradation efficiency [15].

Till date, the effectiveness of an SME-mediator system immobilised onto hydrogel in dye removal has not been studied. As such, this study, conducted in three phrases, aims to develop cost-effective and sustainable spent mushroom extract-alginate beads (SME-AB) and investigate its bioremediation abilities of textile dye pollutant MB, at the same time reducing organic waste in Singapore. Phase 1 compares the effectiveness of MB removal by *P. ostreatus* and *H. erinaceus* SME. Phase 2 involves the synthesis of SME-AB to increase enzyme stability and bead reusability, and investigates MB removal by SME-AB. Phase 3 involves the addition of laccase-mediator ABTS to further enhance decolourisation ability of SME-AB.

2 Materials and Methods

2.1 Phase 1 – Removal of Methylene Blue (MB) by Spent Mushroom Extract (SME)

Preparation of SME Extract. SMC of P. ostreatus and H. erinaceus was obtained from Fogo Fungi. Fungal mycelia, which contain extracellular enzymes, was sectioned and homogenised in cool 50 mM sodium acetate (NaOAc) buffer in a blender for 5 min, in the ratio of 1 g mycelia to 2 mL. Mixtures were incubated at 10 °C with shaking at 150 rpm for 1.5 h. Each mixture was filtered through cheesecloth (20 μm pore size), then centrifuged at 10,000 rpm at 4 °C for 30 min. Supernatants were decanted into a vacuum filter unit with a 0.22 μm pore size membrane, yielding SME.

Preparation of MB. 1000 mg L^{-1} MB (GCE Laboratory Chemicals) was diluted to 100 mg L^{-1} using deionized (DI) water and stored in dark conditions for subsequent use. Varying concentrations were prepared by diluting the 100 mg L^{-1} stock with DI water. A fresh stock was prepared daily.

ABTS Assay. The activity of laccase in *P. ostreatus* and *H. erinaceus* SME was assayed using 0.05mM ABTS as substrate at 25 °C, by adding 1.2 mL of ABTS to 0.3 mL of SME. Absorbance of the mixed solution was recorded every 30 s over 2 min using a UV–Vis spectrophotometer at 420 nm, in triplicate. Average activity was then calculated with one unit of activity defined as the quantity of laccase required to oxidize 1 μM of ABTS per min, using the following equation:

$$\text{Activity (U/L)} = \frac{A \times V \times f_{dilution}}{t \times \varepsilon \times V_0} \times 1000$$

where A is absorbance, V is volume of mixture (mL), t is reaction time (min), ε is the molar extinction coefficient for ABTS oxidation and V_0 is volume of SME (L).

Decolourisation of MB by SME. MB removal by *P. ostreatus* and *H. erinaceus* SME was monitored over 24 h with initial MB concentration of 5 mg L^{-1} and 10 mg L^{-1}, each in triplicate. Each sample containing 12 mL of crude enzyme extract and 4 mL of MB solution was incubated in a 35 °C water bath with shaking. A control was run under identical conditions without SME. Standard MB solutions (2, 4, 6, 8, 10 mg L^{-1}) were prepared to plot a MB calibration curve. Absorbance was measured at 668 nm at fixed intervals, then converted to MB concentration using the calibration curve. Percentage decolourisation of MB over 24 h was calculated as follows:

$$\text{Percentage Decolourisation (\%)} = \left[\frac{C_0 - C}{C_0}\right] \times 100\%$$

where C_0 and C are MB concentrations in the sample at 0 h and 24 h respectively. SME achieved greater MB decolourisation and was chosen for subsequent experimentation.

2.2 Phase 2 – Synthesis and Investigation of Spent Mushroom Extract-Alginate Beads (SME-AB)

Preparation of Alginate Beads. Sodium alginate solution (2.0% w/v) was prepared by dissolving 3 g of sodium alginate (GCE Laboratory Chemicals) in 150 mL of DI water and stirring at room temperature for 4 h. To form SME-AB and sodium acetate-alginate beads (NaOAc-AB), *P. ostreatus* SME and NaOAc buffer were homogenised with an equal volume of alginate solution (2.0 w/v%) respectively for 10 min by magnetic stirring. The solutions were then added dropwise to a 0.2M aqueous $CaCl_2$ solution. After 1h of curing, the beads were rinsed with DI water to remove residual $CaCl_2$ and stored in NaOAc buffer in the fridge for subsequent use.

Characterisation of Alginate Beads. Surface morphologies of SME-AB and NaOAc-AB were analysed using a JEOL JSM-6010LA scanning electron microscope (SEM) in the Science Centre Singapore lab and elemental analysis was carried out. 1.00 g of each type of bead was weighed and dried at 35 °C overnight in triplicates for swelling test. Dried beads were soaked in NaOAc buffer for two days to reach swelling equilibrium. Degree of swelling, DS, was calculated using the following formula:

$$\text{DS (\%)} = \frac{M_S - M_D}{M_D} \times 100\%$$

where M_S is the mass of the buffer-saturated beads and M_D is the mass of the dried beads.

Decolourisation of MB by SME-AB. MB removal capacity of hydrogel-immobilised SME was investigated. Each experimental set-up contained 10g of SME-AB and NaOAc-AB respectively in 200 ml of 20 mg L^{-1} MB. The control contained no alginate beads. Set-ups were placed in a shaking incubator at 150 rpm at 35 ºC. At fixed intervals of 0, 1, 3, 6, 24 h, 0.75 ml aliquots were drawn in triplicates from each setup and mixed with equal volume of DI water. Absorbance was measured at 668 nm and converted to MB concentration using a calibration curve.

Reusability of Alginate Beads. Reusability of alginate beads was tested using identical set ups and method from Sect. 2.2.3. After 24h, absorbance at 668 nm was measured and converted to MB concentrations using a calibration curve. SME-AB and NaOAc-AB were filtered, rinsed with 25 ml DI water, soaked in DI for 5min. This was repeated with NaOAc buffer. Beads were pat-dried between filter papers and added to 200 ml of freshly prepared 20 mg L-1 MB. This process was repeated after 24h. A new calibration curve was plotted each time.

2.3 Phase 3 – Enhancement of Spent Mushroom Extract-Alginate Beads (SME-AB) by Mediation

To study the effect of laccase mediator system on MB removal, a new batch of ABTS-SME-AB and SME-AB was prepared. 9 ml P. ostreatus SME and 9 ml alginate (2%w/v) was mixed with 2 ml 1mM ABTS and 2 ml DI respectively. ABTS-SME-AB and SME-AB were formed and cured in CaCl2 as in Sect. 2.2.1. 4.5 g of each type of alginate bead was added to 150 ml of freshly prepared 20 mg L^{-1} MB. Absorbance was measured at 668 nm at the start of the incubation (0h) and after 14 h and converted to MB concentrations using a calibration curve.

Percentage decolourisation of MB was calculated as follows:

$$\text{Percentage Decolourisation }(\%) = \left[\frac{C_0 - C}{C_0}\right] \times 100\%$$

where C_0 and C are the concentrations of MB in the sample at 0h and 14h respectively.

3 Results and Discussion

3.1 Phase 1 – Removal of Methylene Blue (MB) by Spent Mushroom Extract (SME)

ABTS Assay. The activity of laccase in SME of *P. ostreatus* and *H. erinaceus* was determined to be 2.6U/L and 7.3U/L respectively. This is lower than that reported in a study using similar extraction methods [16], likely due to variations in SMC source and storage conditions (Fig. 1).

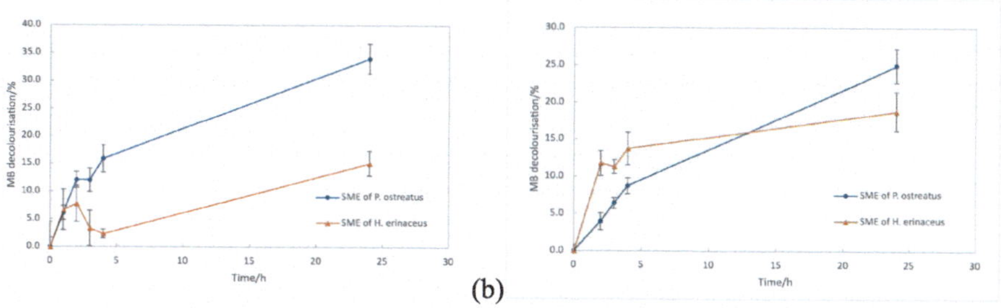

Fig. 1. **(a), (b), (c).** ABTS assay for SME of *P. ostreatus*, **(d) (e), (f)** SME of *H. erinaceus*

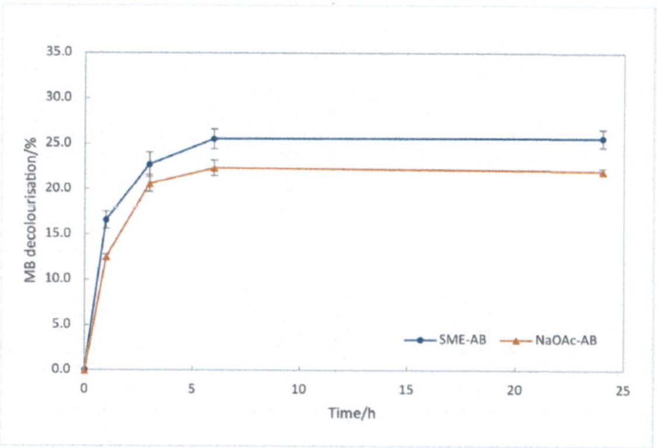

Fig. 2. MB decolourisation by SME of *P. ostreatus* and *H. erinaceus* over 24 h, (a) initial MB concentration 5 mg L^{-1}, (b) initial MB concentration 10 mg L^{-1}.

Decolourisation of MB by SME

At initial MB concentration 5 mg L^{-1}, over 24 h, *P. ostreatus* SME performed consistently better than that of *H. erinaceus*, with a higher percentage of MB decolourised from 2 h onwards. After 24 h, *P. ostreatus* SME achieved significantly greater ($p < 0.05$) MB decolourisation than that of *H. erinaceus*, with 34.0% compared to 15.1% (Fig. 2(a)). At initial MB concentration 10 mg $L^{-}1$, for the first 13 h, *H. erinaceus* SME performed better than *P. ostreatus* and vice versa for the next 11 h. After 24 h, *P. ostreatus* SME decolourised significantly more MB than *H. erinaceus*, with 24.9% compared to 18.8% (Fig. 2(b)).

Laccases in SME oxidise MB by reducing molecular oxygen to water [8]. Although laccase activity of *P. ostreatus* SME was higher than that of *H. erinaceus* SME, *P. ostreatus* SME exhibited significantly greater MB removal potential and was selected for immobilisation in alginate beads and further studies. This could be attributed to a larger variety and quantity of ligninolytic enzymes [11] that are able to degrade MB in *P. ostreatus* SME (Tables 1, 2 and 3).

Table 1. T-test comparing percentage decolourisation of MB by SME from *P. ostreatus* and *H. erinaceus*, (a) initial MB concentration 5 mg L^{-1}, (b) initial MB concentration 10 mg L^{-1}.

(a)

Sample	Mean	Standard Deviation	t-statistic	p
SME of *P. Ostreatus*	0.339	9.40E-03	4.47	1.95E-04
SME of *H.erinaceus*	0.15	6.74E-03		

(b)

Sample	Mean	Standard Deviation	t-statistic	p
SME of *P. Ostreatus*	0.334	5.19E-04	9.63	1.39E-07
SME of *H.erinaceus*	0.188	1.31E-03		

Table 2. T-test comparing percentage decolourisation of MB by SME-AB and NaOAc-AB

Sample	Mean	Standard Deviation	t-statistic	p
SME-AB	0.26	9.44E-05	11.57	2.06E-07
NaOAc-AB	0.22	1.00E-05		

Table 3. T-test comparing percentage decolourisation of MB by ABTS-SME-AB and SME-AB

Sample	Mean	Standard Deviation	t-statistic	p
ABTS-SME-AB	50.3	0.45	16.81	5.82E-09
NaOAc-AB	39.2	3.47		

3.2 Phase 2 – Synthesis and Investigation of Spent Mushroom Extract-Alginate Beads (SME-AB)

Characterisation of Alginate Beads. SEM images showed dried SME-AB to have a rough, irregular surface while dried NaOAc-AB was smoother and more uniform (Fig. 3). This is consistent with findings from previous studies [14, 17, 18]. EDX analysis (Fig. 4) showed presence of C, O, Na and Cl in both samples, confirming successful cross-linking of sodium alginate with $CaCl_2$ to form hydrogel.

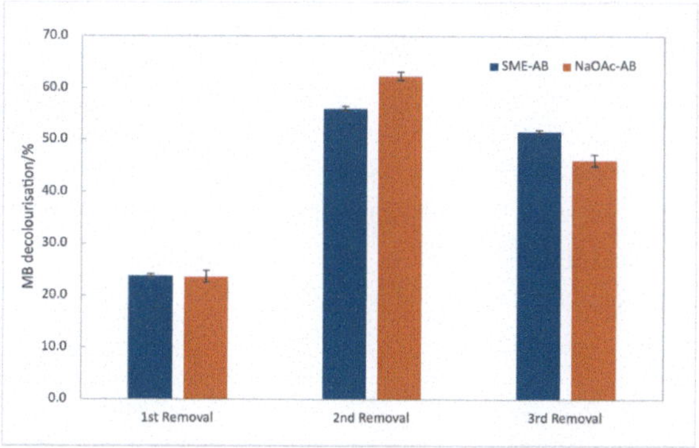

Fig. 3. Surface morphology of (a) dried NaOAc-AB at 200 μm, (b) dried NaOAc-AB at 50 μm, (c) dried SME-AB at 200 μm, (d) dried SME-AB at 50 μm

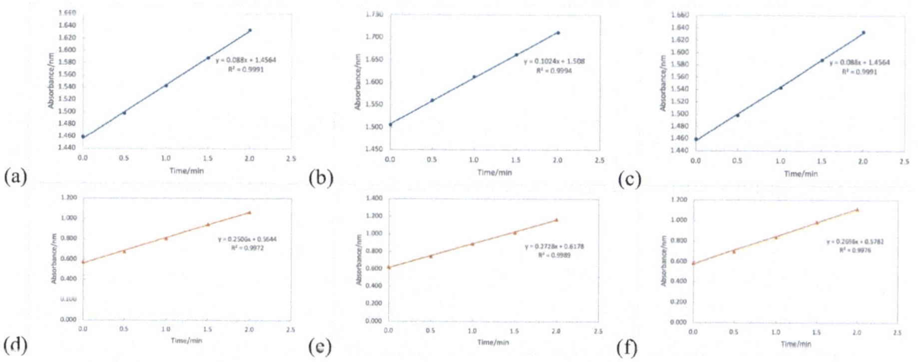

Fig. 4. EDX analysis of (a) dried SME-AB and (b) dried NaOAc-AB

Degree of swelling of SME-AB and NaOAc-AB was 6009.2% and 4210.8% respectively, which is significantly higher than reported in previous studies [14] on laccase hydrogels. This demonstrates mechanical stability as structures of both samples are preserved after drying. As dried hydrogels show better storage stability, maintaining activity for a longer period [19][20], SME-AB showed promise as a viable long-term product.

Decolourisation of MB by SME-AB

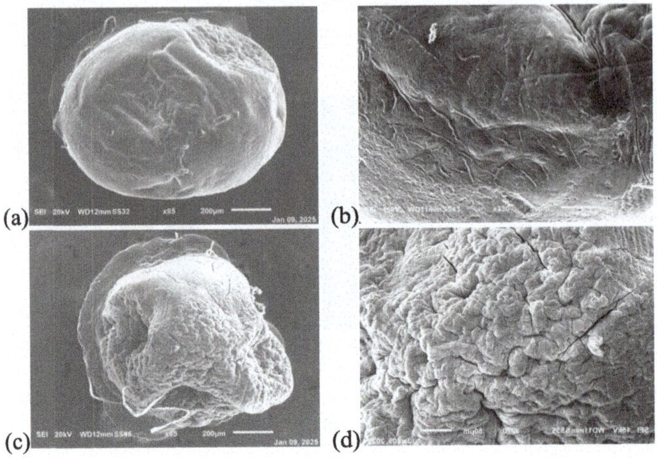

Fig. 5. MB decolourisation kinetics by SME-AB and NaOAc-AB

Kinetic studies (Fig. 5) show increased MB decolourisation but a reduced rate with longer contact time. After 24h, SME-AB achieved significantly greater ($p < 0.05$) MB removal than NaOAc-AB, with 26.0% compared to 22.0%. Initially, both samples follow first-order kinetics, transitioning to zero-order kinetics after 6h as decolourisation plateaus.

The significant removal of MB by NaOAc-AB indicates decolourisation by SME-AB was primarily due to the adsorption of cationic MB onto anionic hydrogel. This accounts for the dark blue stained appearance of the alginate beads after shaking with MB. Several mechanisms take place during the adsorption process, with that of textile dyes including electrostatic interactions, hydrogen bonding, π-π interactions, and surface complexation. Greater MB decolourisation by SME-AB compared to NaOAc-AB shows that immobilised SME was also able to catalyse substantial degradation of MB by enzymatic action.

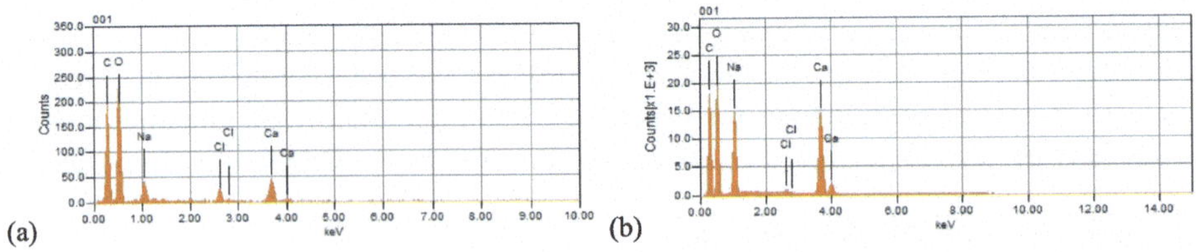

Fig. 6. NaOAc-AB (a) before 1st use and (b) after 2nd use

Reusability of Alginate Beads

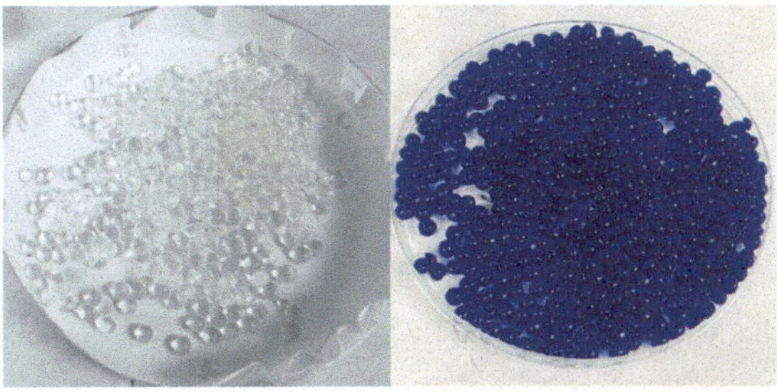

Fig. 7. Reusability of SME-AB and NaOAc-AB for MB decolourisation

After 24h, SME-AB and NaOAc-AB removed 23.8% and 23.6% of MB respectively. In the second removal, substantial increases in MB decolourisation, of 135.3% and 164.0% respectively, were observed. In the third removal, decreases in MB decolourisation, of 8.0% and 26.2% respectively, were observed (Fig. 6).

Even without the addition of activating agents, the smaller decrease in the decolourisation ability of SME-AB is comparable to past studies [21]. This is attributed to the joint activity of biodegradation by SME and adsorption by hydrogel, as the ligninolytic enzymes in SME as reusable. As such, SME-AB demonstrates greater reusability potential.

3.3 Phase 3 – Enhancement of Spent Mushroom Extract-Alginate Beads (SME-AB) by Mediation

After 14h, ABTS-SME-AB achieved significantly higher ($p < 0.05$) MB decolourisation than SME-AB, with 50.3% compared to 39.2%. This was a 28.3% increase in decolourisation ability, showing ABTS mediation was effective (Fig. 7).

This improvement is comparable to past studies conducted on co-immobilised laccase-ABTS systems [22][23]. The mechanism of dye decolourisation by such systems is well-documented. Laccase first oxidises the ABTS molecule to the cationic radical $ABTS^{+}$, then further oxidises it to form $ABTS^{2+}$, which degrades the substrate (Figs. 8, 9, 10, 11 and 12).

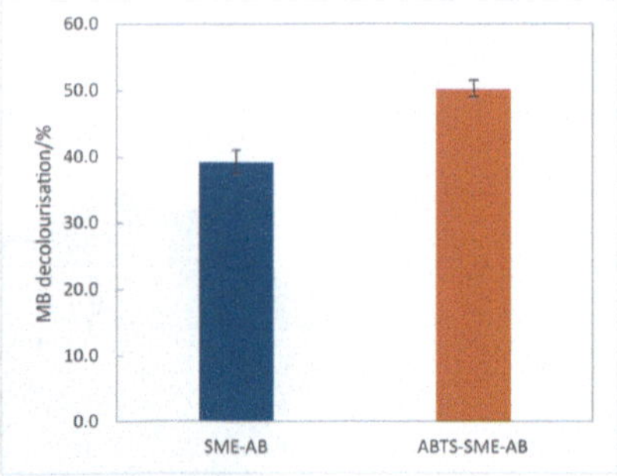

Fig. 8. MB decolourisation by SME-AB and ABTS-SME-AB

Fig. 9. Beads after first reuse

Fig. 10. Beads after second reuse

Fig. 11. Beads being rinsed

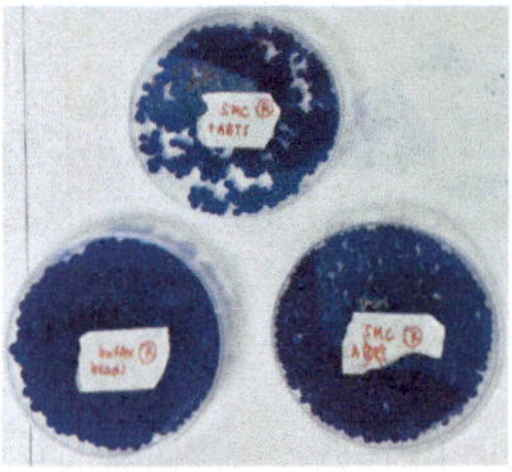

Fig. 12. Beads for reuse and filtered for reuse

4 Conclusion

Reusable spent mushroom extract-alginate beads were successfully synthesised from *P. ostreatus* SMC waste and enhanced by ABTS mediation to effectively remove wastewater pollutant, MB, from aqueous solution. *P. ostreatus* SME was determined to be 2.25 times and 1.32 times more effective in decolourising MB than *H. erinaceus* in initial MB concentrations of 5 mg L^{-}1 and 10 mg L^{-}1 respectively. It was subsequently immobilised in alginate for ease of recovery and improved enzyme stability. SME-AB not only effectively decolourised 26.0% of MB through the combined mechanisms of enzymatic degradation by SME and adsorption by hydrogel, but also demonstrated high reusability potential, with decolourisation ability decreasing by only 8.0% between second and third use. The addition of ABTS further enhanced performance of SME-AB by 28.3%.

As we transition to a circular economy, SMC waste can be repurposed into dried ABTS-SME-AB, a reusable, transportable product, presenting a sustainable solution for textile wastewater treatment, while promoting resource efficiency. Further research can be done to propose a kinetic model for MB decolourisation by ABTS-SME-AB, and to investigate its potential in removing other pollutants.

References

1. Halepoto, H., Gong, T., Memon, H.: Current status and research trends of textile wastewater treatments-a bibliometric-based study. Front. Environ. Sci. **10** (2022). https://doi.org/10.3389/fenvs.2022.1042256
2. Khan, I., et al.: Review on methylene blue: its properties, uses, toxicity and photodegradation. Water **14**(2), 242 (2022). https://doi.org/10.3390/w14020242
3. Sahithya, K., Mouli, T., Biswas, A., Mercy Scorlet, T.: Remediation potential of mushrooms and their spent substrate against environmental contaminants: an overview. Biocatal. Agric. Biotechnol. **42**, 102323 (2022). https://doi.org/10.1016/j.bcab.2022.102323
4. Madzin, Z., Zahidi, I., Talei, A., Raghunandan, M.E., Hermawan, A.A., Karam, D.S.: Optimising spent mushroom compost biochar for heavy metal removal: mechanisms and kinetics in mine water treatment. J. Water Process Eng. **69**, 106829 (2024). https://doi.org/10.1016/j.jwpe.2024.106829
5. Kumar, V., Dwivedi, S.K.: Mycoremediation of heavy metals: processes, mechanisms, and affecting factors. Environ. Sci. Pollut. Res. **28**(9), 10375–10412 (2021). https://doi.org/10.1007/s11356-020-11491-8
6. Alhujaily, A., Yu, H., Zhang, X., Ma, F.: Highly efficient and sustainable spent mushroom waste adsorbent based on surfactant modification for the removal of toxic dyes. Int. J. Environ. Res. Public Health **15**(7), 1421 (2018). https://doi.org/10.3390/ijerph15071421
7. Ghose, A., Mitra, S.: Spent waste from edible mushrooms offers innovative strategies for the remediation of persistent organic micropollutants: a review. Environ. Pollut. **307**, 119285 (2022). https://doi.org/10.1016/j.envpol.2022.119285
8. Lim, S.H., Lee, Y.H., Kang, H.W.: Efficient recovery of lignocellulolytic enzymes of spent mushroom compost from oyster mushrooms, pleurotus spp., and potential use in dye decolourisation. Mycobiology **41**(4), 214–220 (2013). https://doi.org/10.5941/MYCO.2013.41.4.214
9. Akhtar, N., Mannan, M.A.: Mycoremediation: expunging environmental pollutants. Biotechnol. Rep. **26**, e00452 (2020). https://doi.org/10.1016/j.btre.2020.e00452
10. Forootanfar, H., Moezzi, A., Aghaie-Khozani, M., et al.: Synthetic dye decolourisation by three sources of fungal laccase. J. Environ. Health Sci. Eng. **9**, 27 (2012). https://doi.org/10.1186/1735-2746-9-27
11. Saravanan, A., Kumar, P.S., Vo, D.-V.N., Jeevanantham, S., Karishma, S., Yaashikaa, P.R.: A review on catalytic-enzyme degradation of toxic environmental pollutants: microbial enzymes. J. Hazard. Mater. **419**, 126451 (2021). https://doi.org/10.1016/j.jhazmat.2021.126451
12. Chang, B.-V., Fan, S.-N., Tsai, Y.-C., Chung, Y.-L., Tu, P.-X., Yang, C.-W.: Removal of emerging contaminants using spent mushroom compost. Sci. Total. Environ. **634**, 922–933 (2018). https://doi.org/10.1016/j.scitotenv.2018.03.366
13. Lai, Y., et al.: Effective removal of methylene blue and orange II by subsequent immobilized laccase decolourisation on crosslinked polymethacrylate/carbon nanotubes. Mater. Res. Express **6**(8), 085541 (2019). https://doi.org/10.1088/2053-1591/ab26a5

14. Latif, A., Maqbool, A., Sun, K., Si, Y.: Immobilization of Trametes versicolor laccase on Cu-alginate beads for biocatalytic degradation of bisphenol A in water: optimized immobilization, degradation and toxicity assessment. J. Environ. Chem. Eng. **10**(1), 107089 (2022). https://doi.org/10.1016/j.jece.2021.107089
15. Odziomek, K., Drabczyk, A.K., Kościelniak, P., Konieczny, P., Barczewski, M., Bialik-Was, K.: The role of freeze-drying as a multifunctional process in improving the properties of hydrogels for medical use. Pharmaceuticals **17**(11), 1512 (2024). https://doi.org/10.3390/ph17111512
16. Yang, X., Chen, X., Wang, H., Cavaco-Paulo, A., Su, J.: Co-immobilizing laccase-mediator system by in-situ synthesis of MOF in PVA hydrogels for enhanced laccase stability and dye decolourisation efficiency. J. Environ. Manag. **353**, 120114 (2024). https://doi.org/10.1016/j.jenvman.2024.120114
17. Hou, H., Zhou, J., Wang, J., Du, C., Yan, B.: Enhancement of laccase production by *Pleurotus ostreatus* and its use for the decolourisation of anthraquinone dye. Process Biochem. **39**(11), 1415–1419 (2004). https://doi.org/10.1016/s0032-9592(03)00267-x
18. Sunkar, S.: A comparative study of free laccase and laccase immobilized in copper alginate. Indian J. Biochem. Biophys. **59**, 214–223 (2022). https://doi.org/10.56042/ijbb.v59i2.27154
19. Shivakumara, L.R., Demappa, T.: Synthesis and swelling behavior of sodium alginate/poly(vinyl alcohol) hydrogels. Turk. J. Pharm. Sci. **16**(3), 252–260 (2019). https://doi.org/10.4274/tjps.galenos.2018.92408
20. Odziomek, K., Drabczyk, A.K., Kościelniak, P., Konieczny, P., Barczewski, M., Bialik-Wąs, K.: The role of freeze-drying as a multifunctional process in improving the properties of hydrogels for medical use. Pharmaceuticals **17**(11), 1512 (2024). https://doi.org/10.3390/ph17111512
21. Boran, F.: The influence of freeze-thawing conditions on swelling and long-term stability properties of poly(vinyl alcohol) hydrogels for controlled drug release. Polym. Bull. **78**(12), 7369–7387 (2021). https://doi.org/10.1007/s00289-021-03902-8
22. Oktaviani, M., Damin, B.C.S., Suryanegara, L., Yanto, D.H.Y., Watanabe, T.: Immobilization of fungal mycelial and laccase from Trametes hirsuta EDN082 in alginate-cellulose beads and its use in Remazol Brilliant Blue R dye decolourisation. Biores. Technol. Rep. **26**, 101828 (2024). https://doi.org/10.1016/j.biteb.2024.101828
23. Li, Z., Shi, Q., Dong, X., Sun, Y.: Co-immobilization of laccase and mediator into fe-doped ZIF-8 significantly enhances the degradation of organic pollutants. Molecules **29**(2), 307 (2024). https://doi.org/10.3390/molecules29020307
24. Shan, H., Wang, X., Ge, Y., Li, Z.: Homologous amino acids promoted co-immobilization of laccase and mediator onto geopolymer microspheres for enhancing degradation of dyes in water. J. Hazard. Mater. **423**, 127107 (2022). https://doi.org/10.1016/j.jhazmat.2021.127107

Physiological Changes During Aerobic Exercise with Cloth Mask

E. Woo[1(✉)], G. Balasekaran[2], and P. Boey[2]

[1] Hwa Chong Institution, Singapore, Singapore
elenawooyurou@gmail.com
[2] Physical Education and Sports Science, National Institute of Education, Nanyang Technological University, Singapore, Singapore

Abstract. With growing use of reusable face masks, their impact on exercise performance raises concerns for athletes. This study investigates the effect of wearing a Reusable Cloth Mask (RCM) on Ventilatory Breakpoint (V_{pt}), Heart Rate (HR), Oxygen Uptake (VO_2), and Rating of Perceived Exertion (RPE) during aerobic exercise to determine an appropriate exercise intensity level.

A randomised crossover counterbalanced experimental design was conducted with healthy adult females. Participants completed a discontinuous graded submaximal treadmill test until volitional exhaustion with alternate 4-min work and rest stages. Respiratory variables were measured using a metabolic cart, HR was measured by a HR monitor, while participants rated perceived exertion. Data was analysed using a 2-way ANOVA test.

Significant increases in HR (RCM: 166.3 ± 18.9 beats·min^{-1}, non-RCM: 154.2 ± 15.8 beats·min^{-1}, $p = 0.02$) and RPE (RCM: 7.5 ± 1.9, non-RCM: 5.5 ± 1.5, $p = 0.00$) at V_{pt} were observed. No significant difference was observed in VO_2 at V_{pt} (RCM: 1.52 ± 0.28 L·min^{-1}, non-RCM: 1.46 ± 0.22 L·min^{-1}, $p = 0.50$). Significant increases in RMD were observed for overall discomfort (RCM: 7.30 ± 2.26, non-RCM: 4.30 ± 2.31, $p = 0.00$) and hot/stuffiness (RCM: 6.70 ± 2.45, non-RCM: 3.90 ± 2.56, $p = 0.00$).

RCMs significantly increase in HR, higher perceived exertion and discomfort to users but do not greatly impede breathing. Users are advised to wear RCM for exercise only when needed, with consideration of individual comfort. Future studies should evaluate methods to minimise RCM discomfort and explore their potential in enhancing aerobic training.

Keywords: Adult · female · OMNI Rate of Perceived Exertion · Treadmill

1 Background

As the recent Coronavirus disease (COVID-19) pandemic hit countries worldwide, the use of masks has grown in popularity to prevent the spread of airborne viral infections [1]. With growing public awareness about infectious disease prevention, mask-wearing could potentially play a major role in the future. Studies have shown that masks effectively act as a filter to reduce airborne droplet transmission, protecting users from droplets

H. Guo et al. (Eds.): IRC-SET 2025, PTT, pp. 155–166, 2026.
https://doi.org/10.1007/978-981-95-9630-0_14

expelled by others [2]. This is especially so during exercise, due to the forceful release of air emissions propels droplets further than 2 m, including droplets that contain viral particles [3]. Thus, this study is highly relevant in today's world, where mask-wearing is encouraged in high-risk situations.

Considering environmental efforts, reusable cloth masks (RCMs) have become a popular substitute for single-use surgical masks. Yet, it is unclear how wearing RCMs may affect performance and perceived exertion of users during exercise. During the peak of the pandemic, hospitals recommended mask-wearing whenever possible, especially for athletes engaging in team sports [4]. In the 2022 Beijing Winter Olympics, athletes from all countries were also seen donning RCMs. However, the World Health Organisation cautioned against the usage of face masks during exercise in 2020, citing that sweat accumulation could make breathing difficult [5].

Studies have shown conflicting results on the impact of wearing masks on exercise performance, with some citing little to no effects [6–9] and others displaying adverse consequences on physiological indicators [10, 11]. Furthermore, users may experience mask discomfort including heat, humidity, and breathing difficulty, commonly associated with mask-wearing [12, 13]. This could potentially affect perceived exertion levels, especially since it was found that adult females tended to overestimate, while regular exercisers underestimated physical exertion [14].

The assumptions to this study are: (1) RCMs sufficiently impedes airflow transmission between the environment and healthy adults, serving as a basis for comparison between participants wearing and not wearing RCMs, and (2) the specific brand and material of RCMs provided in the study can be a general representation of RCMs commonly used in the market and accessible to an average consumer.

2 Hypothesis/Objectives

This study aims to solidify the hypothesis that RCMs impede airflow transmission between the environment and an individual during exercise, affecting ventilatory breakpoint (V_{pt}), and other physiological variables such as heart rate (HR) and maximal oxygen consumption (VO_{2max}).

The objective of this study is to investigate the (1) perceptual ratings, (2) physiological effects and (3) determine the appropriate exercise intensity level for healthy female adults while wearing RCM during exercise.

3 Methodology

3.1 Subjects

Ten clinically normal female participants were recruited for the study. The inclusion criteria are-age of 21 to 45 years; healthy and have no musculoskeletal injuries in the past 6 months; clinically normal (BMI: 18.5–22.9 $kg \cdot m^{-2}$); at least 150-min of moderate and vigorous intensity physical exercise per week; undertake regular running as part of their exercise regime, with the ability to run continuously for at least 30 min in a single session, at least once per week; should not have worn a mask to exercise for in the past 6 months; and should not be on any medication.

This investigation was approved by the Institutional Review Board (IRB-2021-261). All subjects were informed of the risks and benefits of the study and gave their written consent to participate (Table 1).

Table 1. Participants Characteristics

Variables	
Age (years)	24.30 ± 5.64
Height (m)	1.61 ± 0.07
Weight (kg)	52.18 ± 5.74
Body Mass Index $(kg{\bullet}m^{-2})$	20.13 ± 1.60

All values are presented in mean ± SD. SD refers to standard deviation. m: metres: kg: kilograms.

3.2 Experimental Design

A randomized crossover counterbalanced experimental design was adopted, where participants were randomly assigned into the experimental (with RCM) or control group (without RCM) separated by a washout period of 7 days (Annex A).

3.3 Experimental Variables

This study observed the following experimental variables:

- Rate of perceived exertion (RPE), which is the quantitative measure of the amount of effort taken during physical activity on a scale of 0 to 10.
- Oxygen uptake (VO_2); ($L{\cdot}min^{-1}$), which is the difference between the volume of oxygen inspired and expired within a minute.
- Heart rate (HR); ($beats{\cdot}min^{-1}$), which is the number of times the heart beats in one minute.
- Perceptual ratings of mask discomfort (RMD), which is the various types and intensities of how uncomfortable participants are on a scale of 0 to 10.
- Ventilatory breakpoint (V_{pt}) refers to the instant in which the increase in volume of air expired per unit time exceeds that of VO_2.

3.4 Tests and Measurements

Anthropogenic Measures. The height (cm) and weight (kg) of participants was recorded using a bioelectrical impedance analysis (BIA) scan (Fig. 1).

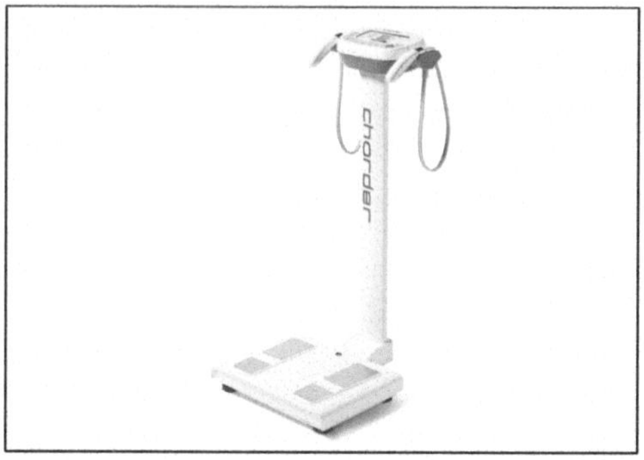

Fig. 1. Bioelectrical impedance analysis (BIA) [15]

Orientation Procedure. As a precautionary measure, participants were given a thorough briefing and walkthrough of full details and procedure of the study. This includes a familiarisation of: (1) Discontinuous graded submaximal treadmill protocol, (2) OMNI RPE scale, (3) RPE rating instructions, (4) HR monitor, (5) Spirometry mask and RCM, as well as other components of the experimental set-up.

Before the tests, participants are anchored on the rating scales from 0 to 10 for OMNI RPE scale to check for their understanding on the various indicators [16]. Usage of the OMNI Pictorial RPE scale (Fig. 2) was explained to participants while they visually examined it. Participants were then briefly tested to rate examples of tasks using the OMNI RPE scale to anchor their understanding. In the following session, participants are briefed again on the scale prior to the test.

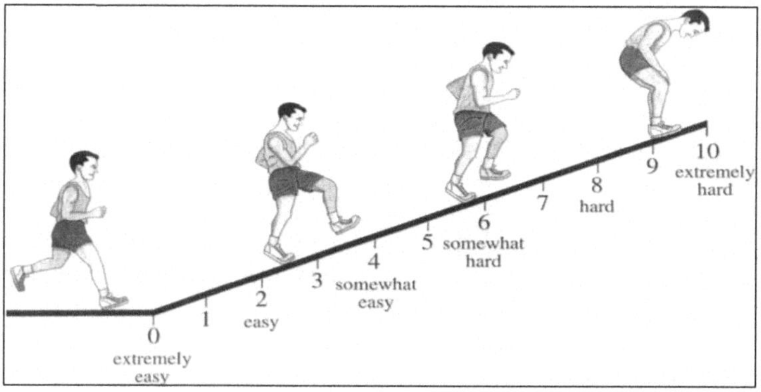

Fig. 2. Adult OMNI pictorial rating of perceived exertion (RPE) scale for running/walking

Protocol. Participants were secured to a safety harness. HR and blood lactate were obtained before the run. Participants strapped on a Bluetooth HR transmitter below their chest level (Fig. 3) (Polar Electro, Singapore). Participants ran a 5-min self-paced warm-up without RCM, followed by the discontinuous graded submaximal treadmill run test with a starting speed of 6 km·h^{-1} and increased by 0.5 km·h^{-1} every work stage. Gradient was maintained at 0% throughout. Test was terminated upon volitional exhaustion, and with the attainment of VO2max criteria with fulfilment of any 2 of the determinants:

- Respiratory Exchange Ratio of > 1.1
- RPE rating of > 9
- Plateau of HR ($\pm$ 5 beats·min^{-1} between stages)
- Plateau of O_2 intake ($VO_2 \leq 150$ mL·min^{-1} between stages)

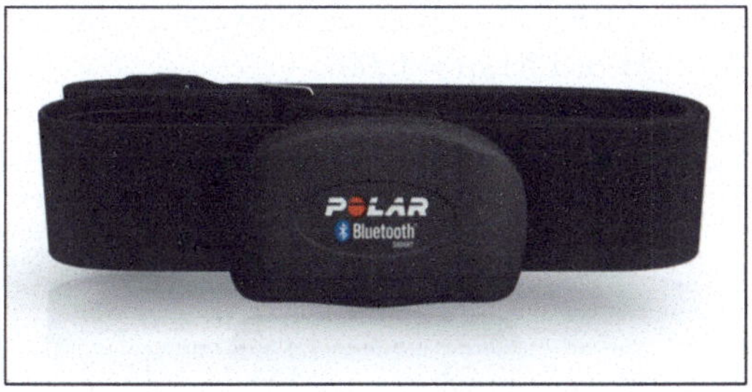

Fig. 3. Polar heart rate sensor and strap

Each work stage lasted 4 min, with a 4-min recovery given between each work stage. RPE and HR were assessed within the last 30 s of each work stage. Participants were shown the OMNI RPE scale and asked to rate their current perceived level by gesturing with their fingers.

3.5 Data Analysis

Descriptive data for anthropogenic measures and dependent variables were calculated as mean $\pm$ SD. Differences between the mask and non-mask tests were observed using a 2-way ANOVA test (Microsoft Excel). Vpt was determined by using a log-log transformation for linear regression. Linear regression was also carried out to determine the HR, RPE and VO_2 at Vpt. RMD ratings were taken based on RPE at Vpt.

4 Results

Listed in Tables 3 and 4 are the mean physiological responses, calculated from Tables 6, 7 and 8 (Appendix) while Table 2 represents mean perceptual ratings calculated from Table 5, as well as their respective p-values. P-values < 0.05 were statistically significant.

Table 2. Rating of Perceived Exertion (RPE) at Ventilatory Breakpoint

Measures	with RCM	without RCM	p-value
RPE	7.48 ± 1.86	5.51 ± 1.51	0.00 *

Values are expressed in mean ± SD. RPE: Rating of perceived exertion, RCM: reusable cloth mask.
*indicates significant difference at $p < 0.05$.

Significant difference in RPE was observed between RCM and non-RCM (RCM: 7.48 ± 1.86, non-RCM: 5.51 ± 1.51, $p = 0.00$).

Table 3. Physiological Variables (VO_2, HR) at Ventilatory Breakpoint (V_{pt})

Measures	with RCM	without RCM	p-value
VO_2 $(L{\cdot}min^{-1})$	1.52 ± 0.28	1.46 ± 0.22	0.50
HR $(beats{\cdot}min^{-1})$	166.27 ± 18.89	154.23 ± 15.83	0.02 *

Values are expressed in mean ± SD. VO_2: oxygen uptake, $L{\cdot}min^{-1}$, RCM: reusable cloth mask.
* indicates significant difference at $p < 0.05$.

No significant difference was observed in VO_2 (RCM: 1.52 ± 0.28 L·min-1, non-RCM: 1.46 ± 0.22 L·min-1, $p = 0.50$). Significant difference was observed in HR between RCMs and non-RCM (RCM: 166.27 ± 18.89 beats·min-1, non-RCM: 154.23 ± 15.83 beats·min-1, $p = 0.02$).

Table 4. Perceptual Ratings of Mask Discomfort (RMD) at Ventilatory Breakpoint (V_{pt})

Measures	with RCM	without RCM	p-value
OD	7.30 ± 2.26	4.30 ± 2.31	0.00 *
BD	6.70 ± 3.27	4.20 ± 2.62	0.09
HS	6.70 ± 2.45	3.90 ± 2.56	0.00 *
O	2.00 ± 2.67	1.40 ± 2.72	0.14
I	1.50 ± 2.27	1.70 ± 3.09	0.74

Values are expressed in mean ± SD. OD: overall discomfort, BD: breathing discomfort, HS: hot/stuffiness, O: odour, I: itchiness, RCM: reusable cloth mask.
*indicates significant difference at $p < 0.05$.

Significant differences for perceptual ratings were observed in both OD (RCM: 7.30 ± 2.26, non-RCM: 4.30 ± 2.31, $p = 0.00$) and HS (RCM: 6.70 ± 2.45, non-RCM: 3.90 ± 2.56, $p = 0.00$). No significant differences were observed in BD (RCM: 6.70 ± 3.27, non-RCM: 4.20 ± 2.62, $p = 0.09$), O (RCM: 2.00 ± 2.67, non-RCM: 1.40 ± 2.72, $p = 0.14$) and I (RCM: 1.50 ± 2.27, non-RCM: 1.70 ± 3.09, $p = 0.74$).

5 Discussion

This study investigated the perceived exertion response and physiological markers, including VO_2 and HR, at V_{pt} during discontinuous graded submaximal treadmill run test under two conditions–with and without RCM.

The RPE score at V_{pt} was significantly higher for RCM (RCM: 7.5 ± 1.9, non-RCM: 5.5 ± 1.5, $p = 0.00$) (Table 2). This is consistent with that of other studies, where RPE was similarly used as an indicator to measure overall psychological discomfort. Previous studies had cited perceived dyspnoea and claustrophobia as a contributing factor to the differences in level of perceived exertion. One of them used a source control management system and found RCM, which had been previously found to block airflow, in the process of other disease particles [17]. In fact, a similar study conducted using surgical face masks on young adults on an incremental treadmill protocol had found that RCM significantly increases dyspnoea much more than surgical cloth masks due to differences in resistance and tightness around the face area [18].

The RMD ratings in this study supports this finding as the ratings for OD and HS were significantly higher for RCM than non-RCM (OD RCM: 7.30 ± 2.26, OD non-RCM: 4.30 ± 2.31, $p = 0.00$, HS RCM: 6.70 ± 2.45, HS non-RCM: 3.90 ± 2.56, $p = 0.00$) (Table 4). Other studies using N95 and surgical masks similarly cited surrounding temperature and humidity to have aggravated the feeling of discomfort for mask users during exercise [19, 19].

This contrasts a similar study which found no significant difference in RPE between surgical and cloth face masks and non-mask conditions after exercise ($p = 0.47$), though it is worth noting the study used a maximal progressive cycle ergometer exercise test, which could have affected the results [9].

5.1 Physiological Outcomes

In this study, HR at Vpt is significantly higher in mask users (RCM: 166.3 ± 18.9 beats·min-1, non-RCM: 154.2 ± 15.8 beats·min-1, $p = 0.02$) (Table 3).

This is consistent with a systematic review by Shawn et al. where HR was higher in those wearing N95 masks ($p = 0.04$) [20]. Yet, it is worth noting that the study found insignificant differences in HR when studies of high risk of bias were removed ($p = 0.45$). However, a standard cycle ergometer ramp protocol conducted on male participants between surgical mask, N95 respirator and no mask conditions showed no significant difference in HR at any stage [21]. Such differences could be attributed to using a different form of aerobic exercise, as well as differences in age group (34 ± 4 years) and BMI (28.72 ± 3.78 $kg{\bullet}m^{-2}$) of subjects.

The VO_2 at V_{pt} was found to have no significant difference between the two groups (RCM: 1.52 ± 0.28 $L{\cdot}min^{-1}$, non-RCM: 1.46 ± 0.22 $L{\cdot}min^{-1}$, $p = 0.49660$) (Table 3). This indicates that there was no hypoxia between participants who did and did not wear masks, which was consistent with another study involving surgical face masks and an incremental treadmill protocol ($p = 0.22$) [18]. VO_2 may not have been compromised as breathing rate may have automatically increased by the body to meet oxygen demand. The effects of RCM may also be felt after $V_{pt,}$ especially once pace gets harder.

However, this data may not seem internally consistent, given that RPE and HR both show positive increases for mask users (Tables 2 and 3). Furthermore, another study involving an incremental ergometer protocol had found ventilation to have significantly decreased from the use of surgical and N95 masks during exercise (131 ± 28 vs 114 ± 23 vs $99 \pm 19\,L{\cdot}m^{-1}$; $p < 0.001$) [19]. Yet, it is worth noting that Shaw et al. determined insignificant effects of masks on physiological variables during submaximal exercise tests, while a small significant reduction in VO_2 (mean difference = − 0.6%, $p = 0.04$) in masked conditions was only observed in maximal exercise tests [20].

5.2 Appropriate Exercise Intensity Level

V_{pt} is known to symbolise the transition from reliance on aerobic respiration to anaerobic respiration, the exercise zones before V_{pt} (Zones 1 and 2) are commonly recommended for sustainable and moderate training (65–75% of maximal HR (HRmax)) [22]. Hence, an appropriate recommended exercise intensity while wearing an RCM is of an RPE rating of 6–7, between 'somewhat hard' and 'hard'.

6 Conclusion

This study has found that reusable cloth masks increase perceived exertion during submaximal exercise, which supports our hypothesis. However, our hypothesis is rejected based on physiological variables, as wearing a RCM had no significant effect on VO_2 at V_{pt}, but a more significant positive effect on HR.

7 Recommendations

Since RCMs can cause discomfort and increased HR during aerobic exercise, mask-wearing is only advised, when necessary, such as for those at high risk of infections or vulnerable to respiratory infections. They can consider using RCMs that have more breathing spaces or fit their facial morphology better to reduce discomfort. To minimise the feeling of humidity in RCMs during exercise, athletes can also switch to different masks over the course of their exercise.

8 Limitations and Improvements

8.1 Limitations of Study

The study assumes that the specific brand and material of RCMs provided can be a general representation of RCMs commonly used in the market. In reality, different brands of RCMs are made of different materials and layers, which could impact breathing and the level of discomfort. Using an airtight silicone mask on top of the RCM may also cause additional discomfort and difficulty breathing, on top of an RCM. This is especially given that one of the significantly impacted indicators for RMD was associated with participants feeling hot and stuffy. However, this is already one of the most accurate and non-invasive methods to collect data on respiratory indicators in an airtight environment.

8.2 Recommendations for Future Work

As masks become an indispensable part of our lives today, further research could be done on ways in which comfort and HR could be less impacted by its use, providing mask wearers with more ease in the long-term. This could be based on the shape and material, as well as the frequency of changing masks. Wearing RCMs during exercise were found to increase HR with VO_2 unchanged, indicating a better circulation of oxygen and nutrients, but a faster depletion of these resources. Thus, this gives rise to the possibility of using RCMs in maximising aerobic training and performance in future studies. Manufacturers of RCMs can also explore ways to make more physiological efficient masks that athletes can use in training.

9 Appendices

10 Annex A: Diagram of Protocol.

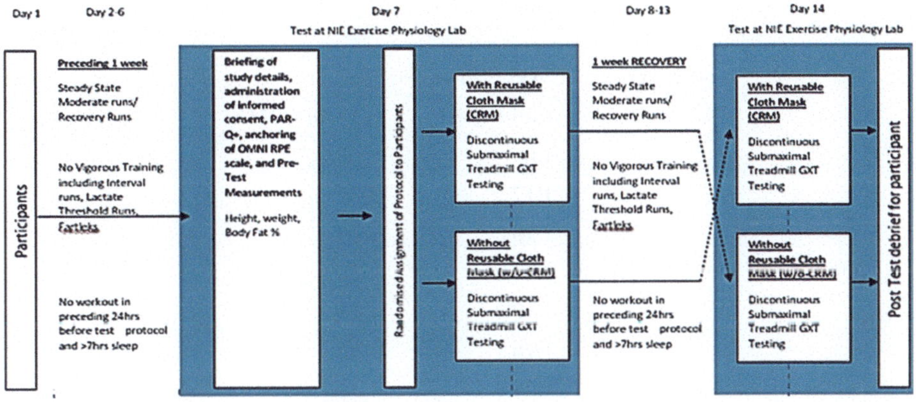

Table 5. Rating of Perceived Exertion (RPE) at Ventilatory Breakpoint for all participants Values are expressed in mean ± SD. RPE: Rating of perceived exertion, RCM: reusable cloth mask.

Participant	RPE	
	with RCM	*without RCM*
1	9.96	7.15
2	8.27	5.90
3	7.72	4.24
4	9.86	6.49
5	6.74	5.44

(*continued*)

Table 5. (*continued*)

Participant	RPE	
	with RCM	*without RCM*
6	4.07	2.83
7	8.45	7.82
8	6.72	4.06
9	7.74	5.15
10	5.26	6.08
Mean ± SD	7.48 ± 1.86	5.52 ± 1.51
p-value	0.001355 *	

* indicates significant difference at $p < 0.05$.

Table 6. Oxygen Uptake (VO_2) at Ventilatory Breakpoint for all participants

Participant	VO_2 ($L{\cdot}min^{-1}$)	
	with RCM	*without RCM*
1	1.05	1.05
2	1.17	1.35
3	1.78	1.29
4	1.79	1.66
5	1.18	1.41
6	1.76	1.32
7	1.64	1.71
8	1.61	1.62
9	1.56	1.48
10	1.64	1.75
Mean ± SD	1.52 ± 0.28	1.46 ± 0.22
p-value	0.496608	

Values are expressed in mean ± SD. VO_2: oxygen uptake, $L{\cdot}min^{-1}$, RCM: reusable cloth mask.

Table 7. Heart Rate (HR) at Ventilatory Breakpoint for all participants

Participant	HR (beats·min^{-1})	
	with RCM	*without RCM*
1	200.40	179.81
2	161.42	167.38
3	158.94	134.16
4	182.77	169.70
5	164.27	151.97
6	145.56	145.76
7	132.59	139.24
8	169.00	143.27
9	175.65	142.05
10	172.10	168.93
Mean ± SD	166.27 ± 18.89	154.23 ± 15.83
P-value	0.024372	

Values are expressed in mean ± SD. HR: heart rate, beats·min^{-1}, RCM: reusable cloth mask. *indicates significant difference at $p < 0.05$.

Table 8. Perceptual Ratings of Mask Discomfort (RMD) at Ventilatory Breakpoint for all participants

Participant	RMD									
	OD		*BD*		*HS*		*O*		***I***	
	with RCM	*without RCM*	*with RCM*	*without RCM*	*with RCM*	*without RCM*	*with RCM*	***without RCM***	***with RCM***	***without RCM***
1	7	3	0	6	7	3	0	0	0	0
2	9	4	9	1	5	2	3	0	3	1
3	10	3	10	3	10	2	0	0	0	0
4	9	6	9	5	8	5	2	0	0	0
5	8	7	9	6	9	7	5	6	3	8
6	5	3	4	2	3	2	0	0	0	0
7	9	8	9	9	9	8	8	7	7	7
8	8	4	8	3	7	4	1	0	1	0
9	3	0	4	1	2	0	0	0	0	0
10	5	5	5	6	5	6	1	1	1	1
Mean ± SD	7.30 ± 2.26	4.30 ± 2.31	6.70 ± 3.27	4.20 ± 2.62	6.70 ± 2.45	3.90 ± 2.56	2.00 ± 2.67	1.40 ± 2.72	1.50 ± 2.27	1.70 ± 3.09
p-value	0.001489 *		0.085224		0.003906 *		0.140453		0.735302	

Values are expressed in mean ± SD. OD: overall discomfort, BD: breathing discomfort, HS: hot/stuffiness, O: odour, I: itchiness, RCM: reusable cloth mask. *indicates significant difference at $p < 0.05$.

References

1. Chiu-Sen, W.: Face masks and prevention of respiratory viral infections: an overview. Asian Aerosol Conference (2022)

2. Dbouk, T., Drikakis, D.: On respiratory droplets and face masks (2020)
3. Nicholas, R.J., Qureshi, Z.U., Temple, R.J., Larwood, J.P.J., Greenhalgh, T., Bourouiba, L.: Two metres or one: what is the evidence for physical distancing in covid-19? (2020)
4. Watkins, K.: Keeping your student athlete safe during COVID-19 (2020)
5. World Health Organisation: Can people wear masks while exercising? (2020)
6. Guardieiro, N.M., et al.: A cloth facemask causes no major respiratory or cardiovascular perturbations during moderate to heavy exercise. J. Phys. Act. Health **20**(1), 35–44 (2022). https://doi.org/10.1123/jpah.2022-0145
7. Ziegler, M.F.: Mask wearing during exercise does not affect breathing or cardiovascular fitness. AGÊNCIA FAPESP (2022)
8. Otsuka, A., Komagata, J., Sakamoto, Y.: Wearing a surgical mask does not affect the anaerobic threshold during pedaling exercise. J. Hum. Sport Exerc. **17**(1), 22–28 (2022)
9. Shaw, K., Butcher, S., Ko, J., Zello, G.A., Chilibeck, P.D.: Wearing of cloth or disposable surgical face masks has no effect on vigorous exercise performance in healthy individuals. Int. J. Environ. Res. Public Health **17**(21), 8110, 107–113 (2020)
10. Chandrasekaran, B., Fernandes, S.: Exercise with facemask; are we handling a devil's sword?"–a physiological hypothesis. Med. Hypotheses **144** (2020)
11. Driver, S., et al.: Effects of wearing a cloth face mask on performance, physiological and perceptual responses during a graded treadmill running exercise test. Br. J. Sports Med. **56**(2), 107–113 (2021)
12. Kampert, M., Singh, T., Sahoo, D., Han, X., Van Iterson, E.H.: Effects of wearing an N95 respirator or cloth mask among adults at peak exercise. JAMA Netw. Open **4**(6), e2115219 (2021). https://doi.org/10.1001/jamanetworkopen.2021.15219
13. Fikenzer, S., et al.: Effects of surgical and FFP2/N95 face masks on cardiopulmonary exercise capacity. Clin. Res. Cardiol. **109**(12), 1522–1530 (2020). https://doi.org/10.1007/s00392-020-01704-y
14. Skatrud-Mickelson, M., Benson, J., Hannon, J.C., Askew, E.W.: A comparison of subjective and objective measures of physical exertion. J. Sports Sci. **29**(15), 1635–1644 (2011). https://doi.org/10.1080/02640414.2011.609898
15. Charder Electric: MA601 information (2024)
16. Utter, A.C., Robertson, R.J., Green, J.M., Suminski, R.R., McAnulty, S.R., Nieman, D.C.: Validation of the Adult OMNI Scale of perceived exertion for walking/running exercise. Med. Sci. Sports Exerc. **36**(10), 1776–1780 (2004)
17. William G.L., et al.: A comparison of performance metrics for cloth face masks as source control devices for simulated cough and exhalation aerosols (2021)
18. Tsz-Chun Poon, E., Zheng, C., Heung-Sang Wong, S.: Effect of Wearing Surgical Face Masks During Exercise: Does Intensity Matter? (2021)
19. Sven Fikenzer, T., et al.: Effects of surgical and FFP2/N95 face masks on cardiopulmonary exercise capacity (2020)
20. Shaw, K., Butcher, S., Ko, J., Zello, G.A., Chilibeck, P.D.: The impact of face masks on performance and physiological outcomes during exercise: a systematic review and meta-analysis (2021)
21. Epstein, D., et al.: Return to training in the COVID-19 era: the physiological effects of face masks during exercise (2020)
22. Mason, T.: Everything you need to know about ventilatory thresholds-VO2 Master (2024). https://vo2master.com/blog/ventilatory-thresholds-explained/?srsltid=AfmBOoqGGS2q-VM099nYDbhIdTxVdVTI_vekUHlCMJqtMp0YWw62s8YX

Biological and Physical Characterisation of a Chitosan Hydrogel for Healing Pressure Sores

Jethro Chai Tsin Kai(✉), Zhu Yijun, and Tan Tai Chuen Gareth

Hwa Chong Institution, Singapore 269734, Singapore
jethrochai85@gmail.com, {211640n,211543b}@student.hci.edu.sg

Abstract. Exposed wounds are vulnerable to infections, oxidative stress and susceptible to trauma. This can slow wound healing, lead to potentially life-threatening complications and result in wound chronicity. Current wound treating methods do not wholly address these issues and can be unsustainable. Chitosan hydrogels may be a suitable and even superior alternative for wound dressing. Hydrogels with varying ratios of the crosslinking agent, glutaraldehyde, to chitosan ratios were synthesised. The antioxidant and antimicrobial properties, and its effect on extracellular matrix formation in *Lumbricus terrestris* wounds were investigated. To ensure that the hydrogel formed is also capable of physically protecting the wound, the Young's modulus, adhesion energy, tensile strength, water absorption capacity and porosity were also investigated. This gel exhibited statistically significant antibacterial and antioxidant properties. Interestingly, the better promotion of extracellular matrix formation could be stimulated by controlling the glutaraldehyde ratio in the gel. There was also a clear trend between concentration of glutaraldehyde and its mechanical properties, as well as the water content and its mechanical properties. These results indicate the potential of this gel to be developed as an improved, possibly customised wound dressing for the treatment and prevention of chronic wounds.

Keywords: Chitosan · Hydrogel · Wound healing

1 Introduction

More than 15 per 1000 people worldwide have an open wound at any time [21]. Among diabetics, these wounds have a 5.6–26% chance of being infected [10], which inhibits healing by causing inflammation and tissue damage (Wound [34]). Among the most concerning types of wounds are pressure sores which commonly occur at areas where bone tissue is close to the skin's surface, particularly the sacrum [2], [24]). Pressure sores are particularly problematic in elderly, bedridden patients.

Oxidative stress [29] and bacterial infection [26] are common inhibitors of wound healing. Current wound dressings are not equipped to reduce the effects of these factors on wound healing. The use of antiseptics requires constant reapplication, may damage skin, and further slow wound healing (Cleveland [3]). Furthermore, most dressings are

H. Guo et al. (Eds.): IRC-SET 2025, PTT, pp. 167–178, 2026.
https://doi.org/10.1007/978-981-95-9630-0_15

made of synthetic, non-biodegradable components such as foam, silicones, polymers, and adhesives, generating 16680 tonnes of waste annually just in the United Kingdom (Willingsford [33]). These dressings are often not specially constructed to withstand the pressures placed on pressure sores, making them ill-equipped for the healing of such wounds.

5-amino-2-(hydroxymethyl)-6-methylhexane-3,4-diol, commonly known as chitosan, is a biocompatible and biodegradable polymer derived from crustacean shells [1], exhibits antibacterial activity against Gram-positive and Gram-negative bacteria [30] and radical scavenging properties (Yan et al., 2008). Synthesised into a hydrogel using glutaraldehyde as a cross-linking agent [14], it promotes extracellular matrix (ECM) formation by supporting glycosaminoglycan (GAG) synthesis [32] and fibroblast production for collagen generation [12]. These properties make chitosan gel a promising candidate for facilitating pressure sore healing. Glutaraldehyde crosslinks effectively in acidic pH [25] and is hence suitable for crosslinking in the acetic acid medium used to dissolve chitosan.

To facilitate the healing process, the Young's modulus of the hydrogel should approximate human tissue to prevent mechanical stress and protect the wound [11]. The hydrogel also needs to have a relatively high tensile strength to prevent hydrogel shedding, bacterial infection and delayed healing. Most importantly, the gel should have a maximum strain that is greater than the strain experienced at common sites for pressure sore formation.

2 Objectives and Hypothesis

This project seeks to formulate hydrogels with chitosan as the active ingredient and adding various concentrations of glutaraldehyde as a cross-linking agent. These hydrogels will be tested for biological efficacy and their abilities to withstand pressures common pressure sore sites.

As a higher glutaraldehyde/chitosan ratio results in more complete cross linking but a lower proportion of biologically active chitosan, we hypothesise that balanced ratios of 1:30 or 1:20 will be most effective regarding of biological and physical efficacy.

3 Materials and Methods

3.1 Materials

Low molecular weight (m.w.) chitosan, 25% glutaraldehyde and 10% acetic acid, methanol, sodium chloride, 1,9-dimethylmethylene blue and 2,2-diphenyl-1-picrylhydrazyl was obtained from Sigma Aldrich Co.. *Lumbricus terrestris* was obtained from licensed breeders. Benedict's reagent and solid gelatine crystals were obtained from GCE.

3.2 Synthesis of the Hydrogel

3.00 g of chitosan was added to 100 ml of the 2.5% acetic acid solution and left under constant stirring at 50 °C to create a 3.0% (w/v) chitosan solution.

1 ml, 2 ml, 3 ml, 4 ml and 5 ml of 1% (v/v) glutaraldehyde solution was directly and separately added from the stock solution to the resultant chitosan solution to initiate gelation. The presence of acetic acid also served to activate the crosslinking ability of glutaraldehyde, catalysing gelation. Each sample was topped up to 25 ml using deionised water to ensure equal chitosan concentration of 2.4% (w/v), forming a glutaraldehyde to chitosan mass ratio of 1:60, 1:30, 1:20, 1:15 and 1:12. The gelation set-ups were left undisturbed until a suitable hydrogel was formed [14]. The hydrogels were then washed in deionised water until the wash solution returned a negative result on a Benedict's test, indicating removal of unlinked glutaraldehyde.

3.3 Well Diffusion Assay

E. coli and *S. Epidermidis* were separately cultured in nutrient broth at 37°C overnight. Freshly grown bacteria were spread evenly onto sterile agar plates. Wells were then punched onto these inoculated agar plates. 80 μL of varying concentrations of glutaraldehyde-chitosan, bleach (positive control) or acetic acid (negative control) were introduced into each well. The set-ups were incubated at 37°C overnight, and the resultant bacterial growth was observed. Resultant zones of inhibition generated were measured [16].

3.4 Diphenyl-1-Picrylhydrazyl (DPPH) Radical Scavenging Activity Assay

1.9 mL of methanol was mixed with 0.1 mL of glutaraldehyde-chitosan samples and 1 mL of 1% (w/v) 2,2-diphenyl-1- picrylhydrazyl (DPPH) solution. Positive control of 1.9 mL methanol mixed with 0.1 mL of 70% ascorbic acid and 1 mL DPPH solution, and negative control of 2 mL methanol and 1 mL of DPPH solution were also prepared. The mixtures were placed in the dark for 10 min before spectrophotometric measurements at 517 nm. Resultant measurements were compared with a calibration curve of DPPH and referenced against the control setup to determine the percentage of DPPH remaining, and hence the antioxidant effects [18].

3.5 *Lumbricus Terrestris* Wound Healing Assay

Lumbricus terrestris were washed with deionised water to remove visual debris. Each *Lumbricus terrestris* had its last three anterior segments surgically removed. The injured organisms were introduced to various chitosan-hydrogel or sterile water, and any rate of wound healing observed after 24h.

3.6 9-Dimethyl-Methylene Blue (DMMB) Glycosaminoglycan Assay

DMMB solution was prepared by adding 1.6mg of DMMB powder in 100mL water containing 0.304g glycine, 0.16g sodium chloride, and 9.5 mL of 0.1M acetic acid [5].

From the same wounds described above, 1g of wound tissue was extracted after 24h of healing and placed in cuvettes containing 3mL of DMMB solution. The samples were incubated in the dark for 10 min and absorbance readings taken at 525 nm. The absorbance readings were compared with a freshly prepared calibration curve of DMMB. This measures the amount of GAGs in the wound tissue, and serves as a proxy for the extent of ECM formation.

3.7 Tensile Strength Test

Hydrogel samples were cut into strips and loaded between the clamps of a tensile strength tester. The machine was first calibrated and the sample was pulled apart by the clamps. The largest magnitude of force that the hydrogel samples withstood without any tear in its structure was recorded and the ultimate tensile strength was calculated using by dividing this reading by the cross-sectional area of the hydrogel samples. The latter were calculated through computational analysis of a photograph of the cross sectional of the hydrogel.

3.8 Bending Assay

A bending assay was carried out to determine the Young's modulus of the hydrogels. Hydrogels were cut into bar shapes, and one end was taped to a retort stand while the other end was left to hang in space. The bending profile of the hydrogel was captured using a camera and the young's modulus was derived by iterative fitting of the Euler Bernoulli beam theory which was solved numerically using the Runge-Kutta method [12].

3.9 Calculation of Strain

Using data from medical journals, pressures at common pressure sore spots were used in the general formula relating to Young's Modulus, Pressure, and Strain. The strain obtained from this formula for every hydrogel was compared with the tensile strength to determine if the strain calculated is within elastic strain limits typical of hydrogels. A mathematical explanation can be found in Appendix B.

3.10 Water Absorption Test

A sample of hydrogel was submerged completely in deionised water and left to absorb water until constant mass. It's mass when fully wet was then measured as m_1. It was then left to evaporate until constant mass and was reweighed as m_0. . The swelling degree and equilibrium water content (EWC) was determined with the following equations respectively [19, 31]:

$$\text{Swelling Degree (\%)}, Q = \frac{m_1 - m_0}{m_0} \times 100\% \tag{1}$$

$$\text{Equilibriium Water Content}, EWC = \frac{m_1 - m_0}{m_1} \tag{2}$$

4 Results and Discussions

4.1 Well Diffusion Assay

All hydrogel samples were significantly effective against *E. coli.* (Fig. 1).

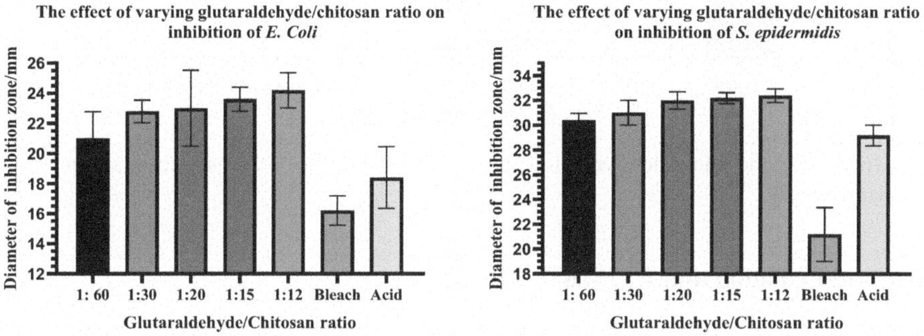

Fig. 1. Hydrogels in all ratios significantly inhibited *E. coli.* (Left) and *S. epidermidis* (Right).

This efficacy can be explained by the ability of chitosan to form complexes with lipopolysaccharides (LPS) in Gram-negative cell walls [9]. These chitosan-LPS complexes disrupt the outer membrane of bacterial cell walls [13], resulting in cell death.

The chitosan hydrogels were similarly but slightly more effective against *S. epidermidis* (Fig. 1). We postulate increased inhibition of Gram-positive bacteria by the hydrogels is due to chitosan interacting electrostatically with negatively charged tei choic acids in the bacteria cell walls, allowing chitosan to enter, causing the leakage of intracellular proteins and disrupting the cell integrity [36].

For both Gram-positive and Gram-negative bacteria, the diameter of the zone of inhibition plateaued at glutaraldehyde/chitosan ratio of 1:20. As such, gels ratios higher than 1:20 were not used for physical tests. Colony counts to quantify the extent of bacterial death was not performed, as bacteriostatic hydrogel would sufficiently inhibit the growth of bacteria, beyond which they can be removed by other immune mechanisms such as phagocytosis. Apoptotic death of bacteria around the wound site might, in fact, elicit a detrimental immune response (Letini et al., 2022).

4.2 DPPH Assay

Hydrogels with glutaraldehyde-chitosan ratios from 1:60 to 1:12 were statistically antioxidant (Fig. 2).

The antioxidative capacity of chitosan can be explained by amino groups donating electrons or hydrogen atoms to neutralise free radicals, and the hydroxyl groups forming hydrogen bonds with reactive oxygen species [1]. Beyond a ratio of 1:20, absorbance values of the substrate did not increase significantly. Hence, only gels of 1:60, 1:30, and 1:20 ratios were subjected to the wound healing assay.

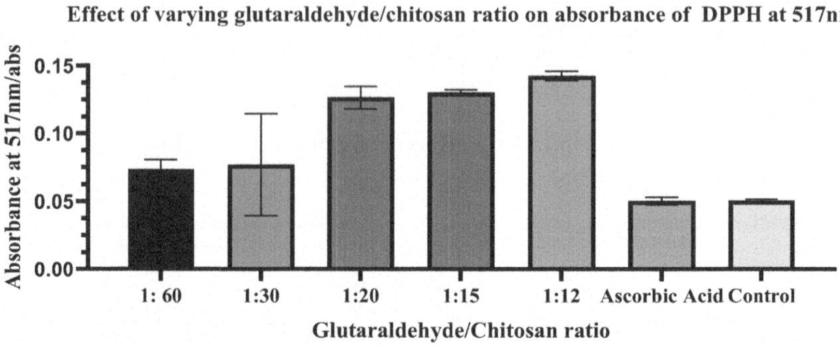

Fig. 2. Increasing the glutaraldehyde/chitosan ratio from 1:60 to 1:15 resulted in an increase in antioxidant activity.

4.3 Wound Healing and DMMB Assay

Gels with glutaraldehyde-chitosan ratios of 1:60, 1:30, and 1:20 recorded absorbance of the chitosan-GAG complex 0.0152, 0.0344, and 0.0586 respectively (Fig. 3).

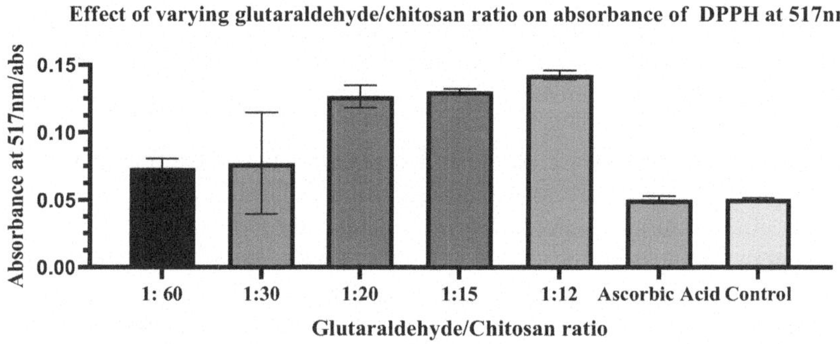

Fig. 3. Hydrogels with glutaraldehyde/chitosan ratios of 1:60, 1:30, and 1:20all stimulated ECM formation.

This suggests that, beyond a glutaraldehyde/chitosan ratio of 1:20, further increasing the ratio has no significant effect on the extent of ECM formation within 24 h of sustaining a wound. Due to the mortality of the worms when cut off from a supply of food and water, limited study can be done on the rate of ECM formation over extended time. Hence, only the 1:60, 1:30, and 1:20 hydrogels were subjected to physical characterisation.

4.4 Tensile Strength, Young's Modulus and Strain

The tensile strength of the hydrogel samples increases when more glutaraldehyde is used in the synthesis of the sample (Fig. 4).

The differences in tensile strength for all three ratios were significant. As the crosslinking degree increases, longer chains of polymers are formed which strengthens the intermolecular bonds in the sample. Hence, a larger force must be applied to

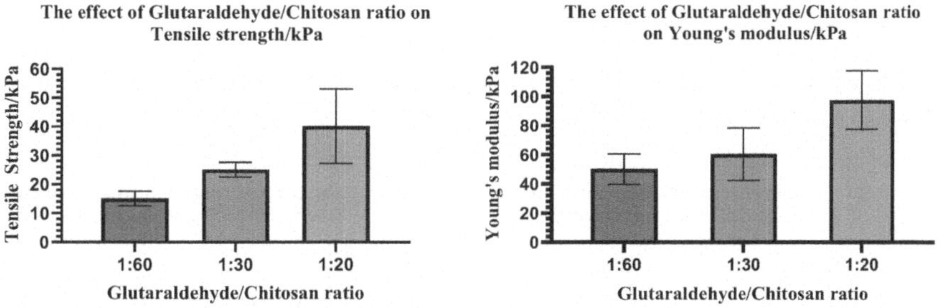

Fig. 4. A greater proportion of glutaraldehyde use on hydrogel synthesis resulted in greater tensile strength (left) and Young's moduli (right).

split the hydrogel sample through tension. This indicates the potential to customise the stiffness of the gel for wound sites of different locations, such as where joint flexibility may be an important consideration to ensure quality of life during the healing process.

Hydrogel that resulted from a higher volume of glutaraldehyde used in their synthesis showed an increase in Young's modulus (Fig. 4). Differences in Young's modulus for all three ratios were significant.

With an increased degree of crosslinking, the extensiveness of the polymers formed during synthesis would also increase, thus it would be more difficult for the polymers to move around each other during flexion, leading to an increase in bending stiffness. Hydrogels with a Young's modulus that lies around the range of human flesh would prevent any mechanical stress acting on the wound when applied, thus speeding up the recovery process of wounds.

Using the method outlined in Appendix, the strain ϵ placed on the hydrogels was calculated, and the ratio ϵ / μ was found to be less than 0.5 for all hydrogels (Fig. 5).

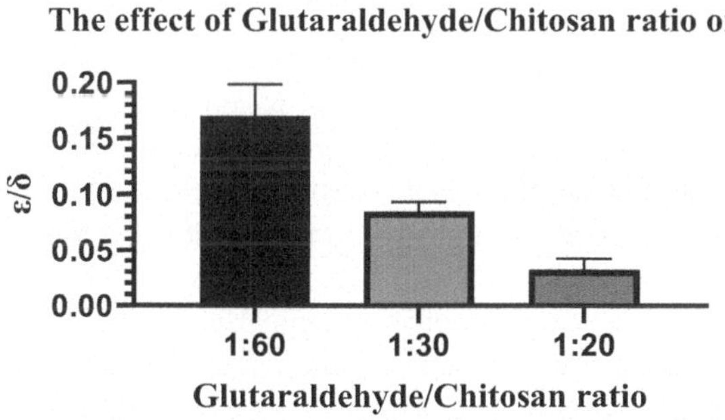

Fig. 5. The ratio ε/μ was found to be less than 0.5 for all hydrogels, suggesting that all the gels can withstand the strains typically associated with pressure sore wounds.

The ratio for the 1:20 gel was 0.03, safely below the elastic strain limit. Given that calculations were done assuming the sores developed at the sacrum of an average human, the 1:20 gel has promise to be used for wounds in other areas where pressures are greater due to lower surface area, such as the spine and shoulder.

4.5 Water Absorption and Swelling

The water absorption capacity of the hydrogels decreases significantly with increasing glutaraldehyde usage (Fig. 6).

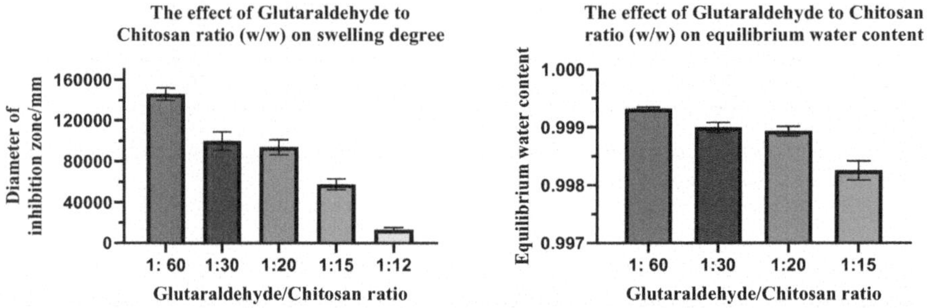

Fig. 6. The relationship between the ratio of glutaraldehyde used to chitosan on swelling (left) and EWC (right), as determined in the water-absorption test. These values suggest that all hydrogels are not likely to dehydrate wound sites.

To absorb water, relaxation of the chitosan polymer chains is required to facilitate the dissociation of intramolecular hydrogen bonds between the amino groups in chitosan. This allows more water into the gel network [27]. However, more glutaraldehyde added leads to greater formation of crosslink structures, restricting the relaxation of the chitosan polymer.

The equilibrium water contents (Fig. 6) of the hydrogels were all significantly above 0.60, higher than the water content values of our body, and indicative of potential biocompatibility in not dehydrating a wound site.

5 Conclusion and Further Work

This study demonstrates the potential of chitosan-based hydrogels cross-linked with glutaraldehyde as effective wound dressings for pressure sores. The hydrogels' antibacterial and antioxidative properties address key challenges in wound healing. Wound healing and the DMMB glycosaminoglycan assay further confirmed the hydrogels' ability to enhance tissue repair and regeneration. Physical characterisation revealed that the hydrogels possess favourable and optimal mechanical properties, ensuring durability and compatibility with human tissue. A glutaraldehyde/chitosan ratio of 1:20 is most optimal for withstanding a wide range of pressures owing to a large margin of error between the strain experienced by the gel and the strain limit.

The chitosan is sourced from prawn shells may include impurities. As a result, our hydrogel may contain traces of allergens that may trigger irritation in users who are allergic to seafood shells, and future work should confirm its safety for patients with relevant allergies [4]. Furthermore, as glutaraldehyde efficacy may not be indefinitely retained beyond the period of a few weeks [23], our hydrogel has potential use in non-vulnerable populations and in cases where freshly produced hydrogels can be used for wound dressings.

While the in vitro results obtained in this study are promising, additional studies are required to assess the material's real-world performance. Notably, human epithelial cell cultures can be used to study more extensively and intensively the effects of wound healing over a longer period. In addition, the re-formation of other layers of skin such as the epidermis can be examined, in addition to the formation of the ECM. Additionally, comparisons could be made to chitosan solutions before gelation to signify the impacts of gelation on its wound healing properties. It would also be useful to compare these hydrogels to currently available wound dressings to evaluate their relative performance.

Appendix A: Mathematical Treatment of Strain Calculation

From the literature review in the introduction, it is found that prolonged lying on the back leads to pressure sores developing most commonly at the sacrum. Modelling the sacrum as a triangular piece of bone, we can calculate the area of the sacrum.

Using the dimensions provided by Kumar *et al.* (2023), we calculate the area of the sacrum as follows:

$$A = \frac{1}{2} \times 0.102\text{m} \times 0.105\text{m} = 0.005355 \text{ m}^2$$

The pressure exerted on a sacrum when the average human weight of 70 kg lies on it is as follows:

$$P = \frac{F}{A} = \frac{mg}{A} = \frac{(70\text{kg})(9.81 \text{ m s}^{-2})}{0.005355\text{m}^2} = 128\text{kPa}$$

The Young's Modulus of a material is defined by the following formula:

$$E = \frac{\sigma}{\varepsilon} \tag{3}$$

where σ is the stress placed on the material and ε is the strain defined as $\Delta L/L$, where ΔL is the change in length, and L is the original length of the material.

Rearranging Eq. (3), the strain placed on a material with a known Young's Modulus when a pressure of is applied can be calculated as follows:

$$\varepsilon = \frac{\sigma}{E} \tag{4}$$

The stress, σ was calculated above to be around 128kPa earlier. Hence, we rewrite (4) as:

$$\varepsilon = \frac{128}{E} \tag{5}$$

For each hydrogel, the strain calculated is compared with the tensile strength as measured experimentally. Lee et al. (2018) estimate that the elastic strain limit of hydrogels is between 30%-50% of the tensile strength μ. Hence, we define that a hydrogel is suitable for use in bedsore patients if the ratio ε/μ is less than 0.5.

Peeling Strength Test Methodology, Results and Discussion

Gelatin-based human flesh model was first created by dissolving 10% gelatine in distilled water with stirring and heating at 50 °C for 2 h [8]. The samples were transferred into a petri dish and dried for 3 days. A strip of hydrogel was loaded onto the gelatine model with one end attached to a string. The string was pulled perpendicular to the surface of the gelatine such that the hydrogel would peel up at a 90° angle. The minimum force to sustain peeling of the hydrogel was measured by reading the mass loaded onto the string, and the adhesion energy derived by dividing the minimum force required to sustain peeling by the width of the sample [6].

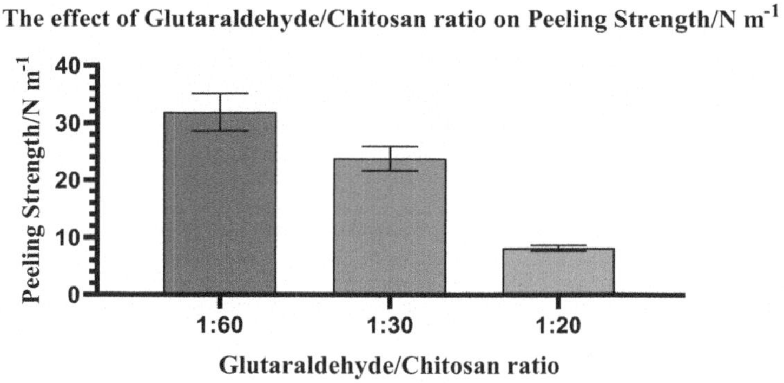

Fig. 7. The relationship between increasing the Glutaraldehyde/Chitosan ratio and the peeling strength of the hydrogel.

Hydrogel with a higher glutaraldehyde to chitosan ratio showed a higher peeling strength (Fig. 7), and the differences in between their peeling strength were significant.

The decreased flexibility in more rigid hydrogels makes it harder for the hydrogel to conform to surfaces, hence decreasing its peeling strength. An optimisation of the glutaraldehyde to chitosan ratio is required to ensure that the adhesion is sufficient to bind on the patient's wound while ensuring it is not too difficult or painful to remove for patients with low pain thresholds or sensitive skin.

References

1. Aranaz, I., et al.: Chitosan: An overview of its properties and applications. Polymers **13**(19), 3256 (2021). https://doi.org/10.3390/polym13193256
2. Bianchetti, A., Zanetti, O., Rozzini, R., Trabucchi, M.: Risk factors for the development of pressure sores in hospitalized elderly patients: results of a prospective study. Arch. Gerontol. Geriatr. **16**(3), 225–232 (1993)
3. Cleveland Clinic: Cuts and grazes. NHS Inform (2022). https://www.nhsinform.scot/illnesses-and-conditions/injuries/skin-injuries/cuts-and-grazes/#
4. Coja, T., et al.: Statement concerning the review of the approval of the basic substances chitosan and chitosan hydrochloride when used in plant protection. EFSA Journal **23**(4) (2025). https://doi.org/10.2903/j.efsa.2025.9318

5. Coulson-Thomas, V.J., Gesteira, T.F.: Dimethylmethylene blue assay (DMMB). Bio-protocol **4**(18), e1236-e1236 (2014). https://doi.org/10.21769/BioProtoc.1236
6. Costa, M.M.: Peel testing: an overview about Material Adhesion Strength. Biopdi (n.d.). https://biopdi.com/peel-testing
7. Cremar, L., et al.: Development of antimicrobial chitosan based nanofiber dressings for wound healing applications. Nanomedicine Journal **5**(1), 6–14 (2018). https://doi.org/10.22038/nmj.2018.05.002
8. Dabrowska, A., et al.: 18A water-responsive, gelatine-based human skin model. Tribol. Int. **113**, 316–322 (2017)
9. Davydova, V.N., et al.: The supramolecular structure of LPS–chitosan complexes of varied composition in relation to their biological activity. Fakhry Carbohydrate polymers **123**, 115–121 (2015)
10. Ding, X., et al.: Challenges and innovations in treating chronic and acute wound infections: from basic science to clinical practice. Burns & Trauma **10**(14) (2022). https://doi.org/10.1093/burnst/tkac014
11. Dong, Z., et al.: High-strength hydrogels: Fabrication, reinforcement mechanisms, and applications. Nano Res. **16**, 3475–3515 (2023). https://doi.org/10.1007/s12274-022-5129-1
12. Fakhry, A., Schneider, G.B., Zaharias, R., Şenel, S.: Chitosan supports the initial attachment and spreading of osteoblasts preferentially over fibroblasts. Biomaterials **25**(11), 2075–2079 (2004). https://doi.org/10.1016/j.biomaterials.2003.08.068
13. Feng, P., et al.: Chitosan-based functional materials for skin wound repair: Mechanisms and applications. Frontiers in Bioeng. Biotechnol. **9**, 650598 (2021)
14. Furuike, T., Komoto, D., Hashimoto, H., Tamura, H.: Preparation of chitosan hydrogel and its solubility in organic acids. Int. J. Biol. Macromol. **104**, 1620–1625 (2017). https://doi.org/10.1016/j.ijbiomac.2017.02.099
15. Gere, J.M., Timoshenko, S.P.: Mechanics of Materials, 4th ed. PWS Publishing Company (1997)
16. Holder, I.A., Boyce, S.T.: Agar well diffusion assay testing of bacterial susceptibility to various antimicrobials in concentrations non toxic for human cells in culture. Burns **20**(5), 426–429 (1994). https://doi.org/10.1016/0305-4179(94)90035-3
17. Izaguirre, N., Gordobil, O., Robles, E., Labidi, J.: Enhancement of UV absorbance and mechanical properties of chitosan films by the incorporation of solvolytically fractionated lignins. Int. J. Biol. Macromol. **155**, 447–455 (2020). https://doi.org/10.1016/j.ijbiomac.2020.03.162
18. Kedare, S.B., Singh, R.P.: Genesis and development of DPPH method of antioxidant assay. J. Food Sci. Technol. **48**, 412–422 (2011)
19. Lange, H.: Determination of the degree of swelling and crosslinking of extremely small polymer gel quantities by analytical ultracentrifugation. Colloid Polym. Sci. **264**(6), 488–493 (1986). https://doi.org/10.1007/bf01422001
20. Lentini, G., et al.: Neutrophils discriminate live from dead bacteria by integrating signals initiated by Fprs and TLRs. EMBO J. **41**(5), e109386 (2022)
21. Maheshwari, G.: Chronic wounds: a rising public health concern. Wounds APAC **7**(1) (2024). Wounds Asia. https://woundsasia.com/journal-articles/chronic-wounds-a-rising-public-health-concern
22. Marya: Blurry SEM Image: Causes and Solutions. Element-Pi (2024). https://elementpi.com/sem-blur/
23. Matei, A., et al.: On the Stability of Glutaraldehyde in Biocide Compositions. Int. J. Mol. Sci. **21**(9), 3372 (2020). https://doi.org/10.3390/ijms21093372
24. Mayo Clinic: Bedsores (Pressure Ulcers) - Symptoms and Causes (n.d.). Retrieved 21 December 2024, from https://www.mayoclinic.org/diseases-conditions/bed-sores/symptoms-causes/syc-20355893

25. Pal, K., Paulson, A.T., Rousseau, D.: Biopolymers in Controlled-Release Delivery Systems. In: Handbook of Biopolymers and Biodegradable Plastics, pp. 329–363. Plastics Design Library (2013). https://doi.org/10.1016/b978-1-4557-2834-3.00014-8
26. Robson, M.C.: Wound infection: a failure of wound healing caused by an imbalance of bacteria. Surg. Clin. North Am. **77**(3), 637–650 (1997)
27. Rohindra, D.R., Nand, A.V., Khurma, J.R.: Swelling properties of chitosan hydrogels. The South Pacific J. Natur. App. Sci. **22**(1), 32 (2004). https://doi.org/10.1071/sp04005
28. Schnabelrauch, M., Scharnweber, D., Schiller, J.: Sulfated glycosaminoglycans as promising artificial extracellular matrix components to improve the regeneration of tissues. Curr. Med. Chem. **20**(20), 2501–2523 (2013)
29. Stadelmann, W.K., Digenis, A.G., Tobin, G.R.: Impediments to wound healing. The American Journal of Surgery **176**(2), 39S-47S (1998)
30. Sudarshan, N.R., Hoover, D.G., Knorr, D.: Antibacterial action of chitosan. Food Biotechnol. **6**(3), 257–272 (1992)
31. Timur, M., Paşa, A.: Synthesis, Characterization, Swelling, and Metal Uptake Studies of Aryl Cross-Linked Chitosan Hydrogels. ACS Omega **3**(12), 17416–17424 (2018). https://doi.org/10.1021/acsomega.8b01872
32. Wang, G., Yang, F., Zhou, W., Xiao, N., Luo, M., Tang, Z.: The initiation of oxidative stress and therapeutic strategies in wound healing. Biomed. Pharmacother. **157**, 114004 (2023). https://doi.org/10.1016/j.biopha.2022.114004
33. Willingsford Limited: Pollution - Plastics, Antimicrobials & Chemicals. Willingsford Healthcare (2020). https://willingsford.com/blog/2020/06/08/pollution-plastics-antimicrobials-chemicals
34. Wound Source: Infected Wounds. WoundSource (2018). https://www.woundsource.com/patientcondition/infected-wounds
35. Xue, M., Jackson, C.J.: Extracellular matrix reorganization during wound healing and its impact on abnormal scarring. Adv. Wound Care **4**(3), 119–136 (2015)
36. Yan, D., et al.: Antimicrobial properties of chitosan and chitosan derivatives in the treatment of enteric infections. Molecules **26**(23), 7136 (2021). https://doi.org/10.3390/molecules26237136
37. Zhang, Z., Feng, Y., Wang, L., Liu, D., Qin, C., Shi, Y.: A review of preparation methods of porous skin tissue engineering scaffolds. Materials Today Communications **32**, 104109 (2022). https://doi.org/10.1016/j.mtcomm.2022.104109

The Effect of Flour Type Used on Spread of Cookies, as Well as Effect of Cooling Cookie Dough and Addition of Solids into Dough on Brittleness of Cookies

Gayatri Sujit, Linda Tang(✉), and Xiaohui Clara Ann Cheong

High School of Math and Science, NUS, 20 Clementi Ave 1, Singapore 129957, Singapore
h2310133@nushigh.edu.sg

Abstract. When cookies are baked, they could come out in different sizes and textures due to the different variables affecting them. Knowing what variables affect the outcome of the cookies and how they affect the cookies could be useful for bakers and other home bakers when it comes to getting favourable outcomes. The main findings of the project are: Firstly, the more protein is present in the flour used, the less the cookies spread due to the formation of a gluten network in the dough which supports the structure of the cookies. Secondly, cooling of dough before baking leads to less brittle cookies. Finally, adding certain solids to the dough leads to more brittle cookies. Two experiments were conducted as part of this research project. The first experiment was to investigate how the type of flour used, affects the spread by comparing the ratio, which was measured by taking the initial ratio of diameter to thickness of cookie, to the final ratio of diameter to thickness of cookie. In the second experiment, whether the cookie dough was cooled or not and the addition of certain solids into the doughs were the independent variables and we found their effects on brittleness of cookies. The brittleness of the cookies was measured by dropping the cookie, in a box, six times from a height of one metre, and then finding the percentage of the cookie's mass that is the largest remaining piece after the drop.

1 Background and Purpose of Research Project

Whether it is baking or purchasing cookies, it was observed that one often gets varying cookie sizes and textures. The definition of a good cookie is different for everyone, as different people have different preferences for texture, depending on how they want the cookies to be presented and eaten. Cookies that are crispy tend to break easily due to their rigid structure, while those that are chewy are usually held together better. There needs to be a right balance of these characteristics of the cookies to achieve the respective texture desired. Hence, an experiment on different variables that would affect these characteristics of the cookie will give us more insight into how one can control and tweak the process to obtain the desired texture.

Two experiments were conducted in 2024, one in April and the other in July. In experiment 1, the independent variable chosen to be investigated was the type of flour

H. Guo et al. (Eds.): IRC-SET 2025, PTT, pp. 179–187, 2026.
https://doi.org/10.1007/978-981-95-9630-0_16

used. The dependent variable was the spread of the cookies. Based on the results in experiment 1, the flour type that resulted in the least spread of cookies was used for the recipe used in experiment 2 to find out more about how variables such as whether the dough was cooled or not before baking and the addition of solids into the dough affect the brittleness of the cookies. If the cookies spread less during the first experiment, it meant that they were denser and firmer, and the structure was least affected. This enabled other variables to be changed on the densest and firmest doughs, which would provide more reliable results, as spread did not affect the structure of these cookies much, and thus this enabled the investigation of a new dependent variable, the brittleness of the cookies.

Knowing how different variables affect the different qualities of the cookies that are baked, such as how much the cookies spread upon baking and how brittle the cookies are, is beneficial, as this helps one to predict and better control the appearance, texture, and size of the cookie. The research done can further add to, or support, these findings already made so that bakeries can benefit from them. Bakeries or home bakers can use these results to bake cookies of desired appearance and texture, which will be more appealing to customers or others.

2 Hypothesis of Research

Based on research done, protein content in flour affects spread. Gluten, formed by two proteins, glutenin and gliadin, gives the cookie shape and structure as a protein chain is formed when gluten proteins are hydrated and bind with each other [1]. This keeps the structure of the cookie and likely makes the cookie spread less [2].

Hence, the hypothesis for experiment 1 was that flour types with more protein content would result in less spread. Flour types in increasing order of protein content are corn starch flour, tapioca starch flour, rice flour, and all-purpose flour (see more under Sect. 4, "Discussion of Results and Implications, Conclusions"). These types of flour were chosen as they were commonly used in baking to make various kinds of food. They are also easily accessible and affordable to many people.

For experiment 2, a few factors that affected the brittleness of baked cookies were researched. The temperature of the fat in the cookie dough is especially important. Fat is one of the basic components of cookie dough. Chilling the dough solidifies the fat, causing a lower rate of spreading in the oven, thus leading to a thicker and more resistant cookie [3]. Another factor that affects the brittleness of a cookie is the addition of solids. By using add-ins like marshmallows and chocolate chips in the cookie dough, this may allow the cookie dough to stick together as the solids melt when baked and solidify once again when cooled, keeping the cookie together and hence leading to less brittle cookies. For experiment 2, the hypothesis was that both cooling the dough, and the addition of solids, results in less brittle cookies.

3 Experimental Methods and Results

3.1 Methodology and Procedures for Experiment 1

1. Make batters excluding flour.

- Whisk together 170g of unsalted butter and 225g of granulated sugar in a bowl using an electric whisk.
- Add ¾ teaspoons vanilla extract to the creamed butter and sugar.
- Add 1 small egg to the mixture and continue whisking the mixture until it is smooth
- Add ¾ teaspoons salt and ¼ teaspoon baking soda to the mixture and mix well

2. Separate batters into 3 equal portions and add 80 g of the respective flours.
3. For each dough type, weigh 3 portions of 15g each, and put them on a baking sheet to bake. Before baking, measure the initial height of the batter with a toothpick stuck into the center of the batter and the diameter of the dough with a ruler. Find the ratio of diameter to height of batter.
4. After baking, measure the height and diameter of the cookie with the same method as measuring the dough. Find the ratio of the diameter to the height of the cookie.
5. Find the spread ratio by dividing the ratio of final diameter to final height by the ratio of initial diameter to initial height.

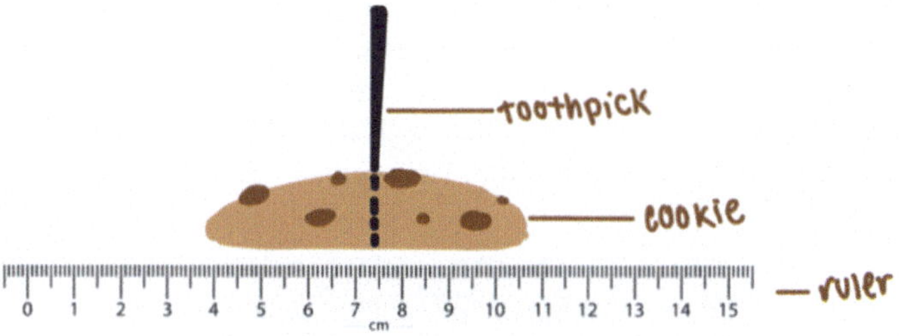

Diagram of experimental set-up when taking measurements

Independent variable: Type of flour used for cookie batter Dependent variable: Spread ratio of the cookie

3.2 Methodology and Procedures for Experiment 2

1. Make cookie dough using all-purpose flour.
2. Separate batter into 2 equal portions, with 1 batch as a control and 1 frozen for a week. Repeat step 1 and similarly, form 2 equal portions for with and without marshmallows, another batch for with and without chocolate chips.
3. Bake each type of cookie dough using same oven setting.
4. After baking, measure the initial weight of baked cookie on a measuring scale.

5. Push plastic box containing cookie off the 1 m tall table. Repeat this for 6 times in total.

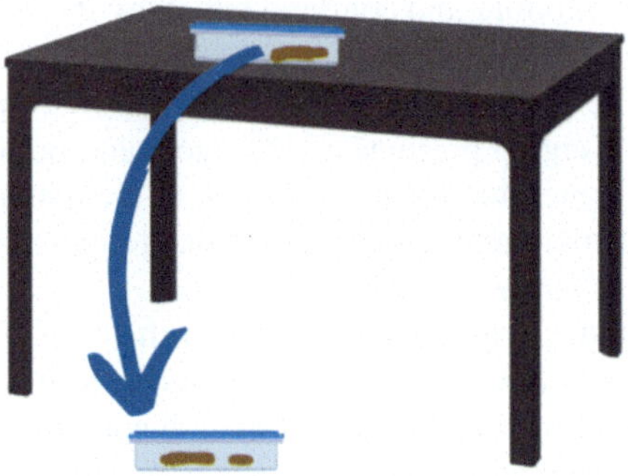

6. Measure the weight of largest leftover piece of cookie after pushing box containing cookie off the table for 6 times. Record down the results. Find the percentage of cookie left by dividing the end weight by the initial weight of cookie.
7. Repeat steps 1 to 6 on another day, to get more results for the same experiment, average the results and ensure reliability of results.

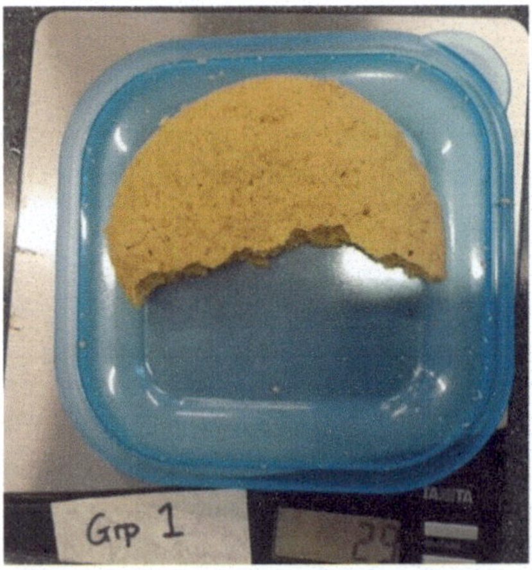

Independent variable: Add in of solids, Temperature of cookie batter Dependent variable: Percentage of cookie left after dropping the box containing cookie for 6 times

4 Discussion of Results and Implications, Conclusions

4.1 Experiment 1

As shown in Fig. 1 (below) the value in the last row and last column is the spread ratio for each flour's cookies, which indicates the average of how much the cookie spreads. The higher the spread ratio, the higher the spread of the cookie.

Based on the USDA website, the protein contents in the flours are 0.3% in corn starch flour [4], 2% in tapioca starch flour [5], 6% in rice flour [6] and 10% in all-purpose flour [7]. The spread ratios of the cookies of the flour types are 8.51, 3.36, 2.78 and 2.62 respectively. This implies that the higher the protein content in the flour, the less the cookies spread.

As mentioned previously, the more protein that is present in the flour, the more structure and shape the cookies would have, as more gluten would be present in the flour, forming a protein chain in the cookies when the gluten proteins are hydrated and bind with each other. Gluten is made up of two different types of protein, glutenin and gliadin. Hydrogen bonds form between the two types of proteins due to the amino acids in the proteins, and this allows for a network to be formed. Hence, this keeps the shape and supports the structure of the cookie and makes the cookie spread less.

These results tally with the hypothesis, which was that the more protein is present in the flour, the less the cookies of that flour spread (Fig. 2).

Corn Starch Flour:	Number 1	Number 2	Number 3	Average
Before (Thickness)/cm	1.30	1.20	0.90	–
After (Thickness)/cm	0.30	0.30	0.20	–
Before (Diameter)/cm	4.30	4.30	4.30	–
After (Diameter)/cm	9.50	7.80	8.30	–
Diameter/Thickness (Before)	3.31	3.58	4.78	–
Diameter/Thickness (After)	31.67	26.00	41.50	–
Spread ratio	9.57	7.26	8.69	8.51

Tapioca starch flour:	Number 1	Number 2	Number 3	Average
Before (Thickness)/cm	0.70	0.60	0.60	–
After (Thickness)/cm	0.30	0.40	0.40	–
Before (Diameter)/cm	4.30	4.30	4.30	–
After (Diameter)/cm	8.20	7.80	8.30	–
Diameter/Thickness (Before)	6.14	7.17	7.17	–
Diameter/Thickness (After)	27.33	19.50	20.75	–
Spread ratio	4.45	2.72	2.90	3.36

Rice flour:	Number 1	Number 2	Number 3	Average
Before (Thickness)/cm	1.00	1.00	0.90	–
After (Thickness)/cm	0.60	0.60	0.50	–
Before (Diameter)/cm	4.30	4.30	4.30	–
After (Diameter)/cm	7.00	6.50	7.40	–
Diameter/Thickness (Before)	4.30	4.30	4.78	–
Diameter/Thickness (After)	11.67	10.83	14.80	–
Spread ratio	2.71	2.52	3.10	2.78

All purpose flour	Number 1	Number 2	Number 3	Average
Before (Thickness)/cm	0.80	0.80	0.90	–
After (Thickness)/cm	0.40	0.50	0.60	–
Before (Diameter)/cm	4.30	4.30	4.30	–
After (Diameter)/cm	6.50	6.60	6.80	–
Diameter/Thickness (Before)	5.38	5.38	4.78	–
Diameter/Thickness (After)	16.25	13.20	11.33	–
Spread ratio	3.02	2.46	2.37	2.62

Fig. 1. Spread of Corn Starch, Tapioca Starch, Rice, and All-purpose flour cookies.

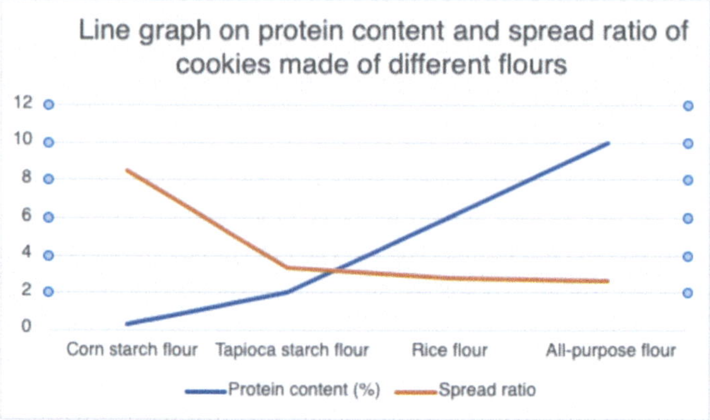

Fig. 2. Graphical representation of Experiment 1 results

4.2 Experiment 2

In experiment 2, for each variable, the experiment was conducted on 2 separate batches of dough. The average of the percentages of the 3 cookies baked from each batch was taken, then the average of the averages of the 2 batches was taken to ensure more reliable results. Figure 3 summarises the averages of batches 1 and 2 for each variable.

Cooled dough resulted in cookies that had a greater percentage leftover, 60%, than dough that was not cooled, 52%. Thus, cooled dough resulted in less brittle cookies. The addition of chocolate chips into dough resulted in a lower percentage left over, 41%, than 54% for the control dough without chocolate chips. The addition of marshmallows into dough resulted in a lower percentage leftover, 46%, than 55% for the control dough without marshmallows. Thus, the addition of marshmallows and chocolate chips resulted in more brittle cookies.

The result for cooled dough supports the hypothesis that cooled dough would result in less brittle cookies. However, the results for the addition of solids, including chocolate chips and marshmallows, do not support the hypothesis, which was that the addition of solids leads to less brittle cookies.

Cooling the dough before baking causes the fats in the dough to solidify, which slows down the rate they melt when baked. This causes less spread of the cookie and a denser texture. This made the cookies less brittle, as the solidification of fats helped create a more cohesive dough, which keeps the cookie moist and flexible, preventing the dry, rigid structure that makes cookies brittle [3].

The addition of chocolate chips made the cookies thick and chewy (as expected). However, it was more brittle. This could have been because chocolate chips disrupted the gluten network in the cookies and weakened the uniform structure, so although it was thick and chewy, the cookie dough was not cohesive as expected and instead led to the cookie breaking apart due to the disrupted structure. The structure was more disrupted than held together by the presence of the chocolate chips, and thus the cookies were crumbly and broke apart easily.

The addition of marshmallows caused cookies to be crispier and more brittle, which was not expected. Marshmallows contain a lot of sugar, so when they melted, they caramalised, and when cooled, they hardened, causing the cookies to have crisp edges and bottoms, contributing to their rigid structure and thus their brittleness [8] (Figs. 4 and 5).

Not Cooled Batch 1	cookie 1	cookie 2	cookie 3	Average
Before weight	42	43	47	–
After weight	22	22	26	–
% of cookie left	52.38095238	51.1627907	55.31914894	52.95429734

Not Cooled Batch 2	cookie 1	cookie 2	cookie 3	Average
Before weight	46	44	45	–
After weight	25	23	22	–
% of cookie left	54.34782609	52.27272727	48.88888889	51.83648075

Cooled Batch 1	cookie 1	cookie 2	cookie 3	Average
Before weight:	53	51	52	–
After weight:	30	31	31	–
% of cookie left:	56.60377358	60.78431373	59.61538462	59.00115731

Cooled Batch 2	cookie 1	cookie 2	cookie 3	Average
Before weight:	50	52	51	–
After weight:	32	31	31	–
% of cookie left:	64	59.61538462	60.78431373	61.46656611

Without Chocolate Chips Batch 1	cookie 1	cookie 2	cookie 3	Average
Before weight	40	43	42	–
After weight	19	24	24	–
% of cookie left	47.5	55.81395349	57.14285714	53.48560354

Without Chocolate Chips Batch 2	cookie 1	cookie 2	cookie 3	Average
Before weight	43	40	48	–
After weight	23	24	25	–
% of cookie left	53.48837209	60	52.08333333	55.19056848

With Chocolate Chips Batch 1	cookie 1	cookie 2	cookie 3	Average
Before weight:	51	52	56	–
After weight:	21	22	23	–
% of cookie left:	41.17647059	42.30769231	41.07142857	41.51853049

With Chocolate Chips Batch 2	cookie 1	cookie 2	cookie 3	Average
Before weight:	54	55	53	–
After weight:	23	22	21	–
% of cookie left:	42.59259259	40	39.62264151	40.73841137

Without Marshmallows Batch 1	cookie 1	cookie 2	cookie 3	Average
Before weight	36	33	35	–
After weight	20	18	20	–
% of cookie left	55.55555556	54.54545455	57.14285714	55.74795575

Without Marshmallows Batch 2	cookie 1	cookie 2	cookie 3	Average
Before weight	42	43	40	–
After weight	23	22	23	–
% of cookie left	54.76190476	51.1627907	57.5	54.47489849

With Marshmallows Batch 1	cookie 1	cookie 2	cookie 3	Average
Before weight:	39	40	42	–
After weight:	18	17	21	–
% of cookie left:	46.15384615	42.5	50	46.21794872

With Marshmallows Batch 2	cookie 1	cookie 2	cookie 3	Average
Before weight:	44	45	44	–
After weight:	20	19	21	–
% of cookie left:	45.45454545	42.22222222	47.72727273	45.13468013

Fig. 3. Brittleness of cookies based on whether dough was cooled or not, addition chocolate chips, addition of marshmallows

Dough used to bake cookies	Average % of cookie that is the largest leftover piece
Not Cooled	52
Cooled	60
Without Chocolate Chips	54
With Chocolate Chips	41
Without Marshmallows	55
With Marshmallows	46

Fig. 4. Average readings for batches 1 and 2 for each variable

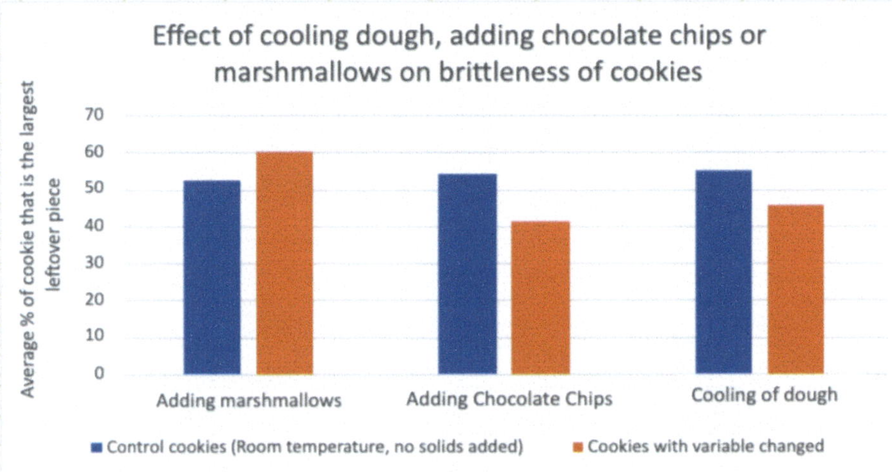

Fig. 5. Graphical representation of Experiment 2 results

5 Limitations and Future Research

There were some constraints in the experiment that restricted the ability to get more consistent results. The methods and procedures used in the experiment are not the best method to measure the independent variable due to a few reasons. Firstly, there was a lack of appropriate machinery to measure the spread and brittleness of the cookies more accurately due to a limited budget. There was also insufficient apparatus to measure the amount of force we used to drop the box containing cookies for the second experiment. There are more accurate and better ways of measuring the independent variables. If constraints were removed, the experiment and its research can be extended to become more advanced and accurate and find out more about how to get the desired appearance of a cookie.

For future research, there are a few possible directions. Firstly, an investigation about how the ratio of different types of flour used can affect the cookie appearance and texture can be carried out, as it was observed that some recipes have tried to use a few types of flour to make the cookie dough. Investigation of how the mixing method and duration of the dough and its ingredients, such as overmixing the dough compared to undermixing the dough, will affect the product can also be carried out. During the experimental phase, it had been observed that when butter is used at different states, such as solid and liquid state, to bake, the cookie had different appearance and texture, such as having more spread. This had led to the interest in wanting to explore more about how fat content in the cookie dough, such as using oil instead of butter, affects the cookie texture and appearance.

References

1. Hill, M.: Attack of the Gluten. American Chemical Society (2012). www.acs.org/education/resources/highschool/chemmatters/past-issues/archive-2011-2012/gluten.html
2. Gray, E.: Science of Flour in Baking. Baking Sense® (2024). www.baking-sense.com/2017/03/29/baking-ingredients-flour/#:~:text=Wheat%20flour%20is%20composed%20of,millions%20of%20tiny%20air%20bubbles
3. Hamel, P.J., et al.: Chilling Cookie Dough: Does It Make a Difference? King Arthur Baking (2017). www.kingarthurbaking.com/blog/2015/05/17/chilling-cookie-dough
4. "Fooddata Central Search Results." FoodData Central (2019). fdc.nal.usda.gov/fdc-app.html#/food-details/169698/nutrients
5. "Fooddata Central Search Results." FoodData Central, fdc.nal.usda.gov/fdc-app.html#/food-details/169614/nutrients. Accessed 25 Sept. 2024
6. "Fooddata Central Search Results." FoodData Central, fdc.nal.usda.gov/fdc-app.html#/food-details/169714/nutrients. Accessed 25 Sept. 2024
7. "Fooddata Central Search Results." FoodData Central, fdc.nal.usda.gov/fdc-app.html#/food-details/168936/nutrients. Accessed 25 Sept. 2024
8. "Why Are My Marshmallows Getting Crunchy after Baking?" Seasoned Advice (1967). cooking.stackexchange.com/questions/120877/why-are-my-marshmallows-getting-crunchy-after-baking

Chemical Engineering

Network Clustering of Drug-Drug Interactions

Paige Zi Ning Gan[1] and Judice L. Y. Koh[2(✉)]

[1] Hwa Chong Institution, Singapore, Singapore
[2] Department of Biomedical Informatics, Yong Loo Lin School of Medicine, National University of Singapore, Singapore, Singapore
judice@nus.edu.sg

Abstract. Drug drug interactions (DDIs) are instances when one drug blunts the effectiveness of another and may lead to unpredictable side effects that threaten patient safety. Elucidating DDIs is paramount to the safety and effectiveness of the drug treatments. In this project, we applied network clustering to identify groups of drugs that tend to interact with one another, and determine potential biochemical similarities underpinning the interactions. Using network clustering - an unsupervised AI/machine learning method involving grouping of drugs based on their interactions within a network, we constructed a DDI network of more than 1 million known DDIs, with nodes depicting the drugs, and the edges depicting evidence of the DDIs. Using the Louvain Community algorithm, we clustered drugs into groups with more frequent interactions. To optimize clustering, inter- and intra-cluster analyses were performed to validate the structure. Categories of the drugs were mapped within these clusters to identify properties influencing DDIs. It was found that some significant determinants of DDIs include the drug structures, the part of the body or type of drug they target. These insights highlight the value of predicting DDIs in preventing adverse side effects. These properties underpin the DDI clusters and are valuable in the predictions of DDIs in the prevention of side effects.

Keywords: Drug-Drug Interactions · Louvain community · network clustering

1 Introduction

Drug-drug interactions (DDIs) occur when a drug affects the ability of another drug to carry out its mechanism of action (MOA) in the body. These interactions could influence how well drugs work and/or result in adverse side effects in patients who take these drug combinations. There are three main types of drug-drug interactions: (1) Duplication where two drugs possess similar MOAs which could lead to an overdose of medication to the patient; (2) Opposition (or Antagonism) where two drugs interact to blunt the effect one or both drugs have on the body; and (3) Alteration where one drug changes the physiology of the patient's body rendering the other drug less effective in carrying out its MOA. Minimizing DDIs is critical to the optimal functioning of the medication and accurately predicting DDIs maximizes patient safety by minimizing adverse effects. Although drug development protocols require drug candidates to be tested extensively

H. Guo et al. (Eds.): IRC-SET 2025, PTT, pp. 191–200, 2026.
https://doi.org/10.1007/978-981-95-9630-0_17

in animals before clinical trials in humans can be carried out, the physiology of animal models such as mice are still sufficiently different from that of humans, and hence drug-drug interactions are often missed in animal testing. Most drug-drug interactions identified during clinical trials are due to unexpected deleterious effects on the tested drugs. It is only when physicians are aware of such interactions will it be possible to avoid the administration of drugs with known DDIs to patients.

This project aims to provide drug developers and clinicians with an additional tool to more accurately predict possible DDIs by predicting DDIs from clustering known DDIs using unsupervised machine learning clustering algorithms. From the clusters, we determine the categories of drugs which are enriched in each cluster to draw insights to the properties underpinning drugs that tend to interact with one another.

2 Materials

The dataset used in our study was obtained from Drugbank. DrugBank is an online database that combines drug data with information on drug targets, mechanisms, and interactions. It is used in pharmaceutical research, bioinformatics, and clinical studies for drug discovery and drug-related biological processes [1].

The initial dataset in the Drugbank database included 1,048,577 different interactions. Due to the network being too dense, the dataset was filtered by removing singletons i.e. drugs with only 1 known DDI and promiscuous drugs with exception high number of DDIs. 470 drugs and their interactions were then randomly selected from the remaining interaction pairs to be used for this study.

3 Methods

3.1 Network Visualization Using Cytoscape

Cytoscape is a software platform that is open source and able to visualize different types of complex networks and integrate any attribute data with it [2] This program is used in bioinformatics and systems biology fields for exploration of complex biological systems and associated connections/interactions. In this project, an initial visualization of the dataset was constructed using Cytoscape version 3.0. The drugs were represented as nodes in the form of a graphical network, and the number of evidences of DDIs between them as edges. The excel file with pairs of interacting drugs was uploaded, with the nodes as individual drugs and the edges as the interactions between the drugs. Markov Cluster Algorithm (MCL) clustering [3] was done on Cytoscape. MCL clustering is a network clustering algorithm that identifies densely connected regions in a network. However, it is better suited for dense network, so for this project, we utilize Louvain for the clustering..

3.2 Network Clustering

To cluster the DDI network, we utilize the community detection algorithm - an unsupervised machine learning methods for identifying modules. NetworkX is a python library intended for the design and investigation of complex networks and graphs [4]. Using NetworkX, various community clustering approaches were explored. For the DDI dataset, the Louvain approach was deemed the most suitable. The communitybased algorithms tested were divisive, k-clique and modularity based communities.

These were found to be not suitable for the dataset as the data required to be inputted did not match the format of the dataset as a position-based community algorithm is required to generate an appropriate network. The Louvain algorithm finds groups that are closely linked together within the given framework. Although the code did not specify the number of clusters, the algorithm selects the optimal number of clusters depending on the structure of the graph and how interconnected its nodes are.

The Louvain algorithm is a method designed for community detection in big networks using modularity optimization [5]. It optimises the degree of modularity which is defined as a measure of the degree to which the network is detachable into a number of smaller parts. It begins by adding each node as their own community and attempts to improve overall modularity by evaluating the addition of a node to a neighbouring community. When no further improvement to the clustering can be achieved, the algorithm compresses the network by gathering all the nodes within a community into a single node, while building a new graph with intact edges connecting those communities. This is done again until no more improvements in the modularity are achievable. The modularity used in the Louvain algorithm can be calculated using a mathematical formula. The formula measures the difference between the actual edges in the network and the expected number of edges if the edges were distributed randomly while preserving the node degrees, normalized by the total edge weight. Higher values of modularity indicate a stronger community structure.

The sequence code that was inputted is shown in the flowchart in Figure 1.

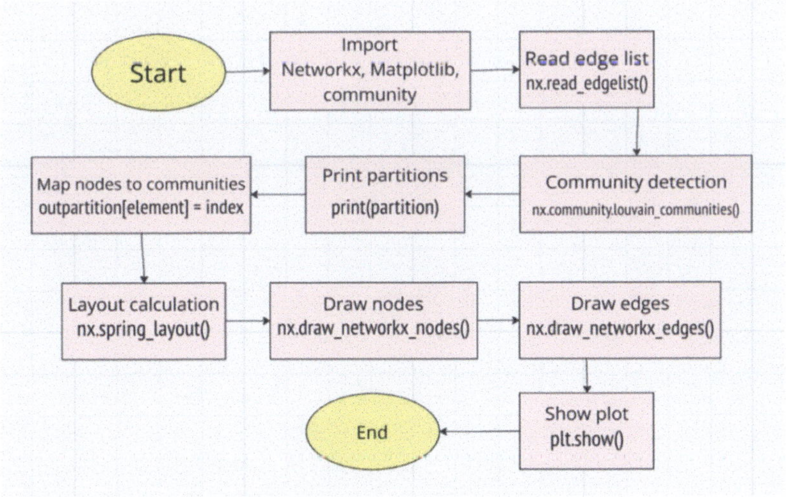

Fig. 1. Sequence of code for clustering dataset

3.3 Inter and Intra Cluster Distance

The code to calculate intra and inter cluster distance is shown in Figs. 2a and 2b respectively.

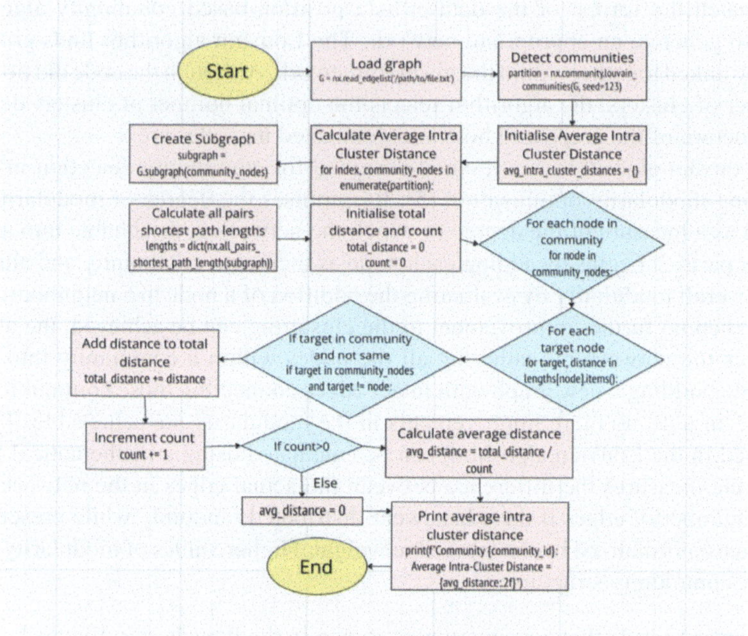

Fig. 2a. Sequence of code for determining intra cluster distance

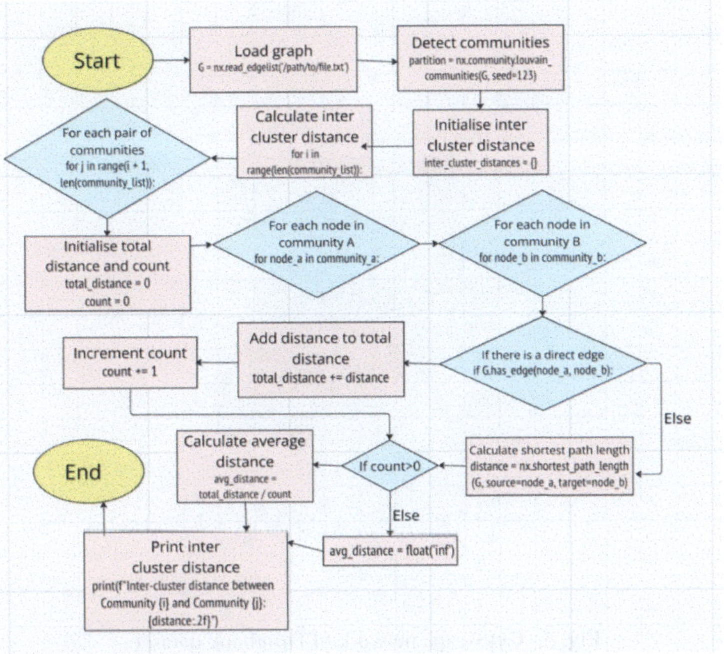

Fig. 2b. Sequence of code for determining inter cluster distance

3.4 Determining Drug Categories

From the drugbank dataset, categories of each drug were listed, as well as other information such as target, enzyme, salt, carrier and transporter (name, polypeptide, ID and mass). Drugs in each cluster from the Louvain network were matched to its category to determine what category of drugs tend to interact with one another.

4 Results

4.1 Cytoscape Clustering

Before using MCL clustering, the network is shown in Fig. 3. It is densely populated with 470 nodes and 12,740 edges.

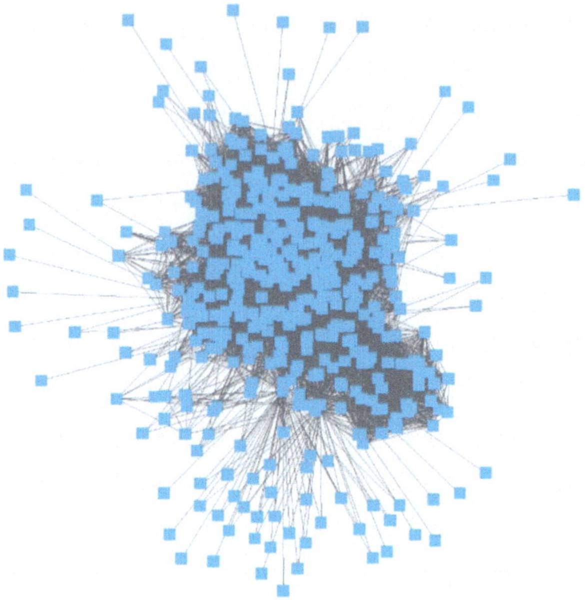

Fig. 3. Cytoscape network of Drugbank dataset

Figure 4 shows the Markov Cluster Algorithm (MCL) clustering done by Cytoscape.

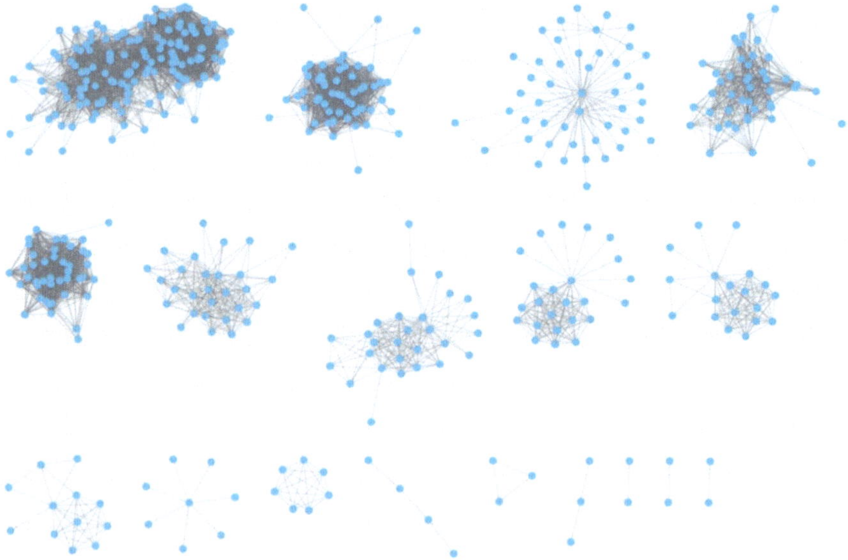

Fig. 4. MCL clustering of dataset

In the MCL clustering network, there are 17 clusters with the maximum number of nodes in 1 cluster being 142 in cluster 1. Of these 142 drugs, there were 97 antibodies, 68 amino acids, 61 peptides and 56 Proteins.

4.2 Network Clustering

Figure 5 illustrates the network cluster generated by the Louvain community-based algorithm. This clustering resulted in the identification of 4 distinct communities, comprising a total of 470 nodes and 12,740 edges. Each node within this network represents a specific drug, while each edge are the interactions between the drugs. The six communities contain between 101 to 134 drugs and reflect groups of drugs that tend to interact more frequently with one another, highlighting the potential of similar chemical properties between them. The structure of the network allows for further analysis of the interconnectedness of these drugs.

The number of nodes in each cluster are 101, 123, 112 and 134 nodes in cluster 0, 1, 2 and 3 respectively.

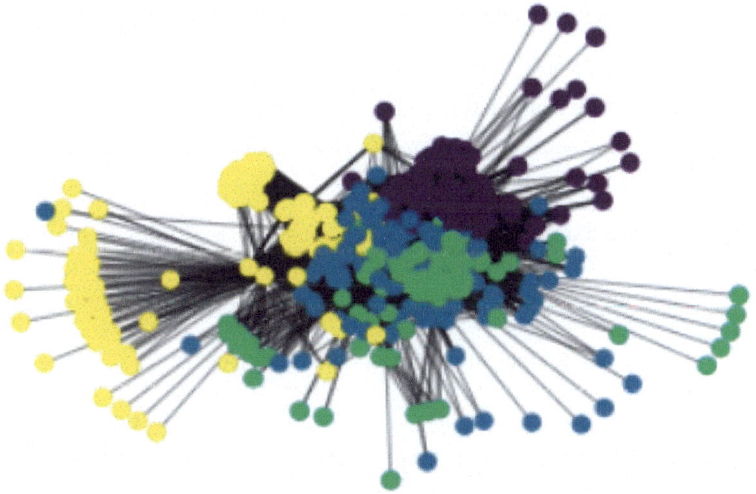

Fig. 5. Network clusters generated by the Louvain communities

4.3 Inter- and Intra- Cluster Distance

To assess the performance of the clustering algorithm, we calculated the intra and inter-cluster distance. Simply for good clustering, the average distance of the nodes within a cluster should be high and that of nodes between clusters should be low.

The intra-cluster distance shows the average distance of points in one cluster from the center measure is relatively low at 1.73. This shows that the points within those clusters are close together and there is high connectivity within the drugs in the DDI clusters, compared to drugs not within the clusters. The measure of inter-cluster distance which shows the average points between the different clusters is higher at 2.46. This suggests that the clusters are far apart from one another. Additionally, Fig. 6 shows that the inter-cluster distance being greater than intra-cluster distance implies that the clusters are well-defined and do not overlap with each other. Therefore, this indicates that Louvain clustering generates well separated clusters.

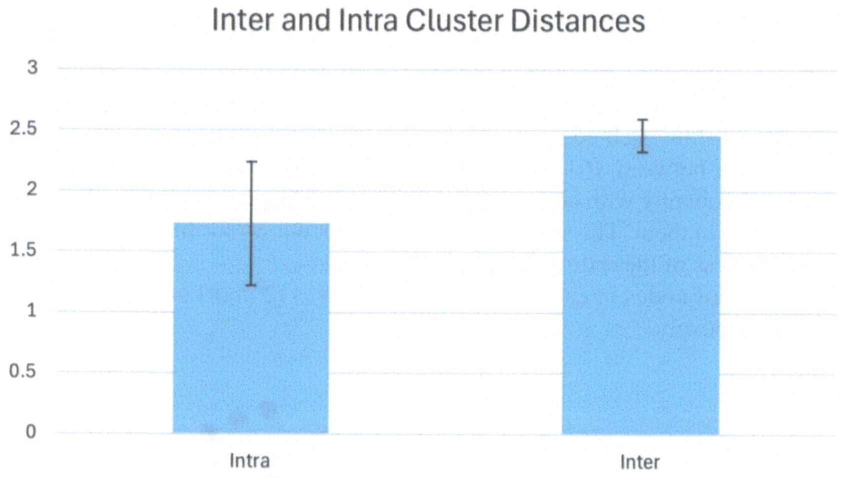

Fig. 6. Bar chart of average inter- and intra- cluster distances

4.4 Identifying Categories from Each Cluster

From the sampled drugbank dataset, the categories that the drugs being investigated belonged to were extracted and compared with other drugs in the same clusters. Table 1 shows the most common categories that the drugs were identified with in each cluster.

Table 1. Table of Categories Matched to the Drugs

Cluster 0		Cluster 1		Cluster 2		Cluster	3
Category	Freq	Category	Freq	Category	Freq	Category	Freq
CNS depressants	48	Cytochrome P-450 Substrates	62	Anti-Infective Agents	40	Antibodies	97
Neurotransmitter Agents	30	Cytochrome P-450 CYP3A Substrates	53	Anti-Bacterial Agents	28	Peptides	62
Nervous System	30	CytochromeP-450 CYP3A4 Substrates	52	Analgesics	18	Amino Acids	57
CNS Agents	22	Cytochrome P-450 Enzyme Inhibitors	39	Anti-Inflammatory gents	17	Monoclonal	53
Antidepressive Agents	21	Cytochrome P-450 CYP3A Inhibitors	23	12 betaLactams	12	Immunoglobulins	46

Cluster 0 shows drugs that target the central nervous system, influencing brain activities and causing changes in behaviour, mood, or physiological processes. Most of these drugs are not fully distinct in their mechanism of action, some fitting into more than one category, such as antidepressants, which also act as CNS agents. Because of their shared targets in the CNS and similar pathways, they may have interactions among themselves that give rise to effects [5].

Cluster 1 share in common their interaction with the cytochrome P450 enzyme system. The drugs either rely on CYP3A enzymes in their metabolism or interfere with the activity of these enzymes. They can interact with multiple drugs that compete for the same enzyme, thus affecting the rate of drug metabolism. CYP3A inhibitors inhibit the activity of these enzymes, leading to the accumulation of substrates and increased risk for side effects or toxicity. However, enzyme inducers accelerate metabolism and reduce drug effectiveness [6].

Cluster 2 drugs often target and eliminate foreign bodies such as bacteria, viruses, fungi, and parasites that cause infections in the body. These drugs may interact because they often share common metabolic pathways, or can affect similar physiological processes like immune response, inflammation, and pain signalling [7].

Cluster 3 drugs are proteins and function through the different levels of their protein structures. These molecules may interact because they often target or influence similar protein structures, cell division & cell death pathways, or receptors in the immune system, potentially affecting each other's effectiveness, metabolism, or immune response.

This is similar to the cluster analysed using Cytoscape [8].

5 Conclusion

This study explores the use of unsupervised machine learning approach for clustering known DDIs. For a dataset of more than 1 million DDIs, we identified 4 clusters using Louvain algorithm. The inter- and intra-cluster distances of the communities were evaluated, suggesting the feasibility of the approach in determining DDI network clusters. Moreover, several categories that the drugs in each cluster were determined. It was found that **drugs that tend to interact often target the same type of molecules, part of the body or are formed from similar materials or structures.** These interactions may affect the drug's efficacy and safety in a patient if not monitored closely.

However, this dataset is small, and more work will have to be done to predict DDIs more accurately. Expanding the dataset or taking into consideration patient's clinical, pharmacokinetic and pharmacodynamic data would allow better prediction of DDIs in the future.

References

1. Wishart, D.S., et al.: DrugBank 6.0: the DrugBank Knowledgebase for 2024. Nucleic Acids Research **52**(D1), D1265–D1272 (2024)
2. Shannon, P., et al.: Cytoscape: a software environment for integrated models of biomolecular interaction networks. Genome Res. **13**(11), 2498–504 (2003); Enright, A.J., Van Dongen, S., Ouzounis, C.A.: An efficient algorithm for large-scale detection of protein families. Nucleic Acids Research **30**(7), 1575–1584 (2002)

3. Hagberg, A., Schult, A., Swart, J.: Exploring Network Structure, Dynamics, and Function using NetworkX. Proceedings of the 7th Python in Science Conference (SciPy2008) (2008)
4. Blondel, D., Guillaume, J., Lambiotte, R., Lefebvre, E.: Fast unfolding of communities in large networks. Cornell University (2008)
5. Low, Y., Setia, S., Lima, G.: Drug–drug interactions involving antidepressants. Neuropsychiatry Dis Treat **14**, 567–580 (2018)
6. Ogu, C., Maxa, J.: Drug interactions due to cytochrome P450. Proc (Bayl Univ Med Cent) **13**(4), 421–423 (2000)
7. Radkowski, P., et al.: Antibiotic– Drug Interactions in the Intensive Care Unit. Antibiotics (Basel) **13**(6), 503 (2024)
8. Kuzmanov, U., Emili, A.: Protein-protein interaction networks: probing disease mechanisms using model systems. Genome Med. **5**, 37 (2013)

Dinoflagellate-Based Biohybrid Mechanolumiscent Transducer for Photonic Communication

Ashlynn Lim Si Han and Ng Zhi Xuan(✉)

Hwa Chong Institution, Singapore, Singapore
ngzhixuann@gmail.com

Abstract. This study explores the development of a novel biohybrid mechanoluminescent device utilising marine dinoflagellates, *Pyrocystis Fusiformis*. The integration of living organisms with synthetic materials in biohybrid systems offers transformative potential for sustainable applications in sensing, monitoring, and communication. Mechanoluminescence (ML), the generation of light triggered by mechanical stress, is particularly suited for low-energy, high-sensitivity biohybrid systems. However, challenges such as material brittleness, environmental sensitivity, and complex fabrication processes limit the scalability and functionality of practical applications of such devices. To tackle these limitations, this study focuses on the feasibility of incorporating dinoflagellates, which exhibit natural mechanosensing capabilities and bioluminescent responses to minimal mechanical stimuli, into biohybrid devices. By leveraging these organisms' natural bioluminescence, this research delves into sustainable alternatives to conventional photonic communication devices. Two distinct fabrication approaches were investigated: 1) Alginate hydrogel composites and 2) Polydimethylsiloxane (PDMS)-encapsulated chambers. Both systems were evaluated for mechanical durability, transparency, bioluminescence, and living viability. The alginate hydrogels, crosslinked with Ca^{2+} ions and reinforced with xanthan gum and elastomer coatings, provided improved structural flexibility and enhanced bioluminescence uniformity. Though challenges remain in maintaining bioluminescent intensity and addressing the manual fabrication processes, the results demonstrated promising potential for versatile, ultra-sensitive applications of the living composites. This study offers a pathway toward sustainable, accessible, and scalable solutions for high-sensitivity mechanosensing and low-energy photonic communication systems through the novel approach of dinoflagellate biohybrid devices.

This project report utilised OpenAI's ChatGPT as a tool for idea drafting text, refining language, and researching common polymer composites used in current ML systems. The AI's contributions were limited to providing conceptual insights and enhancing clarity. All content was reviewed and verified by the authors to ensure accuracy and relevance.

H. Guo et al. (Eds.): IRC-SET 2025, PTT, pp. 201–213, 2026.
https://doi.org/10.1007/978-981-95-9630-0_18

1 Background

In recent times, research has increasingly explored biohybrid strategies, combining living organisms with synthetic materials to harness their natural capabilities. For instance, cell communication, heavy metal sensing [1], and medically-applicable composites that respond to biochemical cues or light [2]. An application of biohybrids is mechanosensing, a fundamental biological phenomenon where organisms detect mechanical stimuli which trigger physiological or functional responses.

Mechanoluminescence (ML), a specific type of mechanosensing, is the emission of light from materials subjected to mechanical forces such as bending, stretching, or compressiont [3]. Mechanical energy converts into visible light through internal structural or chemical changest [4], offering unique advantages for deployment where access to electrical power is limited or unavailable. Dinoflagellates, specifically *Pyrocystis Fusiformis*, embody the ideal biological component for biohybrid systems. These single-celled marine phytoplanktons are known for their distinctive MLt [5], caused by flow or touch, involving a complex intracellular signaling pathway. Their bioluminescence is not merely an aesthetic attribute but a functional response to mechanical agitation, which can emit light with a peak wavelength of near 490 nmt [6]. Unlike inorganic materials, dinoflagellates respond to minimal stress, activating bioluminescence with forces as low as a few Pascals for ultra-sensitive and near-instantaneous reactionst [7]. These characteristics, coupled with their tolerance for encapsulated conditionst [8], as well as high-speed reaction time to stress of 15–20 mst [9], positions them as prime candidates for creating ML biohybrid devices.

Photonic transducers, such as LEDs and laser diodes, convert different forms of energy into optical signals for communication. However, they typically rely on electricity to function, which is not applicable in certain settings due to the dependence on non-renewable energy sources and cumbersome power setups.

As such, ML systems provide a suitable alternative. However, despite advancements in ML technologies, existing systems relying on inorganic solids or polymer composites face significant limitations. Common ML materials, such as $SrAl_2O_4{:}Eu^{2+}$ and $ZnS{:}Cu/Mn^{2+}$, are rigid and brittle, making them incompatible with soft, flexible applications. Photonic communicationt [11] transmits information through light, offering a highly efficient and rapid medium for data exchange, and synthetic photonic devicest [12] include lasers and LEDs. While effective, these systems consume substantial energy and depend on non-renewable materialst [13]. These limitations hinder the scalability, sustainability, and adaptability of current technologies. To address these challenges, this study explores the integration of bioluminescent dinoflagellates into biohybrid ML devices, leveraging their natural mechanoluminescence to create high-sensitivity, energy-efficient mechanical sensors. This approach aims to develop sustainable, low-energy alternatives for real-time mechanical sensing and photonic communication, offering an innovative solution by integrating dinoflagellates into photonic networks as natural transducers.

By advancing the field of biohybrid technologies, this research seeks to open new avenues for applications in resource-limited environments, flexible electronics, and environmentally sustainable designs. The hypothesis of this study is that the incorporation of dinoflagellates in alginate-based hydrogel composites will prove feasible for creating biohybrid ML devices with potential applications in photonic communication, while ensuring sufficient durability, flexibility, and sustained bioluminescent response.

2 Hypothesis

The incorporation of dinoflagellates in alginate-based hydrogel composites will prove feasible for creating biohybrid ML devices with potential applications in photonic communication, while ensuring sufficient durability, flexibility, and sustained bioluminescent response.

3 Methodology

3.1 Culturing of Dinoflagellates

The dinoflagellates were set up as seen in Fig. 1, cultured in artificial seawater enriched with Guillard's F/2 medium which provided a nutrient-rich environment for cellular viability. All apparatus were sterilised using an autoclave to prevent contamination. A 12:12h light-dark cycle was maintained to simulate natural diurnal patterns and regulate its circadian cycle. Tests and observations were all conducted during the dark phase, where peak bioluminescence is achievedt [14]. Feeding protocols were performed routinely to balance cell density and nutrient availability: Maintenance feedings: addition of 7.5% culture volume fortnightly; Expansion feedings: addition of 40% culture volume every 7–10 days. Observations of cultures, included monitoring their bioluminescent intensity during the dark phase of the photoperiod, as seen in Fig. 2, and noting changes in coloration for indicators of cellular health. Artificial seawater [15] was formulated by dissolving 14.16 g, 2.74 g, 3.69 g, 0.555 g, 0.385 g and 0.1 g of $NaCl$, $MgCl_2{\cdot}6H_2O$, $MgSO_4{\cdot}7H_2O$, $CaCl_2{\cdot}2H_2O$, KCl, and $NaHCO_3$ respectively in 500 mL of distilled water. pH of the solution was adjusted to pH8 using 0.01mL HCl/ NaOH to ensure suitability for dinoflagellate survival. The resulting solution of 35% ppt artificial seawater was finally sterilised by autoclaving.

Fig. 1. Culture Set-up in closed controlled environment

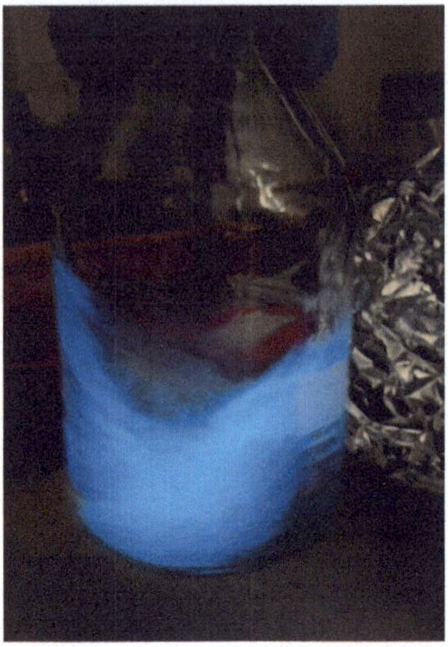

Fig. 2. Dinoflagellate Culture in Dark Phase of 12:12h Light Cycle

3.2 Dinoflagellate-alginate Hydrogel Composite Fabrication

5%, 6%, 7% wt concentration of alginate solution were prepared using alginate powder, recognised for its biocompatibility with cell culturest [16]. Selected samples of alginate precursor were gradually introduced into 5% wt of xanthan gumt [17] to enhance structural integrity while maintaining non-toxicity. Dinoflagellate culture was then gradually mixed in the hydrogel in 1:1 weight ratio under slow, continuous stirring and settled to homogenise for 2h. With reference to Fig. 3, 1 mL of the mixture was then extruded using 3-mL syringes to form roughly spherical shapes on petri dishes. All samples of dinoflagellate-alginate composites were immersed in 5% $CaCl_2$ solution for 30min to facilitate a crosslinking processt [18] between Ca^{2+} ions and alginate chains, fully transforming the solution into a stable hydrogel matrix as seen in Fig. 4a and b.

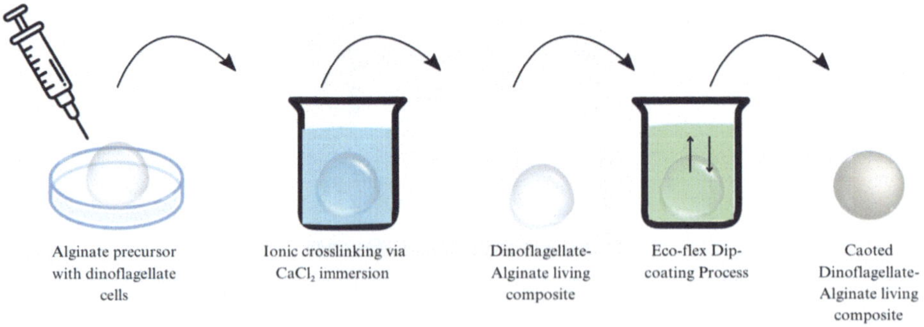

Fig. 3. Fabrication process of dinoflagellate-alginate hydrogel composite

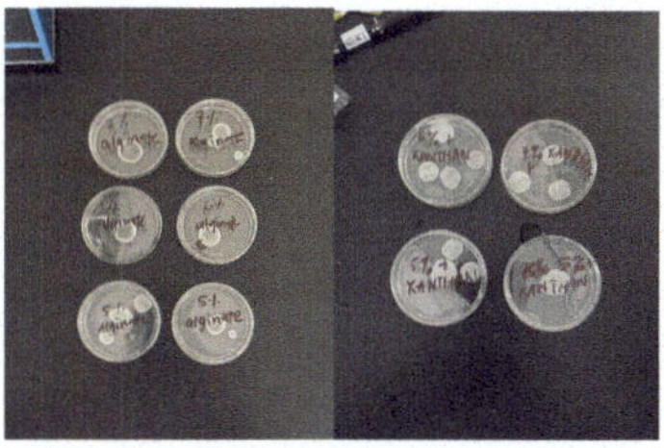

Fig. 4. Dinoflagellate-Alginate Hydrogel Composite: a) 5–7 wt% alginate samples, b) 5–6 wt% with added Xanthan gum

3.3 Ecoflex Elastomer Coating of Hydrogel Composite

To enhance mechanical resilience, an elastomer layer of Smooth-on Ecoflex coating was applied to the samples. Part A and Part B of Ecoflex 00–30 were mixed in a 1:1 wt ratio. The composites were attached to a steel wire and dip-coated in the Ecoflex precursor three times to ensure full coverage and uniformity in structure. Each layer was allowed to cure for 20min under ambient conditions. After air-drying was completed, as seen in Fig. 5, the cured samples were stored in sealed Petri dishes, immersed in nutrient-filled artificial seawater, to prevent water evaporation and maintain cellular viability. To sustain the bioluminescent capabilities of the dinoflagellates, the samples were sustained under their original routine of 12:12h light-dark cycle.

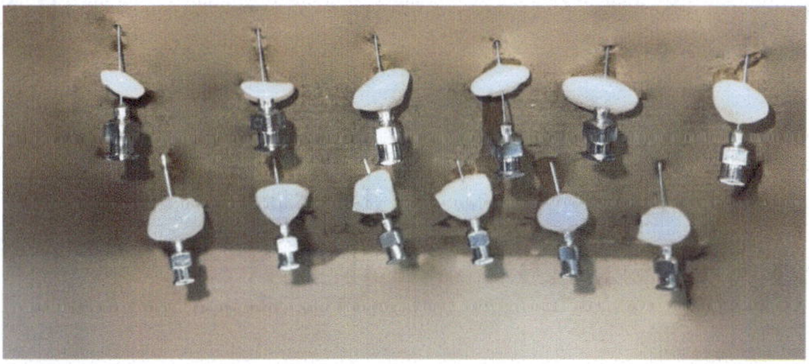

Fig. 5. Curing of elastomer coating of dinoflagellate-alginate hydrogel composites

3.4 Polydimethylsiloxane (PDMS)-encapsulated Chamber Fabrication

To fabricate the biohybrid ML devices, a mold was first designed using SolidWorks software and 3D printed. The dimensions and of molds utilised are specified in Fig. 6. A polydimethylsiloxane (PDMS) precursor was prepared using the base and curing agent of Sylgard 184 (Dow Corning) in 10:1 weight ratio. Referencing Fig. 7, the prepared precursor mixture was mixed and degassed in the Thinky centrifugal mixer for 2 min, then poured into the mold and allowed to cure at 50 °C for 3h. After curing, the structure was bonded to the sealing layers using PDMS precursor as an adhesive, then allowed to cure at 50 °C for 3h. The dinoflagellate culture solution was prepared at the end of its light phase. Dinoflagellate culture was then injected into the PDMS chamber through the pre-punched holes using a 5 mm syringe and the holes were sealed using an Ecoflex precursor prepared in 1:1 ratio. The completed device, as seen in Fig. 8, was then allowed to cure at room temperature, optimal for survival of the dinoflagellates. The biohybrid devices operated as closed systems without refreshing the culture medium. Experimentation occurred at least 4h after the dark phase began to ensure optimal bioluminescent activity.

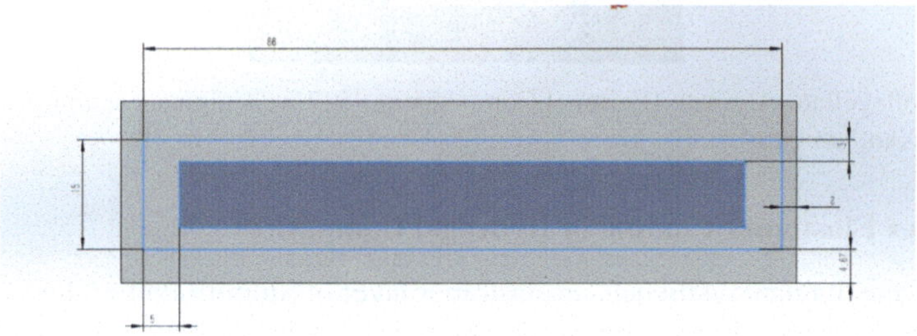

Fig. 6. Dimensions of PDMS Casing

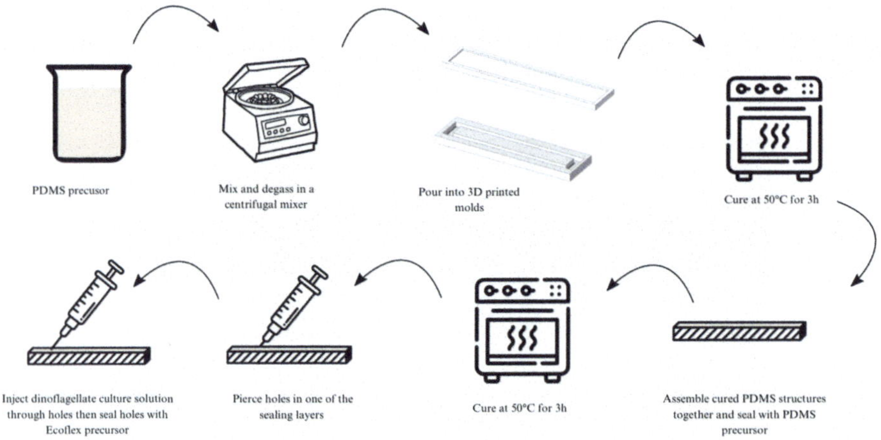

Fig. 7. Fabrication process of Alginate-PDMS device

Fig. 8. Blue-dyed dinoflagellate culture encapsulated within completed PDMS chamber

4 Results and Discussion

1) Alginate hydrogel composite

Table 1. Effect of Alginate Concentration on Survival Duration of Encapsulated Dinoflagellates

Alginate conc./ %	Observations	Survival duration /weeks	Survivability
2	Low structural integrity; prone to degradation.	1	Insufficient matrix stability; rapid nutrient loss.
3	Slightly improved stability; moderate nutrient diffusion.	2	Marginally better survival but still suboptimal.
4	Balanced structure; acceptable nutrient flow.	2	Marginally better survival but still suboptimal.
5	Stable matrix; consistent luminescence observed.	3	Reliable performance but slightly reduced longevity.
6	Optimal balance of stability and nutrient flow.	4	Longest survival; excellent luminescent consistency
7	Overly dense matrix; limited nutrient diffusion	2	Reduced survival due to restricted nutrient availability.

4.1 Environmental Stability Dinoflagellate Survival and Encapsulation Efficiency

Encapsulated dinoflagellates were stored in artificial seawater under a 12:12h light-dark cycle. The survivability of dinoflagellates encapsulated in alginate hydrogel composite was evaluated and shown to be viable and luminous for at least 30 days. These findings suggest that the encapsulation system offers promising breathability and nutrient diffusion for functionality in medium to long-term applications, especially in controlled environments. As shown in Table 1, alginate concentration for dinoflagellate encapsulation was compared between composites of 2–7%. Optimal concentration identified at 6% which supported the survival of dinoflagellates for a minimum of 30 days and balanced structural stability and nutrient diffusion of the composite.

4.2 Compression and Agitation Test

The devices were subjected to consistent mechanical agitation of a laboratory orbital shaker, simulating oscillatory mechanical disturbances. Bioluminescent intensity, upon mechanical stimulation, of respective alginate composites corresponded with the magnitude of agitation, highlighting its sensitivity to real-time indicators of environmental or mechanical disturbances. With reference to Fig. 9a and b, alginate composites lit up intensely when subjected to compression tests.

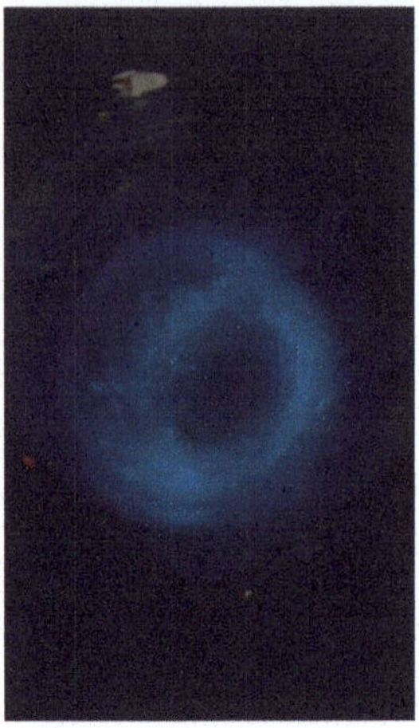

Fig. 9. Compression test results of dinoflagellate-alginate hydrogel composites

4.3 Transparency and Optical Clarity Test

The transparency of the prototypes was evaluated by directing a collimated light source through the samples and measuring the transmitted intensity using a lux meter. Transmittance (T) was calculated using the formula $T = (I / I_0) \times 100$, where I_0 = initial light intensity and I = transmitted intensity. A visual comparison grid provided a complementary qualitative assessment of transparency. Using predictive models and the *Beer-Lambert Law*, alginate-based hydrogels exhibited moderate transmittance of ~ 47%, attributed to increased light scattering within the hydrogel matrix. Despite relatively lower transparency, light diffusion within the alginate matrix creates a more uniform luminescence, reducing localised glare and enabling wide-angle viewing.

4.4 Durability Assessment

Durability was tested via repeated mechanical deformation. In contrast, alginate-based prototypes ruptured after only 5 compression cycles, with strain tolerance at just 4% and a low recovery rate of 25%, highlighting their weak mechanical strength. Failure to endure repeated deformation led to the incorporation of xanthan gum and Ecoflex coatings to enhance the durability.

4.5 Effect of Xanthan Gum Addition

Derived from *Xanthomonas Campestris* through a pure-culture fermentation process and purified with isopropyl alcohol, xanthan gum is a natural polysaccharide with high viscosity and shear-thinning propertiest [19]. It was utilised in this study due to its high biocompatibility and ability to increase the viscosity. Xanthan gum was added at 5% wt concentration to the alginate composites, and showed to increase deformation resistance and reduce likelihood of fracture in compression tests. Composites with added xanthan gum maintained their structural integrity for 50 compression cycles, whereas those without ruptured after 20 cycles. The performance improvement, measured by the fracture resistance ratio (FRR), is the number of cycles before visible degradation. This improvement validates the strengthening effect of xanthan gum, reinforcing the hydrogel matrix by increasing the viscosity of the precursor. Xanthan gum likely facilitates more homogenous ionic crosslinking during $CaCl_2$ immersion, acting as a stabiliser to the hydrogel by reducing excessive water loss, thereby maintaining its elasticity over time.

4.6 Effect of Alginate Concentration on Durability

Alginate concentration was a key factor affecting the mechanical properties of the hydrogel matrix. Concentrations between 2–5 wt% resulted in weak crosslinking density, leading to poor structural integrity and shape retention under compression. In contrast, higher concentrations of 6–7 wt% produced denser and more durable matrices, exhibiting superior durability while maintaining flexibility. This was likely due to strong ionic crosslinking with Ca^{2+} ions which gave rise to a more tightly bound network thus resisting deformation to a greater extent. Although the composites with 6 and 7 wt% alginate displayed similar durability, 7 wt% was significantly stiffer, limiting flexibility and reducing uniform stress distribution. Thus, 6% alginate concentration was identified to have optimal balance between durability and flexibility.

4.7 Elastomer Coating for Durability

To improve the long-term durability of the alginate hydrogel without compromising the flexibility or bioluminescent visibility, a silicone-based coating (Smooth-On Ecoflex 00–30) was applied due to the strength, flexibility (up to 1000% elongation), transparency, and biocompatibility. In addition, the Ecoflex coating is able to cure at room temperature, which is another crucial factor as high curing temperatures of other polymer materials could produce unsuitable living environments [20]. Initial coating attempts resulting in fractured surfaces under compression. After increasing the number of coating layers, it was found the optimal number of times to perform the dip-coating process was 3. It provided uniform coverage, forming a thin, flexible layer that adhered well to the hydrogel surface, maintained transparency for light transmission essential for bioluminescence viewing, and improved structural integrity.

4.8 Crosslinking Efficiency

Table 2. Effect of Crosslinking Duration on Compression Performance of Alginate-Based Composites

Crosslinking duration /min	Observation	Suitability
5	Brittle; easily fractured under mild stress.	Insufficient crosslinking; poor structural integrity.
10	Improved elasticity but still prone to cracking under moderate stress.	Calcium ion diffusion is incomplete for structural support.
15	Moderately brittle; minor fractures observed under repeated stress	Crosslinking insufficient for sustained use.
20	Noticeable improvement in flexibility; some surface cracks.	Near-complete crosslinking for smaller samples.
30	Optimal flexibility and mechanical stability; no fractures observed.	Dense hydrogel matrix; optimal for larger designs.
40	Slightly stiffer texture with no fractures; stable under heavy stress.	Over-crosslinking; unnecessary stiffness observed.

As illustrated in Table 2, the mechanical stability of the alginate-based composites was heavily influenced by crosslinking time. Initial experiments with shorter cross linking periods resulted in brittle composites which fractured under mild stress. This indicated insufficient penetration of calcium ions (Ca2 +) into the alginate matrix, leading to incomplete crosslinking. When crosslinking time was extended to 30min, improved results were obtained. The longer exposure allowed thorough diffusion of Ca^{2+} ions, resulting in a dense and mechanically sound hydrogel network capable of withstanding compression and repeated handling. These results show that crosslinking time is a critical factor, especially important to note for larger, more complex designs.

4.9 Geometries of Dinoflagellate-Alginate Hydrogel Composite

It is able to construct various shapes using dinoflagellate-alginate hydrogel composites, each with distinct advantages and disadvantages. The smooth, symmetrical geometry is beneficial to distribute mechanical stresses more uniformly across the composite, providing consistent activation of bioluminescent response from dinoflagellates. Its absence of corners minimised weak points prone to failure under repeated loading. Hand-drawn patterns created with syringes allowed more intricate designs to be produced, but expectedly suffered from irregularities in shape and inconsistencies in bioluminescent response as shown in Fig. 10a and 10b. This variability underscores the need for greater control during manual fabrication, possibly through semi-automated methods to preserve precision. The alginate composites hold exciting potential for 3D printing due to its tunable concentrations and viscosity. These properties could be adjusted to accommodate a wide range of printer specifications, enabling the creation of more complex shapes and structures.

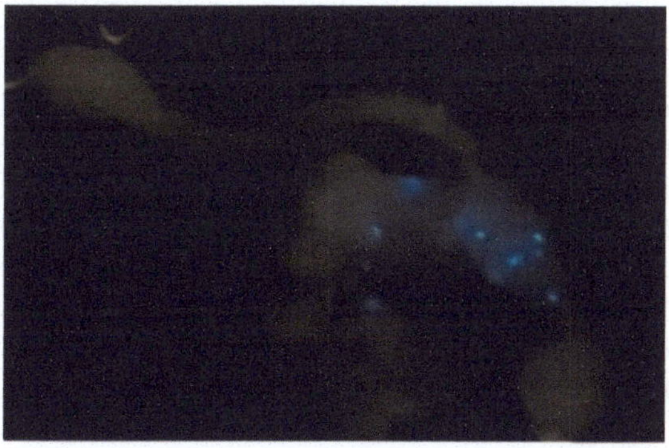

Fig. 10a. Hand drawn patterned dinoflagellate-alginate hydrogel under mechanical stress

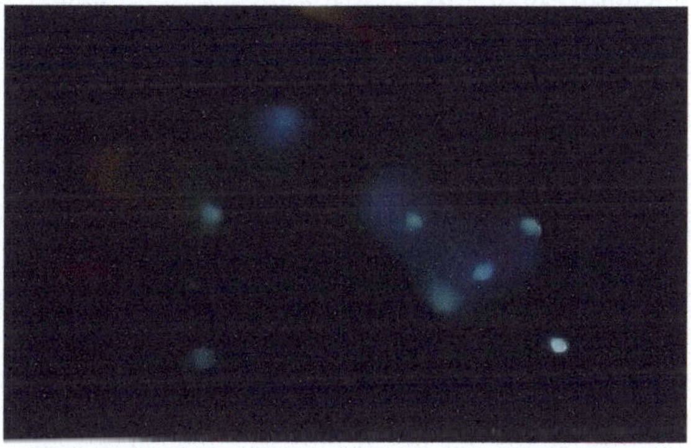

Fig. 10b. Hand drawn patterned dinoflagellate-alginate hydrogel under mechanical stress

2) Polydimethylsiloxane (PDMS)-encapsulated chambers

Initially, PDMS was tested as the encapsulation material, but proved ineffective. This prompted alginate hydrogel as the alternative encapsulation method. High gas permeability of PDMS for gas exchanget [21] supports photosynthesis and nutrient exchange, suggesting its potential to support dinoflagellate survival and functionality under optimised conditions in PDMS chambers. Transparency and optical clarity test estimated PDMS transmittance to be ~ 86%, reflecting high optical clarity and minimal light scattering. In durability tests, PDMS prototypes demonstrated high resilience, with a strain tolerance of 12%, strain recovery rate of 98.6%, and cumulative deformation of 0.5 mm which is within acceptable limits. It remained intact despite stretching, twisting, and bending, indicating high structural integrity. Failure of PDMS may be due to fabrication errors, such as incomplete sealing or improper injection causing air bubbles or contamination, and inadequate control of moisture or nutrient levels.

5 Conclusion

This study establishes the foundational potential of integrating dinoflagellates into ML composites, with results demonstrating feasibility of dinoflagellates encapsulated in alginate-based hydrogel composites and promising bioluminescent response to mechanical stimuli. By optimising composites with 6% wt concentration of alginate, 5% wt concentration of xanthan gum, and Ecoflex elastomer coating, remarkable flexibility, durability, and adaptability of the novel biohybrid device was achieved. Notably, the encapsulated dinoflagellates remained viable for up to 30 days, demonstrating its ability to sustain a living culture over extended periods.

6 Limitations

The bioluminescence of the dinoflagellates was weaker than anticipated, limiting the sensors' immediate applicability in low-light environments. Additionally, the manual nature of certain fabrication steps, such as hand-drawing patterns and dip-coating, introduced variability that hinder scalability and consistency across larger prototypes.

7 Implications/Further Research

The incorporation of alginate hydrogels into biohybrid ML devices shows promise in creating flexible, durable, and adaptable sensors. The tunability of alginate formulations to vary the viscosity allows for 3D printing of more complex designs, such as micro-pillars or lattice structures to tailor to various applications. The ability to create customisable, scalable biohybrids entails applications in areas such as environmental monitoring, structural integrity, and real-time biomedical monitoring. Furthermore, they serve as self-regulating systems, with their biodegradable nature making them more sustainable alternatives to synthetic, non-degradable materials. Advanced spectrophotometric analysis could offer detailed optical clarity characterization and precise optimization for specific applications. Testing under extreme conditions, such as varied pH and temperature, would explore greater robustness and reliability. Future iterations of transparency and clarity experiments should quantify transmission efficiency and assess wavelength-specific clarity to provide a more comprehensive evaluation of optical performance.

References

1. Tang, T., et al.: Hydrogel-based biocontainment of bacteria for continuous sensing and computation. Nature Chemical Biology **17**(6), 724–731 (2021) .https://doi.org/10.1038/s41589-021-00779-6
2. Rivera-Tarazona, L.K., Bhat, V.D., Kim, H., Campbell, Z.T., Ware, T.H.: Shape-morphing living composites. Science Advances **6**(3) (2020). https://doi.org/10.1126/sciadv.aax8582
3. Sun, Q., et al.: Intense and sensitive green mechanoluminescence by Tb3+ doping in Y3GaO6 (2024). https://doi.org/10.1016/j.ceramint.2024.08.289

4. Feng, A., Smet, P.F.: A review of Mechanoluminescence in Inorganic solids: Compounds, mechanisms, models and applications. Materials **11**(4), 484 (2018). https://doi.org/10.3390/ma11040484
5. Valiadi, M., Iglesias-Rodriguez, D.: Understanding bioluminescence in Dinoflagellates—How far have we come? Microorganisms **1**(1), 3–25 (2013). https://doi.org/10.3390/microorganisms1010003
6. Hackett, J.D., Anderson, D.M., Erdner, D.L., Bhattacharya, D.: Dinoflagellates: a remarkable evolutionary experiment. Am. J. Bot. **91**(10), 1523–1534 (2004). https://doi.org/10.3732/ajb.91.10.1523
7. Bioluminescence response of four species of dinoflagellates to fully developed pipe flow. J. Plankton Res. **26**(12) (2004)
8. Da Silva, A.F., Moreira, A.F., Miguel, S.P., Coutinho, P.: Recent advances in microalgae encapsulation techniques for biomedical applications. Adv. Coll. Interface. Sci. **333**, 103297 (2024). https://doi.org/10.1016/j.cis.2024.103297
9. Maldonado, E.M., Latz, M.I.: Shear-Stress dependence of dinoflagellate bioluminescence. Biol. Bull. **212**(3), 242–249 (2007). https://doi.org/10.2307/25066606
10. Xu, C., Watanabe, T., Akiyama, M., Zheng, X.: Direct view of stress distribution in solid by mechanoluminescence. Appl. Phys. Lett. **74**(17), 2414–2416 (1999). https://doi.org/10.1063/1.123865
11. K, A.D.: photonics and optical communication; Illuminating the future of connectivity (2023). https://www.linkedin.com/pulse/photonics-optical-communicationilluminating-future-connectivity-k-rrlpc/
12. Zhou, X., Yi, D., Chan, D.W.U., Tsang, H.K.: Silicon photonics for high-speed communications and photonic signal processing. Deleted Journal **1**(1) (2024). https://doi.org/10.1038/s44310-024-00024-7
13. AzoOptics: Recent Advances in Sustainable Photonics Materials (2022). https://www.azooptics.com/Article.aspx?ArticleID=2189
14. Hastings, J.: Circadian rhythms in dinoflagellates: What is the purpose of synthesis and destruction of proteins? Microorganisms **1**(1), 26–32 (2013). https://doi.org/10.3390/microorganisms1010026
15. Nguyen, T.V.: Preparation of Artificial Sea Water (ASW) for Culturing Marine Bacteria (2018). https://www.researchgate.net/publication/323971616_Preparation_of_Artificial_Sea_Water_ASW_for_Culturing_Marine_Bacteria
16. Augst, A.D., Kong, H.J., Mooney, D.J.: Alginate hydrogels as biomaterials. Macromol. Biosci. **6**(8), 623–633 (2006). https://doi.org/10.1002/mabi.200600069
17. Piola, B., Sabbatini, M., Gino, S., Invernizzi, M., Renò, F.: 3D bioprinting of Gelatin-Xanthan gum composite hydrogels for growth of human skin cells. Int. J. Mol. Sci. **23**(1), 539 (2022). https://doi.org/10.3390/ijms23010539
18. Ionic crosslinking - Labster. (n.d.). https://theory.labster.com/ionic-crosslinking/
19. PubChem: Xanthan Gum. PubChem (n.d.). https://pubchem.ncbi.nlm.nih.gov/compound/Xanthan-Gum#:~:text=Xanthan%20gum%20is%20a%20polysaccharide,by%20recovery%20with%20isopropyl%20alcohol
20. Smooth-On, Inc.: EcoFlexTM Series, Super-Soft, Addition Cure Silicone Rubbers (n.d.). https://www.smooth-on.com/product-line/ecoflex/
21. Pure- and mixed-gas permeation properties of polydimethylsiloxane for hydrocarbon/methane and hydrocarbon/hydrogen separation In: Sciencedirect (2004). https://www.sciencedirect.com/science/article/abs/pii/S0376738804005216
22. OpenAI: ChatGPT (January 15 version) [Large language model] (2025). https://openai.com/chatgpt

In Silico Investigation of Structure-Activity Relationships in Thiophene-Containing BODIPY Photocages

Ngiam Lihong(✉)

NUS High School of Mathematics and Science, Singapore, Singapore
h2010097@nushigh.edu.sg, gnohil88@gmail.com

Abstract. As the development of small molecule drugs continues at rapid pace, off-target side effects and toxicities have emerged as major causes for concern. To this end, a range of photoliable protecting groups (PPGs) have been proposed as possible strategies to mitigate such side effects through selective uncaging only at affected areas. However, many photocages suffer from poor reactivity, toxicity and the need for high-energy ultraviolet radiation for uncaging. In this study, comprehensive structure-activity relationships of an emerging PPG, thiophene-containing boron-dipyrromethene (BODIPY), have been investigated via ab initio Density Functional Theory (DFT) calculations. Singlet-triplet gaps and Spin Orbit Coupling Matrix Elements (SOCME) were calculated for all tested compounds, and the El-Sayed's rule was employed to rationalise differences in singlet-triplet intersystem crossing (ISC) rates between compounds. Through these analyses, we propose optimal thiophene-containing BODIPY structures that exhibit good absorption at longer wavelengths and promising intersystem crossing rates, findings that will certainly aid in the optimisation of future photodynamic therapies.

Keywords: Photodynamic therapy · BODIPY · DFT · PPG · Intersystem crossing

1 Introduction

Off-target toxicity is a major challenge plaguing the advancement of today's modern medicine, with the potential ill side effects of various drugs limiting their applications. To this end, photoliable protecting groups (PPGs) have garnered significant attention in recent years as viable cages that protect drug molecules from unwanted off-target reactivities [7]. Uncaging simply involves UV/Vis/NIR irradiation of the affected area (e.g., tumour location) to trigger drug photorelease and achieveing its desired effect.

One such photocage at the forefront of current research is boron-dipyrromethene (BODIPY) due to its multitude of advantages, including deeper

H. Guo et al. (Eds.): IRC-SET 2025, PTT, pp. 214–221, 2026.
https://doi.org/10.1007/978-981-95-9630-0_19

tissue penetration, greater biocompatibility and relatively simple synthesis [10]. Most commonly, photorelease from BODIPY occurs at the *meso*-methyl position through a photo-S_N1 nucleophilic substition, as shown in Fig. 1(a); and the excited states and key transition steps involved in photorelease are highlighted in Fig. 1(b).

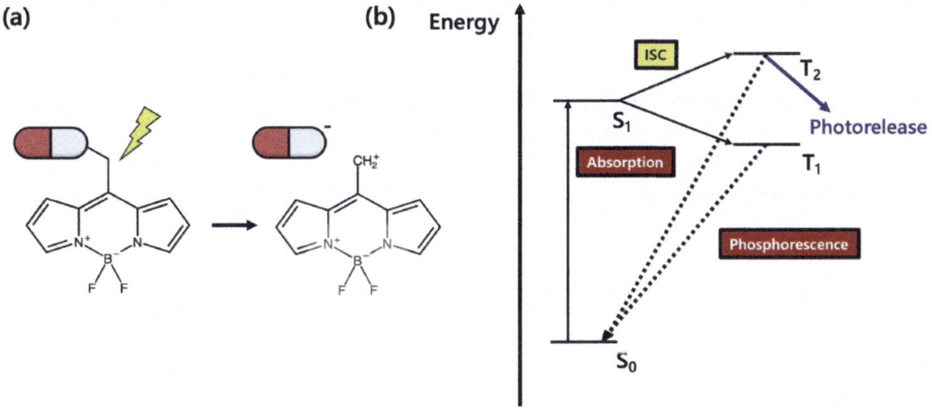

Fig. 1. (a) Mechanism of photorelease from meso-methyl substituted BODIPY photocages. (b) Jablonski Diagram illustrating the excited states and key transitions involved in photorelease.

A key step in photoreactions involves the non-radiative transition from a singlet excited state to a reactive triplet excited state, a process known as Intersystem Crossing (ISC) [3,6]. The rate of ISC is given by Fermi's Golden Rule:

$$k_{ISC} = \frac{2\pi}{\hbar}|\langle\psi_f|\hat{H}'_{SO}|\psi_i\rangle|^2\rho(E_f) \tag{1}$$

The terms $\langle\psi_f|$ and $|\psi_i\rangle$ represent the final and initial wavefunctions respectively, while $\hat{H}'_{SO}$ refers to the spin-orbit interaction Hamiltonian operator governing spin-orbit coupling (SOC). One way to increase ISC is through the well-known heavy atom effect, as investigated and applied to BODIPY in previous studies [14]. However, it has been suggested that the incorporation of heavy atoms in BODIPY (notably halogens - Cl, Br, I) significantly increase their dark-field toxicity [13].

Several groups have proposed the incorporation of thiophene rings to construct feasible heavy atom-free BODIPY for various applications such as imaging and as photosensitizers, showing promising results in preliminary experimental studies (e.g., [1,2,6,13,15,16]). However, few studies have evaluated the potential of these cages as PPGs for use in photodynamic therapy. Hence, in this study, a quantitative evaluation of the ISC and reactivities of different thiophene-substituted BODIPY photocages was conducted *in silico* using ab-initio Density Functional Theory (DFT) calculations, ultimately aiming to elucidate important structure-activity relationships in this important class of photocages.

2 Methods

The structures of all investigated compounds are shown in Fig. 2. These structures were adapted and modified from previously reported designs [1,2,6,13, 15,16] to align this study with current advancements in the field. To serve as benchmarks against which our results were compared against, the parent BODIPY without any activating groups as well as that containing 2 bromine atoms (Parent-Br) were employed. The latter was chosen as Parent-Br is well-known to exhibit high ISC rates owing to the heavy atom effect [3,14]. All DFT calculations were performed using the ORCA 5.0.4 software [12]. The ground state (S_0) and excited state (S_1, T_1, and T_2) structures of all compounds were first optimised using the B3LYP Functional [9], along with the def2-SVP basis set [18] and the def2/J auxilary basis set [17] and the RIJCOSX approxmation [5].

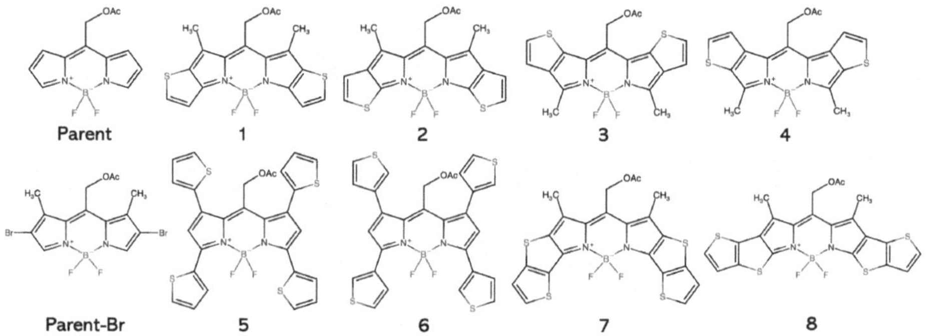

Fig. 2. Structures of all investigated compounds, including the parent BODIPY as a benchmark.

Photorelease was investigated via the release of an -OAc group from the meso position of the BODIPY ring, simulating the release of a COOH-containing drug molecule through a photo-S_N1 nucleophilic substition. To this end, a relaxed potential energy surface (PES) scans were conducted at the B3LYP/6-31+G(d) [8] level of theory along the C-O bond broken to determine the average energy of heterolytic bond cleavage at the S1, T1 and T2 excited states of compound 1.

An unrestricted natural orbital (UNO) calculation was conducted on all compounds at B3LYP/def2-SVP level of theory. All excitation energies, SOC matrix elements and UV-Vis spectra were determined at the S_0 optimised geometries using the B3LYP functional and the cc-PVTZ basis set, as correlation-consistent basis sets are well-known to be generally more accurate at simulating excited states of a wide range of chemical systems [4]. Finally, results on orbital contributions by atom to each of the frontier molecular orbitals were derived from a Löwdin population analysis [11].

3 Results

The UV-Vis spectra of all compounds are shown in Fig. 3 in comparison to that of the parent BODIPY photocage. All thiophene-containing compounds show significant bathochromic shifts in comparison to the parent BODIPY due to their expanded π-conjugation systems.

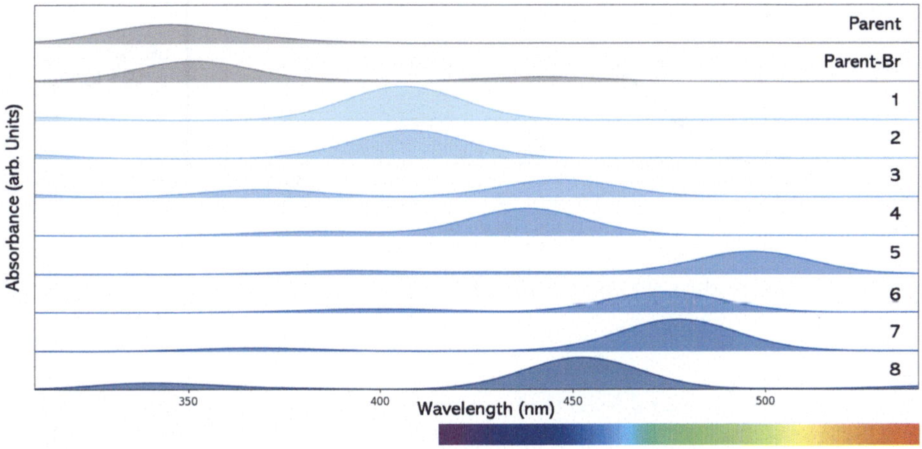

Fig. 3. UV-Vis absorption spectra of all investigated compounds calculated at the B3LYP/cc-PVTZ level of theory.

It is known that the 2 main triplet deactivation channels of the S_1 are the T_1 and T_2 states due to their accessible energy levels. Indeed, the PES scan of compound 1 confirmed that the energy needed for bond cleavage of the C-O bond for photorelease is significantly higher for the lower energy S_1 states as compared to the higher energy T_1 and T_2 states, as shown in Fig. 4. This supports chemical reactivity through the triplet deactivation pathway.

Another key factor underpinning the ISC is the difference in energy, ΔE_{ST}, between the excited singlet and triplet states, where a lower gap generally makes the triplet state more energetically accessible, enabling faster ISC rates. This is referred to as the Energy Gap Law which states that k_{ISC} decreases exponentially with increasing $E_{S}T$. The singlet-triplet energy gaps of all compounds, along with the SOCMEs of their $S_1 \rightarrow T_1$ and $S_1 \rightarrow T_2$ ISC transitions, are shown in Table 1.

Notably, the energy gap between the S_1 and T_2 states of all compounds are significantly lower than those between the S_1 and T_1 states, lending evidence for T_2 being the favoured triplet state for photodeactivation. In particular, compounds 5–6 with 4 Sulfur atoms each exhibit significantly higher $S_1 \rightarrow T_2$ SOCMEs as compared to compounds 1–4, 7 and 8, suggesting that the presence of additional thiophene groups atoms attached in the drives up ISC rates.

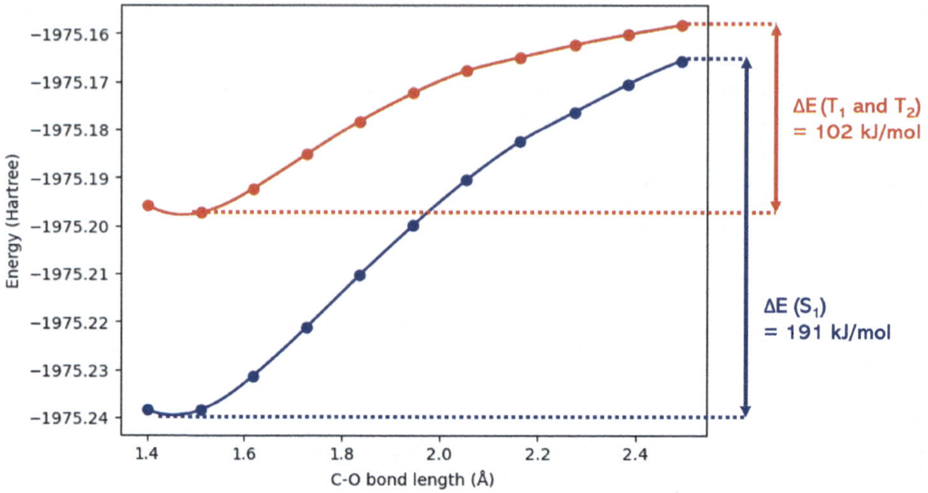

Fig. 4. PES scan of the C-O bond in BODIPY simulating the release of an -OAc group via an SN1-like pathway.

These rates were comparable to that of the Parent-Br, which is known to exhibit high ISC rates due to the well-established heavy atom effect. Thus, our findings lend *in silico* insight into thiophenes being a suitable candidate as a heavy atom replacement in BODIPY photocages.

Table 1. Singlet-triplet excited state energy gaps and SOCMEs of all computed compounds.

Compound		Parent	Parent-Br	1	2	3	4	5	6	7	8
ΔE (eV)	$S_1 \to T_1$	1.587	1.31	1.14	0.97	1.43	1.56	1.25	1.28	1.14	0.91
	$S_1 \to T_2$	0.641	0.48	0.43	0.53	0.03	0.24	0.12	0.07	0.36	0.43
SOCME (cm^{-1})	$S_1 \to T_1$	0.15	1.29	0.29	0.23	0.24	0.22	0.36	0.52	0.06	0.22
	$S_1 \to T_2$	0.95	1.07	0.01	0.10	0.15	0.11	1.63	1.55	0.03	0.03

In order to rationalize differences in ISC, we referred to El-Sayed's rule, which states that ISC transitions are only fast if there is a change in orbital type. In our case, this means that ISC via $n \to \pi^*$ pathways occurs much faster compared to those via $\pi \to \pi^*$ pathways. To investigate this in our molecules, we adapted a procedure from [3], which involves analyzing p orbital contributions from the S atoms to the frontier molecular orbitals (FMOs). The p_z orbitals lie perpendicular to the internuclear axis and are assigned as π-contributing, p_x lies along the CS bond and is σ-contributing, while p_y is non-bonding.

As shown in Fig. 5, the $S_0 \to T_1$ transitions in compounds 5 and 6 have a mostly non-bonding contribution from the lower sulfur atoms (attached with

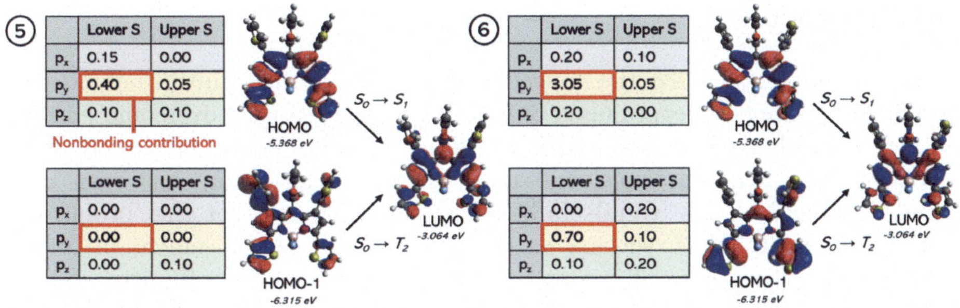

Fig. 5. The MOs involved in the T_1 and T_2 transitions of compounds 5 and 6, as well as Löwdin population analysis on compounds 5 and 6 revealing the sulfur atoms' contributions per MO.

respect to the BODIPY core as displayed in the figure). In contrast, this contribution is absent in the $S_0 \rightarrow T_2$ transition. As such, the nature of these transitions markerdly increases ISC in compounds 5 and 6 according to El-Sayed's rule.

4 Discussion

A key focus of BODIPY design is the development of analogs that absorb in the visible/near-infrared (NIR) region and mitigate the need for potentially harmful ultraviolet (UV) radiation [19]. Indeed, all of our proposed compounds are able to effectively absorb at longer wavelengths to trigger photoreactions, giving rise to improved biocompatibility. Notably, photorelease occurs preferentially via ISC through the $S_1 \rightarrow T_2$ transition across all our proposed compounds as shown by the relatively smaller singlet-triplet energy gap, in line with previous research on similar compounds [3].

The high SOCMEs exhibited by compounds 5 and 6 corroborate previous studies on similar compounds [1], suggesting that the presence of additional thiophene groups attached at the 1, 3, 5 and 7-positions of the BODIPY ring in the drives up ISC rates. However, the low SOCMEs on the other compounds investigated is surprising given their use in earlier studies. We hypothesize that this is due to the fact that all these compounds exhibit thiophene groups connected directly to the BODIPY core which alters Sulfur's contribution to the FMOs of the molecule, giving rise to spin-forbidden transitions as defined by El-Sayed's rule. As such, this study provides a foundation for the application of El-Sayed's rule to this emerging class of compounds.

Moreover, many previous studies have solely focused on using this emerging class of BODIPY as dyes and/or photosensitisers. As such, more in-depth computational (e.g. charge transfer simulations, more accurate multi-configurational self-consistent field (MCSCF) calculations) and experimental (e.g., chemical reactivity, quantum yield and cell inhibition assays) studies are needed to confirm the potential of thiophene-containing BODIPY as a viable PPG.

5 Conclusion

We have explored, *in silico*, the molecular spin-vibronic mechanisms behind several efficient thiophene-containing BODIPY photocages that overcome current limitations of photocages, including UV-induced photorelease and low reactivities. Moreover, a comprehensive structure-acivity relationship has been conducted to elucidate key trends influencing the properties of this promising, emerging class of PPGs, serving as the basis for the advancement of future photodynamic therapies. Ultimately, future research including in-vitro and in-vivo experiments is needed to further validate the effectiveness of our proposed structures.

Acknowledgments. The authors are grateful for Josef Filgas (Department of Physical Chemistry, University of Chemistry and Technology, Prague, Czech Republic) for providing valuable guidance for this project via email correspondence.

Disclosure of Interests. The authors have no competing interests to declare that are relevant to the content of this article.

References

1. Bellier, Q., Dalier, F., Jeanneau, E., Maury, O., Andraud, C.: Thiophene-substituted aza-BODIPY as a strategic synthon for the design of near-infrared dyes. New J. Chem. **36**(3), 768 (2012). https://doi.org/10.1039/c2nj20943h
2. Can, B., Çakmak, Y.: Exploration of two different strategies in near IR absorbing boron dipyrromethene derivatives for photodynamic and bioimaging purposes. ChemistrySelect **6**(33), 8855–8860 (2021). https://doi.org/10.1002/slct.202102508
3. Doležel, J., Poryvai, A., Slanina, T., Filgas, J., Slavíček, P.: Spin-vibronic coupling controls the intersystem crossing of iodine-substituted BODIPY triplet chromophores. Chem. - Eur. J. **30**(4) (2023). https://doi.org/10.1002/chem.202303154
4. Dunning, T.H.: Gaussian basis sets for use in correlated molecular calculations. I. The atoms boron through neon and hydrogen. J. Chem. Phys. **90**(2) (1989). https://doi.org/10.1063/1.456153
5. Helmich-Paris, B., De Souza, B., Neese, F., Izsák, R.: An improved chain of spheres for exchange algorithm. J. Chem. Phys. **155**(10) (2021). https://doi.org/10.1063/5.0058766
6. Ji, S., Ge, J., Escudero, D., Wang, Z., Zhao, J., Jacquemin, D.: Molecular structure–intersystem crossing relationship of heavy-atom-free BODIPY triplet photosensitizers. J. Org. Chem. **80**(11), 5958–5963 (2015). https://doi.org/10.1021/acs.joc.5b00691
7. Klan, P., et al.: Photoremovable protecting groups in chemistry and biology: reaction mechanisms and efficacy. ChemInform **44**(11) (2013). https://doi.org/10.1002/chin.201311243
8. Krishnan, R., Binkley, J.S., Seeger, R., Pople, J.: Self-consistent molecular orbital methods. xx. A basis set for correlated wave functions. J. Chem. Phys. **72**(1), 650–654 (1980). https://doi.org/10.1063/1.438955
9. Lee, C., Yang, W., Parr, R.G.: Development of the Colle-Salvetti correlation-energy formula into a functional of the electron density. Phys. Rev. B **37**(785), 785–789 (1988). https://doi.org/10.1103/PhysRevB.37.785

10. Loudet, A., Burgess, K.: BODIPY dyes and their derivatives: syntheses and spectroscopic properties. Chem. Rev. **107**(11), 4891–4932 (2007). https://doi.org/10.1021/cr078381n
11. Löwdin, P.: On the non-orthogonality problem connected with the use of atomic wave functions in the theory of molecules and crystals. J. Chem. Phys. **18**(3), 365–375 (1950). https://doi.org/10.1063/1.1747632
12. Neese, F., Wennmohs, F., Becke, U., Riplinger, C.: The ORCA quantum chemistry program package. J. Chem. Phys. **152**(22) (2020). https://doi.org/10.1063/5.0004608
13. Qi, S., Kwon, N., Yim, Y., Nguyen, V., Yoon, J.: Fine-tuning the electronic structure of heavy-atom-free BODIPY photosensitizers for fluorescence imaging and mitochondria-targeted photodynamic therapy. Chem. Sci. **11**(25), 6479–6484 (2020). https://doi.org/10.1039/d0sc01171a
14. Slanina, T., et al.: In search of the perfect photocage: structure-reactivity relationships in meso-methyl BODIPY photoremovable protecting groups. J. Am. Chem. Soc. **139**(42), 15168–15175 (2017). https://doi.org/10.1021/jacs.7b08532
15. Sun, Y., Qu, Z., Zhou, Z., Gai, L., Lu, H.: Thieno[3,2-b]thiophene fused BODIPYs: synthesis, near-infrared luminescence and photosensitive properties. Organic Biomol. Chem. **17**(14), 3617–3622 (2019). https://doi.org/10.1039/c9ob00030e
16. Wang, J., et al.: Synthesis, structure and properties of thiophene-fused BODIPYs and azaBODIPYs as near-infrared agents. New J. Chem. **40**(7), 5966–5975 (2016). https://doi.org/10.1039/c6nj01011c
17. Weigenda, F.: Accurate coulomb-fitting basis sets for H to RN. Phys. Chem. Chem. Phys. **8**(9), 1057 (2006). https://doi.org/10.1039/b515623h
18. Weigenda, F., Ahlrichsb, R.: Balanced basis sets of split valence, triple zeta valence and quadruple zeta valence quality for h to rn: Design and assessment of accuracy. Phys. Chem. Chem. Phys. **7**(18), 3297 (2005). https://doi.org/10.1007/s00214-007-0310-x
19. Xiong, H., et al.: Photo-controllable biochemistry: Exploiting the photocages in phototherapeutic window. Chem **9**(1), 29–64 (2022). https://doi.org/10.1016/j.chempr.2022.11.007

Green Synthesis of Ferromagnetic Metal-Doped TiO_2 Nanoparticles Using Spent-Coffee-Grounds for Efficient Degradation of Methylene Blue Dye

Syed Ahnaf Mahmud(✉) and Aravind Balamurugan

NUS High School of Mathematics and Science, Singapore, Singapore
h2010134@nushigh.edu.sg

Abstract. Semiconductor nanoparticles (NPs) serve as promising method to degrade organic dyes via their photocatalytic properties. However, their resource-intensive syntheses, inefficiencies under visible-light, and recovering the NPs remain challenges to viable solutions. Our study combines the benefits of green synthesis and oxygen-vacancy induced ferromagnetism to synthesize Fe-, Ni-, and Co-Doped TiO_2 NPs as effective photocatalysts for degrading Methylene Blue (MB) dye under visible-light. Green synthesis was done using Spent-Coffee-Ground extract, with varying dopant concentrations (0% mol, 2% mol, and 4% mol) and a novel Air-Nitrogen 2-step calcination was employed to produce oxygen vacancies in the NPs. The identities of the metal-doped TiO_2 NPs were confirmed using X-Ray Diffraction. Scanning Electron Microscopy analysis showed our NPs were spherical with 91 nm average diameter. Desorption of MB at the start of light irradiation, an uncommonly reported phenomenon, was observed and analysed in detail. Experimental data was fitted to a pseudo-first order model to reflect both desorption and degradation processes. We found that Ni 2% mol TiO_2 NPs possess the best degradation rate constant of 0.022 min^{-1} with a net removal of 93% of MB dye. Our work highlights the potential for magnetically recoverable metal-doped TiO_2 NPs for a scalable and green solution to critical challenges in wastewater treatment.

Keywords: Photocatalytic Degradation · Doping · Green Synthesis

1 Introduction

Water pollution from textile effluents remains a persistent global problem [1]. The dyeing process, allows 15%–20% of the dye to escape into the environment, releasing aromatic molecules which are toxic, carcinogenic and resistant to breakdown [1–4]. Titanium dioxide (TiO_2) nanoparticles have been used of photocatalytic degradation of organic dyes into non-toxic byproducts such as H_2O and CO_2 while being chemically inert and reusable [5, 6]. However, traditional synthesis of TiO_2 such as sol-gel or hydrothermal routes requires high temperatures, pressures and harmful chemicals. Hence, we have

H. Guo et al. (Eds.): IRC-SET 2025, PTT, pp. 222–233, 2026.
https://doi.org/10.1007/978-981-95-9630-0_20

employed a green synthesis method utilizing bioactive components present in spent coffee grounds (SCG) to facilitate the bio-reduction and capping processes, which is cost-effective and green. SCGs have been extensively researched for the extraction of bioactive compounds such as caffeine, caffeic acid, phenols etc. [7, 8]. These compounds have been suggested to be responsible for bio-reduction and capping processes [9]. Upcycling SCGs is more logistically viable and require fewer preparatory steps than other plant alternatives, as they are already grounded, sterilized and stored independently from other waste in the coffee-making process. Yet, to the best of our knowledge SCGs have not been used to make TiO2 nanoparticles.

Post-treatment recovery of TiO_2 nanoparticles dispersed in wastewater is arduous. After they have been dispersed in wastewater is arduous. We propose making the nanoparticles ferromagnetic and use a magnetic field to control them and recover the nanoparticles. Others have made TiO_2 nanoparticles ferromagnetic by inducing oxygen vacancies by calcining the nanoparticles in an oxygen-deficient atmosphere. We used a novel double calcination method to achieve ferromagnetism. First, nanoparticles were annealed in an oxygen-rich atmosphere to burn away organic residues from green synthesis. Then annealed in an oxygen-poor atmosphere to achieve ferromagnetism. TiO_2 nanoparticles have a high band gap energy of (3–3.2 eV); hence only utilizes the UV region of the sunlight. Thus, we used transition metal doping to lower its bandgap and utilize the visible light region [10–12]. We doped our nanoparticles with Nickel, Cobalt and Iron and found the nanoparticles performance improved similarly.

2 Methodology

2.1 Materials and Chemicals

The materials used in this work were spent coffee grounds (SCGs), which were obtained as a mix of waste from Starbucks, Titanium Butoxide (TTB), ethanol, deionized (DI) water, NaOH, HNO_3, Iron (II) Sulphate Heptahydrate, Nickel (II) Sulphate Heptahydrate, Cobalt (II) Nitrate Heptahydrate, and Methylene Blue (MB) dye. MB was chosen as an organic dye as it is a standard in testing organic dye removal.

2.2 Extraction of Spent Coffee Grounds

SCGs were collected from a Starbucks Coffee Shop outlet, then were dried in the dark using a drying oven, at 80 °C overnight, and subsequently stored in a desiccator to remove any moisture within the SCGs. Following this, 20 g of the SCGs were added to 340 ml of DI water, at a ratio of 1:17 g/ml of solid to liquid which was reported to be optimal for the extraction of phenolic compounds [15]. The mixture was stirred and then heated to 60 °C, before being ultrasonicated at 20 kHz at 60 °C for 40 min in an ultrasonicator bath. After ultrasonic extraction, the mixture was centrifuged at 5000 rpm for 15 min, and the supernatant was filtered using vacuum filtration with a 2.7 μm pore size filter paper to remove solid residue. The prepared SCG extract would be stored at −4 °C.

2.3 Synthesis of Ferromagnetic Doped TiO_2 NPs

The TiO_2 NPs were synthesized using the SCG extract via the following process. 1 drop of 15% HNO_3 was added to 10 ml of ethanol to acidify it. About 1ml of TTB was added to the acidified ethanol and dissolved, to ensure better dispersion of the TTB. The ethanol was acidified to prevent hydrolysis of the TTB into bulk TiO_2 by trace amounts of water in the ethanol. Separately, 1 ml of the dopant precursor (2% mol and 4% mol of Fe, Ni, and Co each) was prepared by dissolving the dopant in 0.25% HNO_3. To 40 ml of the SCG Extract, the 1ml of dopant precursor was added and mixed, afterwards 10 ml of the TTB solution was added dropwise while mixing. The mixture was then adjusted to pH 7 and then mixed for 4h at room temperature. Afterwards, the mixture was centrifuged at 5000 rpm for 15 min to isolate the NPs which were then washed and centrifuged at the same settings with DI water twice and ethanol once to remove residual impurities. The TiO_2 NPs were then dried in a drying oven at 80 °C overnight. Next, the NPs were ground into a fine powder using a mortar and pestle. The powdered NPs were then calcined in air atmosphere at 500 °C for 3 h at a heating rate of 10 °C/min to burn off any organic residue, whilst annealing the TiO_2 NPs into its anatase phase [18]. To make our TiO_2 NPs ferromagnetic they were then calcined again in an oxygen-deficient nitrogen atmosphere to induce oxygen vacancy defects in the crystal structure of the NPs, as per reported in literature. Oxygen vacancies were reported to enhance room temperature ferromagnetism (RTFM [19–24].

2.4 Characterization of the Doped TiO_2 NPs

X-Ray diffraction (XRD) analysis was done using an X-Ray Diffractometer to analyze the crystallinity of the NPs. The XRD analysis was done at a 2θ range of 10° to 80° at normal background noise. Scanning Electron Microscopy (SEM) analysis was done to observe surface morphology of the NPs, obtaining images at a magnification of x25000, accelerating voltage of 15.0 kV, and a working distance of 7.5 mm. The presence of room temperature ferromagnetism was tested in a qualitative approach, where a strong magnetic field was set up to the side of a vial with a suspension of the NPs, and the vial was monitored for congregation of the NPs at the magnetic field.

2.5 Photocatalytic Performance Test

The photocatalytic activity of the NPs was tested by observing the photocatalytic degradation of MB dye using the NPs as photocatalysts under solar lamp simulated visible light. 15 mg of NPs were added to 30 ml of 5 ppm MB solution. Achieving a catalyst concentration of 0.5 g/L. The mixture was then sonicated in the dark at 20 kHz for 30 min to ensure dispersion of the NPs and achieve adsorption-desorption equilibrium before irradiation. The solution was then transferred to a water-cooled photoreactor cell, then placed under a Solar Simulator Lamp at 8 cm above the MB solution level. The water cooling kept the solution at a constant temperature. The photoreactor had its outer walls covered in aluminum foil, to prevent light leakage. The solution was continually stirred to maintain and aerate the solution to increase dissolved O_2 content, which was hypothesized to increase the rate of photocatalytic degradation. At 20 min intervals,

2 ml of the solution was taken out of the photoreactor and centrifuged at 13300 rpm for 4 min to isolate the NPs, from which the absorbance of the MB solution was measured using a UV-Vis Spectrophotometer. Readings were taken over the course of 3h, and the absorbance at 665 nm, the wavelength for peak absorption for MB dye, was measured and plotted (Fig. 1).

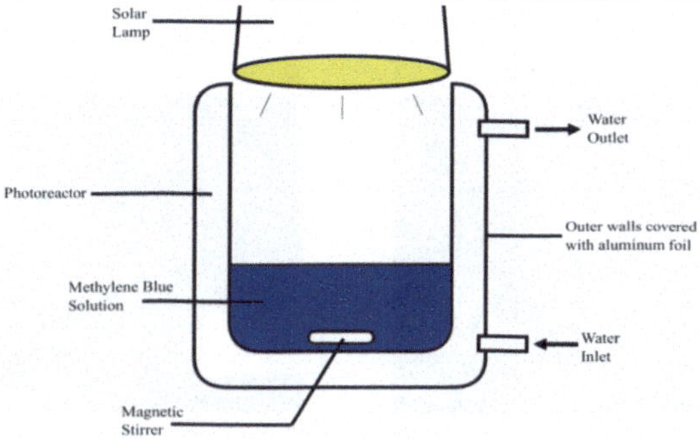

Fig. 1. Schematic Diagram of Photoreactor Setup

2.6 Modelling of Experimental Data

It was observed that the adsorption-desorption equilibrium of MB on the surface of the TiO_2 NPs would shift towards desorption when changed from dark conditions to under visible light solar irradiation. With the amount of free MB in the solution both increasing due to desorption of MB from the NP catalyst surface as well as decreasing due to photocatalytic degradation, the degradation process no longer fits into the usual Langmuir-Hinshelwood model for heterogenous catalyst kinetics. Instead, experimental data was fitted into Eq. (14) (Refer to Sect. 3.5). A python program was used to perform nonlinear regression and evaluate functions to find values that best fit the collected experimental data, as per the theoretical equation. This would give graphs fitted to the experimental data and provide estimates for both degradation and desorption rate constants, allowing both processes to be accounted for (Fig. 2).

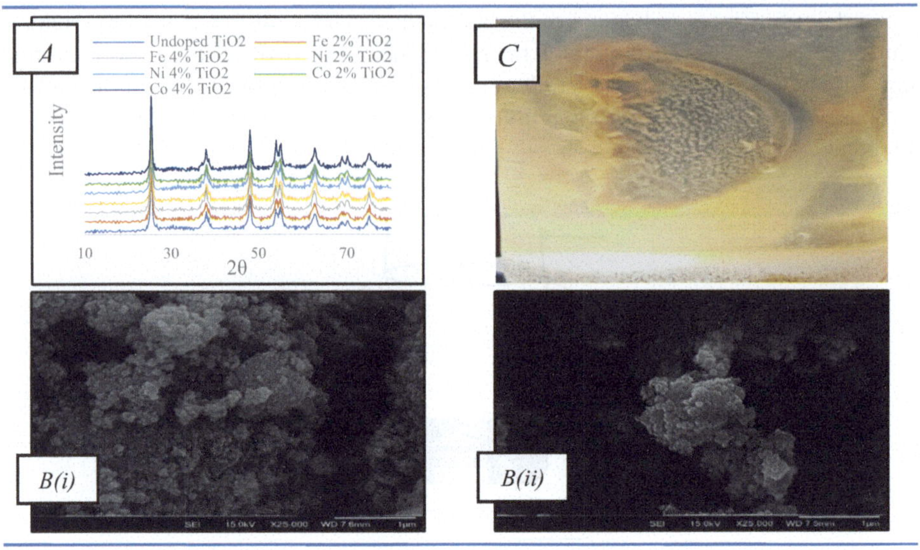

Fig. 2. (A) Compiled XRD Patterns of synthesized NPs, (B) SEM Images of Undoped (i) and Ni2% mol-doped (ii) TiO2 NPs, (C) Image of collection of ferromagnetic NPs using a strong neodymium

3 Results and Discussion

3.1 Characterization

XRD analysis confirmed the anatase phase of all synthesized TiO_2 NPs by matching their diffraction patterns to reference data. As shown in Fig. 3A, all samples demonstrate a high quality of fitting, displaying characteristic anatase peaks at $2\theta = 25.31°$, 37.07°, 37.98°, 38.58°, 48.12°, 53.96°, 55.15°, 62.79°, 68.88°, 70.42°, 75.08°, and 76.13°. Indicating a tetragonal crystal structure. The XRD profiles of all the doped TiO_2 NPs showed no significant signs of phase impurity, indicating that dopants were well-dispersed without forming separate metal oxide phases. SEM analysis was employed to determine surface morphology. Undoped TiO_2 (Fig. 3B(i)) exhibited relatively uniform and fine structure with densely packed agglomerates. In the Ni2%mol-doped TiO_2 (Fig. 3B(ii)), the particles appear rougher, more irregular slightly more agglomerated, with some larger clusters likely due to changes in the crystal growth dynamics or altered surface energy due to Ni-doping. Both samples show spherical nanostructures averaging 91 nm in diameter showing successful synthesis of NPs despite agglomeration.

3.2 Photocatalytic Performance Test

The photocatalytic degradation efficiency of the synthesized TiO_2 NPs was tested against MB dye solution with a concentration of 5 ppm. The initial absorbance of the 5 ppm MB dye solution was measured to be 1.027 Absorbance using a UV-Vis Spectrophotometer. UV-Vis absorbance spectra was recorded at 20 min intervals over 3 h (Fig. 4), and the data were modelled to evaluate photocatalytic activity (Fig. 5). Since absorbance is directly proportional to free MB concentration, it reflects both adsorption-desorption of MB dye onto the NPs and its photocatalytic degradation. Most TiO_2 samples showed a brief absorbance *increase* from 0 to 20 min, indicating dye desorption upon initial illumination. Afterwards, the rate of desorption was lower than degradation, and thus, absorbance decreased with a decreasing gradient. In 2% mol & 4%mol Fe-doped TiO_2 NPs, this initial spike was not present. Additionally, their $t = 0$ absorbance matched the MB blank, suggesting weaker dye adsorption, starting with fewer adsorbed MB molecules, leading to desorption being lower too, causing the degradation pathway to dominate. Fe-doping might have altered the surface characteristics, lowering the adsorption affinity between MB and the TiO_2 NPs. Figure 4 shows occasional shifts in shape or spread of absorbance graph likely due to degradation by-products with differing UV-Vis spectra.

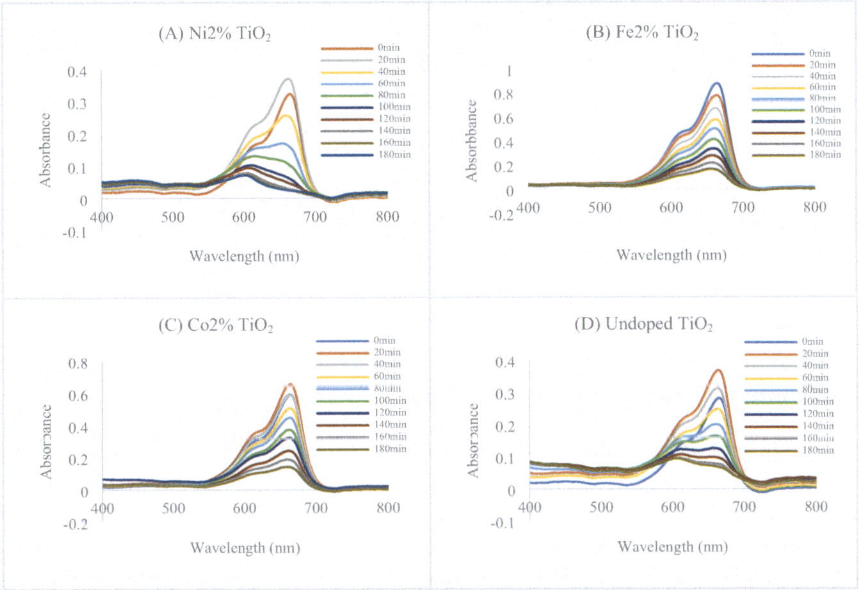

Fig. 3. UV-Vis Absorbance data over time for Ni2% TiO2 (A), Fe2% TiO2 (B), Co2% TiO2 (C), Undoped TiO2 (D)

3.3 Adsorption-Desorption Equilibrium During Degradation

In the dark, MB absorbed onto the NPs would give it a deep blue hue, which will remain so till exposure to solar irradiation, when a brief increase (0–20 min) in absorbance is observed, indicating an increase in free MB concentration resulting from desorption of MB from the surface of the NPs being the dominant process. This slight initial increase is not reported by most of concentration of free MB dye in the first interval of 20 min is a phenomenon that most work in this field. A possible explanation is that the photogenerated holes generated by the NP electrostatically repels the cationic MB, shifting the adsorption-desorption equilibrium towards desorption, which is occurring concurrently with photocatalytic degradation. Over time, as the desorption continually re-equilibrates, the amount of MB molecules released from catalyst surface decreases, while degradation of the MB molecules goes on. This would result in the absorbance readings reaching a local maximum before decreasing continually.

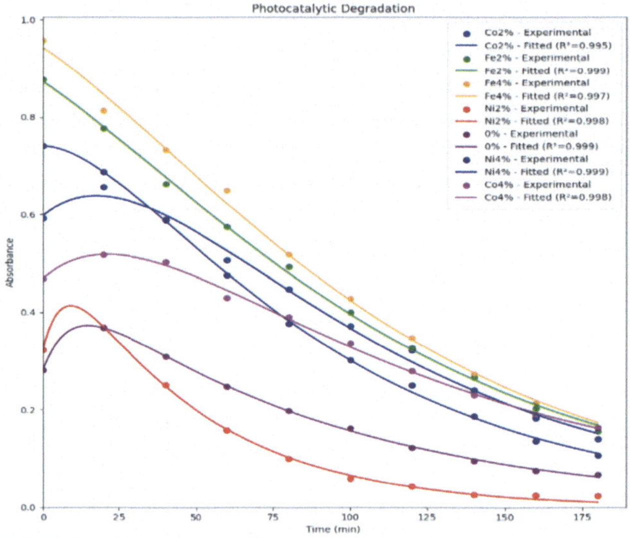

Fig. 4. Graph of Absorbance of MB solution against time for all samples

3.4 Modelling Desorption and Degradation

The change in light absorbance is attributed to both processes of desorption and degradation, due to the changed adsorption equilibrium under light irradiation. Hence, to model the change in absorbance of the MB solutions, both these processes must be considered. Equation (1) describes the Langmuir-Hinshelwood Model for heterogeneous catalysis, which is normally used to determine rate constant, where r is rate of reaction, C is concentration of MB, k is the degradation reaction constant, and K is the adsorption equilibrium constant. The L-H model goes by the assumption that the adsorption equilibrium is achieved much faster than the reaction, while experimental observation showed that the adsorption equilibrium continuously shifts towards the desorption side. Thus, we cannot use Eq. (1) to determine the degradation rate constant [25].

$$-r = \frac{dC}{dt} = \frac{kKC}{1 + KC} \approx kKC \tag{1}$$

Instead, we refer to the following equations. Equation (2) describes a first order model for desorption, where C_a is the concentration of MB adsorbed onto the catalyst surface, C_{a0} is the concentration of adsorbed MB at the start of light irradiation. Equation (3) shows the relationship between the rate of change of C_f, concentration of free MB, with both the rate of degradation r and rate of change of C_a. k_1 is the degradation rate constant and k_2 is the desorption rate constant. C_s would be referring to the catalyst concentration, which would be 0.5 g/L.

$$C_a = C_{a0}e^{-k_2 t} \tag{2}$$

$$\frac{dC_f}{dt} = -r - \frac{dC_a}{dt} = -k_1 C_f + k_2 C_a C_s \tag{3}$$

Solving Eq. (3), we get Eq. (4), where C_{f0} was the initial concentration of free MB dye at the start of light irradiation.

$$C_f = \frac{k_2}{k_1 - k_2} C_s C_{a0} e^{-k_2 t} + \left(C_{f0} - \frac{k_2 C_s C_{a0}}{k_1 - k_2} \right) e^{-k_1 t} \tag{4}$$

We can assume that light absorbance, X, would be linear with respect to both C_a and C_f. This can be written as Eq. (5), where α and β are proportionality constants.

$$X = \alpha C_f + \beta C_a C_s \tag{5}$$

Hence, we can combine Eqs. (4) and (5) to form an overall equation relating k_1 and k_2. It should be noted that both k_1 and k_2 are dependent on light irradiation received by the TiO_2 NP catalyst, and both catalyst concentration and MB concentration would affect the light received by the catalyst.

$$X = me^{-k_1 t} + ne^{-k_2 t} \tag{6}$$

where m and n are each a combination of multiple constants.

$$m = \alpha \left(C_{f0} - \frac{k_2 C_s C_{a0}}{k_1 - k_2} \right) \tag{7}$$

$$n = \left(\frac{\alpha k_2}{k_1 - k_2} + \beta \right) C_{a0} C_s \tag{8}$$

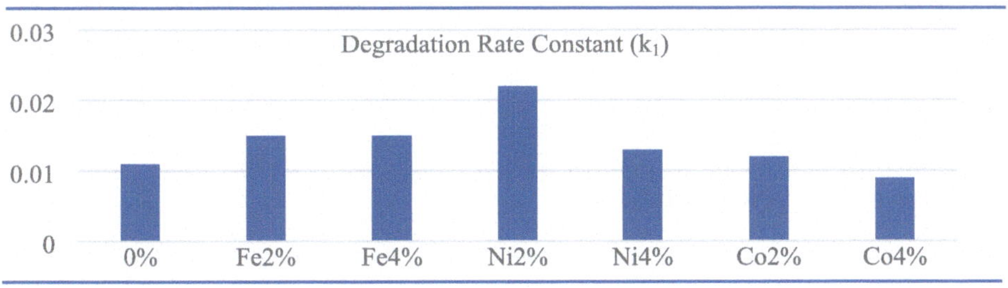

Fig. 5. Comparison of k_1 Values

Experimental data from the photocatalytic performance tests were then fitted to Eq. (6) using nonlinear regression with a python program, to evaluate estimates for the constants in the equation. The estimates for m, n, k1, k2 from the modelling for all synthesized TiO2 NP samples are summarized in Table 1. The slope and R2 values in Table 1 refer to gradient of regression line passing through the origin and its correlation coefficient. All R2 and slopes are close to unity, which would mean that the fitting of the experimental data into Eq. (6) was done relatively well. The fitted graphs are shown in Fig. 5. The k1 of all synthesized TiO2 NPs are shown in Fig. 6 for comparison. It is important to note that the k2 values cannot be directly compared between each other as different NPs had different affinities for adsorption of MB. The % Net removal would refer to the percentage of MB removed from the combined effects of photocatalytic degradation alongside adsorption, by comparing the final absorbance value to the initial absorbance of 1.027 for the 5 ppm MB solution. The values of k1 follow the same trend as that of % Net Removal, which reflects photocatalytic degradation as being the dominant process for removal of dye. Ni2% mol-doped TiO2 NPs were found to have the highest photocatalytic degradation rate constant of 0.022 min^{-1}, whilst being able to remove 92.53% of dye over 3 h. This may possibly be due to formation of NiO structures, a photoreactive material, within the TiO2 crystal. Overall, the addition of dopants increased the photocatalytic activity of almost all synthesized samples, except for Co4%mol-doped TiO2 NPs. This would show that the addition of dopants was likely to have been able to decrease the band gap nearer to the visible light region, as compared to the band gap of normal bulk TiO2 which was reported to be 3.2eV. Among the 3 metals, Co was the least effective dopant. Co4%mol-doped TiO2 NPs alongside being a less effective dopant, had a higher concentration of dopants, which could have acted as electron-hole pair recombination centers, making the reactions involving the electron and hole less probable.

Table 1. Constants from Photocatalytic Degradation and Desorption

Sample	m	n	k_1 (min^{-1})	k_2 (min^{-1})	slope	R^2	% Net Removal
Undoped	0.497	-0.215	0.011	0.104	0.9999	0.9993	75.97
Fe2%	116.262	-115.389	0.015	0.015	1.0039	0.9989	82.07
Fe4%	112.148	-111.204	0.015	0.016	1.0051	0.9966	82.83
Ni2%	0.603	-0.279	0.022	0.138	0.9887	0.9982	92.53
Ni4%	1.127	-0.385	0.013	0.037	1.0011	0.9993	85.53
Co2%	1.469	-0.871	0.012	0.027	1.0033	0.9946	76.23
Co4%	0.867	-0.398	0.009	0.034	1.0006	0.9975	64.97

3.5 Mechanism of Photocatalysis

When the TiO_2 NP absorbs light of sufficient energy, an electron is excited and a hole is generated, these can then react and eventually generate OH• radicals than can attack and degrade the MB molecules, forming byproducts that would be of lower absorbance than MB dye. Generally, photocatalytic degradation of organic pollutants follows the steps below [26]. Starting by the absorption of photon energy and excitation of the electron from valence band to conduction band, to reduction and oxidation via reaction with electron-hole pair, generating reactive oxygen species such as OH• radicals, which then attack and degrade harmful pollutants such as MB dye into less harmful byproducts.

$$TiO_2 + hv \rightarrow TiO_2\left(h_{VB}^{+} + e_{CB}^{-}\right) \tag{9}$$

$$e_{CB}^{-} + O_2 \rightarrow O_2^{\bullet -} \tag{10}$$

$$O_2^{\bullet -} + H^{+} \rightarrow HO_2^{\bullet} \tag{11}$$

$$2HO_2^{\bullet} \rightarrow H_2O_2 + O_2 \tag{12}$$

$$2HO_2^{\bullet} \rightarrow H_2O_2 + O_2^{\bullet -} \rightarrow OH^{\bullet} + O_2 + OH^{-} \tag{13}$$

$$h_{VB}^{+} + OH^{-} \rightarrow OH^{\bullet} \tag{14}$$

$$h_{VB}^{+} + H_2O \rightarrow OH^{\bullet} + H^{+} \tag{15}$$

$$OH^{\bullet} + MB \rightarrow degradationbyproducts \tag{16}$$

4 Conclusions

In our study, undoped, Fe-doped, Ni-doped, and Co-doped TiO$_2$ NPs of varying dopant concentrations were successfully synthesized using spent-coffee-grounds extract, as a reducing, capping, and stabilizing agent. SEM analysis showed that all the synthesized NPs were of 91 nm average diameter spherical crystals that showed some agglomeration XRD analysis showed all synthesized NPs were of anatase phase TiO$_2$ with no significant phase impurities, which would indicate an even dispersion of the dopants. The synthesized NPs were found to be effective as photocatalysts in the degradation of MB dye over a 3 h duration. Desorption of MB dye was observed at the start of solar light irradiation which could have been due to electrostatic forces of repulsion decreasing the adsorption affinity between the NPs and the MB molecules, shifting the adsorption equilibrium towards desorption continuously under light irradiation. The experimental data was fitted to a model that was equated to account for this phenomenon, which allowed for a more accurate analysis of the photocatalytic activities of the synthesized NPs. We found Ni2% mol-doped TiO$_2$ had the largest photocatalytic degradation rate constant of 0.022 min^{-1}, alongside a net removal of 93% of dye. We found Co to be the least effective dopant, with Co4% mol-doped TiO$_2$ having the least degradation rate constant. Green synthesis using SCG offers a solution for scalable green synthesis, and achieved ferromagnetism in our NPs opens new doors for a solution to NP recovery.

5 Future Work

We could not quantify the ferromagnetic properties of our nanoparticles. In the future, we plan on using a Vibrating Sample Magnetometer to measure the hysteresis loop of the synthesized TiO_2 NPs to determine their magnetic coercivity. We could also use UV-Vis Diffuse Reflectance Spectroscopy to measure reflectance of our NPs and thus measure the bandgap of our NPs via a tauc plot, to accurately determine if and how doping has decreased the band gap.

References

1. Halepoto, H., Gong, T., Memon, H.: Current status and research trends of textile wastewater treatments—A bibliometric-based study. Front. Environ. Sci. (2022). https://doi.org/10.3389/fenvs.2022.1042256
2. Kant, R.: Textile dyeing industry an environmental hazard. Nat. Sci. **04**(01), 22–26 (2012). https://doi.org/10.4236/ns.2012.41004
3. Nengsih, S., Nur Abdulmadjid, S., Mursal, M., Jalil, Z.: Photocatalytic performance of Fe3O4-TiO2 in the degradation of methylene blue dye: optimizing the usability of natural iron sand. Mater. Sci. Energy Technol. **7**, 374–380 (2024). https://doi.org/10.1016/j.mset.2024.06.001
4. Goswami, D., Mukherjee, J., Mondal, C., Bhunia, B.: Bioremediation of azo dye: a review on strategies, toxicity assessment, mechanisms, bottlenecks and prospects. Sci. Total Environ. **954**, 176426 (2024). https://doi.org/10.1016/j.scitotenv.2024.176426
5. Armaković, S.J., Savanović, M.M., Armaković, S.: Titanium dioxide as the most used photocatalyst for water purification: an overview. Catalysts **13**(1), 26 (2022). https://doi.org/10.3390/catal13010026
6. Ajmal, A., Majeed, I., Malik, R.N., Idriss, H., Nadeem, M.A.: Principles and mechanisms of photocatalytic dye degradation on TiO2based photocatalysts: a comparative overview. RSC Adv. **4**(70), 37003–37026 (2014). https://doi.org/10.1039/c4ra06658h
7. Al-Dhabi, N.A., Ponmurugan, K., Maran Jeganathan, P.: Development and validation of ultrasound-assisted solid-liquid extraction of phenolic compounds from waste spent coffee grounds. Ultrason. Sonochem. **34**, 206–213 (2017). https://doi.org/10.1016/j.ultsonch.2016.05.005
8. Bouhlal, F., et al.: GC-MS analysis, phenolic compounds quantification, antioxidant, and antibacterial activities of the hydro-alcoholic extract of spent coffee grounds. J. Biologically Active Prod. Nat. **10**(4), 325–337 (2020). https://doi.org/10.1080/22311866.2020.1802333
9. Mustapha, T., Misni, N., Ithnin, N.R., Daskum, A.M., Unyah, N.Z.: A review on plants and microorganisms mediated synthesis of silver nanoparticles, role of plants metabolites and applications. Int. J. Environ. Res. Public Health **19**(2), 674 (2022). https://doi.org/10.3390/ijerph19020674
10. Pandey, P., Gogate, P.R.: Improved synthesis of cobalt-doped TiO2 catalyst using ultrasound and subsequent application for Acid Violet 7 degradation. Asia-Pacif. J. Chem. Eng. **19**(6) (2024). https://doi.org/10.1002/apj.3142
11. Abukhadra, M.R., Ahmed, A.M., Shehata, N., Betiha, M.A., Rabie, A.M.: Ni-doped and Ni/Cr co-doped TiO2 nanotubes for enhancement of photocatalytic degradation of methylene blue. J. Colloid Interface Sci. **555**, 31–41 (2019). https://doi.org/10.1016/j.jcis.2019.07.070
12. Sood, S., Umar, A., Mehta, S.K., Kansal, S.K.: Highly effective Fe-doped TiO2 nanoparticles photocatalysts for visible-light driven photocatalytic degradation of toxic organic compounds. J. Colloid Interface Sci. **450**, 213–223 (2015). https://doi.org/10.1016/j.jcis.2015.03.018

13. Zhang, H., et al.: Room temperature ferromagnetism in pristine TiO2 nanoparticles triggered by singly ionized surface oxygen vacancy induced via calcining in different air pressure. J. Alloy. Compd. **860**, 157913 (2020). https://doi.org/10.1016/j.jallcom.2020.157913
14. Choudhury, B., Verma, R., Choudhury, A.: Oxygen defect assisted paramagnetic to ferromagnetic conversion in Fe doped TiO2nanoparticles. RSC Adv. **4**(55), 29314 (2014). https://doi.org/10.1039/c3ra45286g
15. Li, Q., et al.: Antimicrobial nanomaterials for water disinfection and microbial control: potential applications and implications. Water Res. **42**(18), 4591–4602 (2008). https://doi.org/10.1016/j.watres.2008.08.015
16. Prabhakaran, T., Mangalaraja, R., Denardin, J.C., Jiménez, J.: The effect of calcination temperature on the structural and magnetic properties of co-precipitated CoFe 2 O 4 nanoparticles. J. Alloy. Compd. **716**, 171–183 (2017). https://doi.org/10.1016/j.jallcom.2017.05.048
17. Lal, M., Sharma, P., Ram, C.: Calcination temperature effect on titanium oxide (TiO2) nanoparticles synthesis. Optik **241**, 166934 (2021). https://doi.org/10.1016/j.ijleo.2021.166934
18. Abdullah, J.a.A., Jiménez-Rosado, M., Guerrero, A., Romero, A.: Effect of calcination temperature and time on the synthesis of iron oxide nanoparticles: green vs. chemical method. Materials **16**(5), 1798 (2023). https://doi.org/10.3390/ma16051798
19. Patel, S.K., Kurian, S., Gajbhiye, N.S.: Room-temperature ferromagnetism of Fe-doped TiO2 nanoparticles driven by oxygen vacancy. Mater. Res. Bull. **48**(2), 655–660 (2013). https://doi.org/10.1016/j.materresbull.2012.11.031
20. Motaung, D., Makgwane, P., Ray, S.: Induced ferromagnetic and gas sensing properties in ZnO-nanostructures by altering defect concentration of oxygen and zinc vacancies. Mater. Lett. **139**, 475–479 (2014). https://doi.org/10.1016/j.matlet.2014.10.073
21. Sun, X., et al.: Influence of oxygen vacancies on the structure and magnetic properties of high entropy (Co,Cr,Fe,Mn,Ni) 3O4 ceramics. J. Alloy. Compd. **925**, 166560 (2022). https://doi.org/10.1016/j.jallcom.2022.166560
22. Dependence of structural and magnetic properties on synthesis procedure in silver substituted cobalt ferrite. (n.d.). CoLab. https://colab.ws/articles/10.1063%2F5.0117356
23. Santara, B., Giri, P.K., Imakita, K., Fujii, M.: Evidence of oxygen vacancy induced room temperature ferromagnetism in solvothermally synthesized undoped TiO2 nanoribbons. Nanoscale **5**(12), 5476 (2013). https://doi.org/10.1039/c3nr00799e
24. Li, Y., Lv, G., Liu, H., Liu, X., Liao, L.: Improvement of magnetite adsorption performance for Pb (II) by introducing defects. Front. Chem. (2023). https://doi.org/10.3389/fchem.2023.1137246
25. Xu, C., Rangaiah, G.P., Zhao, X.S.: Photocatalytic degradation of methylene blue by titanium dioxide: experimental and modeling study. Ind. Eng. Chem. Res. **53**(38), 14641–14649 (2014). https://doi.org/10.1021/ie502367x
26. Zhang, L., Wan, J., Hu, Z., Jiang, W.: Preparation and photocatalytic activity of TIO2-Wrapped cotton nanofiber composite catalysts. BioResources **12**(3), 6062–6081 (2017). https://doi.org/10.15376/biores.12.3.6062-6081

Plastic Biodegradation Capabilities of *Zophobas Morio* and *Pseudomonas Putida* in Alginate Beads

Jiang Yuxuan(✉) and Zhao Zilin Wesley

Hwa Chong Institution, Singapore 269734, Singapore
231441B@student.hci.edu.sg

Abstract. In recent years, plastic pollution has plagued ecosystems around the world. Thus, it is increasingly important to research practical and sustainable alternatives to manage such plastic waste. This study assessed the consumption efficiency of extruded polystyrene (XPS), low-density polyethylene (LDPE), polypropylene (PP) plastic strips, and expanded polyethylene (EPE) and expanded polystyrene (EPS) foams over 14 days, by *Zophobas morio*, a common invertebrate commonly known as superworms. Furthermore, the potential of microplastic degradation using *Pseudomonas putida,* a bacterium commonly found in superworm guts, encapsulated in calcium alginate beads was investigated. Our results confirmed that superworms efficiently consume XPS at a rate of 1.56 mg/d/larva, EPE foam at 0.59 mg/d/larva, and EPS foam at 2.82 mg/d/larva. All superworms that advanced into the pupae stage developed into beetles successfully. Fluorescence analysis using Nile red dye and atomic force microscopy (AFM) of microplastics also revealed that *P. putida* encapsulated in calcium alginate beads are capable of degrading microplastics, as indicated by fluorescence analysis indicating a decrease in microplastic concentration and size and AFM analysis indicating a decrease in nanoplastic surface roughness and size. Hence, this study provided insight into the use of superworms and *P. putida* in the management of microplastic waste.

Keywords: Plastic pollution · microplastics · superworm · *Zophobas morio* · *P. putida*

1 Introduction

The persistence of plastic in resisting degradation has made plastic pollution a threat to the environment [1]. The continual stream of plastic pollutants is primarily caused by the increasingly widespread use of plastic worldwide and in various industries, as plastic is lightweight, durable, and inexpensive [2]. The United Nations Environment Programme emphasised this by highlighting that up to five trillion plastic bags are being used worldwide every year around the world [3].

Plastics are inexpensive and widely used organic polymers, but their high durability hinders biodegradation. Hydrolysable plastics, which include polyethylene, polypropylene and polystyrene [4], are more resistant to enzymatic depolymerisation or degradation, taking up to 500 years to fully biodegrade [5]. The conventional methods of physical

H. Guo et al. (Eds.): IRC-SET 2025, PTT, pp. 234–243, 2026.
https://doi.org/10.1007/978-981-95-9630-0_21

degradation of plastics, such as landfilling and incineration, have their limitations. Mass landfilling of plastic waste may lead to the generation and release of volatile organic compounds, such as benzene, toluene and xylene, which are lethal and cause environmental pollution [6]. Furthermore, plastics in landfills experience limited degradation, taking over 20 years to fully degrade. Incineration of plastic leads to the formation of numerous harmful compounds, like heavy metals, toxic carbon and oxygen-based free radicals, most of which are released into the atmosphere [1]. Without a sustainable and efficient method to dispose of plastic, the amount of plastic waste will accumulate in the environment.

Every year, 19–23 million tonnes of plastic waste leaks into aquatic ecosystems, polluting lakes, rivers, and entire oceans [3]. These plastic fragments will gradually break up into microplastics (MPs), which are below 5 mm in diameter. It has been observed that in the last few decades, the mean size of plastic particles in the environment has been reducing, and the abundance and global distribution of microplastic fragments have increased [7].

The management of microplastics is expensive and ineffectual, since their sizes can reach down to a few micrometers in diameter, making detection and disposal difficult. In addition, microplastics adsorb various chemical contaminants and release toxic plastic additives [8], which are detrimental to the environment and human health. Microplastic particles ingested through contaminated food or drinking water can even be found in 15 biological components, including the liver, colon, lung, placenta, and breastmilk. It has been suggested that microplastics can cause damage to human cells, leading to reproductive and developmental issues, neurotoxicity, and even potentially promote the progression of cognitive impairment in Alzheimer's Disease and cancer development [9, 10].

Zophobas morio, or superworms, is a common invertebrate mainly used as fish food due to their rich nutritional value [11]. There has been evidence of superworms being able to consume and digest plastics, having been reported to consume polystyrene foam at an average rate of 0.58 mg/larvae/day [12]. The PS degrading capability was also inhibited by antibiotic suppressions of gut microbiota, indicating that gut microbiota contributed to the degradation of PS foams [13, 14].

Pseudomonas putida has been observed to exhibit successful biodegradation abilities with plastic such as LDPE [15, 16], and PET [17]. Moreover, a high abundance of *P. putida* has been detected in superworm guts [13].

2 Objectives and Hypotheses

This study aimed to investigate the potential of superworms in consuming different types of plastic, and to investigate the potential and effectiveness of microplastic biodegradation with *Pseudomonas putida*.

The hypotheses to be tested were as follows:

1. Superworms will exhibit different levels of potential in consuming different types of plastic.
2. *P. putida* encapsulated in calcium alginate beads will exhibit potential in breaking down microplastic particles.

3 Materials and Methods

3.1 Worm Care

Superworms were purchased from a fish shop in Singapore and kept at room temperature in a plastic container covered with a secure lid poked with holes for air flow. Bran was provided as the initial food source, followed by a 24 h food deprivation period prior to experimentation.

3.2 Consumption of Plastic Strips and Foam Blocks by Superworms

Each experimental setup consisted of 10 superworms with a total mass of 1 g of the respective type of plastic. 2 control setups consisted of either 1 g of plastic or 10 superworms. The PP, LDPE, and XPS pieces were of dimensions 0.5 cm × 1.5 cm, while the EPS and EPE foams were of dimensions 2 cm × 2 cm × 2 cm. All setups were covered with lids poked with holes and left at room temperature. Changes in mass of plastic pieces and foam blocks in each setup were recorded at 2–3-day intervals. Deceased superworms were replaced with new individuals to maintain consistent data collection. The experiment was conducted over a period of 14 days, and the change in the mean mass, cannibalism and pupation rates of worms within the duration of the experiment were recorded.

3.3 Preparation of Bacteria Pre-Cultures

Pseudomonas putida (ATCC 31800) was inoculated into 10 ml Luria-Bertani broth (LB), before being incubated for 24 h in a shaker incubator at 30 °C. The absorbance of the bacteria culture was measured using a UV-VIS spectrophotometer at 600 nm wavelength of light and standardised at 0.6.

3.4 Preparation of Microplastic-Contaminated Water

0.5 g of LDPE plastic bag was placed in a beaker with 200 ml water and microwaved in a 900-W microwave for the release of microplastic particles until only about 1 ml of water remained.

3.5 Biodegradation of Microplastics by *P. Putida* Immobilised in Calcium Alginate Beads

P. putida cultures were centrifuged at 4500 rpm for 10 min, and the collected pellets were thereafter resuspended in 2 ml distilled water before being added into 50 ml 2% (w/v) sodium alginate solution, and mixed at 300 rpm using a magnetic stirrer for 30 min. 20 μl of the sodium alginate and bacteria solution was micropipetted into a beaker containing 1.5% (w/v) calcium alginate solution, to form calcium alginate beads. The formed beads were soaked in the calcium chloride solution for 1 h, and 3 g of calcium alginate beads were incubated with 15 ml microplastic contaminated water for 4 days. The control setup contained microplastic-contaminated water with calcium alginate beads without bacterial immobilisation.

3.6 Fluorescence Microscopic and Spectroscopic Analyses

10 ml of the post-incubation microplastic samples were stained with 5 μl 1 mg/mL Nile red dye (Sigma-Aldrich, #72485) dissolved in acetone and left to rest for about 1 h. The samples were then shone with a blue LED flashlight of wavelength 450 nm and observed under a stereomicroscope. The ImageJ program was used to analyse changes in microplastic particle count and mean diameter.

3.7 Atomic Force Microscopy (AFM) Analysis

The surface roughness of dried microplastic samples ($< 1\,\mu m$) were analysed with atomic force microscopy, and the results were further analysed with the software MountainsSPIP and Gwyddion.

4 Results and Discussion

4.1 Decrease in Mass of Plastic Strips and Foams

Both XPS and EPS were more efficiently consumed by superworms compared to other types of plastic, with XPS at a rate of 1.56 mg/d/larva, PP plastic pieces at 0.059 mg/d/larva, LDPE plastic pieces at 0.098 mg/d/larva, EPS foams at 2.82 mg/d/larva and EPE foams at 0.59 mg/d/larva. The rate of consumption of XPS plastic pieces was similar to some past scholarship, where superworms consumed XPS at 1.41 mg/d/larva. However, superworms consumed slightly more EPE foam than EPS foam as compared to previous similar studies [14, 18]. This difference could be due to different qualities of plastic foams, as EPE foams used in this experiment were obtained from some used packaging. The results are also consistent with some experiments where superworms exhibited limited capability in degrading PP, as well as with significantly lower degradation rates compared to PS and PE, where superworms fed with PP foam consumed it at a rate of 3.1 ± 0.4 mg per 100 larvae per day [19]. The decrease in mass of EPS was the highest in the first four days. It is hypothesised that this was a result of the starvation period of the superworms prior to experimentation, which caused a high amount of consumption of plastics as soon as the experiment began (Fig. 1).

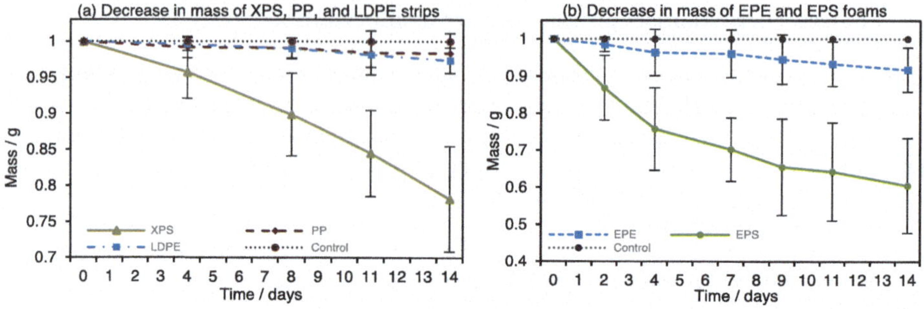

Fig. 1. Decrease in mass of **(a)** XPS, LDPE, and PP stripsA and **(b)** EPE and EPS foams by superworms over 14 days. The respective control setups consisted of 1 g of the respective plastic strips or plastic foams without superworms. The Mann-Whitney U test p-values when compared to the control for PS, LDPE, EPE, and EPS were <0.001 respectively, indicating a significant decrease in mass caused by superworms, while the p-value for PP was >0.05, indicating an insignificant decrease in mass.

4.2 Changes in Superworm Mass After Consumption of Plastic

The percentage decrease in the mean mass of a superworm of each experimental setup, as shown in Fig. 2. Was less than the negative control, which indicates that the superworms consumed at least a portion of the provided plastics. However, the plastic provided insufficient energy for the superworms, which resulted in their weight loss. Superworms in experimental setups shown in Fig. 2 were also observed to exhibit slower movements and reduced activity levels as compared to the positive control fed with bran, suggesting that the plastics provided a diet of little nutrition to the worms, which could explain the overall observed decrease in mass observed in (a) and (b).

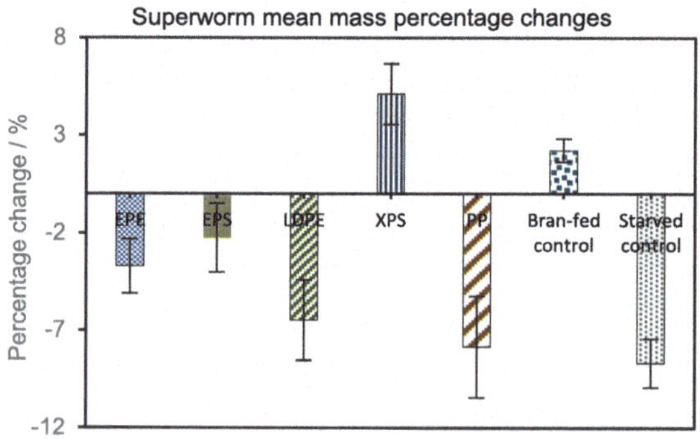

Fig. 2. Percentage changes in mean mass of each superworm when fed with LDPE, XPS, EPE, and EPS plastic foams. The positive control setups consisted of superworms with 5 g bran, and the negative control consisted of superworms without any food source. When compared to the negative control, superworms experienced a significant change in mass when fed with EPS, EPE, and XPS ($p < 0.05$, respectively), whereas this was insignificant for LDPE and PP ($p > 0.05$, respectively).

4.3 Cannibalism and Pupation Rates of Superworms

The number of superworms that developed into pupae and the number of worms that were cannibalised as shown in Fig. 3 corresponded to the amount of consumption of plastic in Fig. 1, with superworms fed with EPS and XPS having a high pupation rate and low cannibalism rate, and superworms fed with PP having a low pupation rate and high cannibalism rate. Superworms typically do not pupate under crowded conditions as a defensive mechanism from being cannibalised by other worms [20], which can be observed in how superworms fed with EPS and EPE tend to burrow into the foam blocks before beginning to pupate.

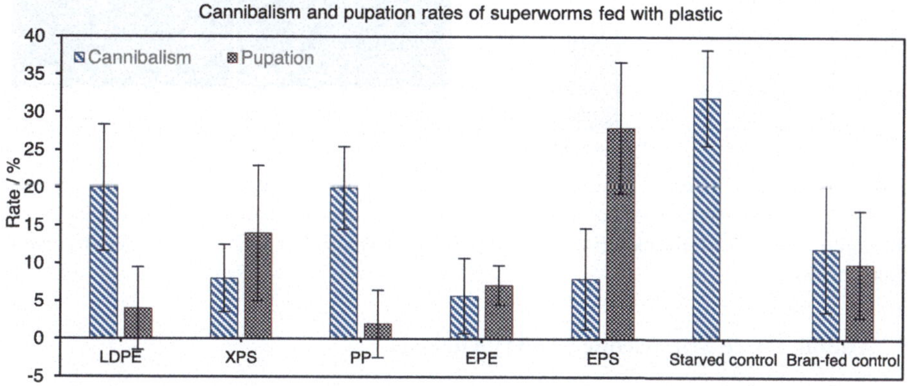

Fig. 3. Cannibalism and pupation rates (per 100 worms) of superworms when fed with LDPE, PS, PP, EPE, and EPS. When compared to the negative control, cannibalism rates of superworms across all experimental setups were significantly lower ($p < 0.05$, respectively). Pupation rates for EPS and XPS were also significantly higher ($p < 0.05$ respectively, contrary to when fed EPE, LDPE, and PP ($p > 0.05$), where there were insignificant differences in pupation rates.

4.4 Fluorescence Microscopy Analysis of Microplastic Concentration After Incubation with *P. Putida*

Nile red-stained microplastics exhibited fluorescence upon shining of 450 nm blue light. Figure 4(A) shows that the concentration of microplastics per cm^3 in the sample decreased after incubation with *P. putida*, with its Mann-Whitney U test p-value being 0.021, indicating that this decrease in concentration is significant. Furthermore, Fig. 4(B) shows a significant decrease in diameter in the experimental setup ($p = 0.019$), suggesting that *P. putida* can partially degrade and reduce the size of such microplastics. Overall, the results agree with the hypothesis that *P. putida* encapsulated in calcium alginate beads can biodegrade microplastics in water.

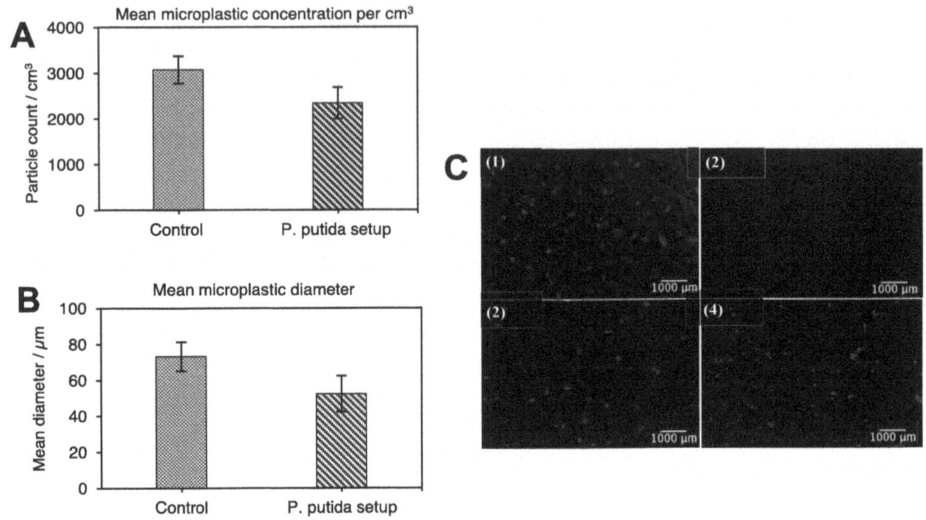

Fig. 4. (A) Graph of microplastic concentrations per cm^3 and (B) graph of mean microplastic diameters. (C) Fluorescence microscopy images of (1) the control setup, (2) the deionised water blank control, and (3), (4) the experimental setups incubated with *P. putida* encapsulated in calcium alginate beads.

4.5 Atomic Force Microscope (AFM) Analysis

A notable decrease in the Sa value could be observed in the setup exposed to *P. putida*, as observed from Fig. 5. This indicates that the bacterial activity has made alterations to the microplastic in water, further reinforcing the hypothesis that *P. putida* immobilised in calcium alginate beads exhibit potential in biodegrading LDPE microplastic particles in water. It is worth noting that the mean roughness values for exposed microplastics are significantly lower ($p = 0.029$), which is unexpected since it was noted in a past study that microplastics that underwent photodegradation are expected to have increased surface roughness [21]. This may require further investigation.

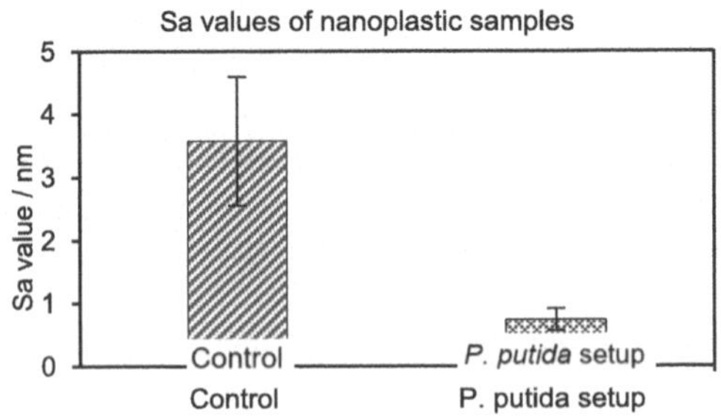

Fig. 5. Mean roughness (Sa) values of control setup and setup incubated with *P. putida*.

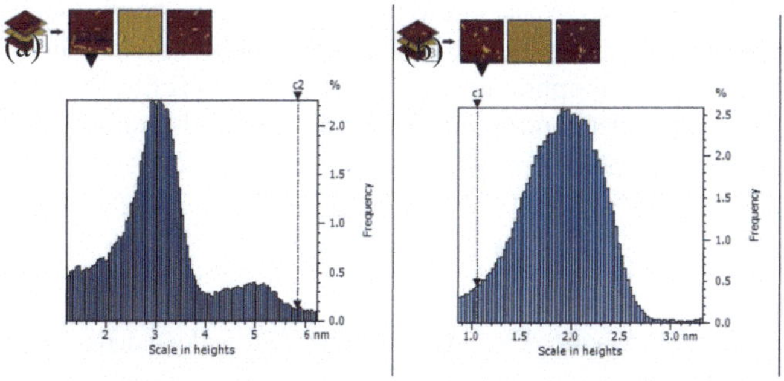

Fig. 6. Histograms showing height distribution of nanoplastics in **(a)** the control setup and **(b)** the experimental setup. The two-sample Kolmogorov-Smirnov test p-value was 0.0064, indicating significant change in height variation.

The overall shift towards lower surface heights in the experimental setup suggests that *P. putida* has effectively reduced the size of the LDPE nanoplastic particles. This corroborates the observed decrease in mean surface roughness seen in Fig. 6, as smaller particles would form a more uniform and smoother surface.

5 Conclusion and Recommendations for Future Work

Superworms consumed plastics at notable rates of 1.56 mg/d/larva (XPS), 2.82 mg/d/larva (EPS), 0.59 mg/d/larva (EPE). This suggests a preference for polystyrene. All superworms that advanced into the pupa stage were able to develop into beetles, confirming that the superworms were able to obtain sufficient nutrients from plastic pieces for progression in the life cycle. However, limited capability was observed in consuming LDPE and PP, with rates of 0.098 mg/d/larva for the former and 0.0593 mg/d/larva for the latter, which was insignificant. In addition, *P. putida* immobilised in calcium alginate beads showed positive results in biodegrading LDPE microplastics, and fluorescence microscopy and fluorescence spectroscopy showing a decrease in the microplastic concentration. The AFM results indicated a decrease in surface roughness and surface heights of exposed microplastics. Thus, both hypotheses that superworms and *P. putida* would exhibit potential in degrading the respective plastic types are supported. While utilising superworms and bacteria in the management of plastic waste on a large scale is challenging at that moment, research might contribute to such a method being used on a smaller scale, to at least mitigate the impacts of plastic pollution on the environment. A possible limitation of this study was the effect of light and humidity in the laboratory on the superworms. Lighting and other conditions in the natural environment of superworms could not be replicated under the laboratory conditions, thus the LED lights in the laboratory might have interfered with the efficiency of consumption of plastic by the superworms. Moreover, fragments of the calcium alginate beads could have broken off, mixing with the microplastic contaminated water, which could have slightly affected experimental results. As future work, the experiment could be further extended to observe whether beetles developed from plastic-fed superworms could reproduce a new generation of superworms that could also consume and digest plastic, investigating if plastic waste could be managed sustainably this way, and could be carried out across generations. Future research might also broaden the scope of plastic types such as polyethylene terephthalate (PET) and polyvinyl chloride (PVC), to provide a more comprehensive understanding of the capabilities and limitations of superworms and bacteria in degrading plastic polymers.

References

1. Webb, H.K., Arnott, J., Crawford, R.J., Ivanova, E.P.: Plastic degradation and its environmental implications with special reference to poly (ethylene terephthalate). Polymers **5**(1), 1–18 (2013)
2. MacLeod, M., Arp, H.P.H., Tekman, M.B., Jahnke, A.: The global threat from plastic pollution. Science **373**(6550), 61–65 (2021)
3. United Nations Environmental Programme Plastic Pollution. https://www.unep.org/plastic-pollution. Last accessed 6 Aug 2024
4. Tokiwa, Y., Calabia, B.P., Ugwu, C.U., Aiba, S.: Biodegradability of plastics. Int. J. Mol. Sci. **10**(9), 3722–3742 (2009)
5. Inderthal, H., Tai, S.L., Harrison, S.T.: Non-Hydrolyzable Plastics–An interdisciplinary look at plastic bio-oxidation. Trends Biotechnol. **39**(1), 12–23 (2021)
6. He, Y., et al.: Current advances, challenges and strategies for enhancing the biodegradation of plastic waste. Sci. Total Environ. **906**, 167850 (2024)
7. Barnes, D.K.A., Galgani, F., Thompson, R.C., Morton, B.: Accumulation and fragmentation of plastic debris in global environments. Phil. Trans. R. Soc. B **364**(1526), 1985–1998 (2009)
8. Carbery, M., O'Connor, W., Thavamani, P.: Trophic transfer of microplastics and mixed contaminants in the marine food web and implications for human health. Environ. Int. **115**, 400–409 (2018)
9. Li, Y., Tao, L., Wang, Q., Wang, F., Li, G., Song, M.: Potential health impact of microplastics: a review of environmental distribution, human exposure, and toxic effects. Environ. Health **1**(4), 249–257 (2023)
10. Ward, P.G., Goff, M., Don ner, M., Kaminsky, W., O'Connor, K.E.: A two step chemo-biotechnological conversion of polystyrene to a biodegradable thermoplastic. Environ. Sci. Technol. **40**(7), 2433–2437 (2006)
11. Araújo, R.R.S., Benfica, T.a.R.D.S., Ferraz, V.P., Santos, E.M.: Nutritional composition of insects Gryllus assimilis and Zophobas morio: Potential foods harvested in Brazil. J. Food Compos. Anal. **76**, 22–26 (2019)
12. Yang, Y., Wang, J., Xia, M.: Biodegradation and mineralization of polystyrene by plastic-eating superworms Zophobas atratus. Sci. Total Environ. **708**, 135233 (2020)
13. Sun, J., Prabhu, A., Aroney, S.T.N., Rinke, C.: Insights into plastic biodegradation: community composition and functional capabilities of the superworm (Zophobas morio) microbiome in styrofoam feeding trials. Microb. Genom. **8**(6) (2022)
14. Peng, B., et al.: Biodegradation of low-density polyethylene and polystyrene in superworms, larvae of Zophobas atratus (Coleoptera: Tenebrionidae): broad and limited extent depolymerization. Environ. Pollut. **266**, 115206 (2020)
15. Montazer, Z., Najafi, M.B.H., Levin, D.B.: Microbial degradation of low-density polyethylene and synthesis of polyhydroxyalkanoate polymers. Can. J. Microbiol. **65**(3), 224–234 (2019)
16. Ward, P.G., Goff, M., Donner, M., Kaminsky, W., O'Connor, K.E.: A two step chemo-biotechnological conversion of polystyrene to a biodegradable thermoplastic. Environ. Sci. Technol. **40**(7), 2433–2437 (2006)

17. Kenny, S.T., Runic, J.N., Kaminsky, W., Woods, T., Babu, R.P., O'Connor, K.E.: Development of a bioprocess to convert PET derived terephthalic acid and biodiesel derived glycerol to medium chain length polyhydroxyalkanoate. Appl. Microbiol. Biotechnol. **75**, 623–633 (2012)
18. Luo, L., et al.: Biodegradation of foam plastics by Zophobas atratus larvae (Coleoptera: Tenebrionidae) associated with changes of gut digestive enzymes activities and microbiome. Chemosphere **282**, 131006 (2021)
19. Yang, S., et al.: Biodegradation of polypropylene by yellow mealworms (Tenebrio molitor) and superworms (Zophobas atratus) via gut-microbe-dependent depolymerization. Sci. Total Environ. **756**, 144087 (2021)
20. Rumbos, C.I., Athanassiou, C.G.: The Superworm, Zophobas morio (Coleoptera:Tenebrionidae): a 'Sleeping Giant' in Nutrient Sources. J. Insect Sci. **21**(2) (2021)
21. Zhu, K., et al.: Long-term phototransformation of microplastics under simulated sunlight irradiation in aquatic environments: roles of reactive oxygen species. Water Res. **173**, 115564 (2020)

EcoHide: Investigation of Properties and Durability of Fruit and Coffee-Ground-Based Biopolymer Composites for Vegan Leather Alternatives

Varshini Arunachalam(✉), Anya Malik, and Yuxin Liu

NUS High School of Mathematics and Science, Singapore, Singapore
{h2310141,h2310005,h2310074}@nushigh.edu.sg

Abstract. In our study, we investigated the mechanical properties and durability of vegan leather composed of biopolymer composites derived from fruit peels (i.e., orange and pomelo) and spent coffee grounds. With increasing environmental concerns and demand for cruelty-free products, sustainable leather substitutes are gaining traction in the fashion and upholstery industries. Composite samples were prepared and tested using cellulose-rich fruit waste and lignin-containing coffee grounds, and their tensile strength, thermal stability, and chemical resistance were analysed under various conditions. Our results show that orange peel-based composites consistently demonstrated the highest mechanical strength and thermal resistance, while pomelo-based composites showed significant improvements with optimised ingredient composition. Coffee-ground-based composites demonstrated good initial strength but were less durable in harsh conditions. None of the tested materials showed electrical conductivity, indicating their potential for applications requiring electrical insulation. By optimising material ratios and processing methods, this study hopes to highlight the potential of low-cost, biodegradable, and scalable vegan leather alternatives matching or exceeding the properties of traditional animal leather in certain applications. This work contributes to the development of sustainable materials for environmentally responsible industries.

Keywords: Vegan leather · biopolymer composites · fruit peel waste · coffee grounds · sustainable materials · sustainability · material science · biochemistry

1 Introduction and Literature Review

1.1 Introduction

In recent years, growing environmental and ethical concerns have driven increasing demand for sustainable alternatives to traditional animal leather. Traditional leather production is associated with high water consumption, toxic chemical pollution, and significant animal welfare issues. As a response, vegan leather (typically made from plant-based or synthetic materials) has emerged as a promising substitute. However, many current vegan leather options are expensive to produce, require specialized equipment, and are often manufactured only in small batches, hindering their accessibility, scalability and affordability.

H. Guo et al. (Eds.): IRC-SET 2025, PTT, pp. 244–252, 2026.
https://doi.org/10.1007/978-981-95-9630-0_22

This study aims to develop a durable, cost-effective, and accessible form of vegan leather using biodegradable biopolymer composites derived from low-cost, sustainably sourced waste materials such as orange peels, pomelo peels, and spent coffee grounds. By modifying the ingredient ratios and production methods, we aim to enhance the tensile strength, flexibility, thermal resistance, and chemical durability of these composites. This also ultimately seeks to demonstrate the feasibility of developing eco-friendly vegan leather rivalling traditional animal-based leather, while remaining economically viable for broader adoption.

1.2 Literature Review

Historical Context and the Emergence of Vegan Leather. The demand for cruelty-free leather has risen exponentially as both consumers and industries aim to reduce their environmental footprints and adhere to ethical standards. Synthetic leather materials such as polyvinyl chloride (PVC) and polyurethane (PU), as shown in Fig. 1, were first introduced in the 20th century as affordable alternatives to animal leather.

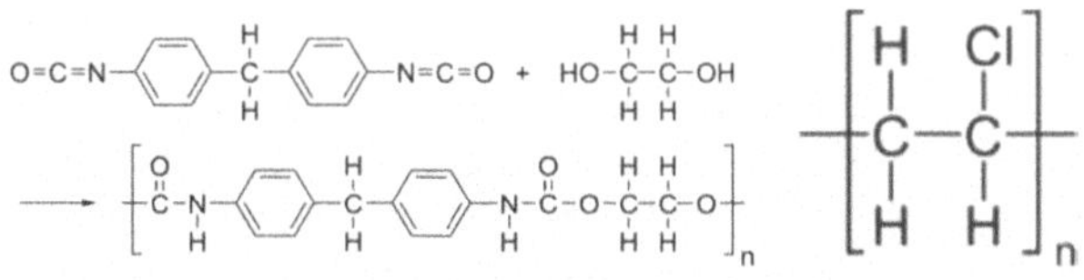

Fig. 1. From left to right, poly (urethane) synthesis and poly(vinyl chloride).

However, these materials have significant environmental drawbacks, including reliance on fossil fuels and the generation of microplastic waste. This has spurred interest in plant-based vegan leathers, which utilize renewable agricultural by-products or waste. Recent innovations include Piñatex®, a leather-like material made from pineapple leaf fibres, and mycelium-based leathers developed by companies such as MycoWorks. However, although these materials offer many ethical and ecological benefits, they still face challenges in cost, durability, and production scalability [1, 2].

Material Properties and Performance. Traditional leather has been valued for its durability and aesthetic qualities, making it the material of choice in a wide range of industries despite its harmful nature. Hence, a key focus of research conducted on vegan alternatives for leather has been their ability to replicate and mimic the physical properties of traditional, animal-based leather, including tensile strength, flexibility, and water resistance, amongst others. Studies on mycelium-based leather have shown it can replicate the texture and aesthetics of animal leather, with its durability and elasticity remaining limited for certain applications. Piñatex demonstrates promising tensile strength, but lacks sufficient water resistance and flexibility for products such as footwear and outdoor equipment. Moreover, the need for expensive equipment and high-quality processing has kept the cost of these materials relatively high, limiting widespread adoption [2, 3].

Environmental Impact. Traditional leather production is resource-intensive, requiring large amounts of water, energy, and chemicals that generate toxic waste [4]. Vegan leather made from agricultural waste such as pineapple leaves, fruit peels, or mycelium can significantly reduce this impact by reusing materials that would otherwise be discarded. For example, Piñatex uses pineapple leaves left over after harvest, minimising land and water usage [5]. Similarly, mycelium can be grown on agricultural waste and is biodegradable. However, both approaches still face challenges, such as high production costs and energy requirements for controlled growth conditions and processing. Moreover, synthetic alternatives continue to dominate a significant portion of the market, despite their reliance on fossil fuels and contribution to plastic waste. [5, 6].

Rationale Behind Ingredients. Fruit peels and coffee grounds are abundant, low-cost waste products rich in biopolymers such as cellulose, hemicellulose, pectin, and lignin. Pomelo peels contain high levels of pectin, which acts as a natural binder, providing elasticity, and hemicellulose, which improves flexibility, while cellulose provides tensile strength. Existing research also shows that the cellulose extracted from pomelo peels has high degrees of crystallinity and that it enhances the tensile strengths of composite materials significantly. Furthermore, pomelo peels contain limonene, a natural solvent and plasticiser, which aids in the breakdown of peels during processing, imparting a certain degree of antimicrobial properties to the pomelo leather. Orange peels offer similar advantages, with additional polyphenols enhancing UV resistance and durability. This increases the lifespan of composites derived from orange peels. Spent coffee grounds contain lignin for rigidity and microbial resistance. Furthermore, they are rich in cellulose, which increases flexibility and tensile strength, and contain residual oils that increase water-resistance. Additionally, the dark colour of coffee grounds can reduce the need for synthetic dyes, further lowering the environmental impact [5–8].

Accessibility and Conclusion. Although current plant-based leathers are innovative, they are often too costly for mass production due to expensive processing and limited scalability. By utilising widely available waste materials with simpler processing methods, this project aims to address cost and accessibility. In doing so, we hope to develop a biodegradable, durable vegan leather that may be produced affordably and at scale, in order to expand sustainable alternatives for industries and consumers.

2 Materials and Methods

2.1 Preparation of Biopolymer Composites with Different Vegan Materials

We prepared various samples of biopolymer composites made from three main organic materials (i.e. orange peels, pomelo peels, and coffee grounds). These materials were selected due to their respective cellulose, hemicellulose, and lignin contents, which are critical factors to the mechanical properties of vegan biopolymer composites. The methodology followed was also standardised across all batches to ensure consistency, allowing for comparative analysis.

Materials. The materials used include dried orange peels, pomelo peels, spent coffee grounds, cotton fibres, calcium chloride, sodium alginate, glycerin, coconut oil, and distilled water.

Methods. For each type of biopolymer composite, fruit peels or coffee grounds were first dehydrated in an oven or food dehydrator and ground into a fine powder. The powdered biomass was blended with sodium alginate, glycerin, coconut oil, and water in a 500mL beaker until a uniform mixture was formed. Cotton fibres were gradually added during blending in order to improve structural integrity and fibre reinforcement. A thin layer of calcium chloride solution was sprayed onto an A4 tray to act as a crosslinking agent. The mixture was spread evenly over the tray, and an additional calcium chloride solution was sprayed over the surface to initiate the biopolymer matrix's crosslinking. The composite was then dehydrated at a controlled temperature until fully dry, ensuring both sides dried evenly to prevent warping or cracking. This process was repeated for orange peel, pomelo peel, and coffee-ground-based composites.

2.2 Preparation of Biopolymer Composites with Different Fibre Content.

To investigate the effect of fibre reinforcement on mechanical properties, separate batches were prepared with varying amounts of cotton fibres (i.e. 1.2g and 1.6g per batch). The same mixing, casting, crosslinking, and drying procedures were followed as described in Section 2.1.

2.3 Testing the Properties of the Biopolymer Composites.

Testing the Tensile Strength. Strips measuring 2 cm × 5 cm were cut from each dried composite sample. Using weight hangers and standard weights (50 g increments), tensile strength was measured by gradually adding weights until the sample tore. The maximum load sustained was recorded. Each test was repeated twice per sample type for reliability, and average values were calculated to two decimal places.

Testing the Electric Conductivity. Each sample was tested for electrical conductivity using a simple closed circuit with a battery, crocodile clips, and a light bulb. If the light bulb was illuminated, the sample was considered conductive. Tests were repeated twice for consistency.

Testing for Chemical Resistance. Strips were submerged in either dilute hydrochloric acid (HCl) or dilute sodium hydroxide (NaOH) solutions for 5 minutes. After chemical exposure, the strips were rinsed, dried, and re-tested for tensile strength and conductivity in order to assess any degradation. Each test was repeated twice per sample type for reliability.

Testing for Thermal Resistance. Samples were submerged in boiling distilled water for 5 minutes to simulate heat exposure. After removal, strips were dried and subjected to the same tensile strength and conductivity tests. Each test condition was verified twice to ensure consistency.

2.4 Data Analysis

Data recorded from the above mechanical, chemical, and thermal tests were tabulated and compared across batches and material types to evaluate relative performance. Furthermore, strength retention under simulated harsh conditions was used to assess durability. The results from this section may be found in Table 1.

3 Results and Discussion

Through our investigation, we have evaluated both the mechanical properties and the typical behaviour of our biopolymer leathers (made from pomelo, orange, and coffee fibres and pulp). We tested the electrical conductivity and mechanical strength of these leather samples under various conditions: normal (untreated), basic (0.1M NaOH), acidic (0.1M HCl), and boiling (100 °C in DI water), designed to simulate a sped-up version of natural wear-and-tear and erosion processes. Results across these three batches have allowed us to compare the strengths and weaknesses of the different leathers to each other and also assess how each type responded to various environmental stresses, allowing us to deduce the "best" fibre for organic biopolymer-based leather.

Table 1. Table of Strengths of Biopolymer Composite Batches under Various Conditions. (Note: > 1.25 indicates maximum weight limit of test setup was reached.)

Batch	Condition	Pomelo/kg	Orange/kg	Coffee/kg	Electrical Conductivity
Batch 1	Normal	0.60	> 1.25	1.10	No
	Base	0.45	0.40	0.50	No
	Acid	0.25	0.90	1.00	No
	Boil	0.70	0.60	0.85	No
Batch 2	Normal	0.80	1.10	1.05	No
	Base	0.70	0.80	0.65	No
	Acid	0.30	0.40	0.30	No
	Boil	0.50	0.80	0.50	No
Batch 3	Normal	> 1.25	> 1.25	> 1.25	No
	Base	> 1.25	> 1.25	0.80	No
	Acid	0.85	> 1.25	0.70	No
	Boil	> 1.25	> 1.25	0.75	No

3.1 Electrical Conductivity

None of the tested biopolymer composites demonstrated electrical conductivity under any condition. This would be due to the organic composition of the leather, as the structural components of the fibres (i.e. cellulose, hemicellulose, and lignin-based matrices)

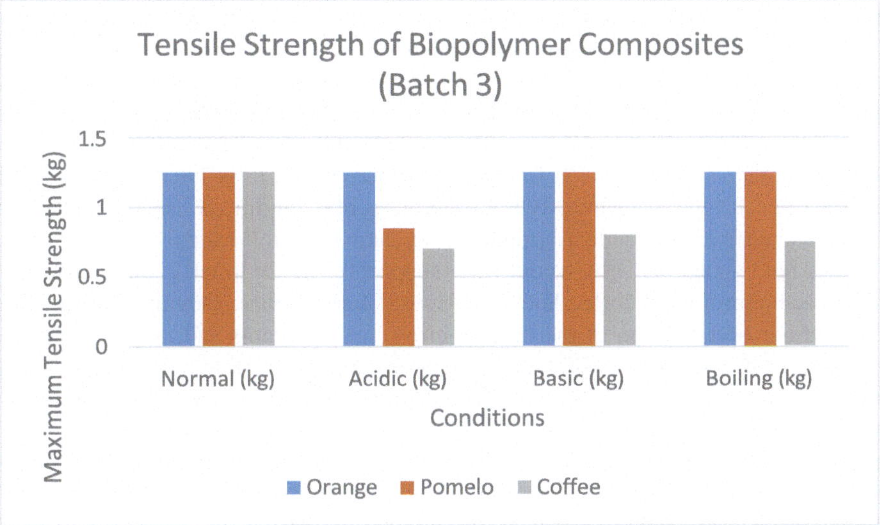

Fig. 2. Comparison of maximum tensile strength of orange peel, pomelo peel, and coffee-ground composites under normal, acidic, basic, and boiling conditions (Batch 3).

are effective insulators of electricity. Moreover, no conductive additives such as metals or ionic polymers were incorporated into the formulations, blocking the facilitation of electron flow. Hence, this may be advantageous for applications requiring electrical insulation, such as protective coverings or wearable technology.

3.2 Mechanical Strength

Batch 1. In initial tests, orange peel-based composites demonstrated the highest strength, especially in normal conditions, supporting weights exceeding 1.25 kg. Coffee leather showed a comparatively good performance, handling 1.10 kg. However, pomelo leather was significantly weaker and could only hold on to 0.60 kg. However, when put under acidic and basic conditions, the mechanical properties of all leathers decreased sharply. Despite this, orange leather was still the second-most resilient and could still support 0.9 kg in acidic conditions, while coffee leather could support 1.0 kg. However, pomelo leather was the least resistant in acidic conditions, with its weight capacity dropping to 0.25 kg. This may be due partly to the higher hemicellulose content in pomelo fibres, which are more sensitive to pH changes. When boiling tests were carried out, orange leather was still heat-resistant and could support 0.6 kg, while coffee leather could support 0.85 kg. However, pomelo leather now demonstrated a higher resistance of 0.7 kg as compared to orange leather. From Batch 1, we could conclude that orange leather steadily showed good properties across all conditions in Batch 1 as compared to pomelo and coffee leather, which appeared relatively strong under certain conditions but weak under others. Overall, pomelo leather demonstrated the weakest and poorest mechanical strength, particularly in acidic, basic, and normal conditions.

Batch 2. Modified fibre content in this batch resulted in higher and more consistent strengths across all materials. The mechanical properties of pomelo leather improved significantly, and coffee leather also showed slightly higher resilience, while orange leather displayed consistent strength. When tested under normal conditions, orange leather could support 0.8 kg, coffee leather 1.1 kg, and pomelo leather showed a significant improvement, now able to support 0.8 kg. In this batch, there were closer intervals between the strengths of each type of leather. When tested in acidic conditions, orange leather still continued to perform best, able to support 0.4 kg, and it was followed by coffee leather at 0.3 kg and pomelo leather, which could also support 0.3 kg. In Batch 2, the performance gap between pomelo and coffee leather narrowed significantly. This may be caused as a result of our optimisation of the ratio of cross-linking agents and may have led to a more uniform structure across all types of leather samples. Next, in boiling tests, orange leather was still the best performer and could support 0.8 kg, while coffee and pomelo leather both supported 0.5 kg. In this batch, the thermal resistance of coffee and pomelo leather dropped. From Batch 2, we concluded that orange leather remained the strongest, especially in extreme conditions like boiling and acidic environments. Although pomelo leather lagged slightly behind orange in strength, particularly in acidic conditions, it still improved significantly. Coffee leather was also weaker in Batch 2 as compared to Batch 1.

Batch 3. Optimising the ratios of sodium alginate, calcium chloride, and glycerin, as in this batch, yielded the strongest results overall. The mechanical strength of all types of leather increased, especially for pomelo and coffee leather. This may be due to our modified formulation, material ratios, and methods. When under normal conditions, all three types of leather were able to support weights of more than 1.25 kg, and this shows a significant increase in strength. Orange leather was still the best performer, but pomelo leather also closed the gap, and this shows that optimising the glycerin, calcium chloride, and sodium alginate ratio enhanced the strength of the leather samples. Next, in acidic conditions, orange leather was still able to support more than 1.25 kg, while pomelo leather improved significantly, able to support 0.85 kg, and coffee leather, while still showing improvements, lagged slightly, and was only able to support 0.7 kg. The stronger crosslinking in the formation of pomelo leather may also have aided in helping it maintain its structure even under acidic conditions, and it performed best in Batch 3. Next, when boiling tests were conducted, orange and pomelo leather could both support more than 1.25 kg and outperformed coffee composites, which could only support 0.75 kg. From Batch 3, we concluded that orange leather remained the best overall performer, but that pomelo leather showed impressive and significant improvements, especially in boiling and acidic conditions. Coffee composites were also less resilient in boiling and acidic conditions in this batch, despite their good performance under normal conditions.

3.3 Comparative Discussions

Across all three batches, orange composites consistently demonstrated the highest mechanical strength and durability, especially under challenging conditions such as acidic or boiling environments. This can be attributed to their naturally higher cellulose and lignin content, which helps provide excellent tensile strength and thermal stability as well. Despite their generally strong performance in normal conditions, coffee composites was more sensitive to acidic and thermal conditions than orange composites. As compared to orange composites, their lower lignin content may explain this, affecting rigidity and resistance to chemical breakdown. However, coffee composites' relatively high cellulose content could have contributed to their initial strength, which was maintained across the batches, despite some minimal susceptibility to acidic conditions. Last, when compared from Batch 1 to Batch 3, pomelo composites showed the most dramatic improvement, despite their initial weak performance. Pomelo composites also bridged the performance gap by Batch 3, mainly in boiling and acidic conditions, similar to orange composites. This can be attributed to the interaction between sodium alginate and calcium chloride, along with the refined concentration of crosslinking agents, which thereby assist in facilitating stronger bonds between the fibers, seen from the "egg-box" model of gelation (Figure 3) [9]. Hence, this, in turn, causes greater mechanical strength and resistance to environmental strength as well, enhancing its performance.

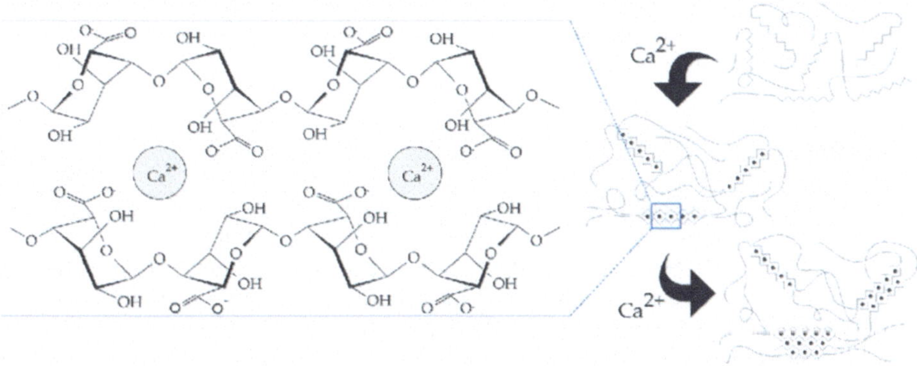

Fig. 3. The "egg-box" model of gelation of alginate by calcium.

4 Conclusion and Future Directions

This study demonstrates the feasibility of producing durable, biodegradable vegan leather alternatives using biopolymer composites derived from fruit peels and spent coffee grounds. By varying the fibre content and crosslinking ratios systematically, we were able to achieve notable improvements in mechanical strength and durability, particularly for orange-peel based composites, which consistently outperformed other materials. Pomelo-based composites showed significant improvements as well with optimised formulation ratios, while although coffee-ground-based composites offered good initial strength, they proved more sensitive to acidic and thermal stress. The main reason for these differences might be due to the biopolymer composition of each fibre, and their cellulose, lignin, and hemicellulose content. Optimising crosslinking agents (i.e. sodium alginate, calcium chloride) can also help improve the mechanical and thermal properties of our biopolymer composites.

Although we were able to enhance the properties of vegan biopolymer composites, further work is required to improve their consistency and replicate the qualities of animal leather. In the future, we plan to conduct further research regarding refining the ratios of biopolymers, fiber reinforcement, and crosslinking agents (e.g., sodium alginate, calcium chloride, glycerin) to maximize strength and thermal resistance, in order to be tailored to specific conditions. Additional biodegradable materials such as banana peels, guava skins, or seaweed will also be investigated to diversify both properties and performance. Furthermore, future work will focus on expanded surface treatments and coatings, such as eco-friendly leather coatings in the form of natural resins, to enhance water resistance, UV durability, and the overall lifespan. Moreover, long-term performance testing will be explored by assessing the behaviour of produced biopolymer composites under real-world conditions, such as exposure to ultraviolet (UV) radiation, humidity fluctuations, and mechanical wear. We would also focus on expanding the incorporation of bacterial cellulose or nanofibers to further increase tensile strength and tear resistance, through the biopolymer matrix.

By addressing these areas, we aim to develop cost-effective, scalable vegan leather alternatives that offer a sustainable, ethical solution for industries traditionally reliant on animal-based leather. This research contributes to the growing field of sustainable consumer materials, focusing on reducing environmental impact through innovative reuse of agricultural and organic waste.

References

1. PETA: Environmental Hazards of Leather. https://www.peta.org/issues/animals-used-for-clothing/leather-industry/leather-environmental-hazards/. (Accessed: July 2025)
2. Håkansson, E., Carter, N., Coen, L., LaBarbera, N.: Leather's Impact on the Planet. Changing MarketsFoundation(2022). https://static1.squarespace.com/static/5f5f02dd9b510014eef4fc4f/t/6386865fa112a35adea84ccd/1669760650422/CFJ+leather%27s+impact+on+the+planet+%28launch%29.pdf
3. Serban-Bondrea, D., Mocanu, R.: the impact of the leather industry left on the environment. J. Environ. Prot. Ecol. **17**(1), 273–280 (2016)
4. Sivaram, N.M., Barik, D.: Toxic waste from leather industries. In: Energy from Toxic Organic Waste for Heat and Power Generation, Woodhead Publishing Series in Energy, pp. 55–67. Elsevier, Amsterdam (2019). https://doi.org/10.1016/B978-0-08-102528-4.00005-5
5. Yaradoddi, J.S., et al.: Bio-based material from fruit waste of orange peel for industrial applications. J. Mater. Res. Technol. **17**, 3186–3197 (2022). https://doi.org/10.1016/j.jmrt.2021.09.016
6. Santos, L.B., et al.: Bioplastics from orange processing byproducts by an ecoefficient hydrothermal approach. Food Packag. Shelf Life **38**, 101114 (2023). https://doi.org/10.1016/j.fpsl.2023.101114
7. Michael-Igolima, U., et al.: Modified orange peel waste as a sustainable material for ad-sorption of contaminants. Materials (Basel) **16**(3), 1092 (2023). https://doi.org/10.3390/ma16031092
8. Chuangchote, S., et al.: Development of biodegradable polymeric leather from pomelo peel. Polymers **14**(3), 518 (2022). https://doi.org/10.3390/polym14030518
9. Yadav, M., Kumar, V., Baweja, M., Shukla, P.: Alginate and Sericin: environmental and pharmaceutical applications. ResearchGate. https://www.researchgate.net/publication/312344348_Alginate_and_Sericin_Environmental_and_Pharmaceutical_Applications

Immobilising Laccase onto *Moringa Oleifera* Seeds and Husks as Novel Bioadsorbents for Acetaminophen Removal

Asnith Muniraju(✉), Daisy Liow Han Yi, Jane Therese Ng, Grace Lim, and Abigayle Ng

Raffles Institution (Students), 1 Raffles Institution Ln, Singapore 575954, Singapore
25YMUNI291F@student.ri.edu.sg

1 Introduction

Paracetamol, also known as acetaminophen (ACT), a widely used pain reliever, has become a concerning pharmaceutical contaminant in water sources [1]. After human excretion, ACT often escapes complete removal in wastewater treatment plants, leading to bioaccumulation in aquatic systems, thus disrupting ecosystems and inhibiting plant and animal growth [2]. Furthermore, toxic byproducts of ACT can also cause adverse long-term health effects from the consumption of contaminated water. [3] Conventional removal methods, such as chlorination or membrane filtration, are often energy-intensive, costly, and unsustainable [4]. Furthermore, rural and underserved areas frequently lack access to such technologies, exacerbating the issue [5]. Referencing a case study on ACT pollution in Indian waters, high concentrations of the drug detected raised concerns about the risks associated with long-term exposure and the impact on agriculture [6]. Given pharmaceuticals' status as emerging contaminants, further research and solutions are needed. Addressing these challenges requires innovative, eco-friendly technologies that can cater to such communities, especially ones that may have limited access to costly industrial equipment.

Immobilising laccase enzymes on natural supports, such as *Moringa oleifera* seeds (MOS) and seed husks (MOH), is one potential strategy to address the challenges above. Native to the Indian subcontinent and well-known for its medicinal properties, *Moringa oleifera* has been highlighted in a number of adsorption studies due to its capacity to remove biological contaminants from water sources [7]. The seed husks are typically thrown away as biowaste, while the fruit and leaves are usually consumed. Additionally, laccase, a multipurpose oxidative enzyme, is found in *Trametes versicolor*, a species of white-rot fungus. Numerous organic pollutants, including ACT, have been shown to be broken down by it into less toxic compounds [8]. In this study, we aim to immobilise laccase onto Moringa seeds and seed husks to not only enhance the enzyme's stability, but also add value to an abundant biowaste material. Furthermore, we aim to investigate such biosorbents' potential as practical and sustainable solutions in removing ACT from polluted water, particularly in underserved areas.

H. Guo et al. (Eds.): IRC-SET 2025, PTT, pp. 253–271, 2026.
https://doi.org/10.1007/978-981-95-9630-0_23

2 Materials and Methods

2.1 Preparation of Moringa-Derived Biosorbents (MDB)

Dried *Moringa oleifera* seeds were acquired from a local supplier (Lao Ban Niang, Geck Huat Trading Pte Ltd), deshelled, washed and oven dried overnight at 40°C. The seeds and seed husks were then ground and sieved to 1000 m. The resulting powders, MOS and MOH respectively, were subsequently stored at 25°C.

2.2 Immobilisation of Laccase onto MDB

To immobilise laccase onto the MDB supports, 40g of MDB (MOS or MOH) was added to 400 mL of stock sodium acetate buffer solution (pH 5, 0.2 M) (Sigma-Aldrich) with 5% glutaraldehyde (Sigma-Aldrich). This mixed suspension was placed in a constant temperature oscillator at 25 °C for 3 h at 150 rpm. Thereafter, 40g of laccase (Creative Enzymes) was added to the suspension before shaking for another 3 h. Mixture was filtered to obtain the residue, which was then washed with buffer. The final solid obtained, laccase-immobilised *Moringa oleifera* seed (L.MOS) or laccase-immobilised *Moringa oleifera* husk (L.MOH) respectively, was dried at 37 °C in an incubator oven overnight and stored at 25 °C.

2.3 Characterisation of Moringa-Derived Biosorbents (MDB)

Surface morphology of the 4 samples was analysed using a JEOL JSM-6010LA scanning electron microscope (SEM) from Science Centre Singapore. BET (Brunauer–Emmett–Teller) analysis of resolution 2–50 nm was done using ASAP 2020 Plus 2.00 analyser in a certified lab CRADLΣ @ Science Centre Singapore.

2.4 Enzyme Assay

The activities of free laccase, L.MOS and L.MOH were determined through a colorimetric assay using a UV–VIS spectrophotometer (Shimadzu UV-1280) at 420 nm. Activity was determined by adding 0.02 g of L.MOS, L.MOH or free laccase to 5.7 ml of stock sodium acetate buffer and 0.3ml of 1mM ABTS (Merk). Absorbance was measured at 2, 4, 6 and 8 min. One unit (U) of laccase was defined as the amount of laccase that can oxidize 1 μmol of ABTS per min under the assay conditions, where the activity is expressed as U/kg. The following equations were used for calculating immobilized and free enzyme activity [9]:

$$\text{Immobilized laccase activity } /\text{U kg}^{-1} = \frac{\Delta \times \text{ab} \times \text{Df} \times \text{Rv}}{\epsilon \times t \times \text{Msupport}}$$

$$\text{Free laccase activity } /\text{mL} = \frac{\Delta \times \text{ab} \times \text{Df} \times \text{Rv}}{\epsilon \times t \times v}$$

where Δab is the absorbance, D_f is the dilution factor, R_V is the reaction volume (mL), ϵ (3.6×10^4 M^{-1} cm^{-1}) is the molar extinction coefficient, *t* is the reaction time (min),

v is laccase amount (mL) and $M_{support}$ is mass of the support (kg). The immobilisation yield [10] is calculated by:

$$Yield\ (\%) = 100 \times \frac{Immobilisedactivityoflaccase}{Activityoffreelaccase}$$

2.5 Adsorption Experiments

Adsorption capacity for all experiments was analysed using a UV-VIS spectrophotometer at 243 nm. Standard ACT solutions (5, 10, 15, 20, 25 mg L^{-1}) (Sigma-Aldrich) were used for a calibration curve.

For kinetic studies, 0.450 g of each MDB sample was added into 4 separate bottles containing 300 mL of 25 mg L^{-1} ACT solution. Samples were shaken at 150 rpm at room temperature, where 2 mL samples were aliquoted from the mixture at 0, 10, 30, 60, 120, 180, 1440 min (MOS, MOH) and 0, 10, 30, 60, 120, 180, 300, 420, 1440, 2880 min (L.MOS, L.MOH) respectively. Samples were centrifuged at 14000 rpm for 5 min and filtered using a 0.20 μ m syringe filter before analysing absorbance.

For batch adsorption studies, 0.450 g of each MDB sample was added into 4 separate bottles containing 300 mL of ACT solution with initial ACT concentrations of 5, 10, 15, 20 and 25 mg L^{-1}. Samples were shaken at 150 rpm at room temperature before aliquoting samples after 24h (MOS, MOH) and 48h (L.MOS, L.MOH).

2.6 MDB Reusability

MDB reusability was analysed over 3 successive adsorption cycles over 4 h using 50 mg L^{-1} ACT dosage. Following each cycle, MDB samples were filtered from the solution, rinsed with standard sodium acetate buffer solution, and dried in an incubator at 37 °C for 24 h before the next use. Percentage adsorption [11] was calculated using the following equations:

$$\text{Percentage Adsorption}(\%) = (\mathrm{C_0} - \mathrm{C_t}/\ \mathrm{C_0}) \times 100\%$$

where C_0 (mg L^{-1}) and C_t (mg L^{-1}) are the initial concentrations of the solution and after time t respectively.

2.7 Data Analysis

Adsorption Kinetics
Kinetic studies were analysed using the following models:

Pseudo-First-Order model (PFO)

$$q_t = q_e (1 - e^{-(k_1 t)})$$

Pseudo-Second-Order model (PSO)

$$q_t = \frac{k_2 q_e^2 t}{1 + k_2 q_e t}$$

Intraparticle diffusion

$$q_e = k_3 \cdot t^{1/2}{}_E + C_1$$

Elovich Model

$$q_t = \frac{1}{\beta_E}(\alpha_E \beta_E) + \frac{1}{\beta_E} lnt$$

Bangham Model

$$log(log\frac{C_o}{C_o}) = log(\frac{k_5 m}{2.303V}) + \alpha logt$$

Avrami Model

$$ln(ln\frac{q_e}{q_e - q}) = n_{av} lnk_4 + n_{av} lnt$$

Where q_t and q_e are the amount of ACT absorbed at time t and equilibrium respectively (mg/g^{-1}) of the MDB samples. For PFO non-linear modelling was employed with a graph of q_t against t. For PSO, a graph of $\frac{t}{qt}$ against t was plotted, where the slope gradient is $\frac{1}{qe}$ and the vertical intercept is $\frac{1}{kqe^2}$. k_1 and k_2 are the rate constants for PFO and PSO reactions respectively. For intraparticle diffusion, a graph of q_t against $t^{1/2}$ was plotted, where the slope gradient is K_3, the intraparticle diffusion constant and the vertical intercept is C [12]. For the Elovich model, a graph of q_t against $ln\ t$ was plotted, where the slope gradient determines α_E, the initial rate coefficient for adsorption and β_E, the rate of desorption. For the Avrami model, *(ln - (ln (1- α))* was plotted against $ln\ t$ to derive k_5, the Avrami rate constant (min^{-1}) from the intercept, and n_{av}, the Avrami dimensionless number from the slope. Lastly, for Bangham kinetics, *log (Co/C0 - mqt)* was plotted against $ln\ t$, to derive k_4, the rate constant of adsorption, and m, the dimensionless number of adsorption intensity [13, 14].

Adsorption Isotherm Studies

The adsorption isotherm studies were analysed using the Langmuir and Freundlich isotherm models.

3 Results and Discussion

3.1 Characterisation of Moringa-Derived Biosorbents

SEM images (Fig. 1) show that MOS and MOH vary in structure: MOS particles are spherical, while MOH has a distinct fibrillar morphology. Laccase immobilisation onto MOS modified its surface texture significantly, causing L.MOS to display striations with irregularly-shaped vessels on its surface. Its surface appears rougher, which can be attributed to the addition of glutaraldehyde during laccase immobilisation [15]. Additionally, laccase immobilisation onto MOH created more porous structures with honeycomb structures.

There was an increase in BET surface area from MOS to L.MOS (Table 1). This may be due to the immobilization of laccase introducing additional pores, thereby increasing available surface area for adsorption [16]. The decrease in surface area from MOH to

L.MOH could be due to enzyme blockage of surface pores from cross-linking with glutaraldehyde, reducing availability of open pores [16]. Pore sizes of MOS and L.MOS vary from 70.6–86.9 nm, indicating a structure of mainly macropores (>50 nm), while that of MOH and L.MOH vary from 12.6–12.7 nm, indicating a structure of mainly mesopores (2–50 nm).

Table 1. BET surface area, pore volume, pore width of MDB incorporated in MOS, L.MOS, MOH, L.MOH

Sample	Surface area/m^2 g^{-1}	Total pore volume/cm^3 g^{-1}	Average pore diameter/nm
MOS	0.0626 ± 0.0207	0.0011	70.5576
MOH	0.6252 ± 0.3640	0.0020	12.6208
L.MOS	0.1080 ± 0.0027	0.0023	86.8694
L.MOH	0.5225 ± 0.2290	0.0017	12.6819

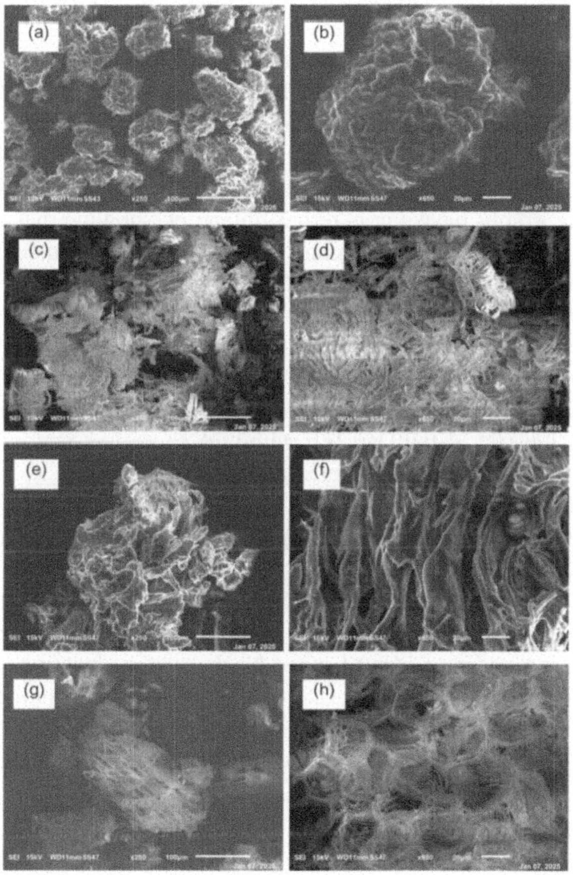

Fig. 1. Surface morphology of MDB incorporated in: (a) MOS at 100 μm, (b) MOS at 20 μm, (c) MOH at 100 μm, (d) MOH at 20 μm, (e) L.MOS at 100 μm, (f) L.MOS at 20 μm, (g) L.MOH at 100 μm, (h) L.MOH at 20 μm

3.2 Immobilization of Laccase on MDB

L.MOH and L.MOS were found to have an activity of 66.667 U/kg and 71.903 U/kg respectively and free laccase activity was calculated to be 144.442 U/kg. (Fig. 2) The immobilisation yield of L.MOH was 46.155% and L.MOS was 49.780%.

3.3 Adsorption Kinetics

The kinetic studies show an increase in adsorption of ACT with increased contact time. Both MOS and MOH plateaued off at 180 min, while L.MOS and L.MOH did not plateau after 2880 min. The adsorption efficiency of the biosorbents were MOH > MOS > L.MOH > L.MOS before 180 min and L.MOH > L.MOS > MOH > MOS after 180 min (Fig. 3).

The adsorption kinetics were analysed by fitting the data to 6 different kinetic models. Kinetic parameters were listed in Table 2. PFO provided a poorer fit ($R^2 = 0.236-0.863$) for MOS, MOH and L.MOS. However, PFO provided an excellent fit for L.MOH ($R^2 = 0.989$), thus showing that its adsorption rate is controlled by the diffusion of ACT on its surface [17].

The application of PSO proved a better fit for MOS and MOH ($R^2 = 1$) along with the comparable experimental and calculated q_e values. This shows that the adsorption rate is controlled by chemisorption[18]. Meanwhile, PSO was an incorrect fit for L.MOH and L.MOS.

L.MOS and L.MOH had a good level of fit for intraparticle diffusion with $R^2 \geq 0.904$. These linear graphs have a non-zero X-intercept, showing that adsorption kinetics is controlled by both film diffusion and intraparticle diffusion sequentially. The C value of 0.176 corresponds to the effect of the boundary layer on adsorptions [19]. The negative C value of -1.330 of L.MOH shows the combined effects of film diffusion and surface reaction control [20]. MOS had multi-linearity and non-zero intercepts, showing that multiple mechanisms controlled the process. In the initial step, external surface adsorption occurs [20] ($R^2 = 0.987$) and in the 2nd step there is gradual adsorption by intraparticle diffusion ($R^2 = 0.953$) [21]. Although there is a new segment observed after 180 min, there was insufficient data to propose another step.

Only MOS yielded good fits for the Elovich and Avrami models. For the Elovich model, MOS had a good fit ($R^2 = 0.974$), showing that the interaction occurs via chemisorption and that the effect of desorption and interaction between adsorbed adsorbents on adsorption is insignificant [21]. It exhibits strong surface heterogeneity [20]. For the Avrami kinetic model, MOS yielded a good fit ($R^2 = 0.870$). Since the value of $n_{av} < 1$, the adsorption process is not limited by the surface reaction [21]. MOH had a good fit for the Bangham model ($R^2 = 0.870$) showing that intra-particle diffusion and pore diffusion are rate determining steps while the poorer fitting of the Bangham model ($R^2 = 0.581-0.841$) for the rest shows that intra-particle diffusion and pore diffusion are not the only rate determining steps [12, 21].

Table 2. (a) Pseudo-Order Kinetic Parameters for ACT adsorption (b) Intraparticle diffusion Kinetic Parameters for ACT adsorption (c) Elovich and Bangham Kinetic Parameters for ACT adsorption (d) Avrami model Kinetic Parameters for ACT adsorption

(a)

Sample	Pseudo-first-order model			Pseudo-second-order model		
	k_1 /min^{-1}	q_e /mg g^{-1}	R^2	k_2 /min^{-1}	q_e /mg g^{-1}	R^2
MOS	0.0352	2.545	0.864	0.0152	2.871	1
MOH	1.946	4.106	0.236	0.303	3.593	1
L.MOS	0.00106	4.482	0.821	0.00032	5.043	0.840
L.MOH	0.00013	35.528	0.989	7.36E-06	-19.139	0.0166

(b)

Sample		Intraparticle diffusion		
		K_3(mg/g/min$^{1/2}$)	C_1	R^2
MOS	First Segment	0.299	0.0456	0.987
	Second Segment	0.121	0.991	0.953
MOH	First Segment	0.45	1.751	0.318
	Second Segment	0.0112	3.186	0.445
L.MOS		0.0768	0.176	0.904
L.MOH		0.213	-1.339	0.928

(c)

Sample	Elovich model			Bangham model		
	α_E (mg g^{-1} min^{-1}	β_E (g mg^{-1})^1	R^2	K_4 /min^{-1}	α_B	R^2
MOS	0.510	2.042	0.974	4.530	0.194	0.841
MOH	-	-	-	1.566	0.165	0.870
L.MOS	0.053	1.715	0.617	0.0167	1.0364	0.776
L.MOH	0.042	0.572	0.666	1.402	0.308	0.581

(d)

Sample	Avrami model		
	K_5 (min^{-1})	α_{av}	R^2
MOS	0.003281	0.2739	0.870
MOH	-	-	-
L.MOS	0.000201	0.318	0.411
L.MOH	7.584 E-05	0.590	0.387

3.4 Concentration Studies

Generally, a positive relationship (Fig. 3) can be established between ACT dosage and MDB adsorption capacity. Best adsorption (1.564–12.019 mg g^{-1}) across all concentrations was achieved by L.MOH, where it peaked at 50 mg g^{-1} ACT dosage.

L.MOS achieved lower adsorption rates (1.083–5.971 mg g^{-1}), likewise peaking at 50 mg g^{-1} ACT dosage. Meanwhile, MOS and MOH's adsorption ranged from 1.000–2.827 mg g^{-1} and 1.039–2.951 mg g^{-1} respectively.

3.5 Adsorption Isotherms

The Freundlich model fits all samples better, with higher R^2 values (0.9098–0.9784) compared to Langmuir (0.036–0.537) (Table 3) using linear regression fitting. This supports the theory that the surface of the MDB samples were heterogeneous and that adsorption of ACT on the MDB samples were multilayered. This was consistent with other studies where ACT adsorption mechanisms followed Langmuir or Freundlich isotherm models [22, 23]. This is corroborated by SEM analysis, which revealed rough porous surfaces on MDB samples, where functional groups are randomly distributed such that adsorption sites possess variable adsorption energies, hence inducing multilayer adsorption [24].

Additionally, the n values of MDB samples were mostly more than 1, implying support of physical process during adsorption so that ACT was easily adsorbed onto the MDB surfaces. For L.MOH however, $n < 1$, which could be due to polar functional groups on ACT molecules competing with water molecules for adsorption sites [25]. k_f values were higher in L.MOH and L.MOS than in MOH and MOS, indicating higher adsorption capacities.

Table 3. Langmuir and Freundlich parameters for ACT adsorption

Sample	Langmuir Model			Freundlich Model		
	k_L /mg L-1	q_m /mg g-1	R^2	n	k_f /mg g-1 (L mg^{-1})$^{1/n}$	R^2
MOS	0.048	1.100	0.392	1.668	3.823	0.911
MOH	0.042	1.178	0.372	1.611	4.108	0.910
L.MOH	-0.010	-23.033	0.537	0.820	6.225	0.978
L.MOS	0.002	90.234	0.036	1.001	6.328	0.965

3.6 MDB Reusability

Over 3 cycles, percentage adsorption by L.MOH was reduced by 82.6%, while that of L.MOS increased by 10.0%. Percentage adsorption of L.MOH from cycles 1 to 3 decreased from 100% to 41.4% to 17.4% (Fig. 4), which could be due to lost biosorbent load during successive adsorption cycles or leaching of laccase from adsorbent during rinsing [16]. In contrast, percentage adsorption of L.MOS from cycles 1 to 3 increased from 100% to 121.9% to 110.0%, which could be attributed to the enhanced stability of laccase after immobilisation, thus increasing catalytic activity and ensuring improved results in subsequent uses [26]. This implies that L.MOS can be effectively reused, enhancing its practicality and cost-effectiveness.

4 Conclusion

Two novel biosorbents were derived from *Moringa oleifera* seeds and husks, enhanced by laccase immobilisation onto the supports. Laccase was successfully immobilised with a 46.2 - 49.8% yield and the bioadsorbents exhibited high reusability, notably L.MOS with increased adsorption by 10.0% over 3 cycles. ACT adsorption was found to fit the Freundlich isotherm model, indicating multilayer adsorption. L.MOH is the most effective adsorbent, with maximum adsorption of 12.0 mg g^{-1}. Both L.MOH and L.MOS exhibited higher adsorption compared to their unmodified counterparts. For kinetics, MOS had IPD, PSO, Elovich and Avrami as best fits, MOH had PSO and Bangham the best fits, L.MOS had IPD and PFO as best fits and L.MOH had IPD as best fit. Thus, seed husk waste can be repurposed as an effective and sustainable biosorbent, which caters to our target communities' needs. Further research can be done to explore such biosorbents' effectiveness in removing other pollutants, or optimising immobilisation methods.

Appendix

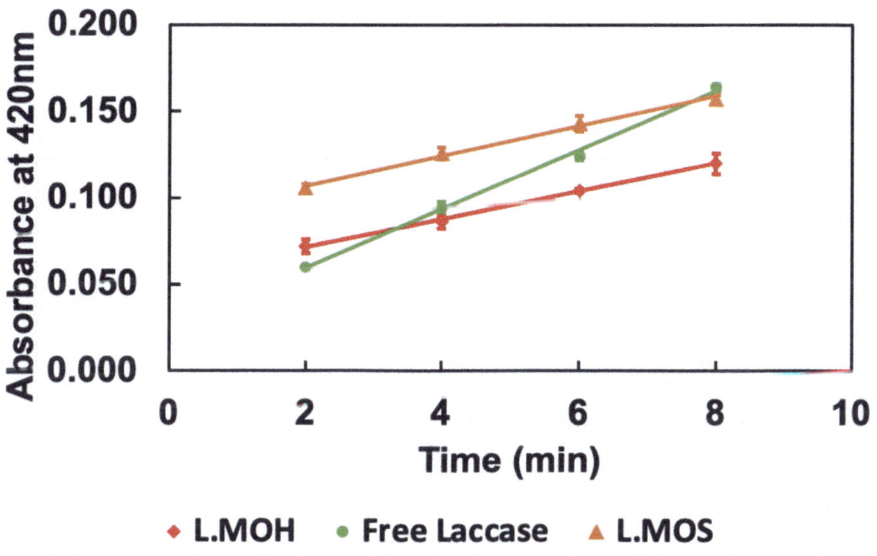

Fig. 2. ABTS enzyme assay change in absorbance at 420 nm over time (error bars are within the points)

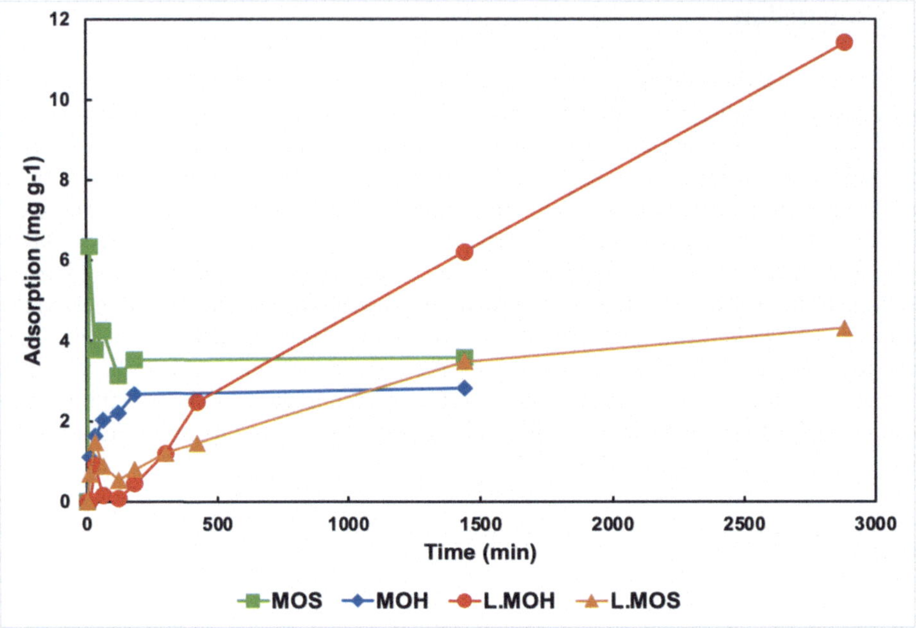

Fig. 3. Graph of raw ACT adsorption kinetics (error bars are within the points)

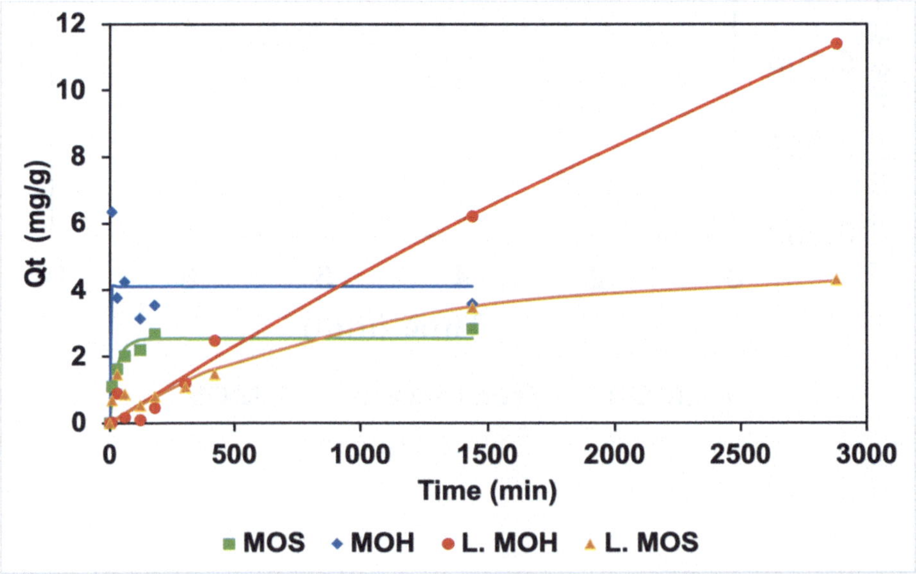

Fig. 4. Kinetics of ACT adsorption using Pseudo-first-order model

See (Fig. 5)

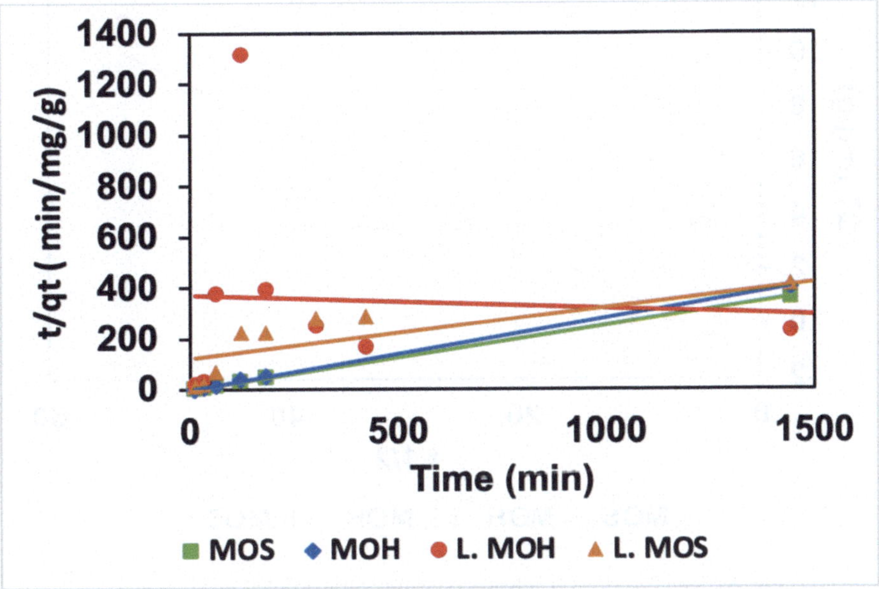

Fig. 5. Kinetics of ACT adsorption using Pseudo-second order model

See (Fig. 6)

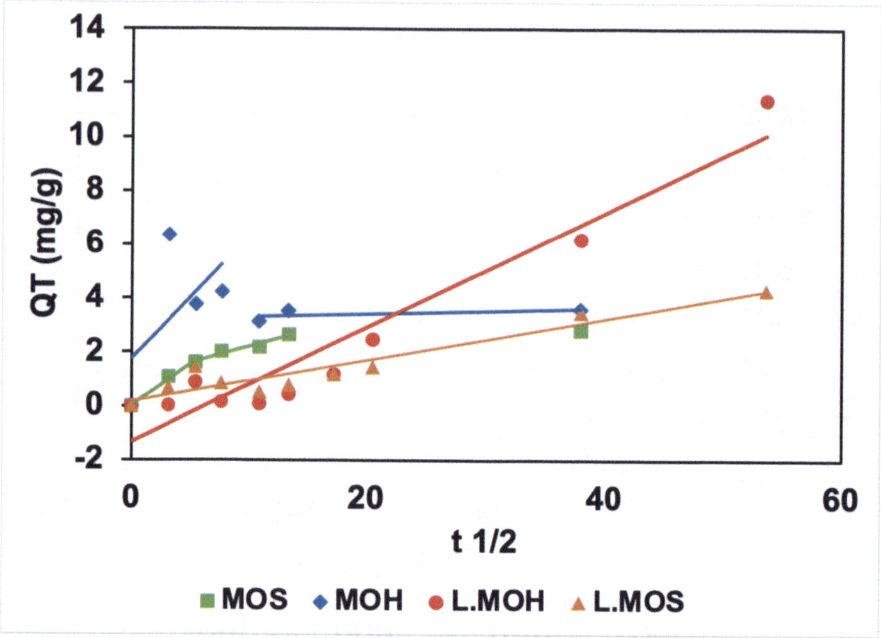

Fig. 6. Kinetics of ACT adsorption using intraparticle kinetic model

See (Fig. 7)

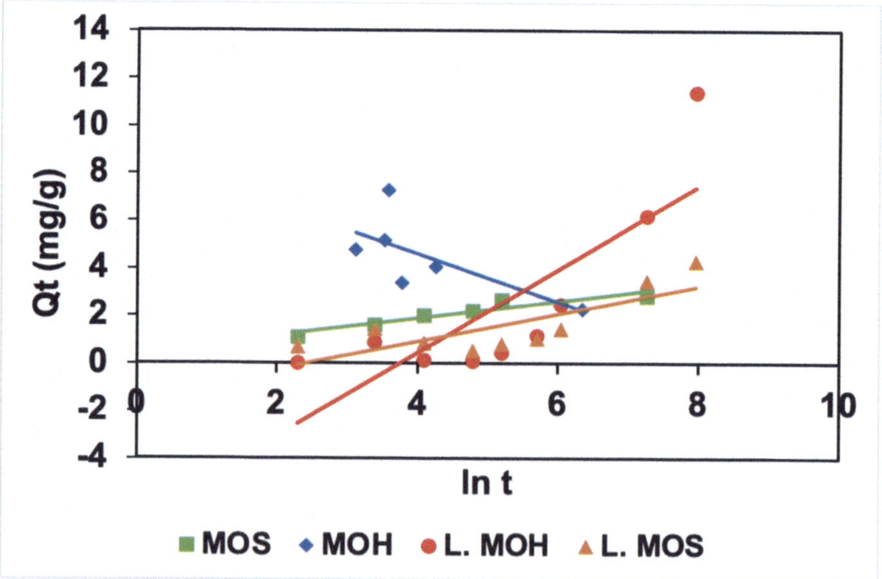

Fig. 7. Kinetics of ACT adsorption using Elovich kinetic model

See (Fig. 8)

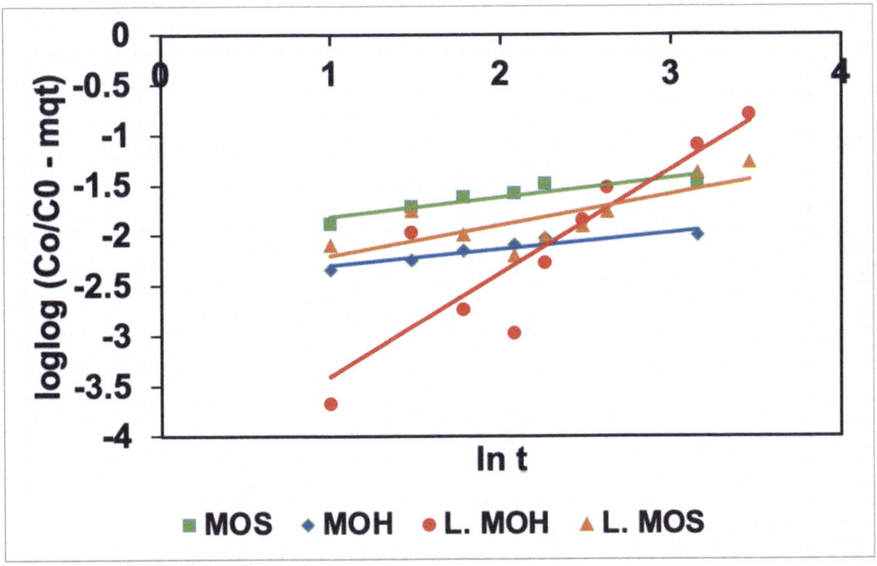

Fig. 8. Kinetic of ACT adsorption using Bangham kinetic model

See (Fig. 9)

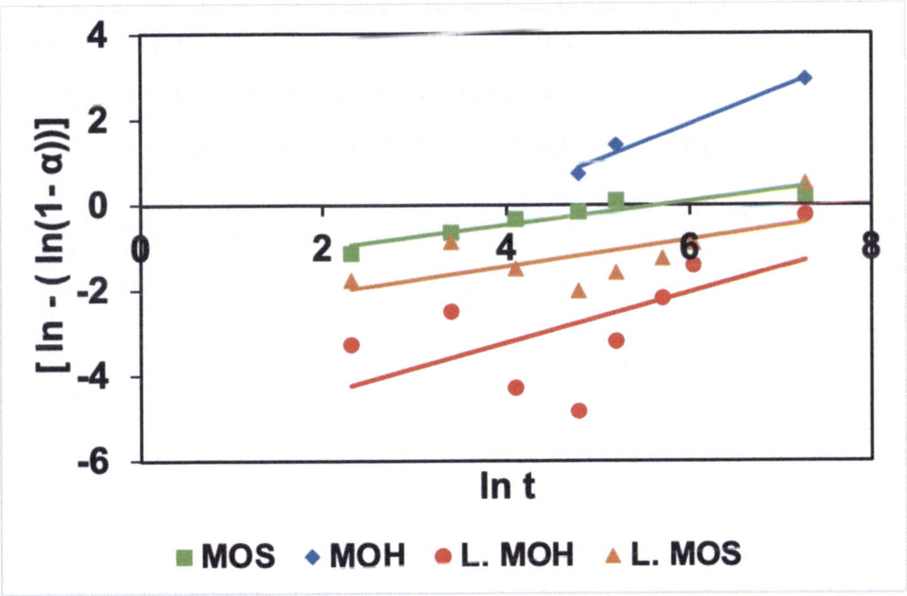

Fig. 9. Kinetics of ACT adsorption using Avrami kinetic model

See (Fig. 10)

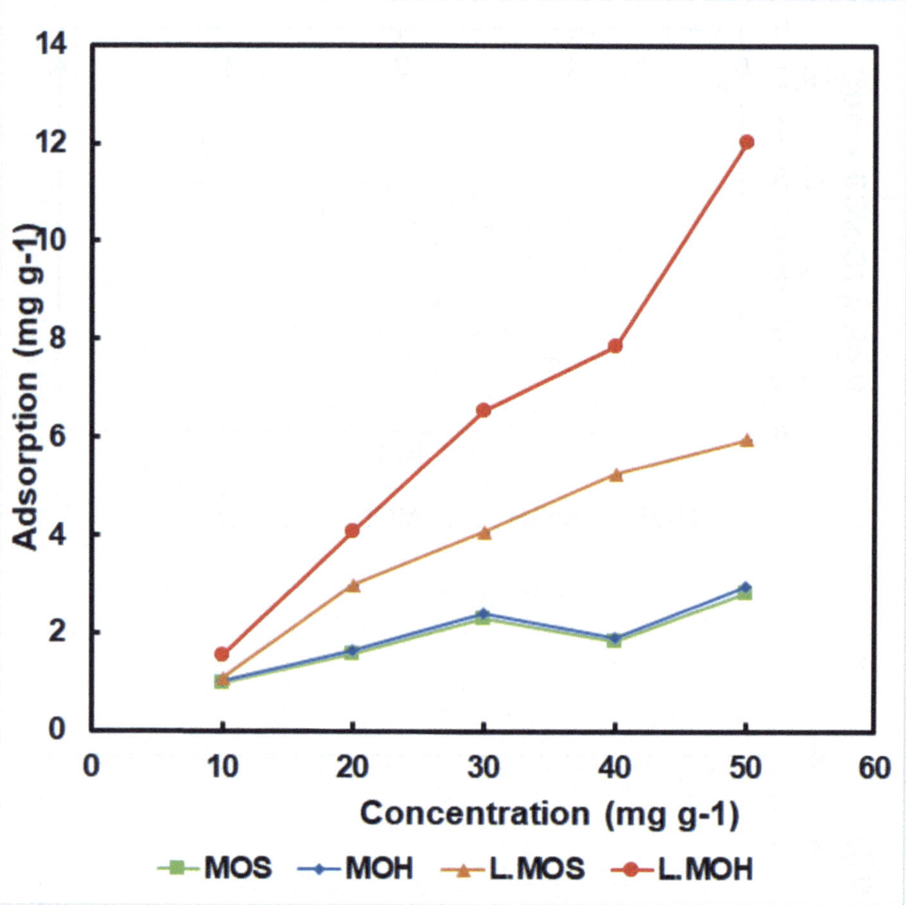

Fig. 10. Graph of ACT batch adsorption using concentrations of 10–50 mg L^{-1} (error bars are within the points)

See (Fig. 11)

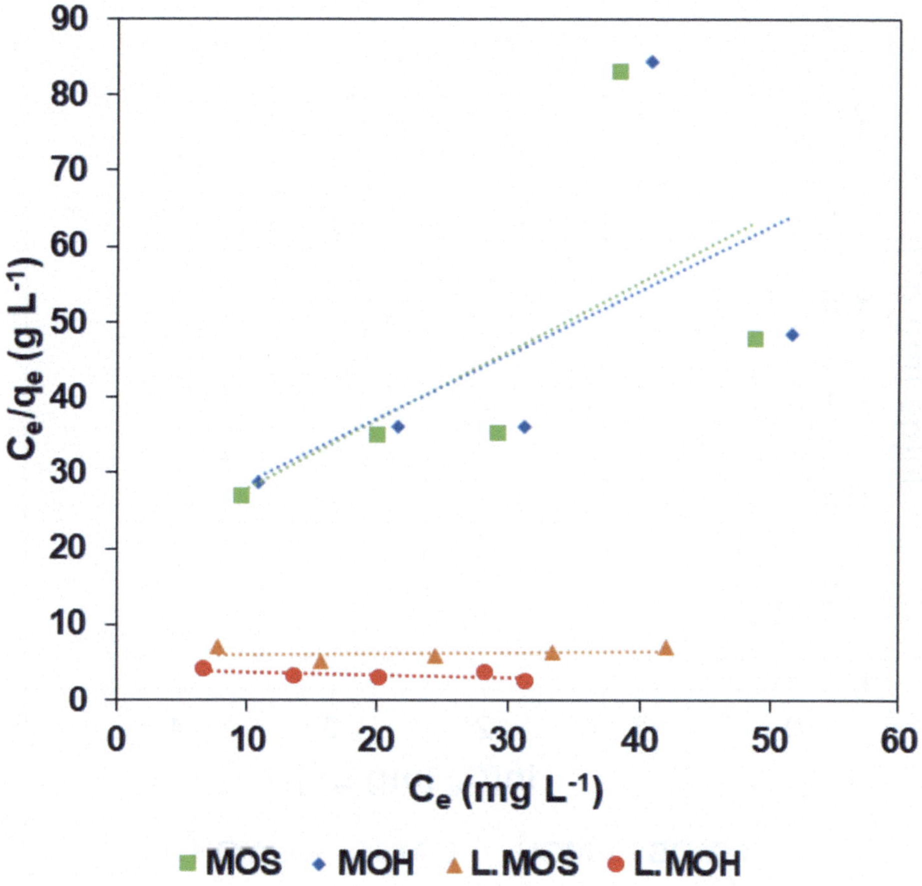

Fig. 11. Modelling of ACT batch adsorption using the Langmuir model

See (Fig. 12)

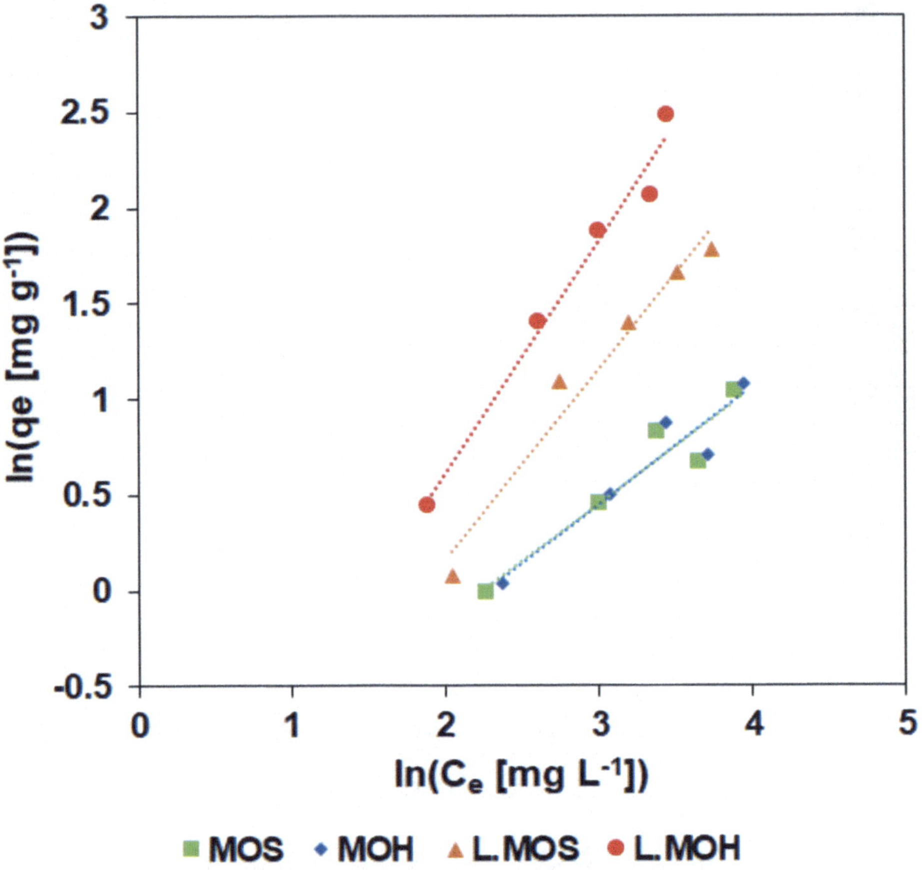

Fig. 12. Modelling of ACT batch adsorption using the Freundlich model

See (Fig. 13)

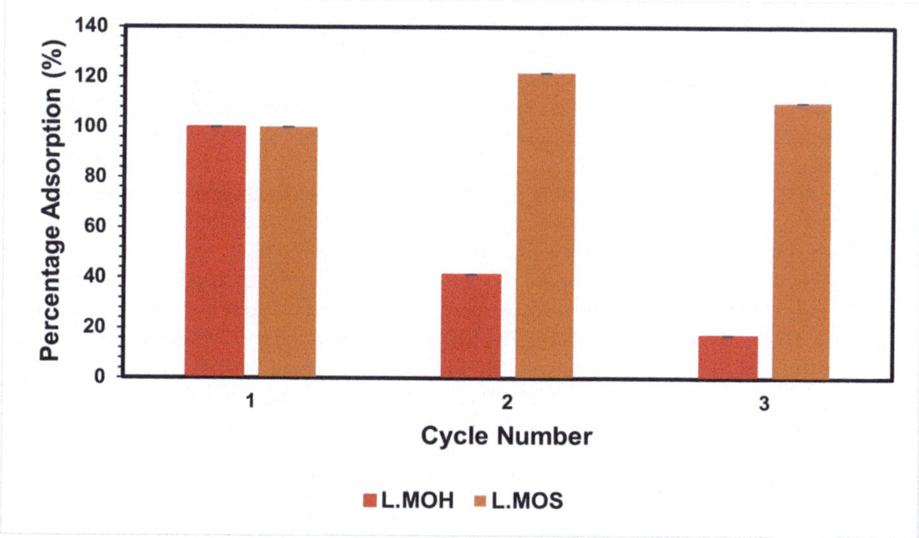

Fig. 13. Reusability of L.MOH, L.MOS adsorbents over 3 successive adsorption cycle (error bars are within the points)

References

1. Munzhelele, E.P., Mudzielwana, R., Ayinde, W.B., Gitari, W.M.: Pharmaceutical contaminants in wastewater and receiving water bodies of South Africa: a review of sources, pathways, occurrence, effects, and geographical distribution. Water **16**(6), 796 (2024). https://doi.org/10.3390/w16060796
2. Pacheco-Álvarez, M., Picos Benítez, R., Rodríguez-Narváez, O.M., Brillas, E., Peralta-Hernández, J.M.: A critical review on paracetamol removal from different aqueous matrices by Fenton and Fenton-based processes, and their combined methods. Chemosphere **303**(Pt 1), 134883 (2022). https://doi.org/10.1016/j.chemosphere.2022.134883
3. Mohammed Al-howri, B., Farhan Azha, S., Sharafee Shamsudin, M., Hamid, N.A., Mubarak Alsobaai, A., Ismail, S.: Paracetamol in diverse water sources: health hazards and treatment efficacy emphasizing adsorption techniques—a review. Int. J. Environ. Sci. Technol. (2024). https://doi.org/10.1007/s13762-024-05581-0
4. Bedner, M., MacCrehan, W.A.: Transformation of Acetaminophen by Chlorination Produces the Toxicants 1,4-Benzoquinone andN-Acetyl-p-benzoquinone Imine. Environ. Sci. Technol. **40**(2), 516–522 (2006). https://doi.org/10.1021/es0509073
5. Healing Waters Media. Water Treatment in Developing Countries. Healing Waters (21 Mar 2024). Healingwaters.org. https://healingwaters.org/water-treatment-in-developping-countries/
6. Al-Kaf, A.G., Naji, K.M., Abdullah, M., Ali, H.: Occurrence of paracetamol in aquatic environments and transformation by microorganisms: a review. **1**(6) (2017). https://www.researchgate.net/publication/322041109_Occurrence_of_Paracetamol_in_Aquatic_Environments_and_Transformation_by_Microorganisms_A_Review
7. Delelegn, A., Sahile, S., Husen, A.: Water purification and antibacterial efficacy of Moringa oleifera Lam. Agric. Food Secur. **7**(1) (2018). https://doi.org/10.1186/s40066-018-0177-1

8. Wang, K., Huang, K., Jiang, G.: Enhanced removal of aqueous acetaminophen by a laccase-catalyzed oxidative coupling reaction under a dual-pH optimization strategy. Sci. Total Environ. **616–617**, 1270–1278 (2018). https://doi.org/10.1016/j.scitotenv.2017.10.191
9. Dos Santos, P.M., et al.: Corn Cob as a Green Support for Laccase Immobilization-Application on Decolorization of Remazol Brilliant Blue R. Int. J. Mol. Sci. **23**(16), 9363 (2022). https://doi.org/10.3390/ijms23169363
10. Kumar, V.P., Sridhar, M.: Sustainable pretreatment method of lignocellulosic depolymerization for enhanced ruminant productivity using laccase protein immobilized agarose beads. Sci. Rep. **14**(1) (2024). https://doi.org/10.1038/s41598-024-76278-0
11. Azad, F.N., et al.: Optimization of the process parameters for the adsorption of ternary dyes by Ni doped FeO(OH)-NWs–AC using response surface methodology and an artificial neural network. RSC Adv. **6**(24), 19768–19779 (2016). https://doi.org/10.1039/c5ra26036a
12. Hamid Raza, H.M., et al.: Chemically Modified biomass from *Triticum aestivum* for Eco-friendly Detoxification of Cd(II)- and Sn(II)-contaminated waters. Univ. Punjab **23**(6), 2163–2174 (2014). https://www.pjoes.com/pdf-89403-23272?filename=Chemically%20Modified.pdf
13. Supriyadi, D., Darmansyah, F.A.C., Sanjaya, A., Soraya, F.: Evaluation of kinetics adsorption models from Lampung ethnic textile industry wastewater for removal chromium onto modified activated sludge and zeolite adsorbent. IOP Conf. Ser.: Earth Environ. Sci. **258**, 012025 (2019). https://doi.org/10.1088/1755-1315/258/1/012025
14. Du, Z.H., Zhang, T.S., Zhu, M.M., Ma, J.: Perovskite crystallization kinetics and dielectric properties of the PMN-PT films prepared by polymer-modified sol-gel processing. J. Mater. Res. **24**(4), 1576–1584 (2009). https://doi.org/10.1557/jmr.2009.0181
15. Lindén, J.B., Larsson, M., Kaur, S., Nosrati, A., Nydén, M.: Glutaraldehyde-crosslinking for improved copper absorption selectivity and chemical stability of polyethyleneimine coatings. J. Appl. Polym. Sci. **133**(37) (2016). https://doi.org/10.1002/app.43954
16. Al-sareji, O.J., et al.: Removal of pharmaceuticals from water using laccase immobilized on orange peels waste-derived activated carbon. Water **15**(19), 3437 (2023). https://doi.org/10.3390/w15193437
17. Tong. W., Minwen. J., Xueling. Y., Na. N., Ligang. C.: Application of lignin adsorbent in wastewater treatment: a review. Sep. Purif. Technol. (2022). https://doi.org/10.1016/j.seppur.2022.122116
18. Bakar, A.A.A., Mazlan, W.N.R.W., Izaham, I.H., Azizan, N.S., Ali, K.A.M., Daud, N.M.: Equilibrium kinetic and isotherm studies of dye colour adsorption on the banana peel. IOP Conf. Ser. Earth Environ. Sci. **920**(1), 012029 (2021). https://doi.org/10.1088/1755-1315/920/1/012029
19. Dharmarathna, S.P., Priyantha, N.: Investigation of Boundary Layer Effect of Intra-Particle Diffusion on Methylene Blue Adsorption on Activated Carbon. Energy Nexus pp. 100294–100294 (2024). https://doi.org/10.1016/j.nexus.2024.100294
20. Tan, K.L., Hameed, B.H.: Insight into the adsorption kinetics models for the removal of contaminants from aqueous solutions. J. Taiwan Inst. Chem. Eng. **74**, 25–48 (2017). https://doi.org/10.1016/j.jtice.2017.01.024
21. Alves, B.R.R., Konzen, R.A., Domingues, R.C.P.R., Bassetti, F.J., L.A., L.A.: Adsorption of methylene blue by seed husks of Moringa oleifera Lam. Desalin. Water Treat. **194**, 280–289 (2020). https://doi.org/10.5004/dwt.2020.25865
22. Skwarczynska-Wojsa, A., Puszkarewicz, A.: Removal of acetaminophen from aqueous solutions in an adsorption process. Materials **17**(2), 431 (2024). https://doi.org/10.3390/ma17020431
23. De Matos, C.F., Leão, M.B., Vendrame, L.F.O., Jauris, I.M., Zanella, I., Fagan, S.B.: Unlocking the paracetamol adsorption mechanism in graphene tridimensional-based materials: an

experimental-theoretical approach. Front. Carbon (2024). https://doi.org/10.3389/frcrb.2024.1305183
24. Chen, Y., Yang, R.T.: Surface and mesoporous diffusion with multilayer adsorption. Carbon **36**(10), 1525–1537 (1998). https://doi.org/10.1016/s0008-6223(98)00144-4
25. ECETOC. Freundlich isotherms-ECETOC (2014). https://www.ecetoc.org/technical-report-123/measured-partitioning-property-data/adsorption-desorption-distribution-kd-and-organic-carbon-water-partition-koc-coefficients/freundlich-isotherms/
26. Adamian, Y., Lonappan, L., Alokpa, K., Agathos, S.N., Cabana, H.: Recent developments in the immobilization of laccase on carbonaceous supports for environmental applications-a critical review. Front. Bioeng. Biotechnol. **9**, 778239 (2021). https://doi.org/10.3389/fbioe.2021.778239

Synthesis and Characterisation of Graphene Oxide for Water Purification

Jethro Chai Tsin Kai(✉), Derrick Lukimin, Chia Ern Kang, and Kong Chiak Wu

Hwa Chong Institution, Singapore 269734, Singapore
211383n@student.hci.edu.sg

Abstract. Industrial wastewater is filled with many pollutants, ranging from chemicals to harmful bacteria. This project aims to investigate the use of Graphene Oxide as an effective way to adsorb heavy metal ions, as well as the possible dye degradation and antibacterial properties of graphene oxide. Graphene oxide was synthesised with the modified Hummers Method. Three different degrees of oxidation of Graphene Oxide were produced. Heavy metal ion adsorption and organic dye degradation was tested using a colorimeter and UV-Vis spectrophotometer respectively. The bacterial tests were conducted using agar well diffusion. It was found that Graphene Oxide with the highest degree of oxidation is very effective in the adsorption of Fe^{3+} with an average of 99% adsorption rate, compared to Zn^{2+} and Cu^{2+} with an average adsorption rate of 60% adsorption and 50% respectively. Graphene oxide was also found to be very effective in degrading organic dyes. But it does not show any antibacterial properties against both gram positive and gram-negative bacteria in the dark. Graphene oxide is thus able to act as a chemical substitute to ultrafiltration and reverse osmosis water purification methods. It has the potential of being incorporated into filtration devices which can assist in curbing water poisoning caused by heavy metal ions and dyes.

Keywords: Graphene Oxide · Water purification · Dye degradation · Adsorption

1 Introduction

98% of water on Earth is seawater, and the other 2% is freshwater. 87% of the freshwater is found in the form of ice, 12% is stored underground, and only 1% can be found in rivers and lakes. Industrial wastewater, which usually contains heavy metal ions and organic dyes, is usually discharged into the waterway in less developed countries. Pollution into waterways renders the water unsafe for consumption. Exposure to heavy metals ions and organic dyes has been linked to chronic and acute toxicity, which may lead to retardation of growth and even death. It may also cause neurotoxicity that can damage the kidneys, lead to the development of different cancers, damage the liver and lungs [8]. Heavy metal ions and organic dyes can be removed using nanotechnology, and one such nanomaterial used is Graphene Oxide, which contains carbonyl, hydroxyl and epoxy groups amongst others [1]. Graphene oxide was known to be effective in adsorbing heavy metal ions [3, 6].

H. Guo et al. (Eds.): IRC-SET 2025, PTT, pp. 272–288, 2026.
https://doi.org/10.1007/978-981-95-9630-0_24

The most used method created is by forming layers of graphene from a carbon source, usually graphite, using the Hummers Method formed by [4]. It consists of adding $NaNO_3$ to graphene powder at 66 °C, then cooling it to 0 °C, where $KMnO_4$ and H_2SO_4 is added. After water is added and the suspension is treated with H_2O_2, the fine suspension is recovered, cleaned and dried. This method mainly involves the direct oxidation of graphite into graphite oxide. Recently, another method, the Modified Hummers Method, has been popularised for improving on this. It modifies the ratios and concentrations of the chemicals used in the synthesis [7].

[2] found that each method and conditions under which the graphene is oxidised would greatly impact the oxidation level, including the oxygen and hydrogen contents per carbon atom. As the oxidation time and amount of oxidant used during the synthesis of GO from graphene is increased, oxygen and hydrogen contents increase as well, which makes the GO more disordered in structure [2]. [5] also cited that photocatalytic graphene oxide has outstanding qualities in water purification. This would affect the physical and chemical properties of graphene oxide, affecting its efficiency in water filtration (among other applications). Thus, our project aims to further investigate how the different amount of oxidants used would affect its efficiency.

2 Objectives and Hypotheses

This study aims (i) to find the optimum oxidation level of the graphene oxide (GO) for the adsorption of heavy metal ions and organic dyes in water, (ii) to investigate how the degree of oxidation affects the antibacterial properties of graphene oxide, and (iii) to investigate the effect of light on the dye degradation capacity of graphene oxide.

It is hypothesised that the higher the degree of oxidation the more effective it is for GO to adsorb heavy metal ions and organic dyes. It is also hypothetical that the higher the degree of oxidation of GO the higher is its antibacterial properties.

3 Materials and Methods

3.1 Materials

For the preparation of GO, graphite was obtained from Sigma Chemical Co.. Potassium manganate (VII) from GCE, 6% hydrogen peroxide from GCE as well as 95% sulfuric acid from Scharlau were used. Rhodamine B, methylene blue, bromothymol blue and methyl orange used for dye degradation tests were obtained from Sigma Chemical Co.. Copper (II) Sulfate, Iron (III) Nitrate and Zinc Nitrate for heavy metal ion adsorption were obtained from GCE. Activated Carbon and Commercial Graphene Oxide was obtained from Sigma Chemical Co.. The bacteria used were *Escherichia coli* ATCC 25922, *Lactobacillus gasseri* ATCC 19992, *Pseudomonas fluorescens* ATCC 948 and *Staphylococcus epidermidis* ATCC 12228.

3.2 Synthesis of Graphene Oxide of Different Degrees of Oxidation

1.0 g of graphite and 30 ml of 95% sulfuric acid were mixed with constant stirring for 10 min in an ice bath. 3.0 g of potassium manganate (VII) was then added with constant stirring for another 10 min. The reaction mixture was then removed from the ice bath, capped and the content was stirred for 18 h.

15 ml of 6% hydrogen peroxide was added to terminate the reaction. The reaction mixture was then added to 600 ml deionised water. It was then filtered and the residue was washed with deionised water until the filtrate has a pH of 7. The residue (graphene oxide) was dried in an oven at 50 °C until constant mass. The Graphene Oxide is ground in a pestle and mortar till fine.

Graphene oxide with different degrees of oxidation were prepared using the same procedure as described above but with varying masses of potassium manganate (VII), namely 2.5 g and 2.0 g. The graphene oxide samples were named 3.0 GO, 2.5 GO and 2.0 GO respectively, which correspond to the mass of $KMnO_4$ used.

3.3 Testing of Heavy Metal Ion Adsorption

50 ppm Cu^{2+}, Fe^{3+} and Zn^{2+} solutions were prepared by dissolving 0.197 g of copper(II) sulfate pentahydrate (M_r 249.5), 0.362 g of iron(III) nitrate nonahydrate (M_r 403.9) and 0.220 g of zinc sulfate heptahydrate (M_r 287.5) separately in 1 L deionised water a 1 L volumetric flask.

The adsorption tests were then carried out by adding 0.100 g graphene oxide sample to 10 ml of each 50 ppm heavy metal ion solution with constant shaking in the orbital shaker for 24 h. The mixtures were then centrifuged at 13500 RPM for 7 min to separate the solution from the graphene oxide sample. The supernatants of the zinc sulfate and copper(II) sulfate were diluted 20 times while the supernatants of the iron(III) nitrate were diluted 10 times before their absorbance were determined by a colorimeter. 5 replicates were conducted for each heavy metal ion test. This was repeated for each sample of graphene oxide as well as activated carbon.

3.4 Organic Dye Calibration Curve

10 ppm organic dyes solutions were prepared by dissolving 0.01 g of rhodamine B., 0.01 g of methylene blue, 0.01 g of methyl orange and 0.001 g of bromothymol blue were dissolved in 1 L deionised water in a 1L volumetric flask. Each organic dye stock solution is diluted to 2, 4, 6 and 8 ppm. The absorbance values of bromothymol blue, methyl orange, methylene blue and rhodamine B were measured at 434 nm, 465 nm, 654 nm and 554 nm respectively. The calibration curves of the dyes are shown in Fig. 11 (Appendix B).

3.5 Testing of the Degradation of Organic Dyes by Graphene Oxide

The adsorption tests were carried out by adding 0.100 g graphene oxide sample to 10 ml of each 50 ppm organic dye solution with constant shaking in the orbital shaker for 24 h. The mixtures were then centrifuged at 13500 RPM for 7 min to separate the solution from the graphene oxide sample. The absorbance of the supernatants was determined by using a UV-VIS spectrophotometer at the λ_{max} of the respective organic dyes. The concentrations of the organic dyes were then determined from the calibration curves. 5 replicates were conducted for each organic dye test. This was repeated for each sample of graphene oxide and the activated carbon.

Due to time constraint, only methyl orange was used to study the photocatalytic properties of graphene oxide. The same procedure was repeated with methyl orange, but the mixture was left to degrade for 24 h in a dark room.

3.6 Antibacterial Tests

1.5 L of M. H. Agar was used as the base agar for agar well diffusion. 100 ml of L. B. Broth was prepared for *E. coli*, *P. fluorescence* and *S. epidermidis*. 50 ml of M.R.S. Broth was prepared for *L. gasseri*. The agar and broths were autoclaved at 121°C. The broths are then stored in the fridge at 4°C for storage. 25 ml of agar are poured into a petri dish. This is repeated for 29 different plates. The plates are then stored in the fridge at 4°C for storage.

The bacteria samples were transferred using a loop and needle. The bacteria sample was placed in 10 ml of the respective broths and placed in an orbital shaker overnight at 37°C.

The agar plates were labelled with 5 spots far enough from each other. The bacteria were swabbed onto the plate in four directions to ensure the bacteria fully covered the plate. 5 wells were created at the marked locations using tips to make the wells. A scalpel is then used to remove the agar inside the well. 0.30 g of graphene oxide is added to 30 ml of water to achieve a dilution ratio of 1g to 100 ml. An ultrasound bath is used to achieve a homogenous suspension of graphene oxide samples. 80 μm^3 of graphene oxide samples and control solutions (10% bleach and sterile deionised water) were then inserted into their respective labelled wells. The plates were then parafilmed before incubating in a 37 °C incubator overnight.

3.7 Characterisation of Graphene Oxide

Graphene oxide sample was sent for scanning electron microscopy (SEM) imaging and FTIR analysis.

4 Results and Discussion

4.1 Characterization of Graphene Oxide

Figure 9 (Appendix B) shows the FTIR spectrum of 3.0 GO obtained in our experiment. The characteristic peaks of the graphene oxide obtained are similar to those reported. [9] The peaks at 3423.02 cm^{-1}, 1725.06 cm^{-1}, 1617.95 cm^{-1}, 1400.53 cm^{-1}, 1226.49 cm^{-1} and 1052.60 cm^{-1} are attributed to the O-H, C = O, C = C, C-O and C-O stretching respectively. The peak at 1052 cm^{-1} is attributed to the C-O-C bending. The O-H and C = O groups contribute to graphene oxide's ability to adsorb heavy metal ions. The O-H groups present are also responsible for any photocatalytic properties observed.

The FTIR scan of the 3.0 GO sample shown in Fig. 9. (Appendix B) resembles scans for graphene oxide found in other references, which suggests that graphene oxide was indeed successfully synthesised. The FTIR scan for the sample of graphene oxide made from the 2.0 sample shown in Fig. 11. (Appendix B) resembles a scan for graphite more than a scan for graphene oxide, suggesting that while graphene oxide was still made, it was less oxidised and hence had a closer nature to graphite than graphene oxide.

Referring to Fig. 4. (Appendix A), the SEM image of graphene oxide showed layers of thickness about 50 nm. While this shows that the graphite had indeed formed layers to make graphene oxide, it also shows that the particles made were not nanoparticles, as they are not small enough, as the lengths of the layers were significantly larger.

4.2 Adsorption of Heavy Metal Ions

The p-value for the tests using the Kruskal-Wallis test are: 0.0793 for Cu^{2+}, 0.00193 for Zn^{2+} and 0.0394 for Fe^{3+}, hence showing the statistical significance of the results except for copper ions. The samples of graphene oxide seemed to primarily adsorb the coloured Fe^{3+} ions, while not adsorbing the Cu^{2+} and Zn^{2+} ions as much. The samples generally seem to follow the trend that a higher oxidation level would lead to higher adsorption levels, although there seems to be an anomaly for Cu^{2+} ions from the 3.0 to the 2.5 sample. Activated carbon, being an allotrope of carbon, seems to have similar adsorption capacities, although it fares much worse for Zn^{2+} ions, not adsorbing most of the ions.

The O-H and $C = O$ groups contribute to graphene oxide's ability to adsorb heavy metal ions. With more oxidation, more of these oxygen-containing functional groups are able to act upon the heavy metal ions, hence resulting in better adsorption.

The percentage adsorption for 3.0, 2.5 and 2.0 graphene oxide of Cu^{2+} ions is 61.6%, 63.9% and 58.1%, Zn^{2+} ions having 65.2%, 57.2% and 44.5%, and Fe^{3+} ions having 97.8%, 97.7% and 96.8%. This high adsorption rate for Fe^{3+} ions could be due to the higher charge on the ion, hence resulting in stronger ion dipole interaction, stronger electrostatic forces of attraction between the ions and the hydroxyl group (Fig. 1).

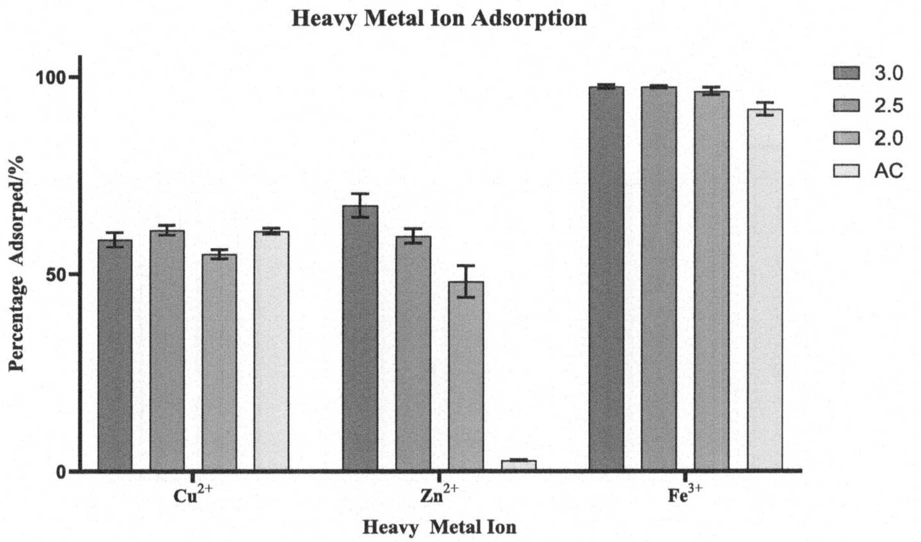

Fig. 1. Graph of heavy metal ion adsorption for different samples of graphene oxide, as well as activated carbon

4.3 Adsorption of Organic Dyes

The results are illustrated in Fig. 2. The results for the 3.0, 2.5, 2.0 GO samples and the activated carbon samples are, respectively, 99.3%, 99.8%, 99.2%, 99.6% for Methylene Blue, 97.8%, 99.6%, 99.4% and 98.0% for Methyl Orange, 99.6%, 99.2%, 97.4%, 89.1% for Bromothymol Blue and 98.6%, 99.8%, 99.8%, 98.4% for Rhodamine B. All our Graphene Oxide samples produced similar results, adsorbing the dye by more than 95% of sample each for all the dyes. This is due to graphene oxide being a photocatalyst, where it effectively removes coloured compounds when exposed to light. The p-value for the tests using the Kruskal-Wallis test are: 0.0272 for Bromothymol Blue, 0.614 for Rhodamine B, 0.442 for Methyl Orange, and 0.673 for Methylene Blue. This shows that the level of oxidation does not significantly affect the photocatalytic adsorption properties of graphene oxide. The only exception is Bromothymol Blue, where the greater the level of oxidation, the greater the level of degradation. The high adsorption rate in the organic dyes can be attributed to their ability to undergo photocatalytic degradation by the GO, which has no capacity and would thus degrade until most of the dyes have been degraded. Thus, the degree of oxidation would not affect its degradation and hence its removal percentage.

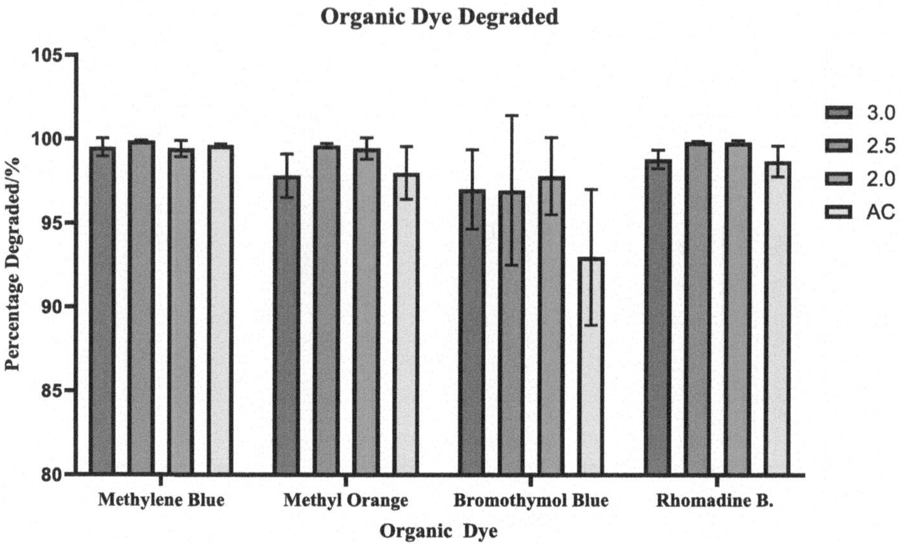

Fig. 2. Graph of Organic Dye adsorption for different samples of graphene oxide, as well as activated carbon

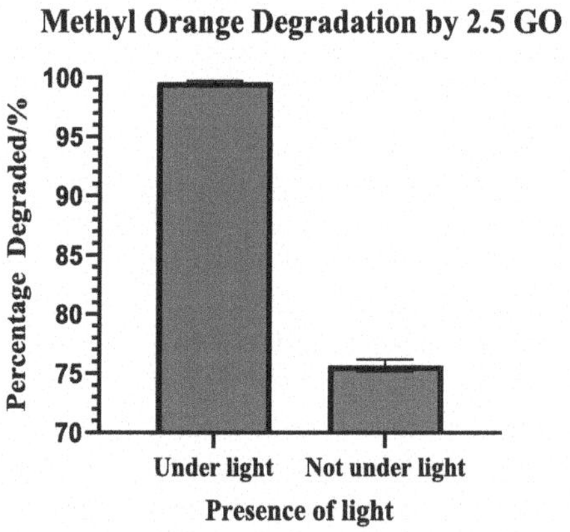

Fig. 3. Graph of Adsorption of Methyl Orange of 2.5 Graphene Oxide Sample in the Presence and Absence of Light

The results of methyl orange adsorption for graphene oxide under light and in the dark (Fig. 3.) suggests that graphene oxide can degrade methyl orange under light asides from just adsorbing it, hence making it a photocatalyst. Using the Mann-Whitney U test, the U statistic is found to be 12.5, which is greater than the U critical of 5. Thus, the test results are significant.

When light is shown onto the graphene oxide, hydroxyl radicals can form, which is able to degrade the dyes faster, hence resulting in more effective dye degradation. This is due to the presence of hydroxyl groups in the graphene oxide, which form hydroxyl radicals under light. These hydroxyl radicals consist of an unpaired electron, making them highly reactive and attracting the organic dyes close towards the graphene oxide for degradation and adsorption.

4.4 Antibacterial Tests

Despite the positive controls and negative controls validating the tests, the zone of inhibition for the graphene oxide samples seemed to be 0.0 mm for all samples, including both gram positive and gram-negative bacteria. This suggests that the graphene oxide synthesised does not have antibacterial properties. One possible explanation is the fact that the graphene oxide made was not in the form of nanoparticles or nanosheets, as most past literature suggests that nanoparticles are the main cause of the antibacterial properties of graphene oxide. The graphene oxide samples produced were not small enough to penetrate the bacteria, migrate internally, and interact with key components of biofilm. Also, the large particle graphene oxide does not have high enough surface-to-volume ratios and multivalent interactions to have antibacterial effects.

5 Conclusion and Recommendations for Future Work

Agreeing with our hypothesis, graphene oxide is also able to act as a photocatalyst, adsorbing and degrading coloured compounds and substances with optimal efficiency under light. It works better, in some cases, than the commercial activated carbon provided by the lab. Graphene oxide is thus able to act as a chemical substitute to the energy intensive ultrafiltration and reverse osmosis water purification methods. It has the potential of being incorporated into filtration devices which can assist in curbing water poisoning in 3rd world countries by acting as a water purifier, especially since this project has shown that even graphene oxide particles outside the nanometer range can efficiently clean water. Scientifically, the test provides an encouraging outlook for the scientific development of graphene oxide and showcases the relationship of the degree of oxidation of the graphene oxide and its effectiveness.

More extensive experimentation can be used to figure out the large-scale synthesis of graphene oxide efficiently. Due to constraints on sample size, more tests were unable to be done.

More experimentation can be used to figure out the heavy metal ion adsorption of graphene oxide for other heavy metal ions, to confirm that the charge of the ion is at play. Future research can be done to more concretely prove the photocatalytic nature of graphene oxide, potentially using terephthalic acid degradation as a test. Notably, graphene oxide nanoparticles could be synthesised to comprehensively compare the efficacy of graphene oxide microparticles with its nanoparticle counterparts, especially to confirm that graphene oxide nanoparticles do have antibacterial properties, as well as to test how degree of oxidation of graphene oxide nanoparticles, and thus their particle size, would influence on the nanoparticle's effectiveness and properties. Finally, a more efficient and safer method can be explored to prepare graphene oxide, as other oxidisers such as persulfates and periodates seem to be unable to oxidise graphite.

Acknowledgements. We would like to thank our mentor and the lab facilitator, Mr Kong and Mdm Xia, from Hwa Chong Institution for their guidance and advice, as well as Hwa Chong Institution for providing us with ample resources and support.

Appendices

Appendix A: SEM Images of Graphene Oxide Sample 3.0

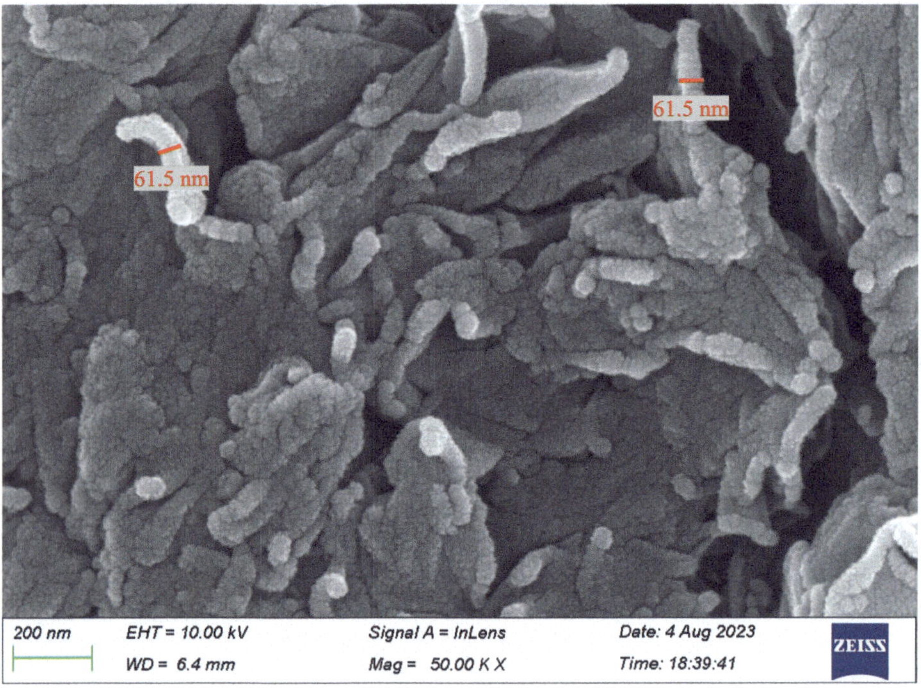

Fig. 4. Annotated scan of graphene oxide under electron microscope

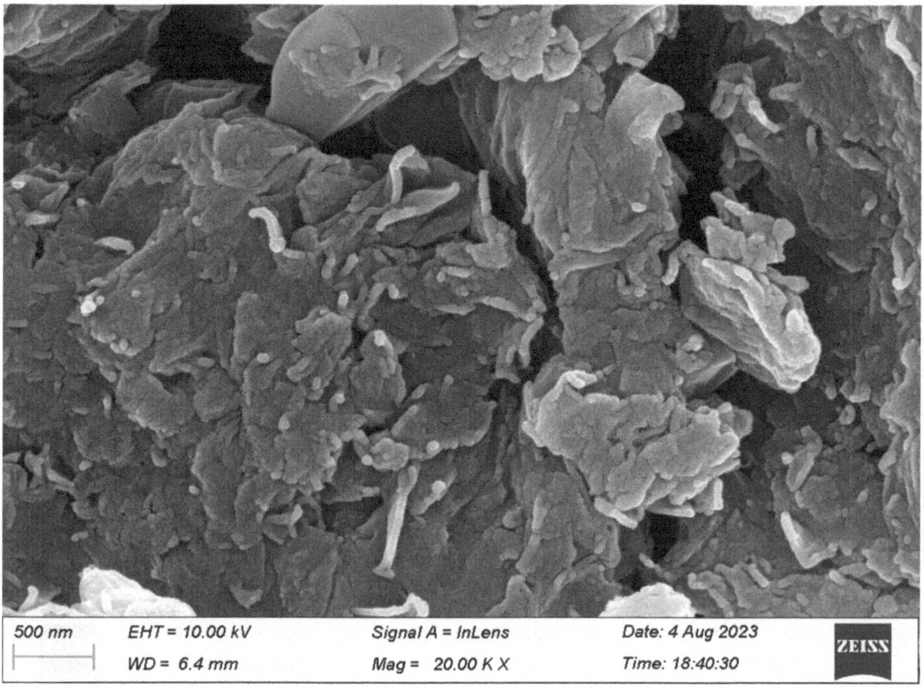

Fig. 5. Scanning electron microscope image of the graphene oxide sample 3 in the 500 nm scale.

Fig. 6. Scanning electron microscope image of the graphene oxide sample 3 in the 5 micron scale.

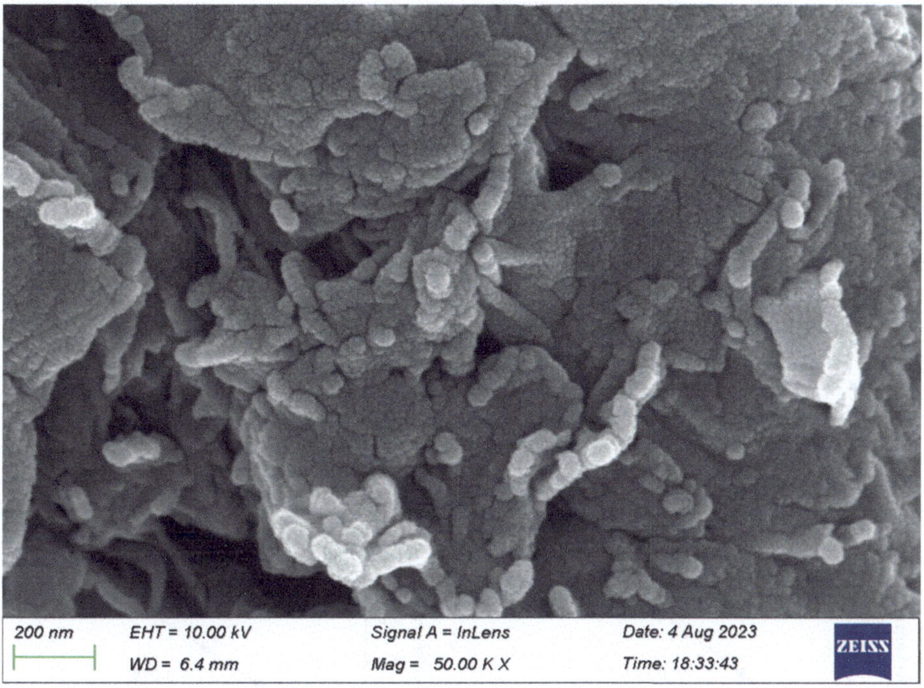

Fig. 7. Scanning electron microscope image of the graphene oxide sample 3 in the 200 nm scale.

Appendix B: Calibration Curves of Organic Dyes

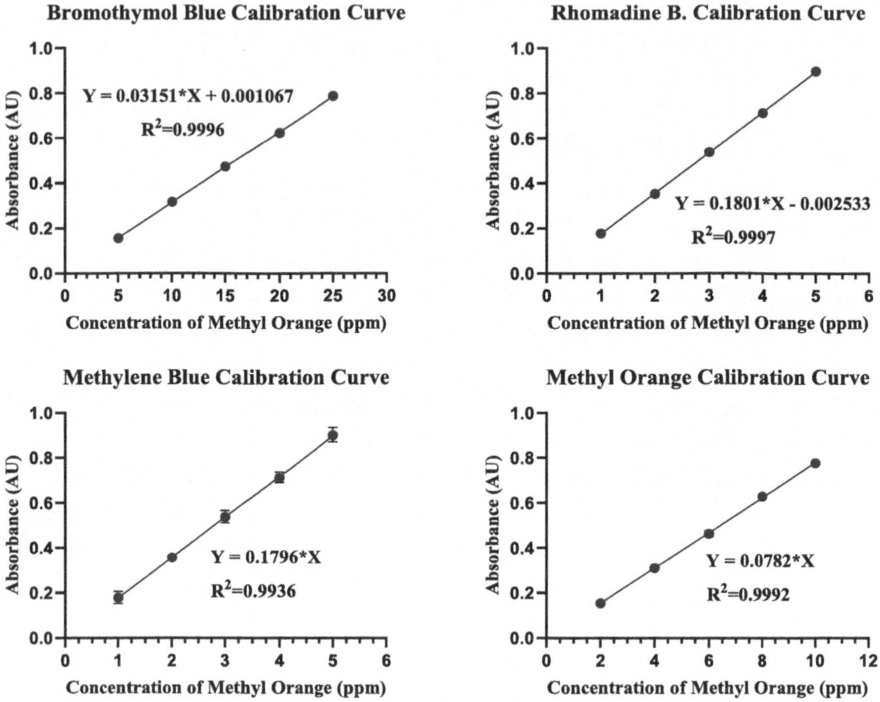

Fig. 8. Bromothymol Blue absorbance calibration curve at 434 nm (Top Left), Rhomadine B. absorbance calibration curve at 554 nm (Top Right), Methylene Blue absorbance calibration curve at 664 m, (Bottom Left), Methyl Orange absorbance calibration curve at 465 nm (Bottom Right).

Appendix C: FTIR

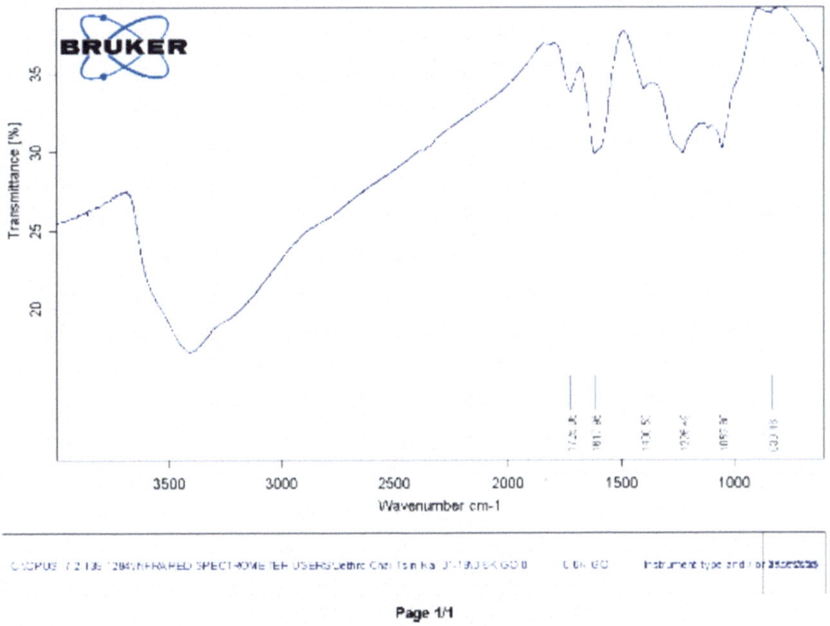

Fig. 9. FTIR Wavelength of 3.0 GO

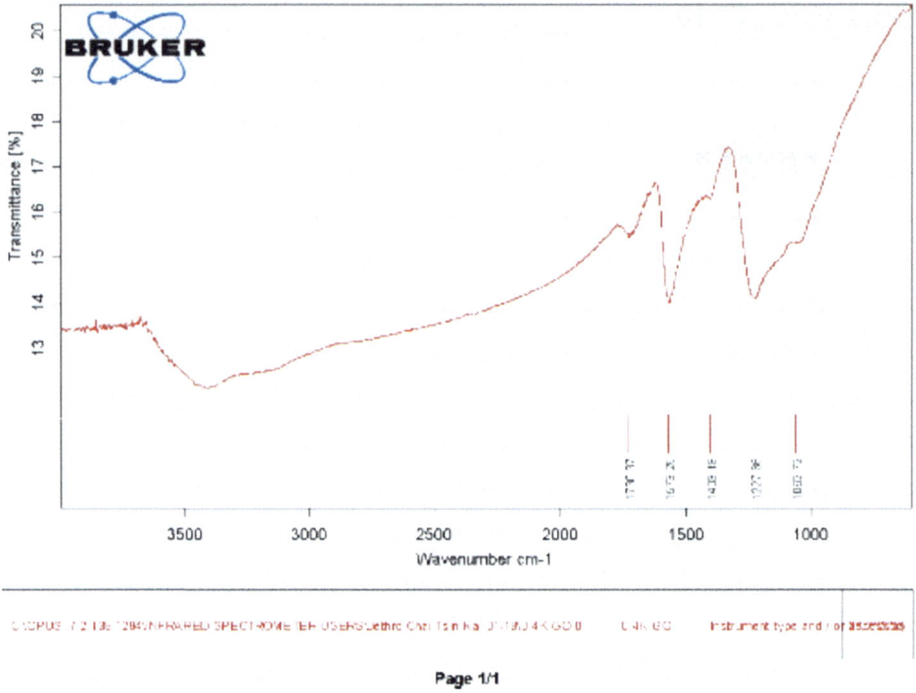

Fig. 10. FTIR Wavelength of 2.5 GO

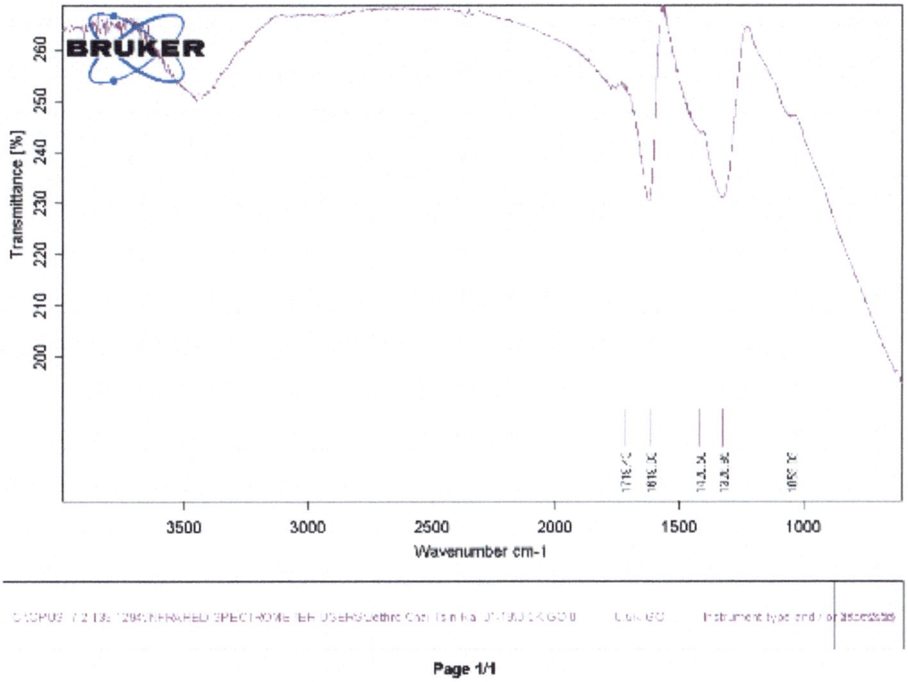

Fig. 11. FTIR Wavelength of 2.0 GO

References

1. Casabianca, L.B., et al.: NMR-based structural modeling of graphite oxide using multidimensional 13C solid-state NMR and ab Initio chemical shift calculations. J. Am. Chem. Soc. **132**(16), 5672–5676 (2010). https://doi.org/10.1021/ja9030243
2. Fan, J., Yang, J., Li, H., Tian, J., Wang, M., Zhao, Y.: Cryogenic mechanical properties of graphene oxide/epoxy nanocomposites: Influence of graphene oxide with different oxidation degrees. Polym. Testing **96**(107074), 9142–9418 (2021). https://doi.org/10.1016/j.polymertesting.2021.107074
3. Guerrero-Fajardo, C.A., Giraldo, L., Moreno-Piraján, J.C.: Preparation and characterization of graphene oxide for Pb(II) and Zn(II) ions adsorption from aqueous solution: experimental, thermodynamic and kinetic study. Nanomaterials (Basel, Switzerland) **10**(6), 1022 (2020). https://doi.org/10.3390/nano10061022
4. Hummers, W.S., Jr., Offerman, R.E.: Preparation of graphitic oxide. J. Am. Chem. Soc. **80**(6), 1269–1522 (1958). https://doi.org/10.1021/ja01539a017
5. Khalid, U., Abraheem, J., Rohana, A., Mohd, R., Nasir, M.: Synthesis, characterization, and photocatalytic activities of graphene oxide/metal oxides nanocomposites: a review. Front. Chem. **9** (2021)
6. Kong, Q., et al.: Relations between metal ion characteristics and adsorption performance of graphene oxide: a comprehensive experimental and theoretical study. Sep. Purif. Technol. **232**, 2022 (2020). https://doi.org/10.1016/j.seppur.2019.115956

7. Nanda, S., Gaur, A., Duchaniya, R.K.: Synthesis, properties and applications of graphene oxide: an overview. World Sci. News **143**(September 2015), 17–27 (2020). http://www.worldscientificnews.com/wp-content/uploads/2020/02/WSN-143-2020-17-27-1.pdf
8. Singh, A., et al.: Heavy metal contamination of water and their toxic effect on living organisms. Toxicity Environ. Pollutants **1**(2), 1–302 (2022). https://doi.org/10.5772/intechopen.98127
9. Zafer, Ç., Nuray, Y., Ayla, Ç.: Investigation of Graphene/Ag nanocomposites synthesis parameters for two different synthesis methods. Fullerenes Nanotubes Carbon Nanostruct. **23**(4), 361–370 (2015). https://doi.org/10.1080/1536383X.2014.894025

Effect of Copper (II) Ions on the Removal of Antibiotics from Wastewater via Electrocoagulation

Yan Hai De(✉), Xu Zixin, Xu Wanjia, Lee Zhiying, Grace Lim, and Abigayle Ng

Raffles Institution, 1 Raffles Institution Ln, Singapore 575954, Singapore
25yyanh632c@student.ri.edu.sg

Abstract. Antibiotic residues such as tetracycline (TC) and ofloxacin (OFL) are frequently found coexisting with copper (II) ions ($Cu2^{+}$) in wastewater, forming stable metal-ligand complexes that hinder effective degradation and removal. This study examines the effect of varying $Cu2^{+}$ concentrations on the efficiency of EC as a treatment method for removal of TC and OFL. Experiments were carried out in a batch electrochemical reactor with Al electrodes at initial pH5.5 and current densities of 12.6 A/m2 for TC and 13.5 A/m2 for OFL. Results obtained using ultraviolet-visible (UV-Vis) spectroscopy indicated that at TC/Cu molar ratio of 1.4298, TC removal reached a maximum of 88.3%, a 103.9% improvement from removal in the absence of $Cu2^{+}$. Meanwhile, at OFL/Cu molar ratio of 1.7585, OFL removal reached a maximum of 75.3%, attaining a 20.5% increase from removal without $Cu2^{+}$. The adsorption kinetics followed the pseudo-second-order model, suggesting that chemisorption was the primary removal mechanism. Scanning electron microscopy (SEM) and elemental analysis by Energy Dispersive X-ray Spectroscopy (EDX) revealed crystalline surface morphology on flocs with optimum $Cu2^{+}$ concentration, characterized by slightly poorer adsorption but heightened retention. This work demonstrates that optimizing the molar ratio of pharmaceutical to $Cu2^{+}$ in EC systems can significantly enhance pollutant removal efficiency while lowering operational costs by reducing energy, electricity, and electrode consumption.

Keywords: Electrocoagulation · Wastewater Treatment · Antibiotic Removal

1 Introduction

The increasing prevalence of antibiotic residues, such as tetracycline (TC) and ofloxacin (OFL), in aquatic systems poses a significant environmental and public health concern due to their potential to promote antimicrobial resistance and disrupt the balance of aquatic ecosystems [1]. Similarly, copper (II) ions ($Cu2^{+}$), widely utilized in industrial processes such as semiconductor manufacturing, electroplating, and mining, are highly toxic to aquatic organisms even at low concentrations [2]. In wastewater systems, the confluence of multiple streams often results in the coexistence of pharmaceuticals and $Cu2^{+}$, particularly in livestock farm effluents where antibiotics and heavy metals

H. Guo et al. (Eds.): IRC-SET 2025, PTT, pp. 289–298, 2026.
https://doi.org/10.1007/978-981-95-9630-0_25

have been extensively used as feed additives [3], and in pharmaceutical industrial discharges [4]. Interactions between TC/OFL and heavy metals such as $Cu2^{+}$ in aquatic environments lead to the formation of metal-ligand complexes, which exhibit recalcitrant behaviors and resistance to degradation and adsorption [5]. Despite the growing prevalence of these composite pollutants, much of the existing research focuses on the removal of single contaminants in testing efficacies of treatment technologies, leaving a critical gap in understanding their effectiveness under simultaneous removal of metal ions and pharmaceuticals.

Electrocoagulation (EC) is an advanced process that utilizes metal electrodes to generate flocs that destabilize and aggregate suspended solids, heavy metals, emulsified oils, and microorganisms, which can then be easily separated and removed from water by filtration or skimming in industrial scale [6]. EC offers several advantages, including efficient contaminant removal, sludge minimization, environmental compatibility, low capital costs, and potential scalability [6, 7]. A study by Lu et al. [5] on the optimization of EC parameters for the removal of TC-Cu complexes demonstrated significant potential for $Cu2^{+}$ in enhancing the removal of TC. Yet, a conclusive trend regarding $Cu2^{+}$ concentration on TC adsorption was unable to be established due to the limited range of TC-Cu ratios explored.

The kinetics of the adsorption process of TC/OFL-Cu complexes onto flocs can be investigated through the application of adsorption kinetic models including the pseudo-first-order (PFO) and pseudo-second-order (PSO) kinetic models [8] or adsorption isotherm models such as the Langmuir and Freundlich models [9]. By evaluating how well these models fit the experimental data, insights can be gained into the dominant adsorption mechanism governing the removal process.

2 Methodology

Experimental procedures were designed to investigate the effect of $Cu2^{+}$ on the removal efficiencies of two common antibiotics, tetracycline (TC) and ofloxacin (OFL), from wastewater using electrocoagulation (EC).

2.1 Preparation of Synthetic Wastewater

Synthetic wastewater was prepared in 1 L batch reactors, each containing 40 ppm of either tetracycline (TC) or ofloxacin (OFL), along with copper (II) ions ($Cu2^{+}$) at varying concentrations of 0, 2, 4, 6, 8, and 10 ppm. Sodium chloride (NaCl) was added at 1.8 g/L as a supporting electrolyte, and deionized (DI) water was used as the solvent. The initial pH of each solution was adjusted to 5.5 using 0.1 M HCl or 0.1 M NaOH. A magnetic stirrer maintained continuous mixing at 300 rpm throughout the experiments. Stock solutions of TC ($C_{22}H_{24}N_2O_8$) and OFL ($C_{18}H_{20}FN_3O_4$) were prepared by dissolving each compound in DI water, while $Cu2^{+}$ was introduced in the form of copper(II) sulfate pentahydrate ($CuSO_4 \cdot 5H_2O$).

2.2 Electrochemical Reactor Configuration

Electrocoagulation was carried out in a batch reactor fitted with four aluminum (Al) electrodes arranged in an internal bipolar electrode (IBPE) configuration. Each electrode measured 10.0 cm × 2.5 cm × 0.1 cm. Prior to experimentation, electrodes were treated by immersion in 1 M HCl for 15 minutes and mechanically polished with fine-grit sandpaper to remove any oxide layers. The electrodes were spaced 1.8 cm apart and submerged to a depth of 3.8 cm in the reactor.

The electrochemical reactions involved are as follows:

Anodic oxidation: $Al \rightarrow Al3^{+} + 3e^{-}$

Cathodic reduction: $2H_2O + 2e^{-} \rightarrow H_2 + 2OH^{-}$

Overall reaction: $Al3^{+} + 3OH^{-} \rightarrow Al(OH)_3$

The in-situ formation of $Al(OH)_3$ serves as the primary coagulant, which adsorbs and destabilizes suspended and dissolved pollutants. Concurrent hydrogen gas evolution aids in the flotation of pollutant-laden flocs to the surface.

2.3 Electrocoagulation Procedure

The reactor was powered by a DC power supply, and both voltage and current were monitored using a multimeter to maintain a constant current density of 13.0 A/m2. For experiments involving TC, 4 mL aliquots were collected at 5-minute intervals over a 45-minute period. Preliminary trials showed that OFL removal was slower; thus, samples for OFL were taken at 10-minute intervals for the first 90 minutes, followed by 30-minute intervals until the change in concentration was less than 1 ppm. All experiments were conducted at room temperature (approximately 25 ± 2 °C).

2.4 Characterization of Generated $Al(OH)_3$ Flocs

Residual TC/OFL concentrations were measured by UV-Vis spectroscopy at the wavelength of 357 nm for TC and 293 nm for OFL, before being converted to concentrations via calibration curves. Removal percentage was calculated with the following expression:

$$Removal(\%) = (C_0 - C_t)/C_0 \times 100 \quad (1)$$

where C_0 is the initial concentration of TC/OFL, and C_t is the concentration of TC/OFL at time t.

2.5 Adsorption Kinetic Modelling of Experimental Data

The removal kinetics were evaluated using pseudo-second-order (PSO) models to determine the dominant removal mechanisms. Scanning Electron Microscopy (SEM) and Energy Dispersive X-ray Spectroscopy (EDX) were employed to examine the morphology and elemental composition of the resulting flocs. This enabled the assessment of structural differences and adsorption properties under varying $Cu2^{+}$ concentrations.

3 Results and Discussion

3.1 Removal Efficiency of Antibiotics against Time

Effectiveness of TC removal was found to vary with initial concentrations of $Cu2^{+}$, with the highest removal of 88.3% being achieved at 4 ppm $Cu2^{+}$ within 45 mins, corresponding to an optimal TC/Cu molar ratio of 1.4298. In comparison, the control (0ppm $Cu2^{+}$) achieved only 43.3%, showing a 103.9% increase in removal efficiency (Fig. 1).

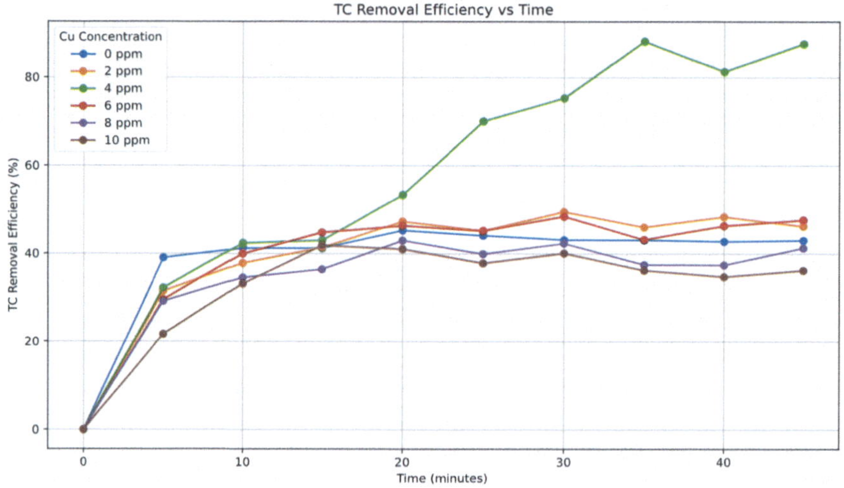

Fig. 1. Removal efficiency of tetracycline (TC) over time at varying $Cu2^{+}$ concentrations using electrocoagulation.

EC alone partially removes TC through standard coagulation mechanisms. Adsorption occurs via the donation of electron pairs from carbonyl (C=O), hydroxyl (-OH), and amino ($-NH_2$) groups of TC to Al to form coordination bonds, and hydrogen bonding. The high surface area and porous structure of $Al(OH)_3$ flocs aid physical entrapment of TC. Under slightly acidic conditions, (-OH) of TC become partially deprotonated, generating localized negative charges that are attracted to $Al(OH)_2{}^{+}$ through electrostatic interactions, leading to aggregation by charge neutralization. Flocculation may also occur when $Al3^{+}$ destabilizes and coagulates colloidal particles and TC.

At low $Cu2^{+}$ concentrations from 0-2 ppm, 2:1 type complexes, such as CuH_2TC_2 and $CuH_4TC_22^{+}$, dominate. These complexes are highly stable due to strong coordination between $Cu2^{+}$ and TC. Multi-substituted complexes, such as CuH_3TC_3, exist in lower quantities as the electron density donated by three or more TC exceeds the electron-accepting capacity of a single $Cu2^{+}$, leading to weaker bonds and instability. In the same vein, complexes such as $Cu_2H_3TC_3{}^{+}$ decompose easily from steric hindrance, where spatial crowding around the copper centers destabilizes the complex and promotes dissociation.

At 4 ppm, the mixture is dominated by 1:1 type complexes, $CuHTC^+$ and CuH_2TC^+ instead. These 1:1 type complexes exhibit greater reactivity than TC alone, as $Cu2^+$ withdraws electron density and makes TC more electrophilic and polarized, weakening its internal bonds. Beyond 8 ppm, removal rates declined progressively below that of the control, as $Cu2^+$ may be preferentially adsorbed onto flocs, interfering with the EC process.

Additionally, a secondary effect of electrocoagulation is the direct oxidation at the anode and reduction at the cathode, which generates reactive oxygen species (ROS), such as hydroxyl radicals (·OH) and hydrogen peroxide (H_2O_2), which may aid in oxidizing TC. In the presence of $Cu2^+$, ROS production is catalyzed, contributing to higher rates of removal observed between 2-6 ppm (Fig. 2).

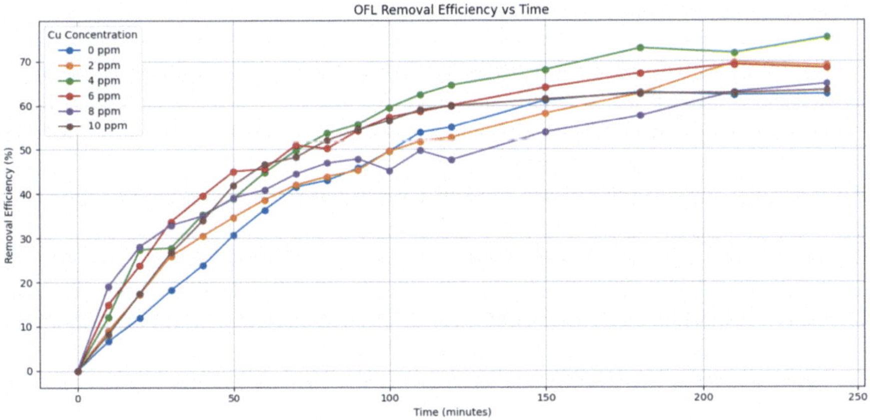

Fig. 2. Removal efficiency of ofloxacin (OFL) over time at varying $Cu2^+$ concentrations using electrocoagulation.

In this study, EC was applied for the first time to remove OFL and $Cu2^+$ simultaneously. OFL removal with respect to initial $Cu2^+$ concentrations was found to follow a similar trend to TC but with less pronounced variations. At OFL absorbance peak 293 nm, the highest removal efficiency of 75.3% was achieved at 4 ppm $Cu2^+$ at 240 min, corresponding to an optimal 1.7585 OFL/Cu ratio, whereas the control achieved 62.5% at the same time, showing a 20.5% increase in removal efficiency. Unlike TC, the final percentage removal for OFL was consistently higher at all tested $Cu2^+$ concentrations than in its absence. $Cu2^+$ addition improved OFL removal by 20.5%, but the relative gain was smaller than that of TC at 103.9%. This milder enhancement for OFL is consistent with its fewer coordination sites and lower tendency to form strong complexes which may be responsible for the increase in removal due to the addition of $Cu2^+$.

Baseline removal of OFL occurs through similar mechanisms of flocculation. For adsorption, at pH5.5, OFL predominantly exists in its zwitterionic form, so its partially deprotonated carboxyl group (-COO-) is able to interact electrostatically with $Al(OH)_2{}^+$, while ketone (C=O) and amine ($-NH_3{}^+$) groups form hydrogen bonds with -OH of flocs.

From 0–2 ppm, 2:1 type complexes $CuOFL_2$, $CuHOFL_2^+$, and $CuH_2OFL_22^+$ dominate, characterized by strong coordination bonds and high stability. At 4 ppm, the mixture is dominated by 1:1 type complexes $CuOFL^+$ and $CuHOFL_2^+$, with lower stability due to the partially filled coordination sphere of $Cu2^+$, and their smaller size enhancing adsorption. As OFL has fewer reactive sites than TC, lower rates of effective collisions for complex formation or degradation lead to less pronounced removal rate improvements, as well as a longer reaction time. At higher $Cu2^+$ concentrations, though $Cu2^+$ competes with OFL for adsorption sites, easily degradable complexes such as $Cu_3H_2OFL_2$ and $Cu_3H_4OFL_22^+$, with high steric hindrance and partially exposed coordination sites, may have been formed, causing removal to persist at a level higher than in the absence of $Cu2^+$.

3.2 Mechanisms of Removal

Both TC and OFL abatement rates declined over time. Kinetic analysis of initial rates revealed $Rate=k[Cu2^+]^{0.189}$ for TC, and $Rate=k[Cu2^+]^{0.292}$ for OFL. The fractional reaction orders suggested multiple removal pathways and indicated that $Cu2^+$-mediated processes were not the rate-determining step.

Additionally, experimental data closely followed the pseudo-second-order (PSO) adsorption kinetic model, with consistently high correlation coefficients, suggesting that chemisorption was the dominant mechanism governing TC and OFL removal, with the rate of saturation of adsorption sites governing the overall kinetics. This means optimal $Cu2^+$ concentration enhances the formation of reactive metal-ligand compounds with stronger binding affinities to flocs (Table 1).

Table 1. Coefficient of determination (R2) values for pseudo-second-order (PSO) kinetic models at varying $Cu2^+$ concentrations

Cu^{2+} concentration (ppm)	TC: R^2	OFL: R^2
0	0.99923	0.94655
2	0.99729	0.97857
4	0.99702	0.98716
6	0.99624	0.99261
8	0.99422	0.98329
10	0.99604	0.99245

3.3 Floc Characterization

Scanning electron microscopy (SEM) was conducted using JEOL JSM-6010LA to characterize the surface morphology of dried flocs, followed by elemental analysis using Energy Dispersive X-ray (EDX) Spectroscopy (Fig. 3).

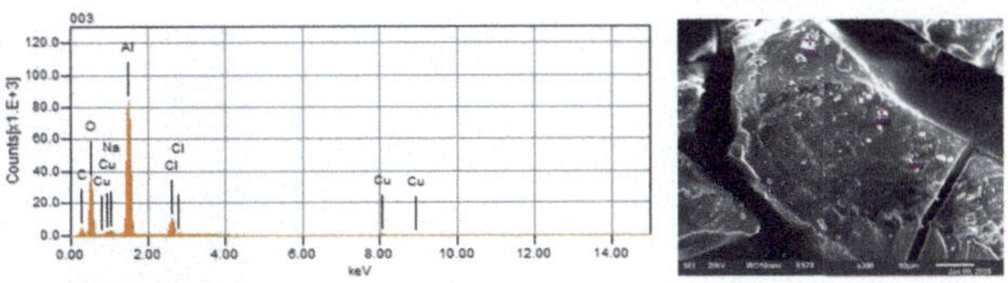

Fig. 3. Counts (x10^3) VS energy (keV) of TC/4 ppm Cu^{2+} Al flocs; Secondary Electron Imaging (SEI) of TC/4 ppm Cu^{2+} Al flocs

TC/4ppm $Cu2^+$ flocs exhibited a more crystalline structure with a distinct surface layer, likely resulting from slower $Al(OH)_3$ precipitation, which enables molecules to migrate and arrange in an orderly, regular lattice. Localized depressions and circular features observed in SEI are attributed to ROS-induced surface modification, as evidenced by the high O content in EDX as compared to TC/6 ppm $Cu2^+$ flocs. Such etching effects expose new active sites and increase the surface area on the floc for adsorption. With PSO adsorption kinetic modelling data confirming adsorption primarily involves molecular interactions at the surface rather than within pores, this combination of greater area and more active sites likely contribute to the superior TC removal rates observed at 4 ppm $Cu2^+$ (Fig. 4).

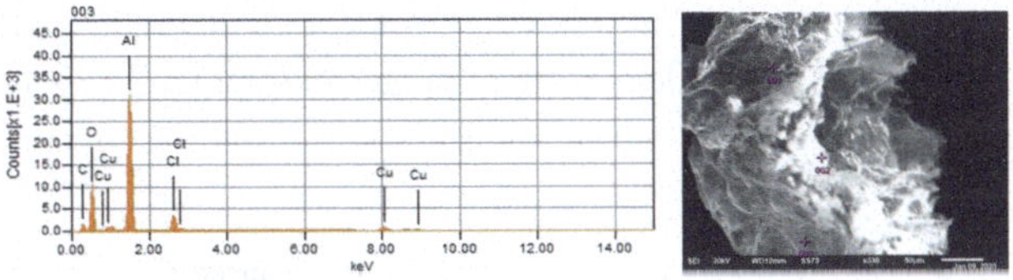

Fig. 4. Counts (x10^3) VS energy (keV) of TC/6 ppm Cu^{2+} Al flocs; Secondary Electron Imaging of TC/6 ppm Cu^{2+} Al flocs

TC/6 ppm $Cu2^+$ flocs exhibit an amorphous structure with rough, layered surfaces, likely resulting from heterogeneous and irregular deposition of flocs and pollutants. While their higher porosity increases physisorption through strong confinement effects in micropores that amplify van der Waals forces, removal was lower than optimal due to the chemisorption-dominated nature of the system. Additionally, reduced structural stability and weaker mechanical strength may make the flocs more prone to breakup and re-release of TC. The increase in alkalinity over reaction duration promoted the formation of soluble $Al(OH)_4^-$, further increasing floc solubility and contributing to TC desorption (Fig. 5).

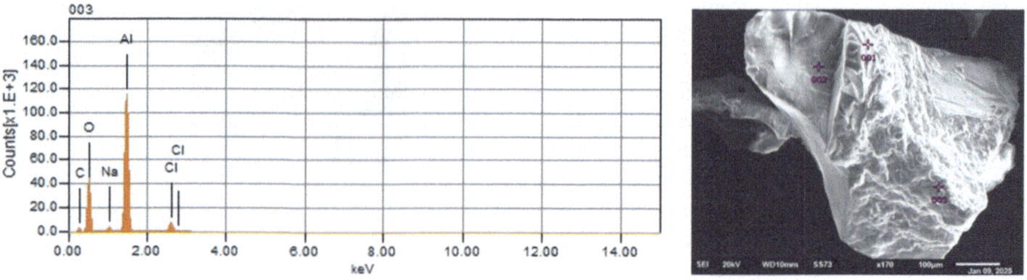

Fig. 5. Counts ($x10^3$) VS energy (keV) of OFL/0ppm Cu^{2+} Al flocs; Secondary Electron Imaging of OFL/0ppm Cu^{2+} Al flocs

OFL/0ppm $Cu2^+$ flocs exhibit a lesser degree of amorphous structure. Relatively high levels of dissolved aluminum species, measured at 21.03% by EDX, result in rapid nucleation at multiple sites. This quick consumption of $Al3^+$ leaves insufficient time for ordered crystal growth, and creates localized variations in $Al3^+$ concentration, leading to voids and aggregated structures which merge into irregular flocs. As pH rises to 7, the solubility of $Al3^+$ decreases drastically to $1.3 \times 10-12M$, potentially causing supersaturation which can further promote the formation of small, disordered floc particles. These amorphous flocs enhance pollutant binding through weaker van der Waals interactions. However, as identified by POS, physisorption, where van der Waals forces dominate, is less significant. Moreover, like TC/6 ppm $Cu2^+$ flocs, they degrade easily under stress from agitation. This suggests that flocs formed at suboptimal $Cu2^+$ concentrations tend to lack long-range order and stability as a prime factor for poorer removal rates (Fig. 6).

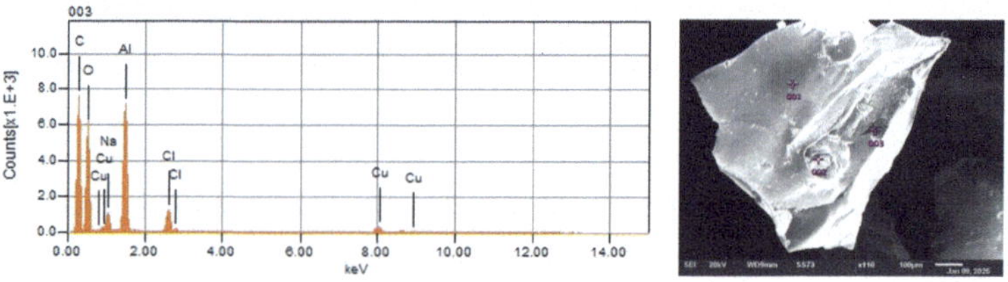

Fig. 6. Counts ($x10^3$) VS energy (keV) of OFL/4ppm Cu^{2+} Al flocs; Secondary Electron Imaging of OFL/4ppm Cu^{2+} Al flocs

OFL/4ppm $Cu2^+$ flocs exhibit a crystalline structure with a compact, flat surface, due to uniform precipitation conditions that promote layer-by-layer addition of $Al(OH)_3$, enhancing structural integrity and settling or flotation rates. This indicates that $Cu2^+$ aids in forming sturdier floc networks for OFL as well, mirroring results from TC.

4 Implications for Wastewater Treatment

Our results highlight the importance of maintaining an optimal metal-to-ligand ratio in enhancing the efficiency of antibiotic removal from wastewater through EC. These findings imply that if the inherent concentration of $Cu2^+$ is insufficient to facilitate effective complexation and subsequent antibiotic removal, a targeted, small dose of $Cu2^+$, or other heavy metal ions, can be added to reach the optimal range that enhances antibiotic removal. Conversely, if heavy metal concentration is too high, pretreatment methods such as selective precipitation, ion exchange, or dilution might be beneficial to avoid inhibiting the rate of antibiotic removal via EC.

With faster removal kinetics at the optimal $Cu2^+$ concentration, effective removal via EC can be achieved in shorter treatment durations, saving energy and electrode material. For example, TC reached 88% removal in 45 min with $Cu2^+$, whereas a longer runtime or a higher current would be needed to approach that efficacy without the addition of $Cu2^+$. This translates to lower electrical energy per volume treated and enables EC treatment to be more cost- and time-effective, which remains one of the most significant drawbacks of EC.

In practice, co-treating different waste streams may present logistical or regulatory constraints that prevent implementation. Yet, in large-scale treatment facilities that receive diverse effluents, it may be feasible and beneficial to intentionally blend streams to approximate the beneficial metal-ligand ratio before EC treatment.

5 Conclusion

This study demonstrates EC's capability to simultaneously remove pharmaceuticals and heavy metals from wastewater and proposes an optimal molar ratio for maximum removal efficiency. The presence of optimal $Cu2^+$ concentrations facilitated the formation of type 1:1 complexes, significantly enhancing the removal of TC and OFL, achieving efficiencies of 88.3% and 75.3% respectively, a marked improvement from removal in the absence of $Cu2^+$ at 43.3% and 62.5% respectively. However, excessive $Cu2^+$ reduced effectiveness by competing for adsorption sites and destabilizing flocs, underscoring the need for precise parameter optimization. Kinetics analysis and pseudo-second-order adsorption kinetic modelling confirmed chemisorption as the primary removal mechanism. SEM and EDX analyses revealed crystalline surface morphology as ideal for satisfactory adsorption and retention of pollutants. These findings establish EC as a promising technology for treating co-contaminated wastewater, contributing to more sustainable and safe water reuse.

References

1. Xuege, G., Junfeng, L., Xueting, F., Wenying, Q., Wenhuai, W., Jiankang, W.: Efficient removal of norfloxacin from water using batch airlift-electrocoagulation reactor: optimization and mechanisms analysis. RSC Adv. **2023**(13), 8944–8954 (2023)
2. Feryal, A., Selva, C.: Copper, chromium and nickel removal from metal plating wastewater by electrocoagulation. Desalination **269**(1–3), 214–222 (2011)
3. Qingdong, Q., Xian, W., Liwei, C., Zhongshuai, J., Yan, X.: Simultaneous removal of tetracycline and Cu(II) by adsorption and coadsorption using oxidized activated carbon. RSC Adv. **2018**(8), 1744–1752 (2017)
4. Hossen, M.A., Sattar, G.S. Mustafa, M.G.: Factors affecting the performance of a pharmaceutical wastewater treatment plant: Characterization of effluent and environmental risk. PMC11015402 (2024)

5. Jianbo, L., et al.: Efficient removal of Tetracycline-Cu complexes from water by electrocoagulation technology. J. Cleaner Prod. **289**, 125729 (2020)
6. Sriram, B., Kolla, A., Jyoti, K., Ruthviz, K., Jayaprakash, V.: A state-of-the-art review of the electrocoagulation technology for wastewater treatment. Water Cycle **4**(2023), 26–36 (2023)
7. Edris, B., Leili, M., Alireza, A., Amir, H.M.: Heavy metals removal from aqueous environments by electrocoagulation process– a systematic review. J. Environ Health Sci. Eng. (2015)
8. Revellame, E.D., et al.: Adsorption kinetic modeling using pseudo-first order and pseudo-second order rate laws: a review. Clean. Eng. Technol. **1**, 100032 (2020)
9. Wang, J., Guo, X.: Adsorption Isotherm Models: Classification, Physical Meaning, Application and Solving Method. Chemosphere **258**, 127279 (2020)
10. Runyu, X., et al.: Removal of microplastics and attached heavy metals from secondary effluent of wastewater treatment plant using interpenetrating bipolar plate electrocoagulation. Separat. Purificat. Technol. **290**, 120905 (2022)
11. Ting, L., et al.: New insights into the effect of pH on the mechanism of ofloxacin electrochemical detection in aqueous solution. Phys. Chem. Chem. Phys. (2019)
12. Sonia, M., Tlili, M., Naceur, J., Abdennaceur, H.: Applicability of electrocoagulation process to the treatment of Ofloxacin and Chloramphenicol in aqueous media: Removal mechanism and antibacterial activity. J. Water Proc. Eng. **49**, 103080 (2022)
13. M. Ebba, P. Asaithambi, Esayas A. 2022. Development of electrocoagulation process for wastewater treatment: optimization by response surface methodology. Heliyon Volume 8, Issue 5, May 2022, e09383.
14. Lu, X., Tang, B., Zhang, Q., Liu, L., Fan, R., Zhang, Z.: The Presence of Cu Facilitates Adsorption of Tetracycline (TC) onto Water Hyacinth Roots. Int. J. Environ Res. Pub. Health **15**(9), 1982 (2018)

Engineering

Biofluorescence of Mussel Shells and its Applications in Water Purification

Josh Sng(✉), Dylan Kai Jie Gan, and Tan Yong Tat

NUS High School of Mathematics and Science, Singapore 129957, Singapore
joshsngg@gmail.com, tanyongtat07@gmail.com

Abstract. Mussels are a widely consumed seafood; yet, most of their shells are discarded as waste. At the same time, pollution of water sources by industrial dyes like Rhodamine B (RhB) poses severe health risks. Our research addresses both issues by upcycling green mussel (*Perna viridis*) shells into cost-effective, reusable, and eco-friendly adsorbents for RhB removal. Using fluorescence microscopy, we characterised the interaction of RhB with the shell layers and found rapid adsorption accompanied by distinctive fluorescence. Further analysis confirmed a combined physisorption and chemisorption mechanism, with calcium carbonate as the primary adsorptive component. Sonication experiments demonstrated the reversibility of adsorption and potential for reuse. Owing to the distinctive fluorescence of RhB-adsorbed shells, our research highlights their applications as RhB biosorbents, visual biosensors, and steganographic materials. Future work will further characterise adsorption dynamics and investigate other pollutants.

Keywords: Shells · Adsorption · Upcycling · Purification

1 Introduction

In a rapidly consumerist and industrialised world, not only does the textile industry continue to grow at expeditious rates, but so do its environmental implications. In particular, Rhodamine B (RhB), whose chemical structure is shown in Fig. 1, is an organic dye often used in textile production that is prone to improper disposal into surrounding waters, especially in developing countries. RhB is a suspected carcinogen and poses a health threat not just to the marine ecosystem, but to humans too via a trickle-down effect [1]. Furthermore, RhB has found its way to foods through its use as a dye, despite being banned for its health risks [2]. The recent banning of RhB usage in cotton candy in India [3] exemplifies how relevant and modern this issue is.

Current methods to remove RhB from water include hydroxyl and sulfate radicals [4], photocatalytic [5], and adsorption [6] treatments, but these remain costly and require extensive processing, which may ironically lead to other environmental implications.

Simultaneously, a major loss in potential value is happening globally due to the annual disposal of 10 million tonnes of shells. These shells, which make up 65–90% of the live weight of shellfish, are discarded [7], either into the sea and landfills in

H. Guo et al. (Eds.): IRC-SET 2025, PTT, pp. 301–310, 2026.
https://doi.org/10.1007/978-981-95-9630-0_26

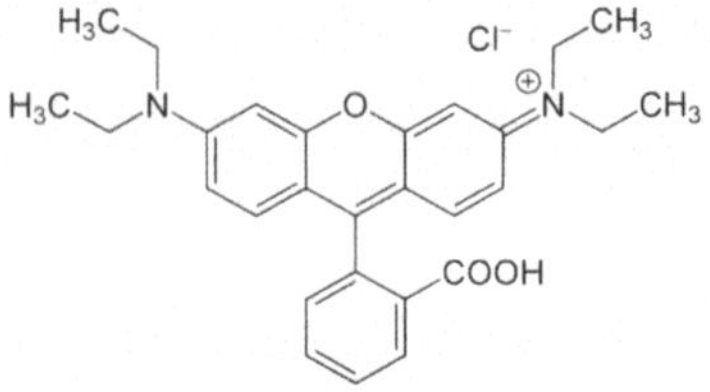

Fig. 1. Chemical structure of Rhodamine B.

developing countries, or via a costly disposal system in developed countries. These discarded shells, which are primarily composed of calcium carbonate, are goldmines of untapped potential, as illustrated by their repurposing into wide-ranging applications such as fertiliser [8] and even adsorbents [9].

As such, this research aims to upcycle shellfish shells into RhB adsorbents and we hypothesise that these shells will be a cost-effective, reusable, and environmentally friendly adsorbent.

2 Materials and Methods

The green mussel (*Perna viridis*) has been selected as the focal shellfish of this research due to its convenience of acquisition and ubiquity in the F&B industry, especially in the Indo-Pacific region. These mussels were acquired from a locally farmed source and deshelled to be cooked and eaten at home, leaving behind shells for us to work on.

To work towards understanding the adsorption mechanism, we investigated how the different layers of the shell (Fig. 2) interacted with the RhB. To expose the prismatic layer, we applied a 2.5% w/w sodium hypochlorite solution to delaminate the periostracum and reveal the prismatic layer.

Further characterization of the mussel shells was done using an Olympus BX51 fluorescence microscope (FM); UV, blue, green and yellow excitations were obtained using U-MWU2 (300–390 nm), U-MWB2 (450–500 nm), U-MWG2 (500–560 nm), and U-MWIY2 (530–580 nm) filter cubes.

To determine the minimum concentration of RhB that could produce detectable fluorescence on the shell, we prepared RhB solutions which varied by one order of magnitude and dipped shell pieces into these solutions for ten minutes. Then, fluorescence spectra of the shells were taken using an Ocean Optics Flame Spectrometer, and we checked if the spectra indicated the presence of RhB through its characteristic peak emission wavelength of 583nm under green excitation.

Next, to determine how long the shell must be dipped to be saturated with RhB, we dipped the shell in a concentration of RhB one order of magnitude higher than the previously determined minimum detectable concentration, then removed the shell at fixed time intervals and recorded the fluorescence intensity at 583nm when the solution was placed under green light excitation. This is repeated with the same shell and solution until the fluorescence intensity remains constant.

We also investigated different adsorption isotherms using concentration data determined by a Shimadzu UV-2600 UV-Vis spectrophotometer.

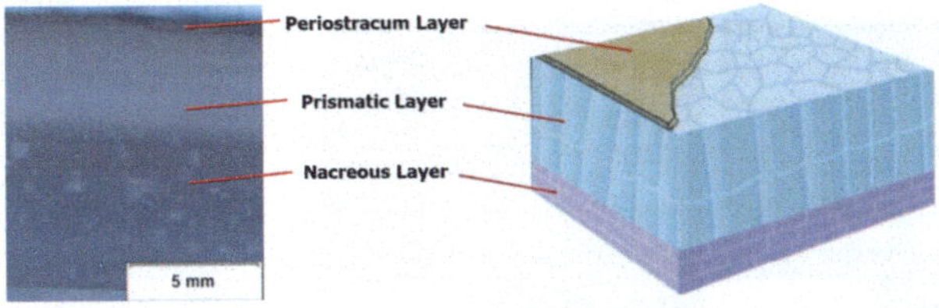

Fig. 2. Cross-section under bright field light (BFL) and diagram of different layers of mussel shell.

Lastly, we attempted to elucidate the mechanism of RhB adsorption onto mussel shells by studying the shell surface using a JEOL JSM-6700F Field Emission Scanning Electron Microscope (SEM) and Oxford Instruments X-MaxN 150 Energy Dispersive X-ray (EDX) detector. We also conducted sonication experiments to determine whether RhB was adsorbed onto the mussel shells through chemical or physical processes, or both. In fact, this can also help to determine the recoverability of our adsorbent.

In this experiment, we placed a RhB-treated mussel shell piece into an Eppendorf tube filled with deionised water and submerged it in an ultrasonic bath. During the sonication process, the ultrasound propagates by a series of compressions. When it passes through the mussel shell, attenuated waves are induced, promoting the 'removal' of the RhB particles adsorbed by the nacreous layer of the mussel shell. We then measure the extent to which RhB is removed by determining the decrease in fluorescence intensity of the nacreous shell layer at 583nm.

3 Results and Discussion

3.1 Fluorescence of Mussel Shells and Rhodamine B

Under the FM, the prismatic and nacreous layers demonstrated fluorescence under UV, blue, green, and yellow light excitations, as seen in Fig. 3(a).

Of note, under green light excitation (500–560 nm), the prismatic and nacreous layers exhibit a red fluorescence with peak emission intensity at 607nm (Fig. 3(b) and (c)). In contrast, our contaminant of study, Rhodamine B, absorbs at a similar range but has a peak emission at 583nm (Inset of Fig. 5(b)). Since both fluorescent materials absorb green light, all experiments henceforth were conducted using green light excitation.

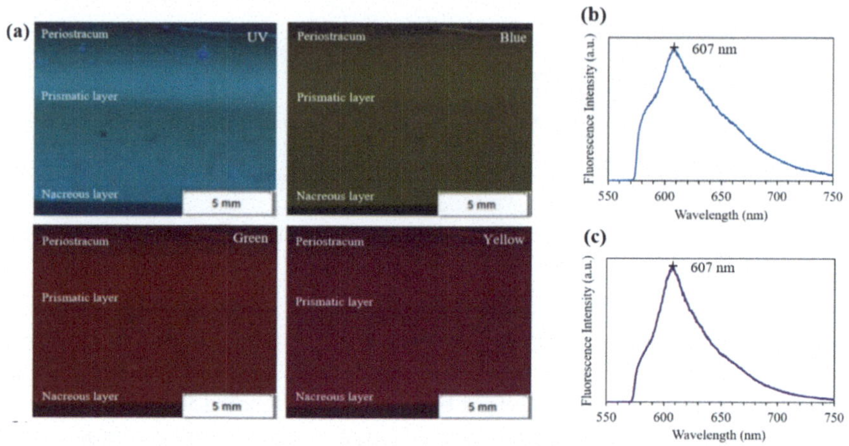

Fig. 3. (a) Fluorescence microscopy images of mussel shell cross-section under UV, blue, green, and yellow light excitations. Fluorescence spectra of (b) prismatic layer and (c) nacreous layer under green light excitation.

3.2 Adsorption of RhB

Figure 4(a) shows a crucial preliminary observation that the shell successfully adsorbs RhB, and only on the prismatic and nacreous layers. RhB adsorption is indicated by a sharp increase of the fluorescence emission peak at 583nm under green excitation, thus appearing bright orange. This allows us to construct mosaics (Fig. 4(b) and (c)) using both pristine and RhB-treated shell pieces: under BFL, the mosaic appears typical, but green light excitation reveals the hidden orange logo formed by RhB-treated shells.

However, exposing the prismatic layer requires bleach treatment, which is antithetical to the broader aim of our research: sustainability. Not only would bleaching require additional processing and incur greater costs, but bleach also poses an environmental threat if disposed improperly. Hence, we focused our subsequent experiments on only the nacreous layer.

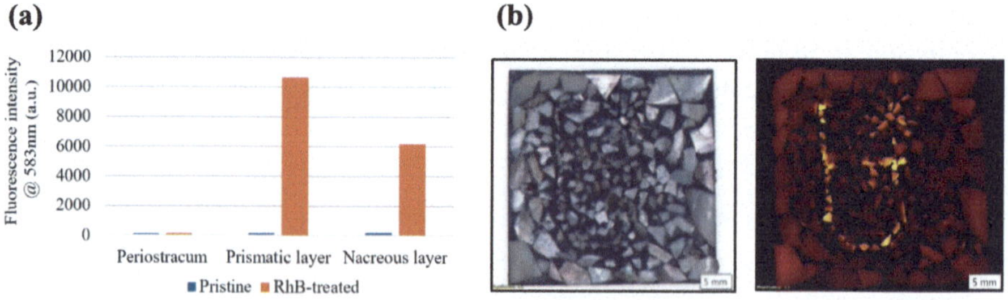

Fig. 4. (a) Fluorescence intensities of pristine and RhB-treated shell layers at 583nm. (b) Mosaic of pristine and RhB-treated mussel shell pieces with a hidden logo under BFL and green light excitation. RhB-treated pieces were used to form the orange logo.

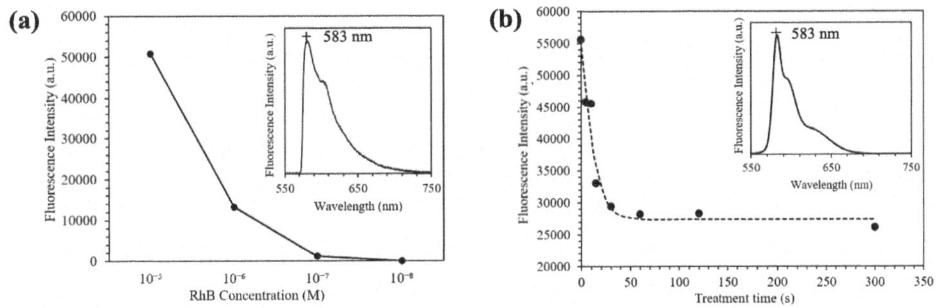

Fig. 5. Peak fluorescence intensity of (a) mussel shell after treatment with varying RhB concentrations and (b) 10^{-6}M RhB solution with varying treatment times. Insets: Fluorescence spectra of (a) RhB-treated mussel shell and (b) RhB solution under green light excitation.

3.3 Limits of RhB Adsorption

The lowest RhB concentration that gave detectable fluorescence is 10^{-7}M (Fig. 5(a)).

According to current literature [10], RhB becomes a threat to aquatic ecosystems at concentrations higher than 140µg/L or 2.92×10^{-7}M. Thus, our results suggest that the shells will be effective in adsorbing RhB at realistically harmful concentrations.

From this experiment, we can also conclude that the fluorescence intensity of the shell at 583nm is positively correlated to the amount of RhB adsorbed onto the shell.

Figure 5(b) shows that the shells take as low as 60 s to be saturated with RhB. This suggests that short treatment times are needed to achieve optimal adsorption, making the shell's applications in chemical cleaning or biosensing even more convenient.

3.4 Adsorption Isotherms

Adsorption isotherms indicate the distribution of solute molecules between the liquid phase and adsorbent surface. Isotherm fitting is important in the design and use of adsorbents. We thus fitted three isotherms to data collected from UV-Vis spectroscopy.

The Langmuir isotherm [11] has four underlying assumptions [12], namely that monolayer adsorption occurs on a structurally homogeneous surface; each adsorption site holds one adsorbate molecule; all adsorption sites are identical and independent; and there are no interactions between adsorbed molecules. The isotherm is given as

$$q_e = \frac{q_{max} K_L C_e}{1 + K_L C_e} \tag{1}$$

where:

q_e = amount of solute adsorbed per mass of adsorbent
C_e = equilibrium concentration of solute
q_{max} = maximum adsorption capacity
K_L = equilibrium constant.

The Freundlich isotherm [13] is an entirely empirical equation is given by Eq. (2).

$$q_e = K_F C_e^{1/n} \tag{2}$$

where:

K_F = distribution coefficient, which is a measure of adsorption capacity

n = dimensionless constant related to the adsorption intensity.

Lastly, for systems where the adsorption enthalpy varies with the surface coverage, the Langmuir isotherm likely cannot be applied as the underlying assumptions are not fulfilled [14]. Thus, the Temkin isotherm was developed to describe a system where adsorption enthalpy decreases linearly with surface coverage. It can be written as [15]:

$$q_e = \frac{q_{max}}{\ln(a_0/a_1)} \ln\left(\frac{1 + a_0 C_e}{1 + a_1 C_e}\right) \tag{3}$$

where:

a_0 = equilibrium binding constant corresponding to maximum binding energy

a_1 = equilibrium binding constant corresponding to minimum binding energy.

We conducted non-linear fitting of our data to the adsorption isotherms using Microsoft Excel's Solver function. The results can be summarized in Fig. 6. With the highest R^2 value of 0.989, the Temkin isotherm is the best fit of our data, although it is only a slightly superior fit to the Langmuir and Freundlich isotherms.

Assuming our system approximates the assumptions made in the Temkin isotherm [16], we can deduce that the mussel shell surface is strongly heterogeneous as $a_0 >> a_1$. That is, there is a large difference between the maximum and minimum binding energies of RhB to the shells.

The good fit to the Freundlich isotherm also suggests that the surface is heterogeneous. On the other hand, the slightly poorer fit to the Langmuir isotherm is expected as its underlying assumptions are not fulfilled. Of note, the mussel shell is undoubtedly structurally heterogeneous, as seen from the SEM images in Fig. 7(e) and (f).

(a)

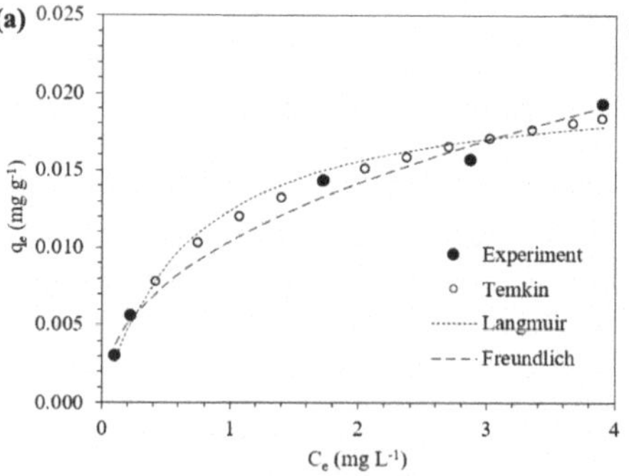

(b)

Langmuir	q_{max} (mg/g)	0.0209
	K_L (L/mg)	1.48
	R^2	0.977
Freundlich	$1/n$	0.442
	K_F ((mg/g)/(L/mg)$^{1/n}$)	0.0104
	R^2	0.987
Temkin	q_{max} (mg/g)	0.0361
	a_0 (L/mol)	7.78
	a_1 (L/mol)	0.00951
	R^2	0.989

Fig. 6. (a) Isotherms and (b) Isotherm parameters obtained using non-linear fitting for the adsorption of RhB onto green mussel shells.

3.5 Adsorption Mechanism

The sonication experiment showed that there is a mean decrease of 41% in the fluorescence intensity after sonication (Fig. 7(a)). We thus infer that the adsorption of RhB onto mussel shells is through both physical and chemical processes. This can be deduced since the energy generated from the ultrasonic bath is insufficient to break the chemical bonds between adsorbent and adsorbate in the case of chemisorption, whereas sonication can effectively remove adsorbate molecules bound through physisorption. This result also shows that these shells can be reused after sonication to adsorb even more RhB.

Additionally, we noticed that the intensity of red fluorescence varies between different shell pieces. One observable physical difference between shells with highly distinct fluorescent intensities was the colour of the nacreous layer under BFL: namely, grey and yellow. Macroscopic and microscopic views of the two differently shaded nacreous pieces are included in Fig. 7(b). In general, the yellow nacreous pieces could adsorb more RhB than the grey pieces, as shown in Fig. 7(c) and (d), where the characteristic orange fluorescence of RhB is more prominent for the treated yellow shell.

We hypothesized that structural differences of the shell surface caused the different appearances and adsorption capabilities of the two different shades of shell. The microscopic views of the two shades of shell in Fig. 7(b) revealed that the grey pieces exhibit a unique iridescence that likely arises from interference of light with the aragonite platelets present in the nacreous layer [17, 18]. This further confirms the hypothesized structural difference between the grey and yellow pieces.

The SEM images (Fig. 8(e) and (f)) of the shells reveal no significant physical difference between the grey and yellow pieces, although the SEM images of the yellow pieces appear darker than those of the grey pieces.

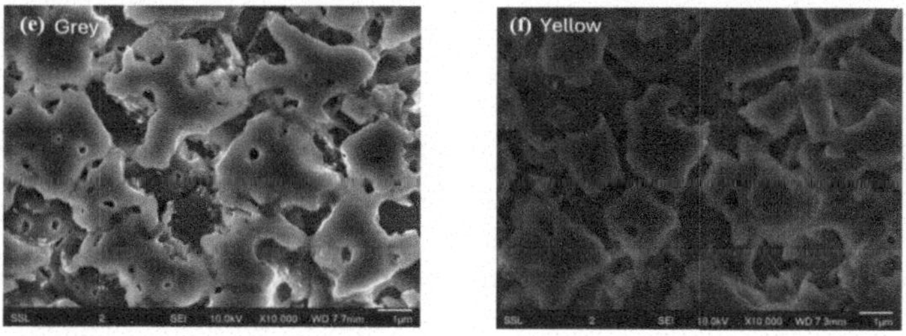

Fig. 7. (a) Median and mean peak fluorescence intensities of mussel shells pre- and post-sonication. (b) Macroscopic and microscopic views of the two shades of nacreous layer. c) FM images of treated and untreated yellow and grey shells and (d) boxplots of peak fluorescence intensities of the two shades of shells after RhB treatment. SEM images of (e) grey and (f) yellow nacreous pieces.

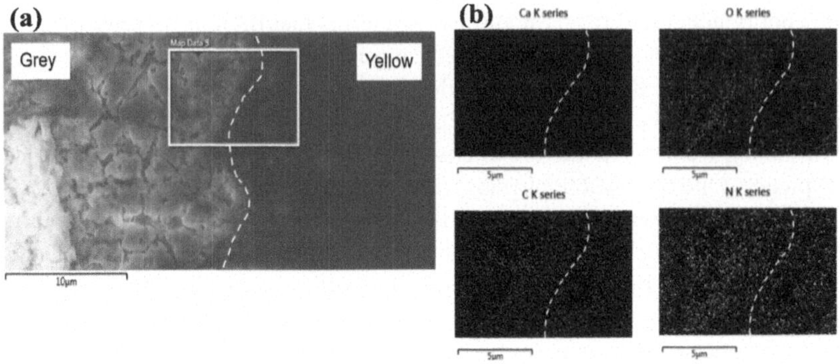

Fig. 8. SEM images of (e) grey and (f) yellow nacreous pieces.

Furthermore, in Fig. 9(b), the SEM-EDX elemental mapping of the interface between the grey and yellow segments of the nacreous layer also show no obvious difference in chemical composition of these two differently shaded areas.

The successful adsorption of RhB by only the calcareous layers of the shell, and not the periostracum, which primarily contains proteins [19], suggests that calcium carbonate plays a major role in the adsorption. This hypothesis is reinforced in Fig. 9(b),

where the elemental map shows that the elements of calcium carbonate (calcium, carbon, oxygen) are present and widespread on the nacreous layer. Fig. 9(b) also confirms this hypothesis as white chalk, which is essentially calcium carbonate, also displays successful adsorption of RhB, as seen from the distinct peak at 583 nm of the FM spectrum taken of RhB-treated chalk.

From Fig. 9, it is apparent that white chalk adsorbs a substantially greater amount of RhB than our shells. However, it would be less appropriate to use white chalk as an adsorbent to remove RhB from contaminated streams for a few reasons: Primarily, the production of chalk is a labor-intensive process beginning with the quarrying of limestone, followed by a multitude of steps that require significant energy. On the other hand, using mussel shells as adsorbents allows this waste material to be upcycled without much labor or processing.

Additionally, when chalk is introduced into water, it disperses into fine particles, creating a cloudy suspension that undermines its utility in the removal of RhB. In contrast, mussel shells offer the advantage in that they can be easily separated after the treatment process, making them a superior alternative to chalk for water purification.

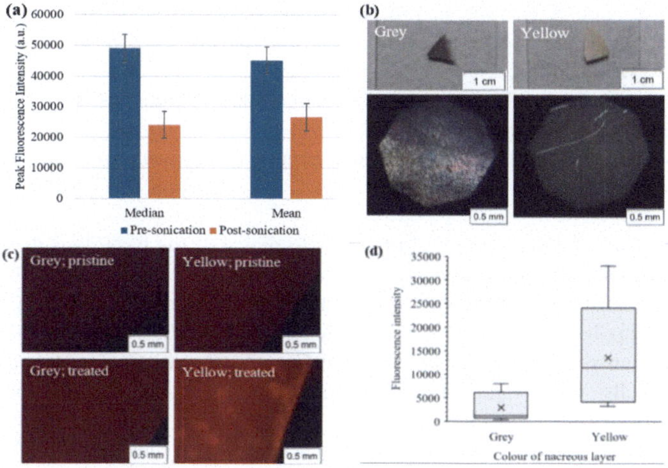

Fig. 9. (a) EDX layered image and (b) elemental map of grey and yellow interface.

4 Conclusion

In conclusion, our mussel shells work as cost-effective, reusable, and environmentally friendly RhB adsorbents. The shells were able to rapidly adsorb RhB at concentrations as low as 10^{-7}M and display the characteristic orange fluorescence of RhB under green light excitation, thus showing great potential as both an adsorbent and a chemical sensor. Adsorption isotherm studies show that the Temkin isotherm adequately describes the adsorption of RhB onto mussel shells, and the isotherm parameters obtained imply a highly heterogeneous adsorption surface. In terms of adsorption mechanism, calcium carbonate is most significant component in the shells that RhB molecules bind to, and the sonication experiment showed that adsorption occurs through both physical and chemical processes.

Furthermore, RhB's distinct bright orange fluorescence, as opposed to our shells' red fluorescence under green light, also allows for steganographic encoding of messages using mussel shells. This expands the applications of these shells beyond just environmental sustainability.

To further characterize the adsorption of RhB onto mussel shells, detailed kinetic and thermodynamic studies can be conducted. Calorimetric experiments can also determine the heat of adsorption, which helps us formally classify the adsorption process as physical or chemical adsorption, or both.

Further investigation is also needed to determine the reason or reasons behind the higher adsorption capacity of the yellow-shaded shell pieces, be it physical or chemical, possibly starting with the differing brightness of the SEM images. This investigation would likely shed more light on the adsorption mechanism of RhB onto mussel shells.

References

1. Saigl, Z.M.: Various adsorbents for removal of rhodamine b dye: a review. Indonesian J. Chem. **21**(4), 1039–1056 (2021)

2. Lin, S., Lin, X., Lou, X.T., Yang, F., Lin, D.Y., Lu, Z.W.: Rapid and sensitive SERS method for determination of Rhodamine B in chili powder with paper-based substrates. Anal. Methods **7**(12), 5289–5294 (2015)
3. Desgupta, D.: Eats not okay: Mayonnaise, cotton candy banned as India cracks down on unsafe street food. The Straits Times (2024)
4. Nidheesh, P.V., Rajan, R.: Removal of rhodamine B from a water medium using hydroxyl and sulphate radicals generated by iron loaded activated carbon. RSC Adv. **6**(7), 5330–5340 (2016)
5. Jain, R., Mathur, M., Sikarwar, S., Mittal, A.: Removal of the hazardous dye rhodamine B through photocatalytic and adsorption treatments. J. Environ. Manage. **85**(4), 956–964 (2007)
6. Li, F., Chen, Y., Huang, H., Cao, W., Li, T.: Removal of rhodamine B and Cr (VI) from aqueous solutions by a polyoxometalate adsorbent. Chem. Eng. Res. Des. **100**, 192–202 (2015)
7. Summa, D., Lanzoni, M., Castaldelli, G., Fano, E.A., Tamburini, E.: Trends and opportunities of bivalve shells' waste valorization in a prospect of circular blue bioeconomy. Resources **11**(5), 48 (2022)
8. Masukume, M., Onyango, M.S., Maree, J.P.: Sea shell derived adsorbent and its potential for treating acid mine drainage. Int. J. Miner. Process. **133**, 52–59 (2014)
9. Topić Popović, N., Lorencin, V., Strunjak-Perović, I., Čož-Rakovac, R.: Shell waste management and utilization: Mitigating organic pollution and enhancing sustainability. Appl. Sci. **13**(1), 623 (2023)
10. Skjolding, L.M., et al.: Assessing the aquatic toxicity and environmental safety of tracer compounds Rhodamine B and Rhodamine WT. Water Res. **197**, 117109 (2021)
11. Langmuir, I.: The constitution and fundamental properties of solids and liquids. Part I. Solids. J. American Chem. Soc. **38**(11), 2221–2295 (1916)
12. Ruthven, D.M.: Principles of adsorption and adsorption processes. John Wiley & Sons (1984)
13. Freundlich, H.: Über die adsorption in lösungen. Z. Phys. Chem. **57**(1), 385–470 (1907)
14. Pursell, C.J., Hartshorn, H., Ward, T., Chandler, B.D., Boccuzzi, F.: Application of the Temkin model to the adsorption of CO on gold. The J. Phys. Chem. C **115**(48), 23880–23892 (2011)
15. Temkin, M.I.: The kinetics of some industrial heterogeneous catalytic reactions. In: Advances in Catalysis, Vol. 28, pp. 173–291. Academic Press (1979)
16. Aharoni, C., Ungarish, M.: Kinetics of activated chemisorption. Part 2.—Theoretical models. J. Chem. Soc. Faraday Tran. 1: Phys. Chem. Condensed Phases **73**, 456–464 (1977)
17. Tan, T.L., Wong, D., Lee, P.: Iridescence of a shell of mollusk Haliotis Glabra. Opt. Express **12**(20), 4847–4854 (2004)
18. Snow, M.R., Pring, A., Self, P., Losic, D., Shapter, J.: The origin of the color of pearls in iridescence from nano-composite structures of the nacre. Am. Miner. **89**(10), 1353–1358 (2004)
19. Bubel, A.: An electron-microscope study of periostracum formation in some marine bivalves. II. The cells lining the periostracal groove. Mar. Biol. **20**, 222–234 (1973)

Developing a Novel Water Filtration System using Carbon-Based Adsorbents

Bhatia Priyansh[1](✉), Joshi Abhilash[1], Vadapalli Dedeep Sai[1], and Valiyaveettil Suresh[2]

[1] School of Math and Science, NUS High, Singapore, Singapore
h2010011@nushigh.edu.sg

[2] Chemistry Department, National University of Singapore, Singapore, Singapore

Abstract. Laundry pollutants including Polymer-Chain-linked Metal ions (PCMs), aromatic compounds, and heavy metal ions critically threaten aquatic ecosystems. Current centralized water reclamation plants suffer from operational inefficiencies and aging infrastructure, preventing optimal pollutant removal. This study proposes a decentralized water filtration system using a gravity-based filtration column with carbon-based adsorbents. Batch extraction experiments revealed that Activated Charcoal (AC) demonstrated superior removal efficiency for both PCMs and aromatic compounds compared to other adsorbents. A gravity-driven prototype was constructed using 39mm PVC pipes with AC and sea sand. Empirical testing determined that a 9g:1.15g sand-to-AC ratio achieved optimal flow parameters while maintaining 94% removal efficiency. Sueded microfiber was identified as the optimal membrane material. Long-term testing with real-life discharge water (443ppm) showed that filters containing 5g of AC maintained effectiveness for 20 hours before dropping below 75% removal efficiency. AC demonstrated regeneration capability under weak acid washing, recovering to 97% removal efficiency after adsorption-desorption cycles, confirming prototype reusability. To address heavy metal ions, a composite polyethylenimine-Alginated-AC (PAA) adsorbent was synthesized and verified for removal efficiency across all pollutant classes. The PAA filter was integrated into a 3D printed mountable casing designed for washing machine output valve applications. This research provides promising groundwork for developing decentralized, scalable AC-based gravity-driven filters for laundry effluent treatment, offering a practical solution to current wastewater management challenges. Limitations and next steps: potential microfiber shedding from the membrane, a hydraulic mismatch versus real washing-machine discharge (≈3–5 gpm, 11–19 L/min), and the need for closed-loop handling of acid-wash effluent during adsorbent regeneration are acknowledged; we outline monitoring protocols, scale-up by empty-bed contact time (EBCT) parity, and neutralization/capture of regeneration liquor for compliant disposal.

Keywords: Activated carbon · Water filtration · UV-vis spectroscopy · Anionic surfactants · Environmental engineering

H. Guo et al. (Eds.): IRC-SET 2025, PTT, pp. 311–321, 2026.
https://doi.org/10.1007/978-981-95-9630-0_27

1 Introduction

1.1 Background

The discharge of pollutants into water bodies poses a critical threat to aquatic ecosystems, with lethal effects on fish and other organisms at concentrations as low as 5–15 ppm [3]. Most laundry pollutants contain significant levels of phosphate salts and complex organic compounds, which inhibit biodegradation and contribute to eutrophication [8].

Conventional centralized wastewater treatment plants often struggle to remove pollutants effectively due to aging infrastructure and high sludge volumes, creating a need for more flexible, decentralized treatment options [2, 5]. Decentralized systems can address wastewater at its source, improving removal efficiency while easing the burden on centralized facilities.

Recent technological advancements have made modern systems particularly effective for treating specific types of wastewaters, such as laundry effluent, where physical-chemical treatments often provide good results (Water, n.d.).

1.2 Scope, Aim and Hypothesis

It has been shown that detergents contain both polymer chain-linked metal ions (PCMs) and aromatic compounds, which are significant pollutants in washing machine discharge [7]. This study focuses on removing three common anionic detergents: sodium n-dodecyl benzene sulphonates (SDBS), sodium n-dodecyl sulphate (SDS), and sodium deoxycholate, all classified as PCMs, with SDBS also acting as a major aromatic pollutant.

Hence, we aim to develop an affordable means of filtering the primary pollutants, the PCMs and harmful aromatic compounds, present in commercially used detergent existing in washing machine discharge. To address this, we turn our attention to four carbon-based adsorbents – Ground Charcoal, Amorphous Activated carbon (AC), graphite, and an AC-containing alginate gel bead (AC-AB) – as potential primary adsorbents. While carbon adsorbents perform well for organic compounds, literature shows that they are less effective for heavy metals often found in industrial soaps and detergents [11]. As such, we then aim to broaden the filtration capacities of our filter by integrating polyethyleneimine (PEI) [9, 10] with the best performing carbon adsorbent, incorporating alginate as the binding medium.

We focus on achieving a prototype with optimal parameters, namely – flow rate, removal efficiency, and breakthrough time. In addition, we treat (i) microplastics control as a co-objective—explicitly minimizing any membrane fibre shedding to the effluent—and (ii) at-source interception under pulsatile, high-throughput discharge conditions characteristic of residential washing machines. Finally, we aim to package the filter as an attachable unit for household/industrial washing units.

We hypothesize that the optimized filter will achieve high removal efficiencies comparable to centralized treatment while remaining practical for household use. We also expect that the filter can be regenerated by mild acid washing and maintain performance over repeated cycles.

2 Materials

Throughout this study, a range of commercially available high-purity carbon-based adsorbents and supporting chemicals were used to develop and test the filtration system. The primary adsorbents included ground charcoal, amorphous activated carbon (AC), graphite, and an alginate-bound AC bead composite (AC-AB). To enhance pollutant binding, polyethyleneimine (PEI) and alginate were employed in synthesizing a novel composite targeting heavy metal ions. Supporting filtration media such as fine sea sand was used to aid dispersion within the column, while detergent solutions were prepared from widely used commercial powders. All chemicals and adsorbents obtained were at purities of $\geq$99% where applicable and were sourced from GCE Lab Chemicals.

3 Methodology and Results

3.1 Materials Characterisation

Prior to determining the efficacies of the different materials in filtering the aforementioned pollutant compounds, we first identify the structural differences between the materials contributing to differences in adsorbing capacities. The Fourier Transform Infrared (FT-IR) spectra of AC, Graphite and ground charcoal within the range of 500-4000 cm^{-1} were obtained using a Bruker ALPHA FT-IR spectrophotometer using KBr as matrix. We then proceeded to obtain a calibration curve for detergent solutions of different concentrations derived for experimental use using UV-vis spectrometry (UVS).

3.2 Determining Efficacies of Various Adsorbents

We used the Minimum Viable Mass Ratio (MVMR) to compare the efficacy of different adsorbents for aromatic compounds and PCMs. The MVMR represents the mass of adsorbent required to fully capture a fixed mass of adsorbate. The filtration process followed a batch extraction procedure, where removal efficiency was measured via UVS at 223nm for aromatic compounds and ICP for sodium ions, ensuring a controlled environment where only the adsorbent type, mass, and adsorbate concentration affected the results.

To optimize testing time, we determined the minimum viable time (MVT) for batch extraction by varying runtimes (15–120 minutes) with 10mg of AC-AB in a 5ml 500ppm solution. The extraction was conducted using a batch extraction machine at 400 revolutions/minute. Three vials were used per condition to ensure reliability and calculate standard deviations. We found that a 30-minute runtime provided the optimal balance of efficiency and time.

For MVMR determination, adsorbents (5mg) were tested with detergent solutions ranging from 50–1500ppm. AC showed the highest MVMR for both aromatic compounds and PCMs, followed by AC-AB, graphite, and charcoal, contradicting our initial hypothesis [6], which suggested AC-AB would perform better for metal ions. Our results indicate that hydrophobic interactions dominate, favoring AC's larger surface area over AC-AB's ionic binding sites. The high correlation between the removal efficiency of aromatic compounds and PCMs is explained by the fact that metal ions in PCMs are linked to large aromatic chains. Therefore, future experiments will focus on the removal efficiency of aromatic compounds as measured by UVS (Fig. 1).

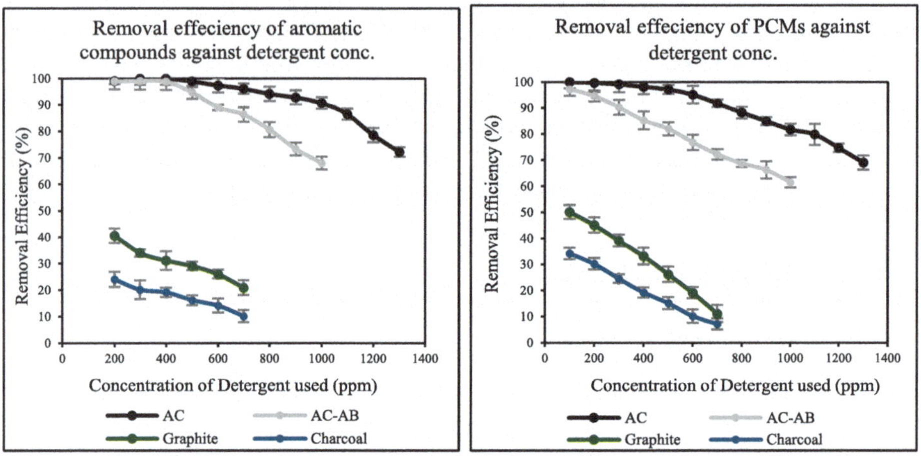

Fig. 1. Graphs of removal efficiencies of aromatic compounds and PCMs via various adsorbents, showing superiority of AC for removal of both types of pollutants.

3.3 Optimising Column Composition

Accordingly, we implement AC as the sole viable packing material for our detergent filter prototype. In order to replicate the gravity-based filter at a smaller scale, we utilised 50 ml syringes for experimentation. The initial steps involved –

i. The arrangements of the packing materials – AC, as well as sea sand to disperse the AC – within the filter (layered or mixed configuration).
ii. Their compositions, specifically the ratios of the packing materials – AC, as well as sea sand to disperse the AC

To determine the optimal packing arrangement, we tested layered versus mixed configurations of AC and sand. Using 5g sand and 1g AC with a 2000ppm detergent solution, the mixed configuration performed better, with a breakthrough time of 36 seconds and a flow rate of 0.32 mL/s, compared to the layered configuration's 67 seconds and 0.12 mL/s. This suggests that AC agglomeration in the layered setup impedes flow. Next, we optimise the sand-to-AC ratio. With 9g sand and varying AC from 0–1.5g, we tested six setups, calculating performance based on breakthrough time (b), flow rate ($1/16f$, where f is the measured quantity in seconds per drop and 16 drops is measured to equal 1ml. **Note that the reciprocal of f is computed to convert the value to drops per second**), and removal efficiency (E), creating a balanced score formula to determine the best configuration (Fig. 2).

$$S = \log_{10} \frac{2^{E-\frac{1}{16f}}}{b} S = \log_{10} \frac{2^{E-\frac{1}{16f}}}{b} S = \log_{10} \frac{2^{E-\frac{1}{16f}}}{b}$$

The scoring formula was based on three parameters mentioned before with removal efficiency weighted most heavily. A logarithmic transformation balanced the parameters. The optimal sand-to-AC ratio was found to be 9g sand to 1.15g AC, achieving a flow rate of 0.021 mL/s, a breakthrough time of 76 seconds, and a removal efficiency of 94%, resulting in a score of 25.59, which closely matched the expected score of 25.58.

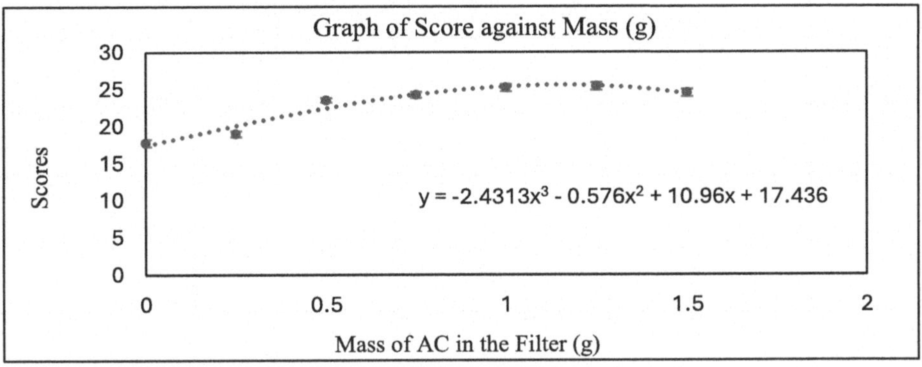

Fig. 2. Graph of score of prototypes against varying mass of AC used, asserting the optimal mass ratio to be 9.1g:1.15g, and validating our score formula as a reliable performance metric.

3.4 Prototyping

After preliminary testing, we switched to 39mm PVC pipes to better simulate real-world use. UVS was used for efficiency testing, as its results aligned closely with ICP trends. Cotton, initially used as a membrane, caused inconsistent filtration due to uneven packing. Tissue paper showed early promise but lacked durability with repeated use, making it unsuitable for long-term application.

To determine a suitable membrane material to be applied at the base of the filter column for our prototype, we tested the efficacies of our other materials as a membrane with 3 main considerations:

i. Pores of the material are small enough to prevent AC from leaking into the output water.
ii. The material should not contribute to the pollution of the output water
iii. The material should not be highly compromising of the flow rate.

These were quantified through evaluating the prototype's score, whose parameters are dependent on the above considerations. Seeing that fabrics are the most likely material to meet all the criteria, we tested our prototype with different types of fabrics as a membrane via filtration of a 2000ppm detergent solution. To confirm the validity of microfiber cloth as a membrane against a large range of AC masses, we also further tested its performance against 0.2-1.5g of AC.

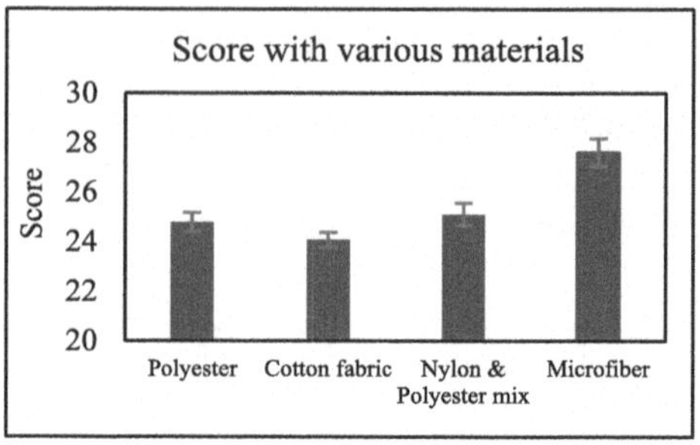

Fig. 3. **(Left)** Graph of score of prototypes against various membrane materials showing microfiber to have the highest score and is the most optimal membrane

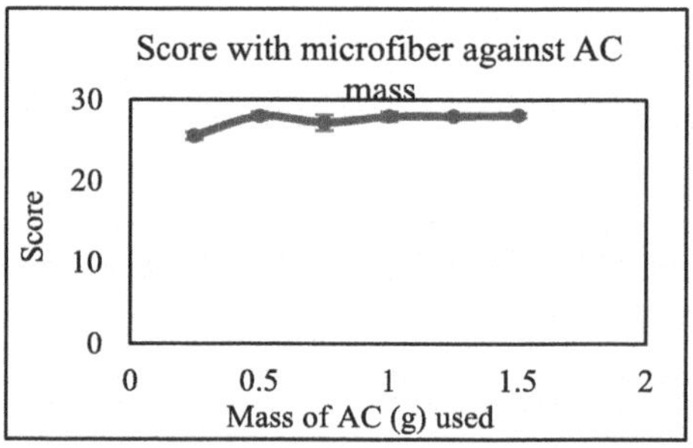

Fig. 4. **(Right)** Graph of score of prototypes against mass of AC, confirming the validity of microfiber as the most optimal membrane by showing a consistent high score

As shown in Fig. 3 below, we observe microfiber to be the display the highest score amongst all tested. From Fig. 4, the score remains relatively high, verifying it to be a suitable membrane that for applications with higher AC masses.

Since the removal efficiency decreases over time, we aim for a prototype that only needs to be replaced each month. Specifically, we designed the prototype to last 16–20 hours, based on a tri-weekly washing load, before its removal efficiency drops below 75%. To simulate real-world conditions, we used discharged water from residential washing machines, which had a concentration of 443ppm. Our benchmark removal efficiency was set at 75%, as levels above 100ppm increase the load on centralized systems [5]. Long-term experiments with varying AC masses showed that 5.0g of AC was sufficient to meet the desired longevity.

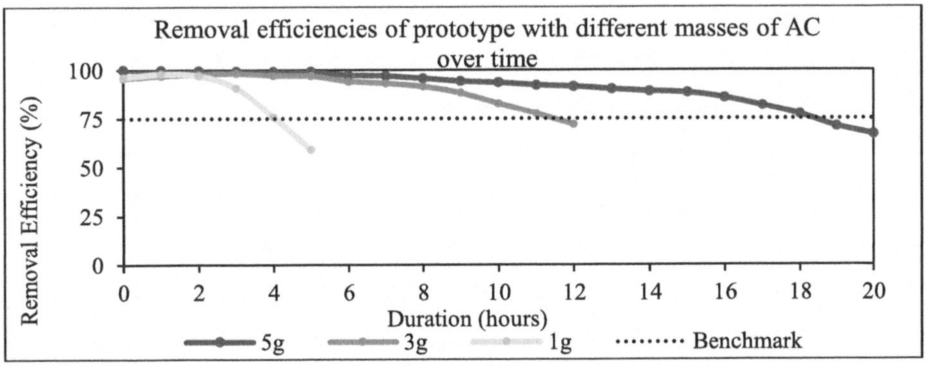

Fig. 5. Graph of removal efficiencies of prototypes against duration of continuous filtration, showing that 5.0g is sufficient for our prototype to have the desired longevity (16-20h)

Next, seeing as we aim for our prototype to be reusable and that acid washing is a common method used to regenerate adsorbents (Jeamichel Puente Torres et al., 2023), we determined the acid exposure (in adsorption-desorption cycles) needed to regenerate our prototype. Here, we first worked with the lower 1.0g AC column to conserve materials. At 5 hours, when the removal efficiency for the column dipped below 75% (see Fig. 5), we allowed the filter to drain and then ran 0.1M HCl through it, taking samples at 2-minute intervals to track the amounts of detergent being discharged, and determined when the column has been regenerated.

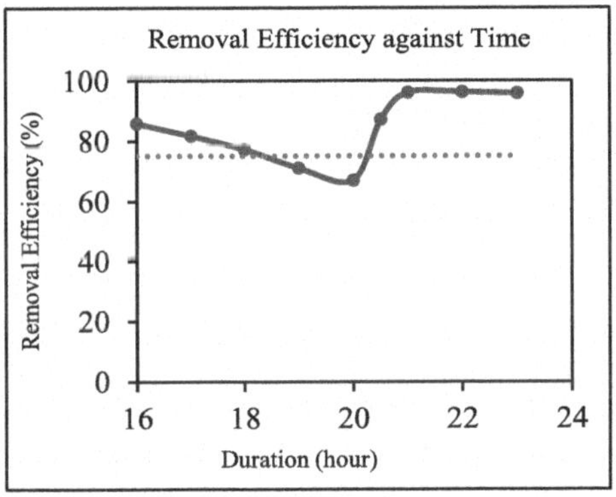

Fig. 6. **(Left)** Graph of removal efficiency of used prototype against duration, showing increase in removal efficiency with 5.0g AC before/after acid washing at 20h

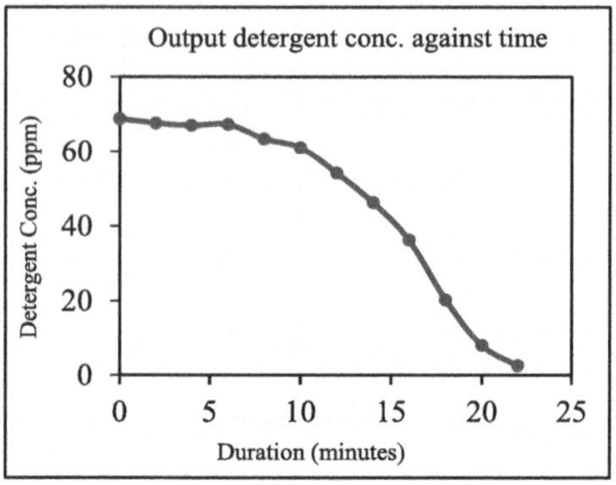

Fig. 7. (Right) Graph of output detergent concentration during acid wash against duration showing how it took 22 minutes to completely regenerate 1.0g of AC

We then repeated this procedure with the 20-hour old 5.0g AC column (as in Fig. 5) to mimic real-world application, carrying out 60 minutes of acid washing at a rate of 100ml/minute until the 21st hour to regenerate it to 97%, as shown in Fig. 6. From Fig. 7, it took 22 minutes at a flow rate of 50ml/minute amounting to roughly 1L of 0.1M acid to completely regenerate 1.0g of AC.

From Fig. 6, we observed the removal efficiency to recover to its initial magnitude (as in Fig. 5) upon acid washing, and it being sustained for 2 hours of filtration afterwards, thus verifying the reusability of our prototype.

4 Further Development of Filter

4.1 Synthesis of Improved Adsorbent

In order to target heavy metal ions, we synthesized a polyethyleneimine-alginate-activated carbon (PAA) composite adsorbent by dissolving 12g alginate in 600mL water (pH 7.0) and 3g PEI in 100mL water (pH 8.0). Amorphous activated carbon (85g) was added to the alginate solution and stirred for 2 hours. The prepared PEI solution was then added dropwise to the alginate-carbon mixture and stirred for another 2 hours. The mixture was poured into 0.1M $CaCl_2$ solution for 2-hour gelation and cross-linking, followed by filtration and washing. The resulting PAA beads were oven-dried and either crushed to powder for better removal efficiency or kept as beads for improved flow rate.

4.2 Determining Efficacy of PAA Composite

We compared PAA with AC and ACAB using batch extractions (10 mg adsorbent, 400 ppm detergent, 30 mins, same methodology as before). All three achieved saturation, with AC performing best (~95% removal), followed by PAA (~85%) and ACAB (~80%), confirming that PAA retains significant efficacy for aromatic compounds. The results demonstrate that all three adsorbents (PAA, AC, and ACAB) achieved saturation levels, with removal efficiencies dropping below 99%. AC exhibited the highest removal efficiency for aromatic compounds at 95%, followed by PAA at 85%. We then compared their removal efficiencies in column filtration.

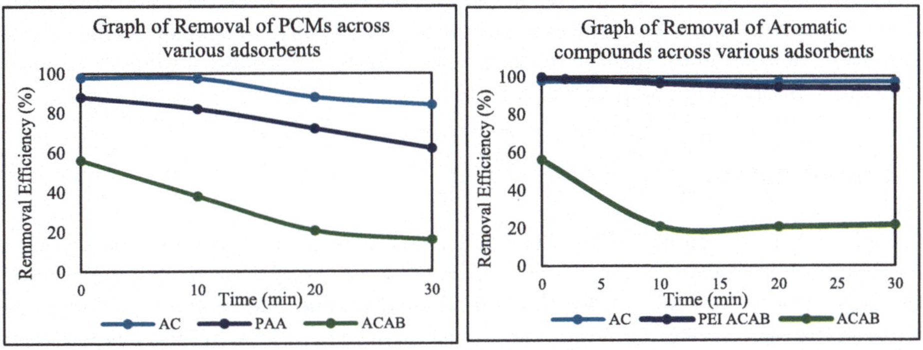

Fig. 8. Graphs comparing the removal efficiency of PCMs and aromatic compounds with PAA against other adsorbents over a long duration

Figure 8 shows that PAA (PEI ACAB) maintains close to 99% removal efficiency for 30 minutes, slightly trailing AC but outperforming ACAB. Over time, PAA shows a minor but acceptable decline compared to AC, remaining above the 75% threshold for a 2.25g mass. This confirms that PAA's performance for both aromatic compounds and PCMs is comparable to AC. We then evaluated PAA's performance against $Pb2^{+}$ ions relative to AC and ACAB.

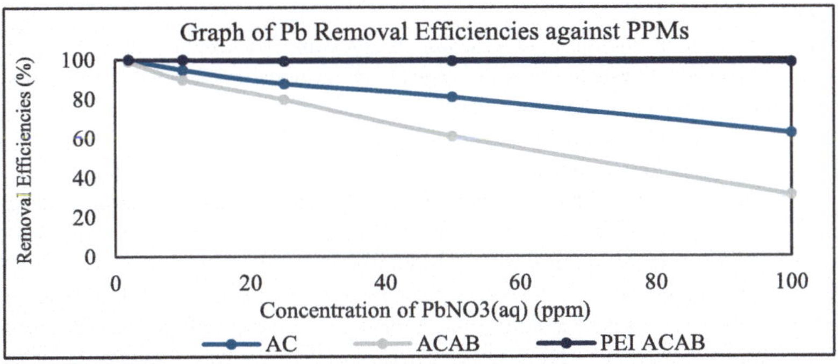

Fig. 9. Graph of removal efficiency of used prototype against duration, showing increase in removal efficiency with 5.0g AC before/after acid washing

From Fig. 9, we observe that PAA significantly outperforms both AC and ACAB in removing Pb^{2+} ions, remaining unsaturated (maintaining 99% removal efficiency) until at least 100ppm, supporting our hypothesis and verifying that PAA is a more effective adsorbent than AC when factoring in heavy metal adsorption.

4.3 3D Printing of Prototype

To develop our prototype into a functional product, we altered its casing from the 39mm PVC pipe used for experimentation into a 3D printed column with a funnel and ball valve attachments at its ends. The design allows for the rate of output to be controlled via the ball valve. Additionally, its threaded ends facilitate easy attachment to the output routes of most household washing machines. Its modular, accessible design also allows for an efficient cleaning and regeneration process.

5 Consolidation

In conclusion, we developed a functional, reusable prototype for decentralized detergent filtration that effectively removes aromatic compounds, PCMs, and heavy metal ions. Activated carbon (AC) proved the most efficient adsorbent, with optimal performance at a 9g:1.15g sand-to-AC ratio and microfiber membrane. A 5g AC filter maintained 97% removal efficiency after regeneration. We also synthesized a novel PEI–alginate–AC (PAA) composite, which matched AC's removal efficiency for organic pollutants and outperformed it for heavy metal removal. Its larger particle size also enables faster flow rates, enhancing practical usability. Our decentralized, scalable solution offers more efficient pollutant removal over existing centralized systems.

References

1. Belhachemi, M.: Adsorption of organic compounds on activated carbons. Sorbents Materials for Controlling Environmental Pollution, 355–385 (2021). https://doi.org/10.1016/b978-0-12-820042-1.00006-7
2. Hallahan, D.: Issues facing decentralized wastewater treatment today (2023). Wwd-mag.com. https://www.wwdmag.com/wastewater-treatment/article/33011011/issues-facing-decentralized-wastewater-treatment-today
3. Lenntech: Effects of detergents on aquatic freshwater life (2019). Lenntech.com. https://www.lenntech.com/aquatic/detergents.htm
4. Moyo, M., Nyamhere, G., Sebata, E., Guyo, U.: Kinetic and equilibrium modelling of lead sorption from aqueous solution by activated carbon from goat dung. Desalin. Water Treat. **57**(2), 765–775 (2014). https://doi.org/10.1080/19443994.2014.968217
5. Muya, A.: Centralized Wastewater Treatment: How it Works? Express Drainage Solutions (2024). https://www.expresswatersolutions.com/waste-management/centralized-wastewater-treatment-how-it-works/Park; Kim, H.G., Chae, T.W., M.Y., Yoo, I.-K.: Activated carbon-containing alginate adsorbent for the simultaneous removal of heavy metals and toxic organics. Process Biochemistry **42**(10), 1371–1377 (2007a). https://doi.org/10.1016/j.procbio.2007.06.016
6. Park, H.G., Kim, T.W., Chae, M.Y., Yoo, I.-K.: Activated carbon-containing alginate adsorbent for the simultaneous removal of heavy metals and toxic organics. Process Biochem. **42**(10), 1371–1377 (2007). https://doi.org/10.1016/j.procbio.2007.06.016
7. Rhazi, M., et al.: Influence of the nature of the metal ions on the complexation with chitosan: application to the treatment of liquid waste. European Polymer Journal **38**(8), 1523–1530 (2002). https://doi.org/10.1016/S0014-3057(02)00026-5

8. Senapati, M.R.: How our detergent footprint is polluting aquatic ecosystems. Down to Earth (2021). https://www.downtoearth.org.in/water/how-our-detergent-footprint-is-polluting-aquatic-ecosystems-77935Water, A. (n.d.). Laundry wastewater treatment plant - AZU Water. www.azuwater.com. https://www.azuwater.com/solution/laundry-wastewater-treatment-plant/
9. Zhang, Y., et al.: Enhanced removal of heavy metal ions using polyethyleneimine-modified adsorbents: Adsorption mechanisms and performance evaluation. J. Environ. Chem. Eng. **10**(2), 107234 (2022). https://doi.org/10.1016/j.jece.2022.107234
10. Liu, H., Wang, K.: Polyethyleneimine-based materials for heavy metal ion adsorption: A comprehensive review. Water Res. **245**, 120567 (2023). https://doi.org/10.1016/j.watres.2023.120567
11. Smith, J.A., Johnson, M.B.: Comparative analysis of water treatment technologies for heavy metal removal. Environ. Sci. Technol. **57**(8), 3245–3258 (2023). https://doi.org/10.1021/acs.est.2c08756

Investigating Potential of AC Electrocoagulation for Removal of Microplastics

Kerthana Shankar, Vidhya Marianne Joseph, Grace Lim(✉), and Lee Zhiying

Raffles Institution, Singapore, Singapore
grace.lim@ri.edu.sg

Abstract. Alternating current electrocoagulation (AC-EC) has been proposed as an alternative to conventional Direct current electrocoagulation (DC-EC) as it mitigates cathode passivation. This is due to the polarity reversal in an AC system preventing the buildup of the unwanted metal oxide layer on the cathode. AC has shown higher removal efficiencies and lower energy consumption in studies involving heavy metals and dyes. This study aimed to replicate those results with microplastics (MPs). A kinetic study of AC-EC with varying NaCl concentrations was conducted to find optimal concentration, which was then used to conduct a DC-EC experiment to compare the removal efficiencies of AC and DC. An AC–DC cycle experiment—run back-to-back without cleaning electrodes—was also performed to compare the effects on cathode passivation. Contrary to expectations, AC showed lower removal efficiency and higher energy consumption. SEM images further revealed greater cathode passivation in electrodes used for AC than DC. A possible reason could be that the frequency of the AC was too high, causing concentration polarization of the coagulants near the electrodes without diffusion into the bulk solution to sufficiently interact with the MPs. Further research should explore AC-EC of microplastics at varying frequencies while optimizing other critical parameters.

Keywords: AC Electrocoagulation · Cathode passivation · Removal efficiency

1 Introduction

Water pollution is an urgent and formidable global concern that presents substantial risks to both ecosystems and human well-being [23]. A vast variety of pollutants, both organic and inorganic, have been wreaking havoc on our water bodies and consequently their ecosystems, some more than others. A recent emergent pollutant of considerable concern these days has been microplastics. Microplastics are defined as plastic particles smaller than 5mm that slough off plastic as it degrades [5]. Microplastics accumulation in marine life and in humans have been reported to have detrimental effects such as inflammation, reduced fertility and mortality in marine life and DNA damage, organ dysfunction, neurotoxicity etc. in humans [14].

Many different wastewater treatment processes (WWTPs) such as adsorption, filtration and chemical coagulation have been investigated for the removal of microplastics

H. Guo et al. (Eds.): IRC-SET 2025, PTT, pp. 322–335, 2026.
https://doi.org/10.1007/978-981-95-9630-0_28

from wastewater. However, electrocoagulation, a recently emerging water treatment process, has been shown to have high MP removal efficiency (>90%) at low operational costs with minimal sludge formation as compared to other WWTPs [6].

Electrocoagulation works by generating heavy metal ions in situ that act as coagulants via electrolytic oxidation and reduction at the electrodes. Coagulants neutralize the electric charge of pollutant particles, allowing them to move close enough together that the Van Der Waals forces takes effect, causing them to clump together and form larger aggregate flocs that are more easily removed [20]. In a conventional electrocoagulation set-up, the electrodes are generally made of either aluminium or iron and the wastewater to be treated serves as the electrolyte. Oxidation occurs at the sacrificial anode to produce the heavy metal ions acting as the coagulants while at the cathode, the water molecule is broken down into OH^- ions and hydrogen gas. The hydrogen gas bubbles formed at the cathode also serve to promote flotation by pushing the flocs formed to the surface to be more easily removed.

Existing studies on the application of electrocoagulation for removal of microplastics focus on using Direct Current (DC) electrocoagulation. A common problem associated with DC electrocoagulation is cathode passivation which occurs when the continuous flow of direct current causes electrochemical reactions which lead to the formation of an impermeable oxide layer on the electrode. This hinders metal dissolution and electron production from the cathode which in turn reduces the number of metal ions formed that can act as coagulants. Hence, cathode passivation reduces the removal efficiency of the pollutant. To solve this problem, some studies have hypothesized that AC electrocoagulation could be a better alternative as the alternating current direction causes the polarity of the electrodes to be reversed periodically thereby preventing the buildup of the oxide layer [7]. Vasudevan and Lakshmi [22] and Othmani et al. [16] have reported AC electrocoagulation to have a higher removal efficiency and lower energy consumption for heavy metal ions and dyes respectively.

However, while polarity reversal is often stated as a solution to mitigate electrode passivation, the scientific literature contains conflicting evidence on the effectiveness of polarity reversal, which has been even reported to show decreased performance under certain conditions and for certain pollutant particles [4]. Hence, this study focuses on the application of AC electrocoagulation for the removal of microplastics specifically and comparing the removal efficiencies and energy consumption of DC and AC electrocoagulation in the context of microplastic removal.

2 Materials and Methods

2.1 Electrochemical Reactor

EC system consists of 4 aluminum plates arranged in monopolar parallel set-up (Fig. S1). The electrodes are of dimensions 10*2 cm^2 and the effective area of each electrode immersed in solution, in a 1L beaker, is 2*4.4 cm^2 (Fig. S1). The distance between the inner electrodes was set at 1cm, while the outer electrode was positioned 1.5cm away from the nearest inner electrodes. The electrodes were connected to a power supply, from which direct current (DC) or alternating current (AC) was drawn. Current was maintained at 0.250A using a variable resistor to maintain a constant current density of 35.5 A/m^2. The voltage was set over a range of 3-5 volts, which varied with NaCl concentration and type of current used. The rotational speed of the magnetic stirrer used was set at 600rpm.

2.2 Wastewater Analogue

The synthetic wastewater solution used for all experiments consisted of 0.5g/L of microplastic spheres, of $(10\text{-}45)*10^{-6}$ m in diameter, suspended in water. 1g/L of detergent (Biograde Fresh Lime Dishwashing Liquid) was added as a surfactant to aid microplastics in forming a uniform suspension. The MPs were made to soak overnight in surfactant to allow them to form a more homogenous solution. A 1mol/L of sodium chloride (NaCl) stock solution was prepared for the electrolyte solution, and an appropriate volume was added to the MP solution the next day after soaking, prior to the experiment. Total volume of the mixture was made up to 1L for each experiment. Initial concentrations of MP and detergent in the wastewater analogue were kept constant at the values given above for all experiments.

2.3 Determination of Optimal Electrolyte Concentration and Current Type

To determine optimal concentration of NaCl for the removal of microplastics using AC, varying concentrations of NaCl at 0.02mol/L, 0.04mol/L, 0.06mol/L, 0.08mol/L and 0.10mol/L were tested. Rotational speed of the magnetic stirrer and electrode arrangement were kept constant at 600rpm and monopolar parallel set up respectively. 1 mL samples were drawn at 0, 2.5, 5, 7.5, 10, 20, 30, 45, 60, 90 and 120 min timepoints for a concurrent kinetics study. An additional experiment was conducted using the same set-up and sampling frequency to compare the removal efficiencies of AC and DC. NaCl concentration was kept constant at 0.08mol/L while current was maintained at 0.250A with a constant current density of 35.5 A/m^2.

2.4 Investigation of Cathode Passivation

To compare energy consumption and impact of cathode passivation when DC and AC were used (The removal efficiency of consecutive runs was expected to decrease more for DC than AC due to greater cathode passivation), a cathode passivation experiment was conducted, where 0.08mol/L of NaCl was used while all other variables were kept constant. 3 back-to-back electrocoagulation runs were conducted, with the electrodes only being rinsed with deionized water in between before being reused, until 100% removal efficiency was attained, which was 75min for AC and 30min for DC. Samples were drawn every 15min for AC, and every 5min for DC due to its shorter duration, starting from 0min.

2.5 Reactor Maintenance

Electrodes were soaked in 0.1mol/L of HCl for 30 minutes, sanded and polished with both rough and smoother sandpapers before being reused. Whereas, for the cathode passivation experiment, HCL soaking treatment for 30minutes was omitted and the electrodes were simply rinsed with deionised water and lightly dabbed dry with paper towels. The duration over which electrodes were kept in air between each experiment was kept constant at 15min for both cycles, so that electrode surfaces are comparable despite atmospheric oxidation of electrodes between experiments, if any.

2.6 Analytical Method

MPs were quantified by manual counting using a hemocytometer. An average of 3 readings were taken for each sample. The formula below was used to calculate the removal efficiency. Removal efficiency (RE) of microspheres was calculated by

$$RE(\%) = \frac{(N_0 - N_t)}{N_0} \chi 100 \tag{1}$$

where N_0 is the average number of microplastics in the sample at timepoint t=0 sample and N_t is the average number of microplastics in the sample at given timepoint.

Energy consumption (E) was calculated by

$$E = vIt \tag{2}$$

where I is the current flowing through the set-up throughout the experiment, which was maintained at 0.250A, v is the voltage required to maintain the current of 0.250A and t is the duration of experiment, which is 120min.

Electrocoagulation Kinetic Study. A kinetic study was also conducted for the AC electrocoagulation with varied NaCl concentration experiments to investigate the mechanism of the adsorption process. This was done by attempting to fit the experimental data into different kinetic models. The two models that were investigated are the pseudo first order (PFO) and pseudo second order (PSO) models. Shown below are the linearized forms of the PFO and PSO kinetic models and how we have fit our experimental data into them.

Pseudo First Order.

$$ln(Ct) = ln(C_0) - kt \tag{3}$$

where Ct is concentration of microplastics at time t, C_0 is initial concentration of microplastics, k is PFO rate constant with units, time^{-1}, t is time in min/hours.

Pseudo Second Order.

$$t/q_t = 1/k2q_e + t/q_e \tag{4}$$

where q_t is amount of adsorbate adsorbed at time t, k2 is the PSO rate constant and q_e is equilibrium amount of adsorbate adsorbed/ maximum amount of adsorbate adsorbed

$$q_t = (C_0 - C) * V/W \tag{5}$$

where V is volume of solution (L), W is weight of adsorbent,

$$W = MIT/nF \tag{6}$$

where M is the molar mass (gmol^{-1}), I is the current(A), t is electrocoagulation time, n is number of electrons involved and F is Faraday's constant (96486 C mol^{-1})

By plotting a graph of ln(Ct) against t and t/qt against t, we would be able to determine how well the experimental data follows PFO and PSO kinetic models respectively by calculating the r^2 value of the plots.

SEM Analysis of Electrode Surfaces and Flocs.
The surface morphology of the Al electrodes and flocs were analyzed using a scanning electron microscope (SEM) and the constituents present on their surface were analyzed by EDS with the SEM made by JEOL (model JSM-6010LA).

3 Results and Discussion

3.1 Effect of NaCl Concentration on Energy Consumption of AC Electrocoagulation

Raising NaCl concentration from 0.02-0.06mol/L, boosts maximum MP removal from 30% to 75%. Removal efficiency plateaued at 90 min, never reaching 100%. At 0.08 and 0.10mol/L, 100% removal was achieved by 120min. Increasing NaCl concentration increases conductivity of the solution, leading to greater removal efficiency and removal rate and this result ties in with a similar study on removal of polyvinyl alcohol using alternating pulse current [26]. Greater conductivity triggers reactions at the electrodes that produce more electrons, which causes the anode to dissolve more and form more coagulants, increasing removal capacity of the process [11]. Additionally, Zhang et al. (2021) [25] also concluded that the Cl^- ions produce ClO^- ions at the anode, which attack the passivated layer on the electrodes, improving removal efficiency of the reactor (Fig. 1).

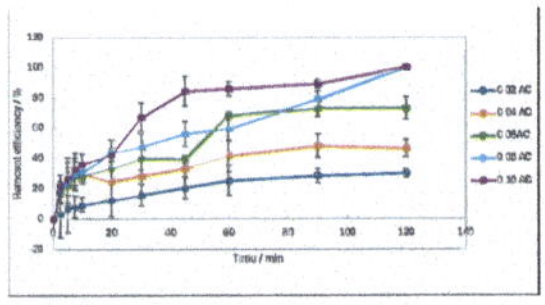

Fig. 1. Graph of percentage removal against time

However, this may not have been the case in this experiment as SEM analysis shows significant passivation on AC electrodes, which is even more than when DC was employed (Fig. S2b,c). This effect of Cl^- ions could have been diminished by the effect of the AC frequency used, as discussed in a later section.

3.2 Kinetics of AC Electrocoagulation

Table 1 shows the r^2 values of the plots for each of the experiments when fit into both pseudo first order and pseudo second order models respectively. The experimental data fit into the pseudo second order model consistently showed higher r^2 values (0.9779, 0.9392, 0.9468, 0.9710, and 0.9426) than those that were fit into the pseudo first order model. This implies that the kinetics of the AC electrocoagulation experiments for removal of microplastics using NaCl as electrolyte can be best explained with a pseudo second order kinetic model. Since the pseudo-second-order model considers chemisorption as the rate-limiting mechanism of the process, it can also be concluded that the adsorption of microplastics onto the $Al(OH)_3$ ions is controlled primarily by chemical reactions such as coordination reactions, hydrogen bonding and π-π interactions and is dependent on the number of unoccupied adsorption sites on the adsorbent surface.

3.3 Effect of NaCl Concentration on Energy Consumption of AC Electrocoagulation

From the graph above, it can be concluded that, increase in NaCl concentration leads to lower energy consumption. When NaCl concentration is increased from 0.02mol/L to 0.10mol/L, energy consumption required to achieve 100% removal of MPs decreases from 0.003461kWh to 0.001941kWh since the voltage required to maintain a current at 0.250A throughout the course of the experiment decreased from 6-7V for 0.02mol/L to 5V for 0.04mol/L and 0.06mol/L, to 4V for 0.08mol/L and to 3-4V for 0.10mol/L of NaCl. This result ties in with a study conducted by Asaithambi et al. (2024) [2] who also concluded that for the AC system, power consumption decreases as solution conductivity increases which caused the higher voltage required for the lower NaCl concentrations in this study (Fig. 2).

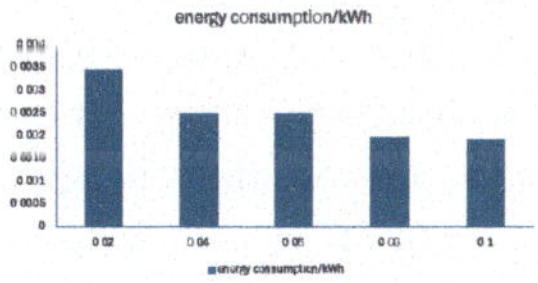

Fig. 2. Graph of energy consumption against NaCl concentration

From the above experiment, the optimal NaCl concentration for AC removal of 0.5 g/L MPs with 1 g/L surfactant and 35.5 A/m2 current density is 0.08 mol/L, achieving 100% MP removal with only slightly higher energy consumption than at 0.10 mol/L.

3.4 Comparison of Removal Efficiency between AC and DC Electrocoagulation

For the DC experiment, 100% MP removal was achieved in 30min and at the 15min time point, over 90% removal. DC is much more efficient than AC in removing MPs as rate of removal is much higher. Maintaining a current of 0.250A, only required 3V, thus energy consumption is much lower, at 0.0001875kWh. This contradicts most existing studies that conclude that AC electrocoagulation offers higher removal efficiency and lower energy consumption for dyes and heavy metal ions [16, 22]. However, dyes and heavy metal ions are small, hydrophilic and electrochemically reactive which allows them to engage in redox reactions, enhanced by AC due to polarity reversal, occurring at the electrodes and not just rely on the aggregation mechanism. Whereas MPs rely mainly on the agglomeration of coagulants for removal due to their poorer electrochemical reactivity. Side redox reactions occurring at the electrodes could have limited release of the coagulants (Al^{3+} ions) reducing AC's removal efficiency. Thus, DC's continuous polarization better promotes $Al(OH)_3$-MP interaction for stable floc formation and higher removal efficiency. Mohammed Tiaiba et al. [21] reported similar findings for Red Nylosan dye removal (Fig. 3).

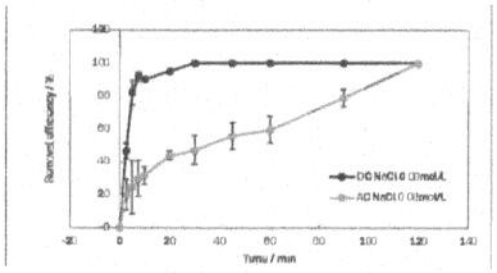

Fig. 3. Graph of removal efficiency against time for AC and DC

3.5 SEM Analysis of Flocs

The SEM images of the flocs produced by DC-EC (Fig. 4a,b) show individual floc groups are more massive. They are tightly packed together and denser. The zoomed-out version, (Fig. 4a), also confirms that the flocs are more compact on a larger scale than the AC flocs (Fig. 4c,d), potentially due to hydrogen gas production at both anode and cathode, which resulted in smaller precipitates that can be packed in a more compact manner [9]. In contrast, the AC flocs were found in numerous distinct clusters. Their structure is more porous and loosely aggregated, largely due to the much shorter intervals between polarity switches. The shorter period results in inadequate time for completion of the 4 stages of hydrogen gas evolution and overall lower Faradaic efficiency , leading to the formation of large bubbles. These bubbles cause the AC flocs to be weakly inter-connected [9]. The observed higher efficiency and speed of DC-EC in removing microplastics, as compared to AC-EC, can be directly attributed to the superior physical characteristics of the flocs it produces. Removal mechanism of pollutants includes destabilization of surface charges, causing the suspended solids to agglomerate and formation of a sludge layer of suspended solid particles [16, 18]. The effectiveness of the flocculation mechanism is dictated by the floc's physical structure and size. The larger DC flocs were more effective as colliding flocs as they sweep a greater volume of water, increasing the probability of encountering microplastic particles. Their considerably more compact nature allows them to form a more effective sludge blanket and trap the suspended microplastic particles better [18]. In contrast, the AC process produces flocs that are significantly smaller in size and have higher porosity with large void spaces. A smaller floc has a lower collision probability, making the second mechanism less efficient. The porous structure was less effective at retaining particles, as smaller microplastics can avoid capture by escaping from the gap between flocs [18] (Fig. 4).

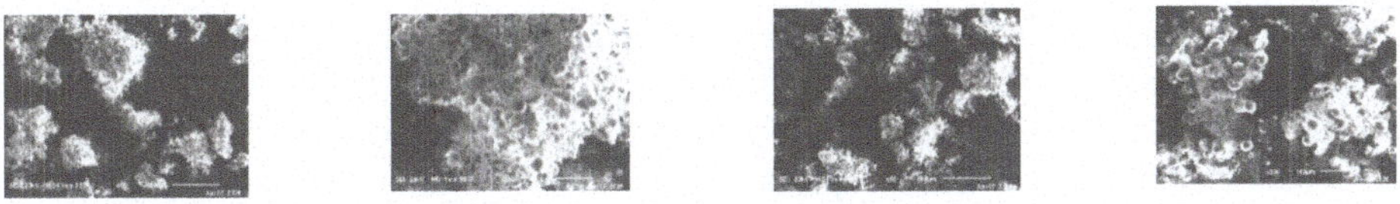

Fig. 4. SEM images of flocs. a) DC flocs 50 times magnified, b) DC flocs 200 times magnified, c) AC flocs 50 times magnified, d) AC flocs 200 times magnified

While the overall process is initiated by the charge neutralization of microplastic particles, it is the superior ability of the DC-generated flocs to physically collide, trap, and rapidly settle that contributes significantly to a higher removal efficiency, especially after the initial phase of charge neutralization, when flocculation and settling appear to be the dominant mechanism of microbead removal [16]. These results also align with the findings of Mohammed Tiaiba et al. [21].

3.6 Electrocoagulation Comparison between AC and DC Cycles

Results of Cathode Passivation experiment show that 3 consecutive experiments is not enough to show any significant decrease in efficiency for both DC and AC. For all 3 consecutive experiments where DC was used, 100% removal efficiency was achieved within 30min, and for AC, within 75min.

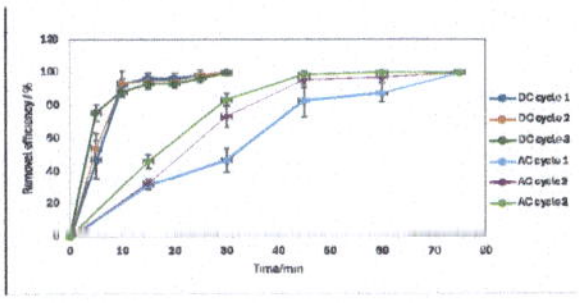

Fig. 5. Graph comparing removal efficiency of AC and DC cycles

3.7 SEM Analysis of Electrode Surfaces

The SEM images of electrodes used for the cathode passivation experiment (Fig. 6c) show that greater passivation has occurred on the AC electrodes. This could also explain the lower efficiency of AC in removing MPs compared to DC. The passivation layer could have reduced electrode dissolution and the formation of $Al(OH)_3$ coagulants, lowering the removal rate of MPs when AC was used (Fig. 3).

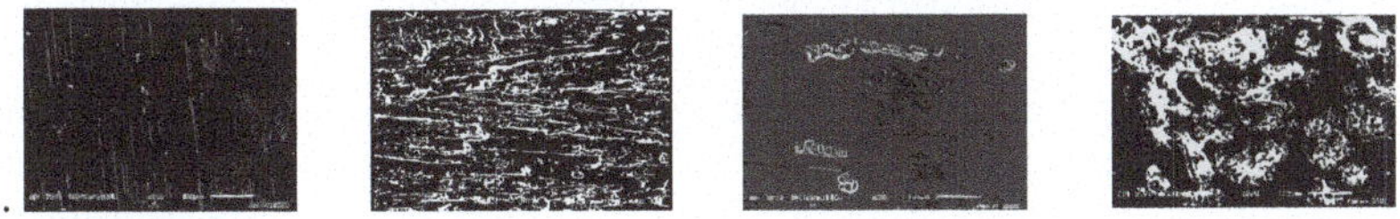

a) b) c) d)

Fig. 6. SEM images of electrodes. a) Surface of unused Al electrode, b) surface of electrode used in experiment to investigate optimal electrolyte concentration, c) surface of electrode used in DC cathode passivation experiment, d) surface of electrode used in AC cathode passivation experiment

Though Vasudevan et al. [23] and Zhang et al. (2021) [26] reported that AC reduces electrode passivation, this study aligns with Chow and Pham (2021) [4] who reported greater passivation for Fe electrodes with 5-minute polarity reversal compared to DC. This could be due the Al^{3+} cations produced by the anode getting reduced by the same electrode after switching to a cathode just 0.0769s later, causing build up and concentration polarization of the ions at the electrode. These cations precipitate as $Al(OH)_3$ on the electrode when they encounter OH^- ions produced by the same cathode, causing electrode passivation. This effect worsens at higher AC frequencies and hence, caused greater passivation in the AC electrodes. Chow and Pham (2021) [4] reached a similar conclusion. Furthermore, the hydrogen gas production in the DC set-up (as mentioned in 3.5) played an important role of dislodging the passivation layer on the Al electrodes [4]. However, insufficient Hydrogen gas was released at the electrodes to do the same in the AC-EC set-up, leading to greater electrode passivation.

3.8 Comparison with Existing Literature

A review of the literature shows that while AC is often reported to be superior to DC by mitigating electrode passivation [7, 8, 12, 15], but its effectiveness is not universal and depends on both the pollutant and the operating conditions. Several studies show that for certain contaminants, such as textile dyes, DC can be far more efficient at removal, suggesting that continuous polarisation is more effective for some interactions [21]. This aligns with our findings for microplastic removal. Furthermore, the argument that AC always prevents fouling is challenged by recent studies that observed significant passivation, even with polarity reversal [3, 4, 15, 21] A critical factor is the frequency of polarity reversal. Research by Chow & Pham [4] and Pi et al. [18] demonstrate that very high frequencies can be detrimental and contribute to electrode passivation, which also supports our observation.

Table 1. Comparison with other studies on AC efficiency

Reference (Author, Year)	Electrodes	Electrode Passivation/Fouling Results	Effect of Period/ Frequency	Author's Justification for Results
[26]	Fe	APC drastically reduces passivation. A 50-cycle test showed DC electrode corroded 4x more than the APC electrode. Current efficiency was 99% vs. 62% for DC.	An optimum exists. Best performance at 30% duty cycle & 500 Hz. Higher frequencies/duty cycles led to passivation or reduced efficiency.	APC's returning current creates a more uniform surface of electrodes, preventing passivation. It also forms porous, spherical flocs with a larger surface area, which is more conducive to PVA adsorption.

(continued)

Table 1. (*continued*)

Reference (Author, Year)	Electrodes	Electrode Passivation/Fouling Results	Effect of Period/ Frequency	Author's Justification for Results
[25]	Al	PR is significantly better. DC produced a hard, crystalline deposit, while PR produced a soft powdery fouling layer. PR at 20s had 63% less fouling than DC.	Shorter periods create softer fouling. Decreasing the period from 60s to 10s resulted in a softer fouling layer but also reduced the filterability of the final sludge.	The periodic change of polarity promotes a faster conversion of hydrated metal to form coagulant species. This prevents the formation of hard, dense, glassy-type precipitates that form under the stable pH conditions of DC.
[18]	Al	PREC effectively retards passivation. The rate of resistance increase for DC was over 3x higher than for PREC.	An optimum exists. The optimal cycle was 15s. Longer cycles (30s) allowed passivation to occur between switches. Shorter cycles (2s) were found to be detrimental to removal efficiency.	Periodic reversal inhibits cathodic polarization and reduces anodic passivation. However, The shorter the cycle of periodic electrode reversal, the less time for anodic dissolution.
[7, 8]	Al	APC overcomes passivation. The drop in DC efficiency is attributed to cathode passivation.	The periods tested (3-15 min) did not significantly affect removal efficiencies. [8]	Polarization via APC shortens the electrolysis time required for metal hydroxides species for removing the pollutants. Cyclical energy supply delays passivation.
[12, 15]	Pb	AC prevents oxide layer formation that hinders DC performance.		The periodicity of this current prevents the oxide layer formation and accelerates the reaction speed.

(*continued*)

Table 1. (*continued*)

Reference (Author, Year)	Electrodes	Electrode Passivation/Fouling Results	Effect of Period/ Frequency	Author's Justification for Results
[4]	Al, Fe	AC is better for Al, not for Fe. PR diminished Al fouling but did not reduce Fe fouling.	Frequency is critical for Fe. Increasing PR frequency decreased coagulant production. At higher frequencies, the coagulant production decreased dramatically. Usage of rapid 0.5-min reversal period decreased the Faradaic efficiency.	For Al, PR is beneficial because chemical dissolution of the Al cathode helps dislodge scale. For Fe, PR is detrimental because Fe ions get stuck and precipitate near the electrode after a switch in polarity instead of migrating into the solution.
[21]	Al	DC was far more efficient. DC achieved steady-state color removal of 90-97%, while AC's highest removal was only 52.5%.		Dyes needed a stable, non-reversing charge on the floc that leads to more efficient polarization of the solid $Al(OH)_3$ interface, for efficient interaction with the dye molecules.
[3, 14]	Fe [3]		AC/PR did not mitigate fouling. This study observed the formation of thick fouling layers in PR-EC	CaCO3 ppt formed on the cathode & trapped the dissolved iron coagulant, which ended up not getting effectively released into the solution

4 Conclusion

Using DC is better than using AC to remove MPs as it results in both greater removal efficiency and lower energy consumption and this is possibly due to MPs' unique surface properties which prevents them from actively participating and benefiting from the enhanced redox reactions caused by using AC. Instead, such chemically inert pollutants depend more on the topological characteristics of the flocs formed since the main removal mechanism other than charge neutralization is flocculation rather than participation in side-reactions. Thus, for such cases, DC proves to be more useful as it forms more compact and denser flocs that aid the above-mentioned removal mechanism. Furthermore, using AC does not always prevent cathode passivation, especially when the current frequency is high enough to prevent the cation from migrating into the bulk solution. This also suggests that the benefits of polarity reversal on fouling are highly dependent on the wastewater matrix and target pollutant and are not universally guaranteed. Further studies can be conducted to determine the optimal frequency of AC current for MP removal by taking into consideration type of contaminants, buffer capacity of solution and other pertinent factors.

Appendix

See Fig. A1 and Table 2.

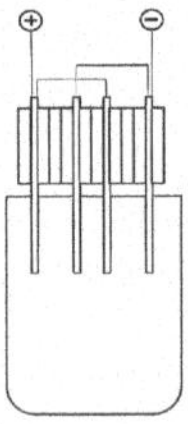

Fig. A1. Schematic diagram of EC reactor set-up

Table 2. r^2 values obtained when experimental data was fitted into pseudo 1st and 2nd order

NaCl Concentration	Pseudo first order r^2 values	Pseudo second order r^2 values
0.02 mol/L	0.9663	0.9779
0.04 mol/L	0.7576	0.9392
0.06 mol/L	0.8976	0.9468
0.08 mol/L	0.6164	0.9710
0.10 mol/L	0.7446	0.9426

References

1. Abdollahi, J., Alavi Moghaddam, M.R., Habibzadeh, S.: The role of the current waveform in mitigating passivation and enhancing electrocoagulation performance: A critical review. Chemosphere **312**, 137212 (2023). https://doi.org/10.1016/j.chemosphere.2022.137212
2. Asaithambi, P., et al.: Hybrid Industrial Wastewater Treatment Using a Sono-Alternating Current-Electrocoagulation Technique with Power Usage Estimation. Chem.Electro. Chem. **11**(13) (2024). https://doi.org/10.1002/celc.202400039
3. Ashraf, S.N., Rajapakse, J., Dawes, L.A., Millar, G.J.: Electrocoagulation for the purification of highly concentrated brine produced from reverse osmosis desalination of coal seam gas associated water. J. Water Proc. Eng. **28**, 300–310 (2019)
4. Chow, H., Pham, A.L.-T.: Mitigating electrode fouling in electrocoagulation by means of polarity reversal: the effects of electrode type, current density, and polarity reversal frequency. Water Res. **197**, 117074 (2021). https://doi.org/10.1016/j.watres.2021.117074
5. Dutchen, B.S.: Microplastics everywhere. Harvard Medicine Magazine (2024). https://magazine.hms.harvard.edu/articles/microplastics-everywhere#:~:text=Where%20they%20are,liver%2C%20kidneys%2C%20and%20placenta
6. Elkhatib, D., Oyanedel-Craver, V., Carissimi, E.: Electrocoagulation applied for the removal of microplastics from wastewater treatment facilities. Sep. Purif. Technol. **276**, 118877 (2021). https://doi.org/10.1016/j.seppur.2021.118877
7. Eyvaz, M.: Treatment of Brewery Wastewater with Electrocoagulation: Improving the Process Performance by Using Alternating Pulse Current. Int. J. Electrochem. Sci. **11**, 4988–5008 (2016)
8. Eyvaz, M., Gürbulak, E., Kara, S., Yüksel, E.: Preventing of cathode passivation/deposition in electrochemical treatment methods – a case study on winery wastewater with electrocoagulation. In: Aliofkhazraei, M. (ed.) Modern Electrochemical Methods in Nano, Surface and Corrosion Science. InTech eBooks (2014). https://doi.org/10.5772/58580
9. Eyvaz, M., Kirlaroglu, M., Aktas, T.S., Yuksel, E.: The effects of alternating current electrocoagulation on dye removal from aqueous solutions. Chem. Eng. J. **153**, 16–22 (2009)
10. Fuladpanjeh-Hojaghan, B., Shah, R.S., Roberts, E.P.L., Trifkovic, M.: Effect of polarity reversal on floc formation and rheological properties of a sludge formed by the electrocoagulation process. Water Res. **242**, 120201 (2023)
11. Husein, S., Slamet, N., Dewi, E.L.: The Impact of Sodium Chloride (NaCl) Concentrations on Electrocoagulation for Simultaneous Tartrazine Dye Removal and Hydrogen Production. MDPI, 4 (2024). https://doi.org/10.3390/engproc2024067004
12. Ifill, R.O.: Alternating Current Electrocoagulation (AC/EC) of Fine Particulate Suspensions. Ph.D. Thesis, Department of Chemical and Materials Engineering, University of Alberta, Edmonton, Canada (2010)
13. Keshmirizadeh, E., Yousefi, S., Rofouei, M.K.: An investigation on the new operational parameter effective in Cr(VI) removal efficiency: A study on electrocoagulation by alternating pulse current. J. Hazard. Mater. **190**(1–3), 119–124 (2011). https://doi.org/10.1016/j.jhazmat.2011.03.010
14. Li, Y., Tao, L., Wang, Q., Wang, F., Li, G., Song, M.: Potential Health impact of Microplastics: A review of environmental distribution, human exposure, and toxic effects. Environment & Health **1**(4), 249–257 (2023). https://doi.org/10.1021/envhealth.3c00052
15. Müller, S., Behrends, T., van Genuchten, C.M.: Sustaining efficient production of aqueous iron during repeated operation of Fe(0)-electrocoagulation. Water Res. **155**, 455–464 (2019)
16. Othmani, A., Kesraoui, A., Seffen, M.: The alternating and direct current effect on the elimination of cationic and anionic dye from aqueous solutions by electrocoagulation and coagulation flocculation. Euro-Mediterranean Journal for Environmental Integration **2**(1) (2017). https://doi.org/10.1007/s41207-017-0016-y

17. Perren, W., Wojtasik, A., Cai, Q.: Removal of Microbeads from Wastewater Using Electrocoagulation. ACS Omega **3**(3), 3357–3364 (2018)
18. Pi, K.W., Xiao, Q., Zhang, H.Q., Xia, M., Gerson, A.R.: Decolorization of synthetic Methyl Orange wastewater by electrocoagulation with periodic reversal of electrodes and optimization by RSM. Process Saf. Environ. Prot. **92**, 796–806 (2014)
19. Shen, M., et al.: Efficient removal of microplastics from wastewater by an electrocoagulation process. Chem. Eng. J. **428**, 131161 (2022). https://doi.org/10.1016/j.cej.2021.131161
20. Tech, G.W.: Coagulants used in water treatment: Optimizing process efficiency. Genesis Water Technologies (2024). https://genesiswatertech.com/blog-post/coagulants-used-in-water-treatment/#:~:text=in%20Water%20Treatment-,What%20are%20the%20most%20common%20coagulants%20used%20in%20water%20treatment,and%20organic%20matter%20from%20water
21. Tiaiba, M., Merzouk, B., Amour, A., Mazour, M., Leclerc, J.-P., Lapicque, F.: Influence of electrodes connection mode and type of current in electrocoagulation process on the removal of a textile dye. Desalin. Water Treat. **73**, 330–338 (2017)
22. Vasudevan, S., Lakshmi, J.: Effect of alternating and direct current in an electrocoagulation process on the removal of cadmium from water. Water Sci. Technol. **65**(2), 353–360 (2012). https://doi.org/10.2166/wst.2012.859
23. Vasudevan, S., Kannan, B.S., Lakshmi, J., Mohanraj, S., Sozhan, G.: Effects of alternating and direct current in electrocoagulation process on the removal of fluoride from water. J. Chem. Technol. Biotechnol. **86**(3), 428–436 (2011). https://doi.org/10.1002/jctb.2534
24. Yadav, S.C.: Water pollution: the problems and solutions. Science Insights **44**(2), 1245–1251 (2024). https://doi.org/10.15354/si.24.re905
25. Yasri, N.G., et al.: Investigation of electrode passivation during electrocoagulation treatment with aluminum electrodes for high silica content produced water. Water Sci. Technol. **85**(3), 925–942 (2022)
26. Zhang, J., Li, J., Ma, C., Yi, L., Gu, T., Wang, J.: High-efficiency and energy-saving alternating pulse current electrocoagulation to remove polyvinyl alcohol in wastewater. RSC Adv. **11**(63), 40085–40099 (2021). https://doi.org/10.1039/d1ra08093h

AI and Cybersecurity

GM-TI: GAI Model for Threat Intelligence

Charisee Zhi Ling Yip(✉), XingXing Chen, Tian Le Chua, Jia Ye Goh, Zheng Jie Ong, Zhen Yu Tan, and Huaqun Guo

Infocomm Technology Cluster, Singapore Institute of Technology, Singapore, Singapore
{2203215,2200626,2200520,2201457,2200605, 2200877}@sit.singaporetech.edu.sg, huaqun.guo@singaporetech.edu.sg

Abstract. This paper focuses on the development of a domain-specific Generative AI (GAI) model specialized in threat intelligence, while providing the literature research and experiments using different available models. Our research and evaluation result led to the usage of Mistral 7B as the baseline model and incorporating Retrieval-Augmented Generation (RAG). Datasets from frameworks like MITRE ATT&CK, CVE repositories, and open threat intelligence platforms are integrated, allowing the model to address the challenges of this rapidly evolving threat landscape. This solution helps to speed up and simplify the process of threat detection while helping analysts to identify their corresponding mitigation processes conveniently and quickly. The result shows the capability of the model in cybersecurity applications, with significant improvements in accuracy and contextual relevance compared to traditional models.

Keywords: LLM · Mistral · AI · Threat Intelligence · RAG · APT · MITRE ATT&CK · TTP · IoCs · Mitigations · Detections

1 Introduction

Organizations today face a significant challenge in extracting actionable intelligence promptly and in the most efficient manner. The generative AI model is adapted based on a well-known Large Language Model (LLM), Mistral [1], a general-purpose model that has been specifically enhanced to focus on the area of threat intelligence. By aligning with the core concepts of Tactics, Techniques and Procedures with references to well-known frameworks such as MITRE ATT&CK [2] and Cyber Kill Chain by Lockheed Martin [3], this model will facilitate threat analysis, identification of Indicators of Compromise (IoCs) and suggestion of effective mitigation strategies by leveraging on advanced AI technologies that aim to facilitate quicker and more effective responses to emerging threats.

H. Guo et al. (Eds.): IRC-SET 2025, PTT, pp. 339–349, 2026.
https://doi.org/10.1007/978-981-95-9630-0_29

2 Literature Review

2.1 Mistral for Threat Intelligence

Mistral has demonstrated its fine-tuning capabilities for domain-specific tasks as showcased in [4] where it outperformed baseline models in adaptability to specific domains, highlighting its potential for accuracy and precision. This is evident from its achievement of over 81.5% accuracy rate as compared to GPT-4. Recent studies on specialized training approaches in LLMs have shown that precision rates can be improved significantly through targeted domain-specific fine-tuning, often yielding up to a 10% increase in task accuracy compared to general models. Therefore, Mistral's fine-tuning can yield high precisions for threat intelligence, allowing it to capture critical nuances in cybersecurity language with notable accuracy improvements [5].

2.2 Challenges and Limitations of Current LLMs in Cybersecurity Applications

Generic LLMs exhibit limited contextual accuracy in cybersecurity, with false positive rates often exceeding 20%. On the other hand, Mistral's specialized training can reduce false positives by up to 15%, where the context-awareness of the model is significantly enhanced. These improvements demonstrate Mistral's potential to address the limitations in generalized LLMs for precise threat detection and assessment in real-world scenarios [6].

2.3 Integration of Mistral with Existing Security Infrastructure

The use of Mistral can streamline and automate threat intelligence workflows as its integration can allow for seamless augmentation of Security Information and Event Management (SIEM) and Security Orchestration, Automation, and Response (SOAR) systems, enhancing overall efficiency in threat detection and incident response. Research supports that integrating LLMs into cybersecurity frameworks can significantly improve real-time threat analysis and event correlation, reduce false positives and enable faster decision-making within SIEM environments [7].

2.4 Reducing Alert Fatigue Through LLM-driven Analysis

LLMs like Mistral have shown its potential to reduce threat alert fatigue by providing rapid and context-driven analysis that significantly aids security analysts and prevent the overwhelming high volume of alerts. LLMs can swiftly interpret and analyze vast amounts of threat data, identifying patterns to provide actionable insights that enable analysts to identify true threats more efficiently. Studies indicate that LLM-driven analysis can reduce non-critical alerts by up to 87%, drastically minimizing the time spent on manual identification of alerts and freeing analysts to focus on high-priority incidents. Moreover, the implementation of LLMs for initial threat assessments has been shown to save an average of 90 hours per week in manual review, enhancing both response time and accuracy in threat detection. This capability to perform quick, automated analyses empowers organizations to handle real-time threats more effectively and to streamline their cybersecurity operations [8].

2.5 Importance of RAG in LLMs for Domain-Specific Tasks

Traditional LLMs are limited by their static training data, which may lead to outdated or generalized responses, especially in rapidly evolving fields like threat intelligence. RAG addresses this limitation by incorporating non-parametric retrieval, allowing the model to pull the latest information from specialized databases, ensuring that the generated outputs are up-to-date and domain relevant [9]. Moreover, the utilization of RAG can reduce hallucination. This is because, one of the challenges in using LLMs is their tendency to "hallucinate" or generate inaccurate information. By leveraging RAG, models can ground their responses in retrieved, factually accurate content from trusted sources, thereby reducing the incidence of hallucinations. This is crucial in domains where misinformation can lead to significant security risks, such as cybersecurity and healthcare [10]. Finally, RAG-enabled LLMs can improve response quality and context utilization by generating more specific and contextually accurate content compared to standard LLMs, especially when applied to complex fields like medical education and legal compliance. This approach has been shown to improve the quality of generated outputs by integrating domain-specific content dynamically, leading to a 71% boost in performance for challenging tasks like multiple-choice questions and factual Q&A [11].

3 Methods

3.1 Dataset Used

The dataset for training the GAI model was meticulously compiled from several authoritative sources to ensure comprehensive coverage and relevance. This includes public threat databases such as MITRE ATT&CK, CVE, GitHub compiled databases and other relevant sources discussed in the following sections.

MITRE ATT&CK Framework. The key knowledge base of the GAI model for threat intelligence lies in the MITRE ATT&CK framework. It is thus crucial to obtain information on the tactics, techniques and procedures performed by various known Advanced Threat Persistent (APT) groups as it explains the attack flow of adversaries. This information is obtained from a publicly available GitHub repository [12] that is frequently updated. The information from both the Enterprise and Mobile domains have been obtained as part of the dataset. The various components of the MITRE ATT&CK framework are explained with more details in the following subsections.

Tactics. The MITRE ATT&CK framework covers 14 different tactics, each for both the Enterprise and Mobile domains. Tactics represent the why of an attack's technique or sub-technique. The relevant information of the various tactics will help a threat analyst to understand the goal of the adversary when performing a particular tactic.

Techniques and Sub-techniques. Each tactic comprises many techniques, where each technique may comprise one or more sub-techniques. Techniques spell the method utilized by an adversary to achieve a particular goal described by the tactic.

Procedures. Information on the procedures of APT groups or software to abuse the techniques and/or sub-techniques are obtained. This information allows users to have a specific understanding on the approach taken by APT groups or malicious software when applying the techniques and/or sub-techniques.

APT Groups. Information on APT groups is key to understanding a known adversary. It helps users to understand the behavior and profile adversaries based on their modus operandi. The APT group is linked their observed behaviors which are techniques and/or sub-techniques utilized in the past as well as software they have leveraged on.

Software. Software, specifically attack tools and malware, are obtained as part of the dataset to provide more insight into the techniques performed using software.

Detections. Information on detection mechanisms that can be employed by threat analysts to detect IoCs of known malicious activity are obtained.

Mitigations. Information on the mitigations that can be employed by threat analysts to better secure their systems against these known techniques are obtained.

CVE Details. As many of the adversary techniques exploited known vulnerabilities, it is important to obtain up-to-date information on Common Vulnerabilities and Exposures (CVEs). This information is obtained from a publicly available GitHub repository [13] that provides immediate and automatic updates and stores all known CVEs that have been published since 1999.

Indicators of Compromise (IoCs). IoCs are key to allow threat analysts to identify signs of exploit infection in their system. Apart from the detections obtained from the MITRE ATT&CK framework, these include known malicious URLs as well as hashes of known malware samples. This information is adapted from URLhaus [14], an open-source and up-to-date threat intelligence database that stores all known records of malicious URLs and payloads.

Malpedia. Malpedia [15] is a resource that provides information such as articles of APT group activity and malware in the wild. Specifically, it provides for rapid identification and actionable context when investigating malware. It is a platform that allows for open sharing of information and knowledge retrieval.

3.2 Data Pre-processing

For MITRE ATT&CK framework and CVE details, the data are pre-processed by extracting relevant portions of information from the datasets obtained from online sources. For Malpedia, the articles, as well as their metadata are recursively retrieved from the Malpedia library. The articles are then grouped and sorted according to their malware families and/or threat actors respectively. Articles without any tags are grouped together. The reformatted dataset is then stored into a jsonl (JSON Lines) file.

The data from the various datasets are then parsed into the all-MiniLM-L6-v2 [16] sentence transformer model, which is used to create the embeddings from the datasets by mapping sentences and paragraphs into a 384-dimensional dense vector space. Thereafter, Txtai [17] is used as an all-in-one embeddings database meant for semantic search. It will be used to store, index and search the embeddings that have been created.

3.3 Initial Approach

The initial approach taken by the team was to generate question and answer pairs using a set of question-and-answer template, to cover the various query permutations and use them to fine-tune the Mistral-7B model. However, using a script to generate Q&A pairs often results in them following a rigid format, making the training data highly predictable. Humans rarely express the same question in the exact same way. Therefore, the model that has been trained with the Q&A pairs will become overly dependent on these rigid patterns and struggle to interpret questions that are not included. As a result, the model generated inaccurate or unrelated responses. Therefore, RAG was used as it addresses the key limitation of models like Mistral-7B.

3.4 Data Retrieval

Rather than performing data correlation on the dataset used to build and train a custom model or performing fine-tuning of a model, the approach chosen was to implement Retrieval-Augmented Generation (RAG) [18]. Most pre-trained models have differing datasets that were used to train which results in varied predictions. RAG is a technique that enhances the accuracy and reliability of prediction by including additional context and information from external sources. This is achieved by having a vector database that the model can refer to and extract new information like a database query. The model would then formulate a response based on the retrieved information.

4 Comparative Model Analysis

As part of the approaches taken when developing a GAI model for Threat Intelligence, the team did a detailed research and analysis with experimentation on various well-known LLM models via trial-and-error. The comparative analysis discusses about the different models experimented as well as their respective outputs and accuracy achieved. The rationales supporting the final decision to use Mistral as the baseline model, based on the findings observed, are detailed accordingly.

4.1 GPT2 by OpenAI

GPT2 [19] is OpenAI's model that was made available and open sourced on GitHub. It allows for building custom models that can be trained with custom sets of data to suit different use cases and needs of AI implementations. It was released to the public in 2019 but has made no further improvements due to deprecation in favor of their proprietary models. However, GPT2 does pale in comparison to more modern models such as Llama or Mistral since there were no planned upgrades unlike other open source models. The maximum parameters are 1.5B yet it is still smaller than other models that generally have at least 7B in size. Running GPT2 with the custom dataset is tricky as there are many cases of hallucinations which will require a more detailed and feature engineered dataset or detailed configurations of parameters. Due to time constraints and better alternatives, GPT2 was not chosen as the base model to implement the GAI model for Threat Intelligence.

4.2 Llama models by Meta Llama

Llama models created by Meta Llama [20] offer great performance which surpasses the capabilities of GPT2 and through various integration of custom datasets, was able to produce moderately accurate results with occasional hallucinations. 7B parameters in size was found to be the most stable size for building and running a model on a consumer Graphics Processing Unit (GPU) without too much cost [21] and can respond with 81.58% to 82.66% accuracy [4].

While Meta allows users to adapt their open-source Llama models for use and redistribution, both Llama 2 [22] and Llama 3 [23] models have certain license agreements to be complied with. Therefore, Llama 2, Llama 3 and any of its other variants serve as a great model for selection but the considerations of the license agreements will be tricky to handle if the model in use is to be at production level. NousResearch [24] offered Llama 2 and Llama 3 as its own distributions, but the license will still follow the parent license created by Meta. In other instances of experimentation and/or for internal use without interactions with public users, Llama will be very practical.

4.3 Mistral7B-Instruct by MistralAI

Building upon the findings of Llama 2 experiments, Mistral7B-Instruct was another alternative that offered similar performance. In fact, while Mistral 7B is under the Apache 2.0 license, it can be used without restrictions, including for commercial purposes [25]. Mistral's model has easy parameterization to be trained with custom datasets and throughout the testing, Mistral has performed similarly to Llama 2 with approximately 79.60% accuracy [4].

Moreover, the team identified through experimentations that using RAG with Mistral is one of the most important techniques to be incorporated into the final model building. Using RAG enables the model to incorporate additional data into its response generation process, enhancing its accuracy and relevance. This concludes the findings that Mistral has great performance comparable to Llama 2 and without the complex license agreements, making it the chosen baseline model.

5 Model Evaluation Metrics

5.1 ROGUE and BLEU score

ROUGE and BLEU are basic and traditional for evaluating modern transformer-based LLM models. They both measure text overlap between a generated candidate (model's response) and a reference text and rely on matching n-grams to evaluate similarity. However, they lack in terms of only being able to measure surface-level lexical similarity and are unable to measure semantic meaning of sentences. This is supported by a paper that states that both ROGUE and BLEU do not effectively align with how humans perceive the quality of open-domain dialogue systems [26].

5.2 BERTScore

BERTScore [27] is able to understand semantic and contextual nuance matter, allowing for the flexibility of word choice and phrasing, as long as the meaning of the model response remains similar. BERTScore computes a similarity score for each token in the generated candidate sentence with each token in the reference sentence. However, instead of exact matches, the token similarity is computed using contextual embeddings. Hence, BERTScore is able to correlate better with human judgments as compared to ROGUE and BLEU that rely on simpler word-matching techniques. Moreover, it provides for stronger model selection performance than existing metrics by being able to better distinguish and evaluate models accurately. We adopt the use of BERTScore to evaluate our GAI model for Threat Intelligence.

6 Result and Insight

This section details the outcomes and evaluation results obtained from the development of a GAI model for Threat Intelligence.

6.1 Model Query API

The method to query the model uses a popular communication format of JSON API. In order to achieve this, the model is wrapped with a Flask web server to run the model querying process. Since most of the model is coded using Python, Flask is a rapid web development framework that allows the server to be written with minimal lines of code. When the API endpoint is called upon with a prompt question, the web server will direct the request to the model to generate a response and that will be used to reply back to the caller in JSON format.

6.2 Model Evaluation

The GAI model for Threat Intelligence is evaluated using BERTScore, which leverages the contextual embedding from the BERT model to compute Precision, Recall, and F1-scores. The model was evaluated with three distinct scenarios to determine the effectiveness of the implemented approach: (1) the baseline model without domain knowledge, (2) the model fine-tuned with domain knowledge but without RAG, and (3) the model with domain knowledge integrated with RAG.

Table 1 illustrates the raw metric scores achieved by the model in 3 different scenarios. It can be seen that throughout the different metrics, a score in the high 80s range has been achieved. It is clearly evident that the Mistral model that is integrated with domain knowledge via RAG performed the best among the 3 scenarios, demonstrating that the team has adopted the right approach in fine-tuning the model.

It is also evident that the baseline model performed better than the model fine tuned with domain knowledge, but without the use of RAG. Our team concluded that this is most probably due to the fact that the model was initially fine-tuned using a set of question and answer pairs and repetitive responses. This approach as explained under

Table 1. Model Evaluation Results.

BERTScore Metrics	Without Domain Knowledge (Baseline Model)	Fine-tuned with Domain Knowledge (without RAG)	With Domain Knowledge and RAG
Precision	0.8411	0.8338	0.8585
Recall	0.8920	0.8603	0.8924
F1-Score	0.8657	0.8465	0.8746

Initial Approach, led to overly specific patterns in training, which causes the fine-tuned model to struggle with interpreting queries that deviated from the training data format. Hence, it resulted in a decrease in performance scores.

The difference in scores between the baseline model and the model that has integrated with domain knowledge via RAG is minimal, with at most a 0.02 score improvement across all metrics for the model enhanced with RAG. When examining the responses produced by both models in detail, it was observed that while the baseline model achieved a high-performance score in the various metrics, its responses were outdated for some instances. A sample of a response from the baseline model is as shown in Fig. 1.

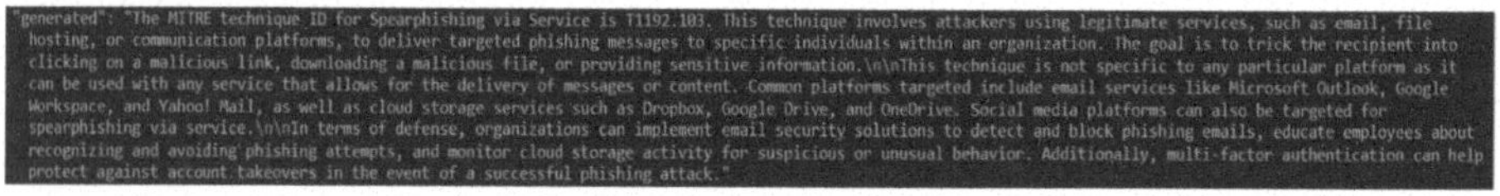
"generated": "The MITRE technique ID for Spearphishing via Service is T1192.103. This technique involves attackers using legitimate services, such as email, file hosting, or communication platforms, to deliver targeted phishing messages to specific individuals within an organization. The goal is to trick the recipient into clicking on a malicious link, downloading a malicious file, or providing sensitive information.\n\nThis technique is not specific to any particular platform as it can be used with any service that allows for the delivery of messages or content. Common platforms targeted include email services like Microsoft Outlook, Google Workspace, and Yahoo! Mail, as well as cloud storage services such as Dropbox, Google Drive, and OneDrive. Social media platforms can also be targeted for spearphishing via service.\n\nIn terms of defense, organizations can implement email security solutions to detect and block phishing emails, educate employees about recognizing and avoiding phishing attempts, and monitor cloud storage activity for suspicious or unusual behavior. Additionally, multi-factor authentication can help protect against account takeovers in the event of a successful phishing attack."

Fig. 1. Baseline Model Response

The technique ID listed in the response from the baseline model no longer exists on the MITRE ATT&CK website and has been redirected as T1566 instead of T1192. This demonstrates that even though the baseline model may achieve a high-performance score, it does not necessarily produce accurate or quality outputs, as compared to the model enhanced with RAG. This undermines the importance of retrieving accurate data which is especially crucial for a domain like threat intelligence. Moreover, new threat intelligence information can simply be added to the RAG database on top of the existing data, eliminating the need for frequent fine-tuning of the model with each update. This approach ensures the model remains up to date.

6.3 Comparison with Output from GPT-4o

Our team used OpenAI's ChatGPT-4o proprietary model as the baseline for comparison of outputs produced by our GAI model for Threat Intelligence.

Fig. 2 illustrates a screenshot output from ChatGPT-4o. The length of the output was longer than one screen height (screenshot has been truncated in this report).

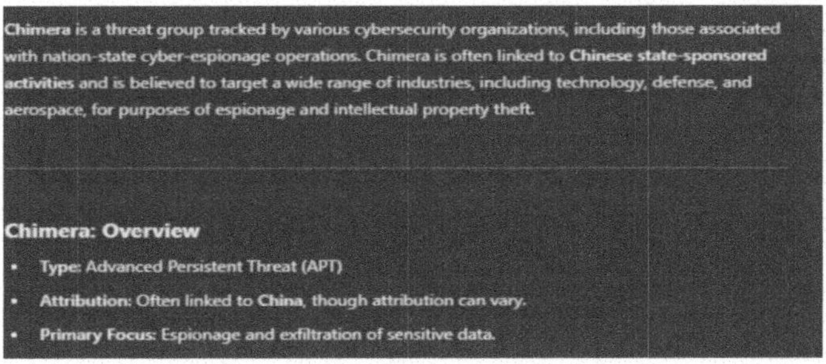

Fig. 2. ChatGPT-4o response on Chimera

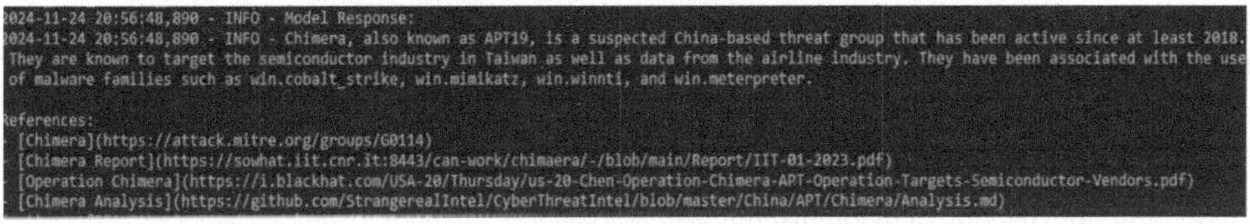

Fig. 3. Threat Intelligence GAI response on Chimera

Fig. 3 illustrates the screenshot output from our GAI for Threat Intelligence. While the output is shorter than the one produced by ChatGPT-4o, it provides a more direct response to the user. References to the MITRE ATT&CK website and other sources of information describing the operations of Chimera are provided, providing a more concise yet complete response.

7 Conclusion

A coherent and comprehensive GAI model for threat intelligence has been successfully developed, with focus on the MITRE ATT&CK framework, Cyber Kill Chain, CVE details, Techniques, Tactics and Procedures (TTPs) as well as IOCs, threat analysis and mitigation strategies. Throughout the research and implementation of the GAI model, the team explored various techniques when it came to handling the dataset and compared existing models such as GPT-2, Llama and Mistral before coming to the conclusion on the approach that would yield the best outcome for the eventual GAI model in terms of computational resources, license agreements and most importantly, accuracy of model responses.

One potential future work would be to implement a pipeline that automatically detects and retrieves the new dataset from GitHub for both the MITRE ATT&CK framework and CVE details. Given that the GitHub repositories for both datasets are updated on a timely and automatic basis, the GitHub repositories that the team obtained our data from are good sources of information to continuously retrieve the latest information given the rapid evolution in the domain of threat intelligence.

Acknowledgement. The authors deeply acknowledge A/P Dr. Guo Huaqun from the Singapore Institute of Technology (SIT) for her help and support in the research and development of this project. The authors would like to declare that there is no conflict of interest.

References

1. "mistralai/Mistral-7B-Instruct-v0.3 · Hugging Face." https://huggingface.co/mistralai/Mistral-7B-Instruct-v0.3
2. "MITRE ATT&CK®." https://attack.mitre.org/
3. Cyber Kill Chain. [Online]. Available: https://www.lockheedmartin.com/en-us/capabilities/cyber/cyber-kill-chain.html
4. Rebei, A.: Fine-Tuning Language Models for Context-Specific SQL Query Generation (2023). arXiv: arXiv:2312.02251
5. Fu, X., et al.: Lightweight Automatic Modulation Classification Based on Decentralized Learning. IEEE Trans. Cognit. Comm. Netw. **8**(1), 57–70 (2022)
6. Elin, M.N.Z.: Comparative Analysis of Decision-Making Efficiency of Large Language Models. IJFMR – Int. J. Multidiscipl. Res. **5**(3)
7. Sapegin, A., Jaeger, D., Cheng, F., Meinel, C.: Towards a system for complex analysis of security events in large-scale networks. Computers & Security **67**, 16–34 (2017)
8. S. McElwee, J.H., Fraley, J., Cannady, J.: Deep learning for prioritizing and responding to intrusion detection alerts. In: MILCOM 2017 - 2017 IEEE Military Communications Conference (MILCOM), pp. 1–5 (2017)
9. Lewis, P., et al.: Retrieval-Augmented Generation for Knowledge-Intensive NLP Tasks (2021). arXiv: arXiv:2005.11401
10. Asai, A., Wu, Z., Wang, Y., Sil, A., Hajishirzi, H.: Self-RAG: Learning to Retrieve, Generate, and Critique through Self-Reflection (2023). arXiv: arXiv:2310.11511
11. Soman, K., et al.: Biomedical knowledge graph-optimized prompt generation for large language models. (2024). arXiv: arXiv:2311.17330
12. Mitre-Attack: GitHub - mitre-attack/attack-stix-data: STIX data representing MITRE ATT&CK. GitHub. https://github.com/mitre-attack/attack-stix-data
13. CVEProject: GitHub - CVEProject/cvelistV5: CVE cache of the official CVE List in CVE JSON 5 format. GitHub. https://github.com/CVEProject/cvelistV5
14. URLhaus - Malware URL exchange. https://urlhaus.abuse.ch/
15. Fkie, F.: Malpedia (Fraunhofer FKIE). https://malpedia.caad.fkie.fraunhofer.de/
16. sentence-transformers/all-MiniLM-L6-v2 · Hugging Face (2001). https://huggingface.co/sentence-transformers/all-MiniLM-L6-v2
17. "txtai." https://neuml.github.io/txtai/
18. Merritt, R.: What Is Retrieval-Augmented Generation aka RAG | NVIDIA Blogs. NVIDIA Blog (2024). https://blogs.nvidia.com/blog/what-is-retrieval-augmented-generation/
19. Openai: GitHub - openai/gpt-2: Code for the paper 'Language Models are Unsupervised Multitask Learners,' GitHub. https://github.com/openai/gpt-2
20. "meta-llama (Meta Llama)." https://huggingface.co/meta-llama
21. Witt, A.: Llama 2 and Llama 3.1 Hardware Requirements: GPU, CPU, RAM. Hardware Corner (2024). https://www.hardware-corner.net/guides/computer-to-run-llama-ai-model/
22. "meta-llama/Llama-2-7b · Hugging Face," (2001). https://huggingface.co/meta-llama/Llama-2-7b
23. "meta-llama/Llama-3.1-8B · Hugging Face," (2001). https://huggingface.co/meta-llama/Llama-3.1-8B

24. "NousResearch (NousResearch)," (2001). https://huggingface.co/NousResearch
25. Ai, M.: Mistral 7B. Mistral AI | Frontier AI in Your Hands (2024). https://mistral.ai/news/announcing-mistral-7b/
26. Ghazarian, S., Hedayatnia, B., Papangelis, A., Liu, Y., Hakkani-Tur, D.: User Response and Sentiment Prediction for Automatic Dialogue Evaluation (2021). arXiv. https://doi.org/10.48550/ARXIV.2111.08808
27. Zhang, T., Kishore, V., Wu, F., Weinberger, K.Q., Artzi, Y.: BERTScore: Evaluating Text Generation with BERT (2019). arXiv. https://doi.org/10.48550/ARXIV.1904.09675

AI-Driven MOSFET Bypass Systems and Computer Vision for Mitigating Partial Shading Losses in Photovoltaic Systems

Rui Siang Lim[1(✉)], Tianyin Li[1,2], Yang Zhi Pang[1,2], Rahul Bhujade[2], and Sanjib Kumar Panda[2]

[1] Hwa Chong Institution, Singapore 269734, Singapore
limruisiang2007@gmail.com

[2] Department of Electrical and Computer Engineering, National University of Singapore, Singapore 117583, Singapore

Abstract. This study investigates the potential of AI-driven MOSFET bypass systems and computer vision to mitigate partial shading losses in photovoltaic (PV) systems. Partial shading, particularly due to cloud cover, can significantly reduce the power output of PV systems. As commonly used bypass diodes have limitations due to power loss and hotspots, this study explores the use of MOSFETs as an alternative, paired with a convolutional neural network (CNN) based computer vision-based cloud detection system. The CNN model is trained to detect cloud edges and track their movement, enabling dynamic adjustment of the MOSFET bypass system. The results show that the MOSFET bypass system outperforms traditional bypass diodes in energy efficiency, particularly in medium to low intensity static shading conditions. The study also highlights the potential of predictive analysis to improve the accuracy of detection and MOSFET control for dynamic shading conditions. Findings suggest that AI-driven MOSFET bypass systems and computer vision can be an effective solution for mitigating partial shading losses in PV systems, particularly in large-scale solar arrays. Future studies can focus on refining the CNN model, exploring alternative control methods, and further investigating the energy efficiency of the system under various shading conditions.

Keywords: Partial shading · MOSFET · neural network

1 Introduction

1.1 Maximum Power Point Tracking

Photovoltaic (PV) systems are one of the most researched and used methods for solar energy production, providing 5.5% of the world's energy in 2024 [1]. While maximum cell efficiency has reached 47.6% [2], challenges with commercial implementation persist, such as the availability of third-generation cells [3].

Maximum Power Point Tracking (MPPT) aims to maximise the power output most commonly by adjusting the voltage across the load, relying on inputs such as voltage,

H. Guo et al. (Eds.): IRC-SET 2025, PTT, pp. 350–359, 2026.
https://doi.org/10.1007/978-981-95-9630-0_30

current or external inputs like irradiance or temperature [4]. Increasingly AI models have been used to enhance MPPT convergence, stability and efficiency [5]. As suggested in [6], choosing an algorithm depends on partial shading conditions (PSC), occurring when different panels receive different irradiance, resulting in the entire array current to decrease to the output of the shaded segment [7]. This causes significant power losses and damage to panels due to the hot spot effect. While PSC has been studied thoroughly [8, 9], a subset of PSC, cloud cover, is not commonly investigated. Although clouds do not cause large irradiance drops, for large strings with high energy outputs, a small drop in irradiance in a small segment can more than proportionately decrease power output for the whole string. Non-linear power-voltage relationships resulting from partial shading make it hard to predict the MPP [10] and dynamic nature of clouds can cause frequent power losses that add up over time.

1.2 Bypass Diodes

Bypass diodes are commonly used to mitigate partial shading conditions (PSC) in solar systems by allowing current to bypass shaded panels when voltage drops are significant [11]. However, traditional bypass diodes cause power losses due to forward voltage drops and introduce multiple peaks in PV curves, requiring complex, energy-intensive maximum power point tracking (MPPT) techniques [12, 13]. MPPT methods, such as Fuzzy Logic Control, Support Vector Machines, and regression models, face challenges including high computational cost and the need for extensive training data [14–17]. Without such MPPT algorithms available, frequent PSC, especially in large arrays affected by moving clouds, worsens inefficiencies.

Smart bypass diodes, such as Texas Instruments' SM74611 employs an N-Channel Metal Oxide Field Effect Transistor (MOSFET) with a controller and charge pump to minimize power dissipation, although its complexity raises costs [18]. Similarly, [19] demonstrated a design using a floating bulk NMOS, achieving up to 5.61% power saving compared to diodes by minimizing voltage drop. However, heating issues still contribute to power loss, while reliance on periodic switching of the NMOS to detect voltage drop introduces noise and inefficiency. Thus, there is still a need for a simple, cost-effective alternative to such smart control systems, which is the aim of the study.

This study investigates the use of AI-driven control systems paired with MOSFETs to improve the performance of photovoltaic (PV) systems under PSC. By exploiting the simple to track PV curve and low voltage drop across MOSFETs, there is significant potential for energy savings for medium-to-large scale PV systems.

2 Methodology

2.1 Static Shading Conditions

Static shading conditions were simulated in MATLAB/Simulink, with the circuitry illustrated in Fig. 1. For finding the MPP of various configurations, a controlled voltage source was connected to both ends of the 20-panel array and swept through a range of voltage, recording the power at each time step.

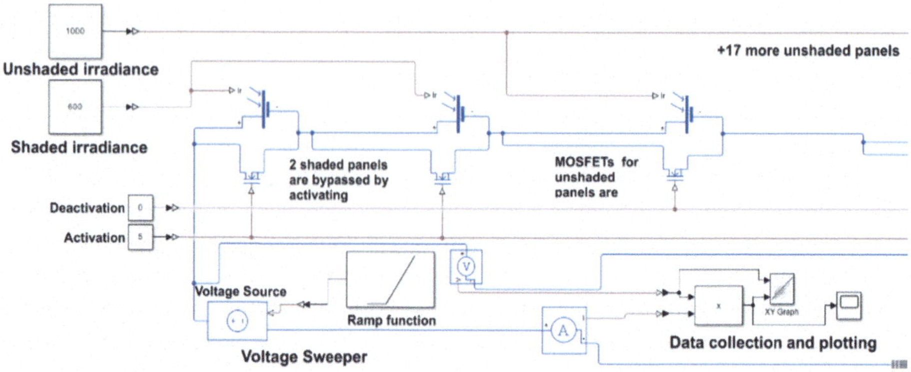

Fig. 1. Circuitry for static PSC simulation. Irradiance values for shaded panels were adjusted between 200–800 W/m^2 in steps of 200 W/m^2, while the irradiance of unshaded panels remained at 1000 W/m^2. The number of shaded panels was varied, with the connection of the MOSFETs being adjusted accordingly.

2.2 Detection of Cloud Front

A convolutional-neural-network (CNN) segmentation model was built using YOLOv8 with 3901 and 1672 images for training and validation respectively. CNN ensures consistent detection under various contrasts, accounting for different shading intensity or field color. Images were obtained in the lab with the setup in Fig. 2. To ensure consistency, OpenCV findContours method was calibrated to fit accurately to the highest contrast shadow. The same set of contours was then superposed upon other shading contrasts to ensure consistency. Which is enabled by linear actuator.

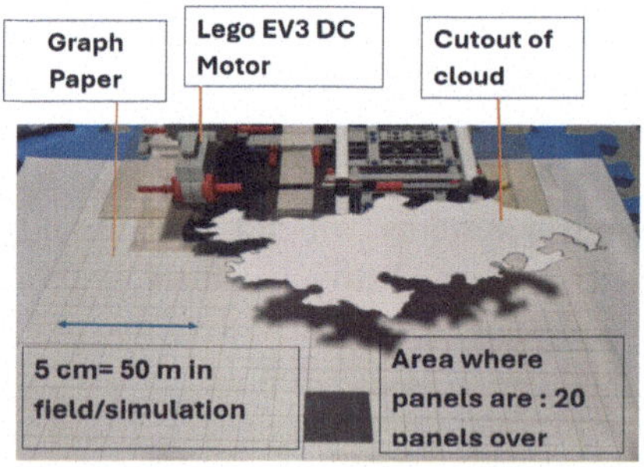

Fig. 2. Set-up for collection of data. Videos were taken from a stationary camera of shadows created by cutouts of cloud shapes. Different lighting conditions were used to vary contrast.

2.3 Dynamic Shading Conditions

For dynamic shading conditions, the circuitry in Fig. 3 was used. The Perturb and Observe (P & O) algorithm was used for the MOSFET bypass system as it is the simplest and most commonly used [20]. P&O steps the voltage up or down and observes the change in power, subsequently adjusting the voltage in the direction of increased power. To simulate real world settings, 1cm on the graph paper is set to be the same as 10m in the "field" frame. This conversion is done through a homography matrix.

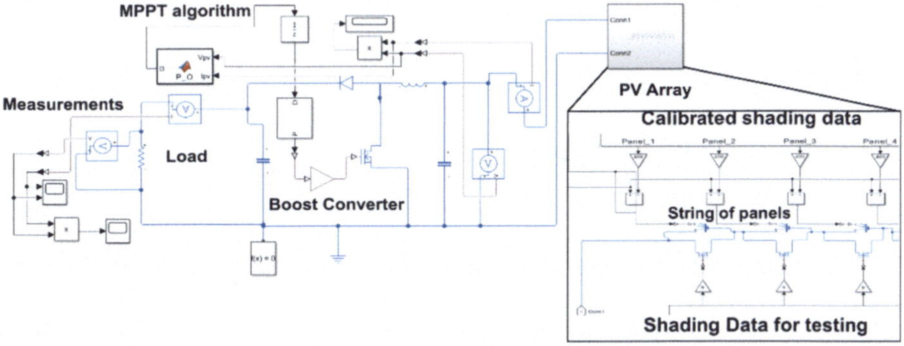

Fig. 3. Circuitry for collection of power data under dynamic PSC. Each panel was assigned a position on the "field", with the coordinates (in meters) highlighted.

2.4 Video Analysis

Two analysis methods were used, with logical flow shown in Fig. 4. In frame-by-frame analysis, every frame was processed. In predictive analysis, the x coordinate of the "cloud front"—mean x coordinate of the points on the contour with the highest values, was recorded for the first 30 frames, before extrapolating using simple linear regression.

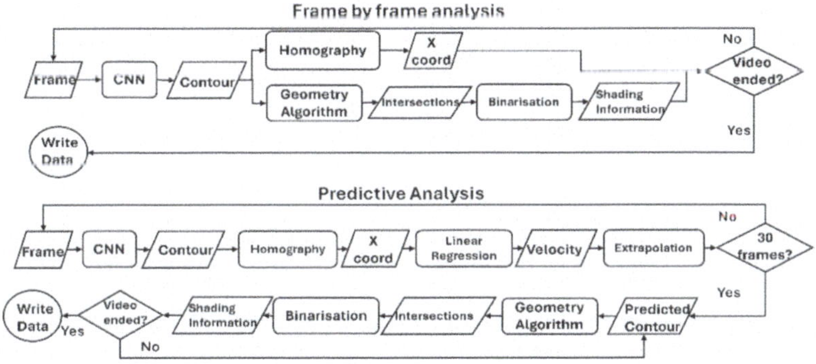

Fig. 4. Flowchart of steps for video analysis

Shaded and unshaded panels were given Boolean values of 1 and 0 respectively per frame. Imperfect shading data from CNN-detected cloud shadows was used as input for MOSFET control signals, to analyse the inaccuracies in detection. To activate idle state in the MOSFET, a separate Support Vector Regression (SVR) model was trained using data on shading intensities and the corresponding maximum power outputs when the MOSFETs were idle. SVR is a machine learning model that finds a hyperplane to predict continuous outputs by maximizing the margin around the training data points [21]. The

SVR compared the predicted idle state power with the power generated when bypass is activated. If the idle state power was higher, the MOSFETs were turned off by setting all panel values to 0. The resulting MOSFET signals were input into Simulink, while the unprocessed shading data was fed into the solar arrays.

3 Results and Discussion

3.1 Comparison of Power Output

Figure 5 presents the PV curves of diode and MOSFET bypass systems. Notably, the MOSFET bypass and idle systems have a single peak in their PV curves, simplifying MPPT algorithms and reducing energy consumption needed for complex algorithms. At certain higher shading intensities (800 W/m^2), the MOSFET bypass achieves significantly higher maximum power compared to the idle bypass and diode systems. This is not so at lower shading intensities (400 W/m^2). Figure 6 compares power output across bypass methods at two shading intensities. The power output for activated MOSFET systems only depends on the unbypassed panels. In contrast, the idle MOSFET system follows a complex relationship due to its non-linear IV behavior when transitioning from the triode to saturation region. The bypass diode follows a piecewise-polynomial relation, with shading thresholds determined by the voltage required to activate reverse bias.

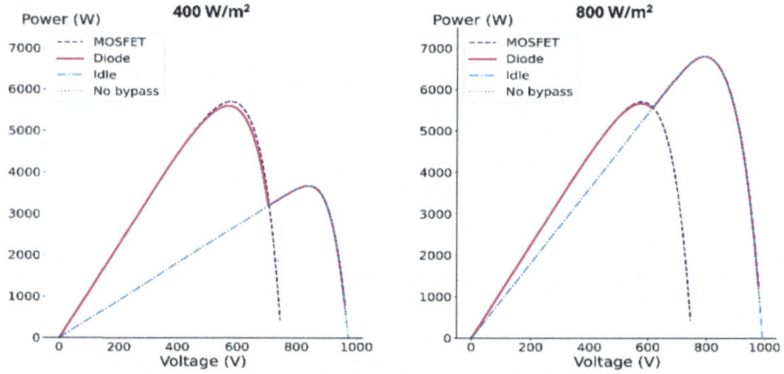

Fig. 5. PV **curves for systems with diode and idle/activated MOSFET bypass systems**. Notice that activated MOSFETs have peaks higher than the global MPP for diodes, indicating greater energy efficiency. After the threshold, where bypass diodes are activated, the idle MOSFETs have almost identical curves to diodes.

For higher shading intensities, the MOSFET bypass shows greater power savings compared to the diode system. Beyond a certain shading percentage, however, bypass diodes produce more power than activated MOSFETs but slightly less than idle MOSFETs. This behavior arises because, under these conditions, the short-circuit current (I_{sc}) of the shaded cells limits the entire string's current. As a result, the string's operating voltage drops to 0V. Consequently, the reverse voltages across the shaded cells must collectively equal the operating voltage of the unshaded cells at Isc. When shading exceeds the threshold, the reverse voltage across the shaded cells becomes too low to activate the bypass diodes, rendering them inactive.

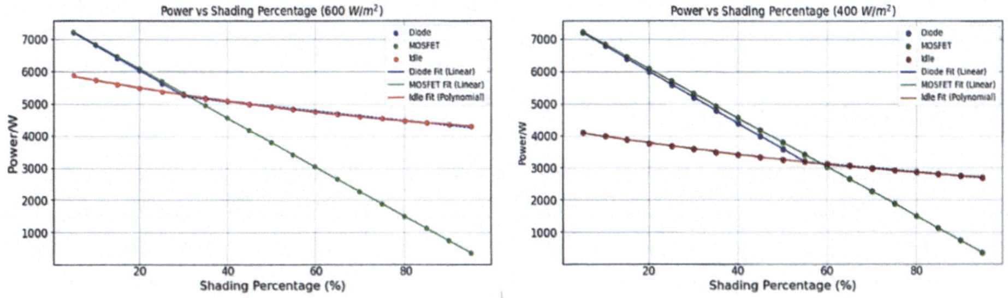

Fig. 6. Relationship **between power saved and the shading percentage.** The energy saved is more visible here. Under darker shading intensities, the power saved is greater.

3.2 Energy Efficiency

To quantify the efficiency of MOSFETs compared to diodes, the maximum power of the activated MOSFETs and diodes systems were compared with the idle-state MOSFET systems maximum power. Figure 7 illustrates these results, with the percentage of power saved given by $\frac{P_{d/m}-P_i}{P_i}$ where $P_{d/m}$ and P_i represent the maximum powers of the diode/MOSFET and idle state respectively. After the threshold voltage, the idle-state MOSFET system behaves identically to bypass diodes, so the diode curves drop to zero and flattens horizontally whilst the activated MOSFETs lose power after crossing the x axis.

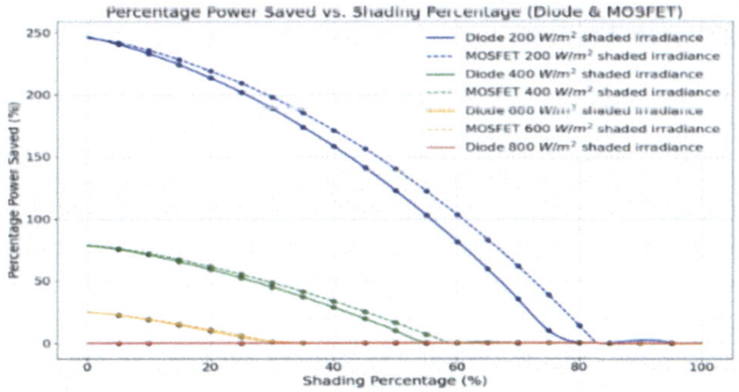

Fig. 7. Comparison of percentage of power saved against shading percentage between diode and MOSFETs. Since most of the power saved using MOSFETs (both idle and activated) occurs above the x axis, negative values are omitted for simplicity.

Above the axis, higher shading intensities result in greater power savings, indicated by the larger area difference between the MOSFET and diode curves. During partial shading, the string's power output is limited by the lowest current. In darker shading, the current in shaded panels drops significantly, causing a larger power loss compared to lighter shading. Bypassing these panels yields greater energy gains under darker conditions. The power saved by activated MOSFETs relative to diodes depends on how much power is saved by bypassing the shaded panels. Therefore, in darker shading intensities, the power saved is greater and the power saved by activated MOSFETs relative to diodes is compounded, resulting in the higher power saved.

3.3 CNN Tracking Accuracy

Subsequently, the CNN model was applied to footage of moving shadows cast by an unseen cloud shape. Figure 8 demonstrates the model's effectiveness in tracking the cloud edge, with a deviation of less than 2m across all contrasts. However, for high contrasts, overfitting may cause clustering of points, while training on lower contrasts may lead to detecting areas within the shadow that still have irradiance. These factors likely contributed to underprediction. Accurate tracking is critical for predictive analysis, as it forms the basis for calculating cloud velocity.

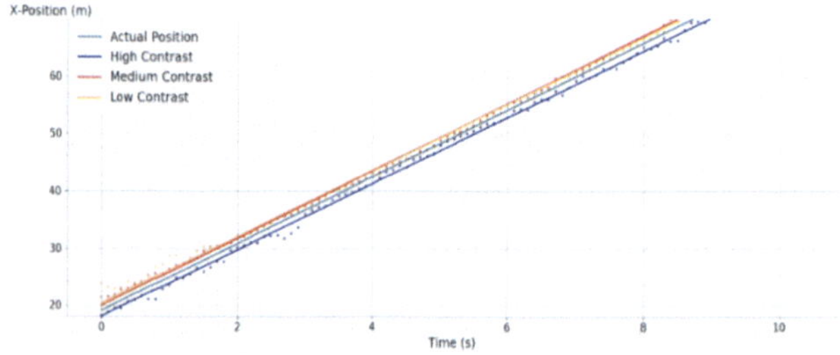

Fig. 8. Graph representation of tracking accuracy

3.4 Frame-By-Frame Analysis

As a preliminary investigation, the CNN model was applied to unseen footage of a moving cloud shadow. Detected cloud edges were processed frame by frame to estimate the system's maximum bypass efficacy independent of time. For standardisation, the shaded panel irradiance was fixed at 600 W/m2 regardless of contrast so that a fair judgement of how good the CNN model was at detecting shadows could be made. The cloud positions were then processed in MATLAB/Simulink. The power output over 4 s is plotted in Fig. 9.

The black curve represents the ideal case where MOSFETs are activated immediately upon shading. This was achieved by using shading data as MOSFET signals so that the time of activation and shading was synced. The green curve shows the power output of a diode system using the P&O, which is the most commonly used system. As the P&O could not converge to the global maximum due to double peaks, the power decreased more significantly compared to the MOSFETs, proving that given the same MPPT method, MOSFETs are indeed the better option. Subsequently the power of the MOSFETS for all contrasts converged, proving how in idle mode, the power drop solely depends on the number of panels if the shading intensity is the same. As seen from Figure 10, medium and low contrasts had best results, which was unexpected. At high contrasts, the model may have underestimated the cloud edge position as discussed, leading to early MOSFET activation. Due to the double effect of an early MOSFET activation and misdetection of shadow, the power loss is exacerbated and the MPPT requires more time to converge, explaining the significant fluctuations. Hence, the training process must be refined further for better results.

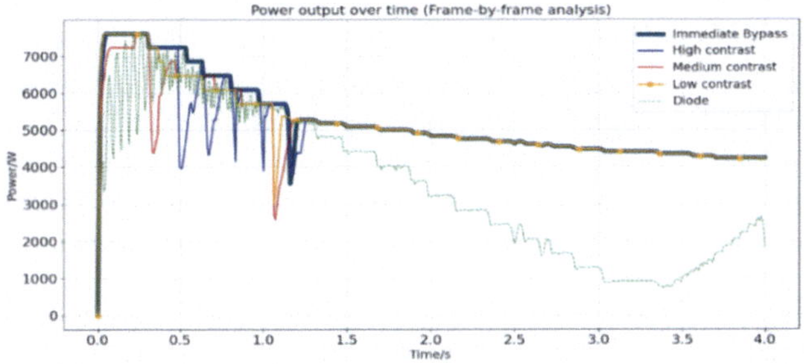

Fig. 9. Comparison of power output over time (Frame-by-frame analysis). The significant drop at 1.3 s indicates the activation of idle mode. It is caused by a large change in the voltage at MPP, resulting in a significant longer convergence time.

Type	Bypass Diode	Low Contrast	Medium Contrast	High Contrast	Actual
Energy output	10255	14025	14180	13912	14219

Fig. 10. Comparison of total energy output (Frame-by-frame analysis)

3.5 Predictive Analysis

For predictive analysis, the same procedure was used, except for only the first 30 frames were used to track the cloud front to find the velocity. The power and total energy output over 4 s are shown in Figs. 11 and 12 respectively.

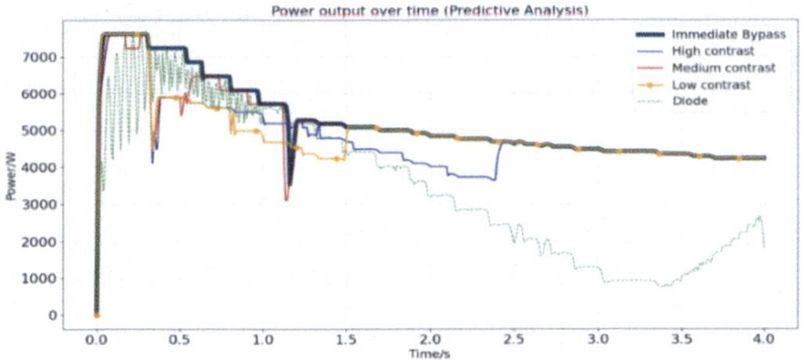

Fig. 11. Comparison of power output over time (Predictive analysis). Significant fluctuations compared to the frame-by-frame analysis due to greater error in measuring further distances, causing greater mismatch in the signal and actual shading and hence greater power loss.

Type	Bypass Diode	Low Contrast	Medium Contrast	High Contrast
Energy output	10255	13443	13861	13187

Fig. 12. Power output over time for different contrasts in comparison to diode bypass system (Predictive analysis).

The predictive analysis also has good tracking of the power after the idle mode is activated, though experiencing deviations for high contrast. This could be due to the propagation of a misdetected contour, amplifying shadow size and causing early activation of MOSFETs. The CNN showed a lag in detection for low contrasts but still had good tracking subsequently, proving short term linear prediction is a reliable method for predicting shading conditions since the cloud shadow can be approximated in short time scales to have a constant velocity without significant deformations in shapes.

4 Conclusion

In this paper, MOSFETs were found to have greater power efficiency than diodes for PSC. A control system was proposed that leverages CNN to track cloud shadows and SVR to optimize the MOSFET control. This removes the need for complex MPP algorithms as there will only be one peak at any point in time. For future work, this control system could be tested on real clouds, to confirm the increase in efficiency gains under small intensity differences. More research could be done to map the intensity of pixels detected to the actual irradiance, to predict the approximate voltage. As the control system performed best under more extreme shading conditions, the control system could be applied to urban settings where buildings can potentially shade PV arrays. This is important for cities aiming to build PV farms in urban areas.

References

1. Lempriere, M.: Wind and solar are 'fastest-growing electricity sources in history.' Carbon Brief (2024). https://www.carbonbrief.org/wind-and-solar-are-fastest-growing-electricity-sources-in-history/
2. Schygulla, P., et al.: Quadruple junction solar cell with 47.6 % conversion efficiency under concentration. In: International Conference on Metal Organic Vapor Phase Epitaxy 2022 (2022)
3. Nagar, H.: Feasibility of emerging technologies in solar panel. Zenodo (CERN Eur. Organ. Nuclear Res.). (2023). https://doi.org/10.5281/zenodo.7851436
4. Katche, M.L., Makokha, A.B., Zachary, S.O., Adaramola, M.S.: A comprehensive review of Maximum Power Point Tracking (MPPT) techniques used in solar PV systems. Energies **16**, 2206 (2023). https://doi.org/10.3390/en16052206
5. Yap, K.Y., Sarimuthu, C.R., Lim, J.M.-Y.: Artificial Intelligence based MPPT techniques for solar power system: a review. J. Mod. Power Syst. Clean Energy **8**, 1043–1059 (2020). https://doi.org/10.35833/mpce.2020.000159
6. Boubaker, O.: MPPT techniques for photovoltaic systems: a systematic review in current trends and recent advances in artificial intelligence. Discov. Energy. **3** (2023). https://doi.org/10.1007/s43937-023-00024-2

7. Giyantara, A., Wisyahyadi, Rizqullah, R.B., Kusuma Priyanto, Y.T.: Analysis of partial shading effect on solar panel power output. J. Phys.: Conf. Ser. **1726**, 012022 (2021). https://doi.org/10.1088/1742-6596/1726/1/012022
8. Mamun, M.A.A., Hasanuzzaman, M., Selvaraj, J.: Experimental investigation of the effect of partial shading on photovoltaic performance. IET Renew. Power Gen. **11**, 912–921 (2017). https://doi.org/10.1049/iet-rpg.2016.0902
9. Alsayid, B.A., Alsadi, S.Y., Jallad, J.S., Dradi, M.H.: Partial shading of PV system simulation with experimental results. SGRE. **04**, 429–435 (2013). https://doi.org/10.4236/sgre.2013.46049
10. Sera, D., Baghzouz, Y.: On the impact of PSC on PV output power. In: Proceedings of RES'08. WSEAS Press (2008)
11. Aghaei, M., et al.: Review of degradation and failure phenomena in photovoltaic modules. Renew. Sustain. Energy Rev. **159**, 112160 (2022). https://doi.org/10.1016/j.rser.2022.112160
12. Teo, J.C., Tan, R.H.G., Mok, V.H., Ramachandaramurthy, V.K., Tan, C.: Impact of partial shading on the P-V characteristics and the maximum power of a photovoltaic string. Energies **11**, 1860 (2018). https://doi.org/10.3390/en11071860
13. Worku, M.Y., et al.: A comprehensive review of recent maximum power point tracking techniques for photovoltaic systems under partial shading. Sustainability. **15**, 11132 (2023). https://doi.org/10.3390/su151411132
14. Craciunescu, D., Fara, L.: Investigation of the partial shading effect of photovoltaic panels and optimization of their performance based on high-efficiency FLC algorithm. Energies **16**, 1169 (2023). https://doi.org/10.3390/en16031169
15. Takruri, M., et al.: Maximum power point tracking of PV system based on machine learning. Energies **13**, 692 (2020). https://doi.org/10.3390/en13030692
16. Farayola, A.M., Sun, Y., Ali, A.: Optimization of PV systems using Linear Interactions Regression MPPT techniques. 2022 IEEE PES/IAS PowerAfrica. pp. 545–550 (2018). https://doi.org/10.1109/powerafrica.2018.8521064
17. Ishrat, Z., Gupta, A., Nayak, S.: A Comprehensive Review of MPPT Techniques Based on ML Applicable for Maximum Power in Solar Power Systems. JREE **11** (2024). https://doi.org/10.30501/jree.2023.385661.1556
18. Texas Instruments: TI introduces smart bypass diode with industry's lowest power dissipation. Texas Instruments (2013). https://www.ti.com/about-ti/newsroom/news-releases/2013/2013-03-12-ti-introduces-smart-bypass-diode-with-industry-s-lowest-power-dissipation.html
19. Bauwens, P., Doutreloigne, J.: Reducing partial shading power loss with an integrated Smart Bypass. Sol. Energy **103**, 134–142 (2014). https://doi.org/10.1016/j.solener.2014.01.040
20. El-Khozondar, H.J., El-Khozondar, R.J., Matter, K., Suntio, T.: A review study of photovoltaic array maximum power tracking algorithms. Renewables **3** (2016). https://doi.org/10.1186/s40807-016-0022-8
21. Smola, A.J., Schölkopf, B.: A tutorial on support vector regression. Stat. Comput. **14**, 199–222 (2004). https://doi.org/10.1023/b:stco.0000035301.49549.88

Social Sciences

Factors Influencing the Calorie Intake from the Selection and Consumption of Beverages Among Youths

Chong Jie Clement Chua[1](✉), Joy Le Xuan Choo[1], Ding Xuan Ng[2,3], Mabel Qi He Leow[2,3,4], and Ngiap Chuan Tan[2,3,4]

[1] Hwa Chong Institution, 661 Bukit Timah Road, Singapore 269734, Singapore
clementchuacj@gmail.com, itsjoychoo@gmail.com

[2] Primary Care Research Institute, 167 Jalan Bukit Merah Connection One (Tower 5) #15-10, Singapore 150167, Singapore
{ng.ding.xuan,mabel.leow.q.h,tan.ngiap.chuan}@singhealth.com.sg

[3] SingHealth Polyclinics, 167 Jalan Bukit Merah Connection One (Tower 5) #15-10, Singapore 150167, Singapore

[4] Duke-NUS Medical School, 8 College Road, Singapore 169857, Singapore

Abstract. Beverage consumption augments youths' calorie intake, increasing their susceptibility to Non-Communicable Diseases. Factors such as knowledge, attitude, perceived norms and personal agency may influence their beverage choices. This study aims to determine the factors that correlate with beverage consumption habits, in terms of daily calorie intake, among youths. The secondary aim of the study is to identify the barriers and enablers to healthier beverage consumption.

From 2023 to 2024, a self-administered e-survey was distributed to participants across Singapore aged 15–24 years. Data on their demographics and beverage consumption habits were collected through the secured electronic platform FormSG. Linear regression was used to assess the factors influencing personal beverage choice and consumption based on the Integrated Behavioral Model (IBM) framework.

A total of 352 participants were analyzed, and their median daily calorie intake was 182 kcal (IQR = 96–331) from beverages, with 43% attributed to Sugar-Sweetened Beverages (SSBs). Linear regression revealed that females ($\beta = -45$, 95%CI = −82 to −8, p = 0.018) and a higher Personal Agency score ($\beta = -57$, 95%CI = −88 to -26, $p < 0.001$) were significantly associated with lower SSB consumption. The Personal Agency construct had the greatest influence (standardized $\beta = -39$) on SSB consumption, while the Knowledge, Attitude and Perceived Norm constructs were not significant. Participants reported taste (60%) and cost (32.7%) as the major factors in selecting healthier beverages. Awareness of SSBs' negative health impact (71%) and concern over future health (70.7%) were the main enablers.

SSBs contributed substantially to youths' daily beverage calorie intake. Personal Agency emerged as a key determinant of participants' SSB consumption habits. Taste preferences, cost considerations and SSB literacy were identified as primary influences on the selection of healthier beverages.

H. Guo et al. (Eds.): IRC-SET 2025, PTT, pp. 363–373, 2026.
https://doi.org/10.1007/978-981-95-9630-0_31

Keywords: Sugar-Sweetened Beverages · Non-Communicable Diseases · Calorie Intake · Integrated Behavioral Model · Personal Agency · Youth Dietary Behavior · Public Health Policy

1 Introduction

Sugar-Sweetened Beverages (SSBs) are defined as drinks that contain added sugars such as sucrose, brown sugar, and high-fructose corn syrup [1]. One key driver of SSB consumption is taste preference [2]. SSB consumption is prevalent globally, with the mean SSB intake among adults and children at 2.7 [3] and 3.6 [4] servings per week (1 serving = 248 g) respectively. However, high SSB intake is a concern with its rising prevalence, from 6.6% to 11.3% over the past two decades [5]. In Singapore, SSBs contributed to more than half the average daily sugar intake of 60 g among residents [6].

Excessive SSB consumption poses a significant threat to public health as high sugar intake is a major risk factor for various Non-Communicable Diseases (NCDs), including type 2 diabetes mellitus and cardiovascular disease [7]. The prevalence of diabetes in Singapore is projected to reach 20% by 2035, an increase from 12% in 2024, with economic cost amounting to US$1,867 million by 2050 [8, 9].

Youths represent a pertinent group of interest as studies have found SSB consumption to be highest among late teens and early adults [10]. Youths' high SSB intake during their formative years might translate into unhealthy dietary practices later in life, increasing their risk of NCDs. In recent years, there has been growing interest in tackling the high SSB consumption levels. The Singapore government has implemented policies such as sugar level restrictions on beverages sold in schools [11] and the Nutri-Grade scheme. Implemented by the Ministry of Health in 2022, Nutri-Grade is Singapore's front-of-pack nutrition labelling scheme for pre-packaged beverages which was later expanded to include freshly-brewed beverages in 2023. The scheme's grading, primarily based on drinks' sugar and saturated fat content, classifies beverages from grade 'A' (most healthy) to grade 'D' (least healthy). The scheme also restricts advertising for beverages graded 'D', as part of Singapore's broader efforts to combat diabetes and promote healthier lifestyles among its population. However, youth-targeted studies about the factors influencing beverage consumption habits remain limited. This study thus aims to identify the factors affecting the SSB calorie intake among youths as well as the barriers and enablers to healthier beverage consumption [12].

2 Methods

2.1 Study Design, Study Site and Participants

This was a cross-sectional study aimed at recruiting participants aged 15–24 years from August 2023 to November 2024 through convenience sampling. The STROBE [13] guideline was utilized as a checklist to ensure standardized reporting. Youths of all ethnicities were eligible to participate in the study. Participants serving National Service were excluded from the analysis to provide a more accurate reflection of youths' consumption habits as those serving National Service are likely to have controlled diets.

To encourage honest and unbiased responses from youths, an e-survey was utilized, allowing them to complete it anonymously at their convenience.

2.2 Data Collection and Sample Size Calculation

The survey collected anonymized data and was administered via the secure electronic platform FormSG. A QR code for the e-survey link was generated, with the study team disseminating it through social media platforms and word of mouth. Prior to starting the e-survey, participants were informed of the study's purpose and assured of confidentiality. Completion and submission of the e-survey implied consent to data collection. Data collected include age, gender, ethnicity, BMI, housing type, educational institution, and diabetic status. Other data collected include participants' knowledge, attitudes and practices pertaining to SSBs as part of the Integrated Behavioral Model (IBM) framework. Participants also completed the Patient Health Questionnaire-2 (PHQ-2) [14].

According to Chen et al., the proportion of youths who consumed one or less SSB per week is 33.2% [15]. With a 5% precision and 95% confidence interval, a sample size of 341 is needed. Given that the average response rate for an e-survey is 44% [16], the e-survey should be distributed to at least 777 participants.

2.3 Integrated Behavioral Model (IBM)

To analyze the factors influencing participants' beverage consumption habits, this study utilized the Integrated Behavioral Model (IBM), which is a psychological model used to understand human behavior (Fig. 1). The IBM summarizes the factors shaping one's intention to perform certain behaviors, while examining other relevant factors influencing behavior [18].

The validated questionnaire BEVQ-15, which pertains to beverage consumption habits, was incorporated to measure participants' consumption patterns by assessing the frequency and amount of calorie intake for different beverage types [19]. This study analyzed 4 of the constructs shaping one's intention to perform a behavior – Attitude, Perceived Norm, Personal Agency and Knowledge. For the Attitude construct, participants answered 4 questions relating to how they perceived the consumption of SSBs, healthy lifestyles and Nutri-Grade labels. For the Perceived Norm and Personal Agency constructs, participants answered 6 questions per construct about social pressure and their confidence in avoiding SSBs respectively. For each of these constructs, participants rated their answers on a 5-point Likert scale (1-Strongly Disagree to 5-Strongly Agree), and their mean score was used as their construct score. Separately, for the Knowledge construct, participants answered 4 factual questions about SSB consumption guidelines and the definition of SSBs. Participants were then assigned a Knowledge score based on the number of correctly answered questions.

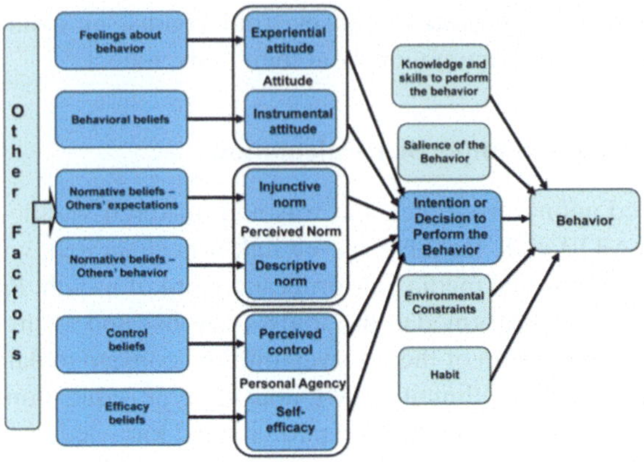

Fig. 1. The IBM provides a framework for understanding human behavior [17].

2.4 Patient Health Questionnaire-2 (PHQ2)

The self-administered validated tool Patient Health Questionnaire-2 (PHQ-2) was used to screen for depressive symptoms among participants. The scores for the questionnaire range from 0 to 6, with scores of 3 and above indicating a likelihood of depressive disorder. Participants who scored 3 or more were advised to seek immediate medical assistance via a pop-up message.

2.5 Statistical Analyses

Descriptive statistics were employed to summarize the demographic characteristics, calorie intake from beverages, and scoring of constructs within the IBM framework. Categorical variables were presented as frequencies and percentages, while continuous measures were reported as mean and standard deviation or median and interquartile range, depending on the normality of the data. To identify significant factors affecting youths' calorie intake from SSBs, linear regression analysis was conducted. A p-value $\leq$ 0.05 was deemed significant. All analyses were performed using R version 4.4.2, SPSS version 27.

3 Results

A total of 830 questionnaires were distributed and 354 were completed by participants (response rate = 42.7%). Participants serving National Service (n = 2) were excluded and 352 participants were analyzed. Most youths were aged between 15 to 19 (86.9%), female (57.4%), Chinese (87.5%), and had a normal BMI (50.9%) based on the guidelines for Asians. Most participants resided in HDB 4–5 room flats (48.9%). Over half the youths were studying at Junior Colleges/Millennia Institute (50.3%) and were unlikely to have depressive disorder (77.8%). All but 2 participants do not have a history of diabetes (Table 1).

Table 1. Demographics, constructs, and daily calorie intake from beverages of participants.

	Categories	Overall
N		352
Daily Calories (median [IQR])		182.00 [95.75, 331.25]
Sugar Sweetened Beverage Daily Calories (median [IQR])		79.00 [23.00, 165.75]
Age (%)	15–19	306 (86.9)
	20–24	46 (13.1)
Gender (%)	Female	202 (57.4)
	Male	150 (42.6)
Ethnicity (%)	Chinese	308 (87.5)
	Malay	19 (5.4)
	Indian	14 (4.0)
	Others	11 (3.1)
BMI (%)*	Underweight (<18.5)	97 (27.6)
	Normal (18.5–22.9)	179 (50.9)
	Overweight/Obese (≥23.0)	76 (21.6)
Housing (%)	HDB 2–3 room flat	20 (5.7)
	HDB 4–5 room flat	172 (48.9)
	Condominium/Executive	128 (36.4)
	Landed property	28 (8.0)
	Others	4 (1.1)
Educational Institution (%)	Upper Secondary	81 (23.0)
	ITE	9 (2.6)
	Junior College/Millennia Institute	177 (50.3)
	Polytechnic	42 (11.9)
	University	42 (11.9)
	Others	1 (0.3)
PHQ-2 (%)	Likely Depressive Disorder	78 (22.2)
	Unlikely Depressive Disorder	274 (77.8)
Diabetes (%)	No	350 (99.4)
	Yes	2 (0.6)
Integrated Behavioral Model Construct		
Knowledge Score (mean (SD))	1.67 (0.96)	
Attitude Score (mean (SD))	3.77 (0.72)	

(continued)

Table 1. *(continued)*

	Categories	Overall
Perceived Norm Score (mean (SD))	3.06 (0.42)	
Personal Agency Score (mean (SD))	3.79 (0.69)	

* BMI was computed in accordance with the Asian BMI guidelines [20].

Youths had a median calorie intake of 182 from beverages, with calories from SSBs forming 43.4% (79 cal). Youths scored an average of 1.67 out of 4 for the Knowledge domain on beverage consumption. Among constructs, Perceived Norm and Personal Agency were the lowest and highest scoring respectively, with a mean score of 3.06 ± 0.42 and 3.79 ± 0.69 respectively (Table 1).

Table 2. Linear Regression for SSB daily calorie intake.

	Unstandardized β	95% Confidence Interval	Standardized β	P-value
Gender				
Male	Ref	-	-	-
Female	−45	[−82, −8]	-	0.018
Ethnicity				
Chinese	−24	[−114, 65]	-	0.595
Malay	−35	[−149, 79]	-	0.548
Indian	Ref	-	-	-
Others	72	[-60, 203]	-	0.286
BMI				
Underweight	−37	[−88, 14]	-	0.151
Normal	−13	[−58, 32]	-	0.581
Overweight/Obese	Ref	-	-	-
Housing				
HDB 2–3 room flat	Ref	-	-	-
HDB 4–5 room flat	−19	[−97, 59]	-	0.633
Condominium/Executive	−3	[−84, 78]	-	0.944
Landed property	1	[−96, 98]	-	0.982
Others	27	[−153, 208]	-	0.766
Educational Institution				
Upper Secondary	Ref	-	-	-

(continued)

Table 2. (*continued*)

	Unstandardized β	95% Confidence Interval	Standardized β	P-value
ITE	38	[−77, 153]	-	0.514
Junior College/Millennia Institute	1	[−45, 48]	-	0.952
Polytechnic	51	[−16, 119]	-	0.135
University	56	[−8, 121]	-	0.088
Others	−23	[−347, 301]	-	0.890
PHQ-2				
Unlikely Depressive Disorder	Ref	-	-	-
Likely Depressive Disorder	31	[−11, 74]	-	0.147
Integrated Behavioral Model Construct				
Knowledge Score	−6	[−25, 13]	−6	0.521
Attitude Score	−14	[−42, 13]	−10	0.309
Perceived Norm Score	−2	[−45, 42]	−1	0.945
Personal Agency Score	−57	[−88, −26]	-39	< 0.001

Linear regression analysis revealed that females ($\beta = -45$, 95%CI = −82 to −8, p = 0.018) and a higher score for the Personal Agency construct ($\beta = -57$, 95%CI = −88 to −26, $p < 0.001$) were associated with lower calorie consumption for SSBs. Personal Agency was the most influential construct contributing to participants' SSB consumption, with a standardized β of −39. The 3 other constructs of Knowledge, Attitude and Perceived Norm were not significant in influencing calorie consumption for SSBs (Table 2).

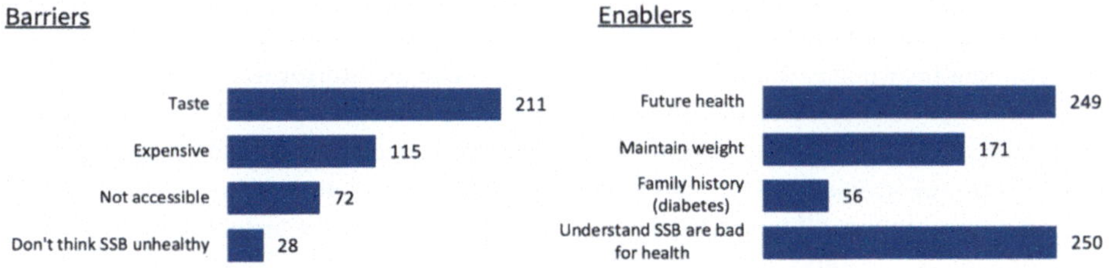

Fig. 2. Barriers to healthier beverage consumption (left) and enablers to healthier beverage consumption (right).

Most participants identified taste as the primary barrier (60%) to choosing healthier beverages. On the other hand, several factors such as worries about long-term health outcomes (70.7%) and increased awareness in understanding that SSBs are bad for health (71%) were the most commonly reported enablers (Fig. 2). Additionally, most participants disagreed with the statement that they have "little access to low- or no-sugar beverages" (65.8%), with purchases made mostly in schools (27.7%) and shopping malls (27.1%).

4 Discussion

To understand youths' consideration in beverage selection, this study utilized the IBM framework, which explores behavioral factors influencing choice. This study found that the significant factors influencing youths' beverage consumption habits were personal agency and gender. Knowledge of the health risks associated with SSB consumption and higher health consciousness were key enablers to healthier beverage choices, while taste was a key barrier.

This study found that the most significant behavioral factor influencing beverage selection and consumption among youths is personal agency. Those who were confident in their ability to avoid SSBs or choose healthier options were significantly more likely to have a lower calorie intake from SSBs. This corroborates various studies on adolescents which have found a positive correlation between greater personal agency, in the form of self-efficacy and perceived behavioral control, and healthier beverage consumption habits [21, 22]. These findings suggest the need to facilitate youth access to healthier beverages or introduce behavioral support programs that empower youths to make healthier beverage consumption choices.

Youths' consumption patterns were also affected by their gender, with females more likely to have a lower SSB calorie intake than males. This aligns with the findings of multiple international studies showing that among youths, males tend to consume more SSBs than females [23, 24]. Possible reasons for this include gender-specific socio-cultural factors leading to differences in dietary habits [25], and females' lower energy requirements and greater consciousness about their appearance causing lower SSB intake [26]. Further research is required to better understand the interplay of biological, socio-cultural and psychological factors responsible for the gender-based differences in SSB consumption.

Taste was found to be the most common barrier to healthier beverage consumption. Most participants indicated taste as their primary consideration when selecting beverages, and that they perceived healthier beverages' taste as inferior to that of SSBs. This taste preference has been well documented among youths [27, 28], who have consistently indicated that taste is a key determinant of their beverage choice. Youths may, potentially because of hearsay or limited exposure to well-formulated alternatives, also hold the misconception that all healthier beverages come with inferior taste. Given that SSBs are engineered to be hyperpalatable [29], this could have made it difficult for youths to associate the less intense tastes of healthier beverages with pleasure. Addressing this barrier may require a multi-pronged approach, including the reduction of youths' exposure to SSBs, the reformulation of healthier drinks to improve their palatability, and the implementation of educational programs to help youths develop an appreciation for less intensely sweetened beverages.

Conversely, an understanding of SSBs' adverse health effects and health consciousness emerged as key enablers to healthier beverage consumption. However, these enablers vary across studies. While some studies reflect an understanding of SSBs' adverse health effects and health consciousness as key enablers [30, 31], others have found no effect of these factors on the enabling of healthier beverage choices [32]. Further research is needed to conclusively determine the main enablers to healthier beverage consumption.

This study has several limitations. Due to the self-reporting of SSB calorie intake by participants, responses may lack objectivity in the actual consumption patterns reported. Desirability bias may have influenced responses, with participants potentially underreporting their SSB intake due to awareness of health implications. However, the anonymous nature of data collection may have mitigated this bias and encouraged more accurate reporting.

5 Conclusion

Youths with greater personal agency were associated with healthier beverage choices and lower SSB consumption. The primary barrier against healthier beverage consumption was the perceived poor taste of healthier options, while the primary enablers were an understanding of SSBs' adverse health effects and health consciousness. The involvement of multiple stakeholders including beverage vendors, schools, and the government is essential in cultivating healthy drinking habits from a young age, laying the foundation for healthier lives and lower NCD risk.

References

1. Get the facts: Sugar-sweetened beverages and consumption. https://www.cdc.gov/nutrition/data-statistics/sugar-sweetened-beverages-intake.html
2. Lee, S.H., et al.: Occasions, locations, and reasons for consuming sugar-sweetened beverages among U.S. adults. Nutrients **15**(4), 920 (2023)
3. Lara-Castor, L., Micha, R., Cudhea, F., et al.: Sugar-sweetened beverage intakes among adults between 1990 and 2018 in 185 countries. Nat. Commun. **14**, 5957 (2023)
4. Lara-Castor, L., et al.: Intake of sugar sweetened beverages among children and adolescents in 185 countries between 1990 and 2018: population based study. BMJ (2024)
5. Ge, C., Xiong, J., Zhu, R., et al.: Global burden of high sugar-sweetened beverage consumption among young adults. Diabetol. Metab. Syndr. **17**, 259 (2025)
6. Public Consultation on Measures to Reduce Sugar Intake from Pre-packaged Sugar-sweetened Beverages (2018). https://www.moh.gov.sg/newsroom/public-consultation-on-measures-to-reduce-sugar-intake-from-pre-packaged-sugar-sweetened-beverages/
7. Hu, F.B., Malik, V.S.: Sugar-sweetened beverages and risk of obesity and type 2 diabetes: epidemiologic evidence (2010)
8. Wong, L.Y., Toh, M.P., Tham, L.W.: Projection of prediabetes and diabetes population size in Singapore using a dynamic Markov model (2017)
9. Png, M.E., et al.: Current and future economic burden of diabetes among working-age adults in Asia: conservative estimates for Singapore from 2010–2050. BMC Public Health **16**, 1 (2016)

10. Joh, H.-K., et al.: Simple sugar and sugar-sweetened beverage intake during adolescence and risk of colorectal cancer precursors. Gastroenterology **161**, 1 (2021)
11. Update on HPB's efforts to reduce sugar intake of our population, https://www.moh.gov.sg/newsroom/update-on-hpb's-efforts-to-reduce-sugar-intake-of-our-population. Last accessed 11 July 2025
12. Measures for Nutri-Grade. https://www.hpb.gov.sg/healthy-living/food-beverage/nutri-grade. Last accessed 11 July 2025
13. von Elm, E., et al.: The strengthening the reporting of observational studies in Epidemiology (strobe) statement: guidelines for reporting observational studies. PLoS Med. **4**, 10 (2007)
14. Kroenke, K., et al.: The patient health questionnaire-2. Med. Care **41**(11), 1284–1292 (2003)
15. Chen, L., et al.: Associations of body roundness index and sugar-sweetened beverage consumption with psychological symptoms in adolescents: A Multicenter cross-sectional survey based on Chinese adolescents aged 12–18 years old. Front. Nutrit. **12** (2025)
16. Wu, M., Zhao, K., Fils-Aime, F.: Response rates of online surveys in published research: a meta-analysis (2022)
17. Ajzen, I., Fishbein, M.: Integrated Behavior Model (2000). https://www.besci.org/models/integrated-behavior-model
18. Fishbein, M., Ajzen, I.: Predicting and changing behavior: the reasoned action approach (2011)
19. Hedrick, V.E., et al.: Development of a brief questionnaire to assess habitual beverage intake (BEVQ-15): sugar-sweetened beverages and total beverage energy intake. J. Acad. Nutr. Diet. **112**(6), 840–849 (2012)
20. World Health Organisation: The Asia-Pacific Perspective: Redefining Obesity and its Treatment. Health Communications Australia (2000)
21. Zhang, C.-Q., et al.: Adolescent sugar-sweetened beverage consumption: an extended health action process approach. Appetite **141**, 104332 (2019)
22. Wang, C.-W., Chen, D.-R.: Associations of sugar-sweetened beverage knowledge, self-efficacy, and perceived benefits and barriers with sugar-sweetened beverage consumption in adolescents: a structural equation modeling approach. Appetite **168**, 105663 (2022)
23. Gui, Z.-H., et al.: Sugar-sweetened beverage consumption and risks of obesity and hypertension in Chinese children and adolescents: a national cross-sectional analysis. Nutrients **9**(12), 1302 (2017)
24. Skeie, G., et al.: Intake of sugar-sweetened beverages in adolescents from Troms, Norway—the Tromsø Study: fit futures. Nutrients **11**(2), 211 (2019)
25. Ge, C., et al.: Global burden of high sugar-sweetened beverage consumption among young adults. Diabetol. Metab. Syndr. **17**, 1 (2025)
26. Fontes, A.S., et al.: Demographic, socioeconomic and lifestyle factors associated with sugar-sweetened beverage intake: a population-based study. Revista Brasileira de Epidemiologia. **23** (2020)
27. Block, J.P., et al.: "If it tastes good, I'm drinking it": Qualitative study of beverage consumption among college students. J. Adolesc. Health **52**(6), 702–706 (2013)
28. Kourouniotis, S., et al.: The importance of taste on dietary choice, behaviour and intake in a group of Young Adults. Appetite **103**, 1–7 (2016)
29. Moss, M.: Salt, sugar, fat: How the Food Giants hooked us. Random House Trade Paperbacks, New York (2014)
30. Park, S., et al.: Associations between knowledge of health risks and sugar-sweetened beverage intake among us adolescents. Nutrients **15**(10), 2408 (2023)

31. Tee, W.K., Rahamat, S., Darwish, M.: Consumers' Intention to purchase healthful foods and beverages from vending machines in a healthcare institution: an application of theory of planned behaviour and health consciousness (2024)
32. Lundeen, E.A., et al.: Adolescent sugar-sweetened beverage intake is associated with parent intake, not knowledge of Health Risks. Am. J. Health Promot. **32**(8), 1661–1670 (2018)

EEG Hyperscanning to Evaluate Team Dynamics and Cooperation Strategies Among Secondary Students

Xue Yuran Yolanda[1(✉)], Gan Li Ting[1], Seow Zi Yi Olivia[1], and Aung Aung Phyo Wai[2]

[1] Hwa Chong Institution (College), Singapore 211706, Singapore
x@student.hci.edu.sg
[2] Nanyang Technological University, Singapore, Singapore
apwaung@ntu.edu.sg

Abstract. Face-to-face communication remains essential for inter-personal connections even in an increasingly digital world. Understanding the neural mechanisms of group interactions can inform strategies to enhance learning and interpersonal communication. While behavioral and selfreports showed coordination patterns among learners, these methods lack temporal precision with objective measures. This study investigate team dynamic patterns among secondary students through neurophysiological analysis using EEG hyper-scanning. Twenty eight female secondary school students were recruited and organized into seven groups of four participants each. Instead of a separate control group, baseline measures and within-subject task contrasts were employed to isolate the neural and behavioral effects of different social interaction contexts. Each group completed two main tasks: (1) a mixed cooperation-competition task, and (2) a purely collaborative task. We used continuous and simultaneous EEG recordings to evaluate behavioral and group-level task performance. A positive correlation was found between inter-brain coherence in the θ frequency band and team's task performance ($R = 0.73$, $p < 0.01$). This implied the neural synchrony was associated with better collaborative outcomes. Increased frontal band-power energy was significantly associated with higher team connectivity ratings ($\beta = 0.63$, $p < 0.05$), suggesting that effective team interactions resulted in enhanced frontal neural activity. However, complex relationships emerged between mental states and group performance; mental attention not always conducive to task performance but also requiring adaptability. Our findings contribute to the evidence-based understanding of interpersonal communication for improving team-based interactions in educational contexts.

Keywords: EEG Hyperscanning · Team Dynamics · Cooperative Learning · Inter-brain Synchrony

H. Guo et al. (Eds.): IRC-SET 2025, PTT, pp. 374–382, 2026.
https://doi.org/10.1007/978-981-95-9630-0_32

1 Background

Understanding students' collaboration, cooperation and communication is vital for their gradual development of social and cognitive skills according to modern synthesis of 21st century skills [8, 18, 27]. The empirical evidences showed that effective collaboration enhances both cognitive outcomes (achievement) and social skills (positive relationships) lead to higher academic achievement, more positive interpersonal relationships, and greater psychological well-being [27]. Structured engagement in collaborative tasks enhances teamwork, communication, and problem-solving skills, fostering emotional intelligence and leadership abilities [19]. The holistic development with problem-based learning prepares students for success in both academic and real-world contexts [16]. In current technology-driven world, proficiency in face-to-face communication remains crucial for staying connected. Expanding our knowledge on team dynamics is integral to promoting physical interactions and enhancing interpersonal skills [22, 25].

The literature review across 104 different studies showed that active participation in collaborative tasks provides students with opportunities to develop vital emotional intelligence, teamwork and leadership capabilities [6]. By working together on group tasks and engaging in meaningful discussions, students enhance their skills in collaboration, communication, and problem-solving [12]. These experiences are foundational for holistic development, equipping students with the competencies required for academic success, social integration, and future professional environments [19, 25]. Furthermore, effective teamwork experiences during schooling can positively shape students' social and cognitive development, preparing them to navigate complex interpersonal relationships and contribute constructively to society [16, 18].

In today's rapidly evolving world, characterized by increased reliance on technology and digital communication, face to face interactions remain irreplaceable for meaningful human connection and development [25]. While technologymediated communication tools have grown prevalent, proficiency in face-to-face communication remains critical for nurturing authentic relationships and maintaining connections with the real world. Thus, we employ EEG hyperscanning techniques to examine and evaluate the neural correlates of team dynamics and cooperation strategies among secondary school students [2]. In hyper-scanning research, it is common to employ baseline measures and condition contrasts rather than separate control groups, as this approach controls for inter-individual variability and maintains ecological validity [1, 13]. Our design follows this practice, comparing cooperative and competitive social contexts within the same participant cohort.

In this report, we present the design of conducting group scenario experiment with secondary school students and preliminary findings from analysis of the team dynamics among individuals working comfortably from EEG and behavioral measurements among during group tasks [2, 10]. By analyzing inter-brain synchrony and communication patterns during collaborative tasks, this study aims to provide actionable insights into enhancing cooperative educational practices that balance technological advancements with interpersonal skill development [2]. This report will present our hypothesis, methodology, and Findings, as well as a discussion about findings of our experiment. Finally, we will also discuss possible courses of action moving forward to improve the quality of group tasks for every student.

2 Objective and Hypotheses

This study aims to investigate how team dynamics and cooperation strategies among secondary school students are influenced by different types of social interaction, such as cooperative and competitive tasks, within four-member teams [7]. By analyzing behavioral and neural data, our research seeks to investigate the effectiveness of team-based learning environments that promote positive relationships, and inclusive participation [13, 19]. The study also examines how personality traits, based on the Myers-Briggs Type Indicator (MBTI), influence group cohesion and interpersonal engagement [26]. So, we can contribute to design effective educational settings with technological insights on the teamwork, particularly for students who may be less socially inclined [8, 10].

We hypothesize that cooperative tasks will result in higher inter-brain synchrony and stronger team cohesion than competitive tasks, with each participant's baseline measures serving as the reference in evaluation [1, 25]. Extroverted individuals are expected to demonstrate more effective communication and engagement, which in turn will lead to improved team dynamics relative to introverted groups [3, 9, 29]. Furthermore, we hypothesize that teams comprising one or two extroverted members will be most effective in cooperative settings due to balanced leadership and participation, by fostering higher levels of engagement, assertiveness, and responsiveness among members [4, 9].

3 Materials and Methods

We obtained ethical approval from relevant Institutional Review Board to ensure compliance with research studies involving minors. We recruited 28 female secondary students, (age 13–16) voluntarily with their parental consents and personal assents. These participants were randomly assigned to seven four-person groups. We employed a within-subject design where each team completed both cooperative and mixed cooperative–competitive tasks [1]. This design controls for inter-individual variability, optimizes statistical power in a small sample, and allows direct comparison between social contexts.

As shown in Fig. 1, each session begins with a baseline and task-practice block, then moves into two 20-min group games (cognitive and social-deduction, ended by 30 s rest epochs and brief surveys), and concludes with a postexperiment questionnaire. The baseline block consists of four tasks for personalised cognitive-affective baselines to interpret the neural and behavioral dynamics observed in the subsequent group-task blocks.

First baseline task is visual stroop test to capture individual's baseline attentional capacity, later to determine the focus levels of playing the group games [21]. Second task is emotional arousal using sequence of affective-image slideshow sequence to establish individual arousal reactivity, providing a reference for engagement fluctuations during gameplay [11]. The third visual spatial awareness test asks the spatial position of the target item in the stimulus image to measures spatial or visual awareness for tracking hidden-role clues in the game [23]. The final task, Prisoner's Dilemma, is a single-round cooperation–defection probe reveals baseline social-strategy preferences that inform interaction styles in social deduction game [24].

The main group tasks involve cognitive game and social deduction Game as shown in Table 1. The first game is to evaluate Cognitive Collaboration where each team tackle—a

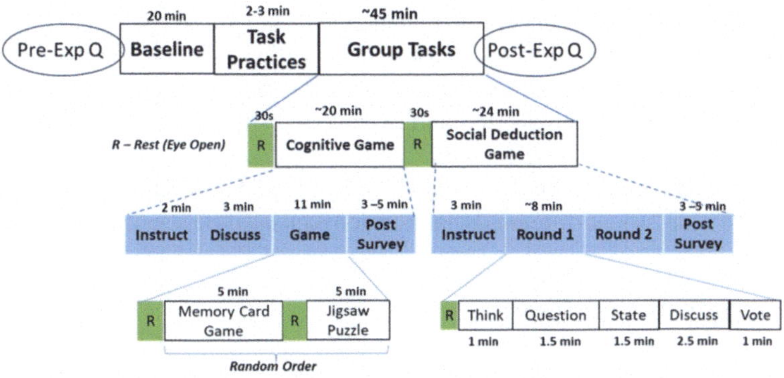

Fig. 1. Experiment Design and Scenario

Table 1. Comparison of Spyfall and Jigsaw Puzzle & Memory Game Tasks

Parameter	Spyfall	Jigsaw & Memory
SkillsIn-volved	Social deduction, Critical thinking, Public Speaking Confidence	Problem Solving, Spatial reasoning, Strategic planning
TeamDy-namics	Information sharing, trust building, verbal communication patterns	Role assignment, Collective goal, Resource sharing
Cooperation Mechanism	Questioning strategies, Collective deduction, Knowledge sharing	Mutual support, Collaborative assembly, Task distribution
Communication needs	Social Cue interpretation, Active verbal interaction, Key words selection	Strategic communication, Role coordination, overall progress updates
Cognitive Demands	Information analysis, Pattern recognition, Deductive reasoning	Visual-spatial Processing, Patience, Systematic organization
Social Interaction	Hybrid competitive cooperative	Cooperative only
Relevant to Students	Inference skills, social awareness, confidence building	Teamworkpromotion,strategic thinking,collaborativeproblem solving

memory-card match and a jigsaw puzzle assembly—within a shared 10-min window. Members negotiate who does what, reassign roles on the fly, and synchronously solve the problems, giving us a controlled lens on real-time division of labour and adaptive cooperation [14]. The second game is adapting the party game called Spyfall, three players receive the same hidden location while a fourth is briefed as the "spy." Over a 5-min questioning round and a 2-min voting phase, teammates must expose the spy before the spy can deduce the location, creating a mixed cooperative–competitive setting that stresses strategic questioning, deception management, and group consensus [15]. Spyfall requires the abstract thinking and social awareness while jigsaw puzzle promotes strategic collaboration. The dual-task design accounts for diverse personality types within teams. Spyfall allows both introverted and extroverted students to contribute meaningfully, while jigsaw puzzles provide opportunities for students with different cognitive strengths and working preferences to excel. So these tasks in our experiment provide comprehensive assessment of team dynamics across varied personality compositions.

4 Data Analysis and Results

In our EEG analysis, raw four-channel data are first segmented, then proceed along two streams. In the first stream, signals are band-pass–filtered (0.5–45 Hz), six band-power features are extracted, and a simple binary classifier assigns each epoch to one of four mental states (attention, arousal, spatial awareness, decision-making); the resulting model (MDL) is saved for later inference. In parallel, band-power features from segmented data are fed into an exploratory data analysis branch for visualization and inter-brain synchrony analysis. Outputs from both streams converge in the statistical-testing, mental-state metrics and synchrony measures are evaluated against our hypotheses to provide the final research findings.

As shown in Fig. 2, cooperative tasks yielded a higher mean phase-locking value ($PLV \approx 0.43 \pm 0.01SEM$) than competitive tasks ($PLV \approx 0.29 \pm 0.01SEM$). A paired-sample t-test on the 10 four-person teams would return $t \approx 9.6$, $p < 0.001$, supporting hypothesis that cooperation promotes greater neural coupling. Teams completed cooperative puzzles faster ($\approx 75s$) than competitive ones ($\approx 95s$).

The scatterplot in Fig. 2 illustrates a negative relationship between interbrain PLV and completion time ($r \approx -0.62$, $p = .03$): greater synchrony aligns with quicker task resolution, linking the neural measure to an observable behavioral benefit. Cooperation seems to foster shared attentional focus and real-time information alignment, reflected in elevated theta synchrony over the Muse $2's$ frontal–temporal sensors. In contrast, competitive framing fragments attention, lowering synchrony and prolonging task execution. These findings mirror real classroom hyperscanning findings and reinforce the value of cooperative pedagogy for both neural engagement and efficiency [13].

Furthermore, Fig. 3 demonstrates that EEG classification performance is lower than subjective task performance across all measures. The EEG performance ranges from 64.44% to 76.89%, while subjective one achieve 82.90% to 86.84%, showing a performance gap of 10–20 percentages. The spatial awareness task show clear visual separation between objective neural measures and subjective behavioral assessments, challenges with EEG signals to predict subtle behaviors.

5 Evaluation and Discussion

As shown in Fig. 4, participants reported positive game-play experience, noting that they clearly understood the importance of effective communication, could adopt adaptive strategies with ease, and felt comfortable throughout the experiment [17]. These

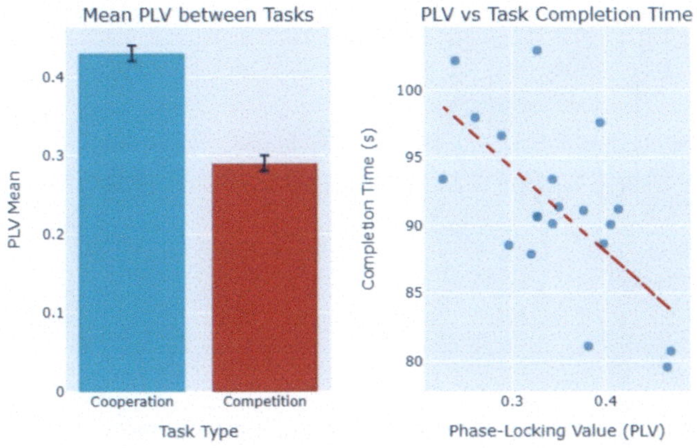

Fig. 2. Comparison of Phase Locked Value (PLV) Analysis and Task completion time.

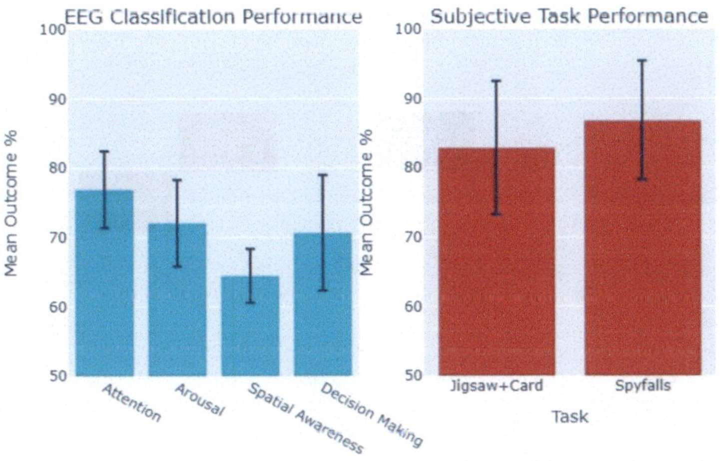

Fig. 3. Comparison between EEG classification and Subjective Performance.

subjective ratings indicate that our game-based training paradigm successfully delivered the intended blend of enjoyment, sustained engagement, and strategic interaction within a game framework [5]. The below plot 4 measures five key aspects of team dynamics and social interaction on a rating scale on playing social deduction game.

However, objective comparisons revealed no definitive relationship between EEG-derived individual mental-state indices like attention/focus and team-based task performance [28]. Elevated attention scores did not uniformly translate into higher outcomes, suggesting that tasks demanding cognitive flexibility—such as Spyfall's rapid role switching—may benefit from alternating, rather than continuous, attentional focus [20]. Likewise, arousal exhibited a bidirectional effect: heightened arousal was detrimental during the socially dynamic Spyfall rounds yet advantageous during memory-

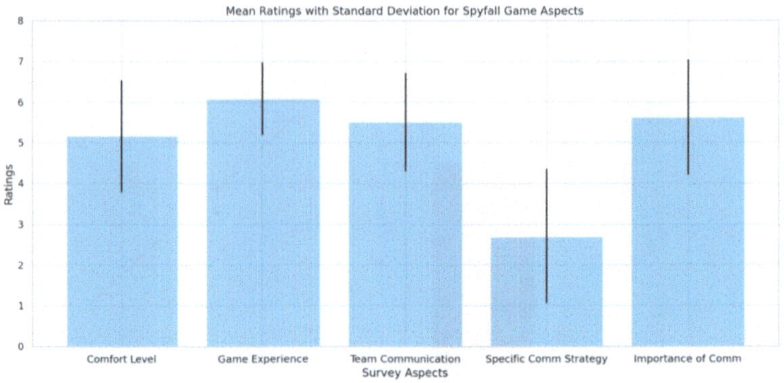

Fig. 4. Summary of survey responses for playing spyfall game

and puzzle-oriented challenges. This divergence underscores that the influence of individual mental states on collaborative success is highly task-dependent and cannot be captured by a single unidimensional metric [28].

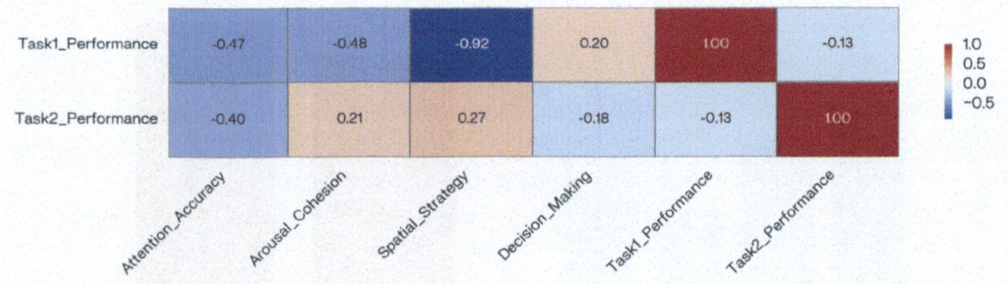

Fig. 5. Correlation between mental health and task performance

The correlation heatmap in Fig. 5 reveals a strong negative relationship between Spatial Awareness and Jigsaw+ card Performance (−0.92) indicating that heightened spatial processing may impair collaborative puzzle-solving tasks. Arousal Cohesion shows opposite effects across task types: hindering structured problem-solving (−0.48) while benefiting social deduction games like Spyfall (0.21). Our systematic and objective approach enables quantification of both individual cognitive contributions and collective neural coordination during teambased activities.

A limitation of this study is the absence of a separate control group performing non-interactive or neutral tasks. This choice was intentional, following established within-subject hyper-scanning designs that leverage baseline tasks and conditions. Future research should include diverse demographic groups to enhance generalizability. Additionally, longitudinal studies examining how neural patterns develop over time could provide deeper insights into the maturation of cooperative behaviors. The data-driven empirical framework in this study offers a foundation for investigating team dynamics across various educational and social contexts, contributing towards efective collaborative learning environments.

6 Conclusion

This study demonstrated the utility of EEG hyperscanning combined with behavioral analysis to investigate team dynamics and cooperation strategies among secondary school students. The neurophysiological data analysis revealed distinct patterns of brain activities that correspond to different cooperation and communication strategies during group tasks. The within-subject design, using baseline and contrasting task conditions as controls, provided robust insights into team neural synchrony while maintaining ecological validity. These findings indicate that neural activities and patterns serve as indicators of effective group communication, with higher synchrony observed during successful collaborative interactions compared to competitive scenarios. These results contribute to the growing body of evidence supporting the neural basis of interpersonal communication and offer practical implications for educational settings. The study's findings can inform evidence-based strategies for enhancing group learning experiences and improving social interaction skills among secondary school students.

References

1. Balconi, M., Vanutelli, M.E.: Cooperation and competition with hyperscanning methods: Review and future application to emotion domain. Front. Comput. Neurosci. **11** (2017). https://doi.org/10.3389/fncom.2017.00086
2. Balconi, M., Vanutelli, M.E.: Eeg hyperscanning and behavioral synchronization during a joint actions. Neuropsychol. Trends **24**, 23–47 (2018). https://doi.org/10.7358/neur-2018-024-balc
3. Barry, B., Stewart, G.L.: Composition, process, and performance in self-managed groups: the role of personality. J. Appl. Psychol. **82**(1), 62–78 (1997). https://doi.org/10.1037/0021-9010.82.1.62
4. Bradley, B.H., Baur, J.E., Banford, C.G., Postlethwaite, B.E.: Team players and collective performance: how agreeableness affects team performance over time. Small Group Res. **44**(6), 680–711 (2013). https://doi.org/10.1177/1046496413507609
5. Connolly, T.M., Boyle, E.A., MacArthur, E., Hainey, T., Boyle, J.M.: A systematic literature review of empirical evidence on computer games and serious games. Comput. Educ. **59**(2), 661–686 (2012). https://doi.org/10.1016/j.compedu.2012.03.004
6. Coronado-Maldonado, I., Benítez-Márquez, M.D.: Emotional intelligence, leadership, and work teams: a hybrid literature review. Heliyon **9**(10), e20356 (2023). https://doi.org/10.1016/j.heliyon.2023.e20356
7. Corrégé, J.B., Michinov, N.: Group size and peer learning: peer discussions in different group size influence learning in a biology exercise performed on a tablet with stylus. Front. Educ. **6** (2021). https://doi.org/10.3389/feduc.2021.733663
8. Council, N.R.: Education for life and work: developing transferable knowledge and skills in the 21st century. The National Academies Press, Washington, DC (2012). https://doi.org/10.17226/13398
9. Curşeu, P.L., Ilies, R., Vîrgă, D., Maricuţoiu, L., Sava, F.A.: Personality characteristics that are valued in teams: not always "more is better"? Int. J. Psychol. **54**(5), 638–649 (2018). https://doi.org/10.1002/ijop.12511
10. Czeszumski, A., et al.: Hyperscanning: a valid method to study neural inter-brain underpinnings of social interaction. Front. Hum. Neurosci. **14** (2020). https://doi.org/10.3389/fnhum.2020.00039

11. Dan-Glauser, E.S., Scherer, K.R.: The geneva affective picture database (gaped): a new 730-picture database focusing on valence and normative significance. Behav. Res. Methods **43**(2), 468–477 (2011). https://doi.org/10.3758/s13428011-0064-1
12. Delaherche, E., Dumas, G., Nadel, J., Chetouani, M.: Automatic measure of imitation during social interaction: a behavioral and hyperscanningeeg benchmark. Pattern Recogn. Lett. **66**, 118–126 (2015). https://doi.org/10.1016/j.patrec.2014.09.002
13. Dikker, S., et al.: Brain-to-brain synchrony tracks real-world dynamic group interactions in the classroom. Curr. Biol. **27**(9), 1375–1380 (2017). https://doi.org/10.1016/j.cub.2017.04.002
14. Fissler, P., Küster, O.C., Laptinskaya, D., Loy, L.S., von Arnim, C.A.F., Kolassa, I.T.: Jigsaw puzzling taps multiple cognitive abilities and is a potential protective factor for cognitive aging. Front. Aging Neurosci. **10** (2018). https://doi.org/10.3389/fnagi.2018.00299
15. Foasberg, N.: Spyfall : information games and scholarly conversation. College Res. Libr. News **78**(4), 218–219 (2017). https://doi.org/10.5860/crln.78.4.9653. https://crln.acrl.org/index.php/crlnews/article/view/9653
16. Hmelo-Silver, C.E.: Problem-based learning: what and how do students learn? Educ. Psychol. Rev. **16**(3), 235–266 (2004). https://doi.org/10.1023/b:edpr.0000034022.16470.f3
17. IJsselsteijn, W., {de Kort}, Y., Poels, K.: The game experience questionnaire. Technische Universiteit Eindhoven (2013)
18. Johnson, D.W., Johnson, R.T.: Making cooperative learning work. Theory Into Pract. **38**(2), 67–73 (1999). https://doi.org/10.1080/00405849909543834
19. Johnson, D.W., Johnson, R.T.: An educational psychology success story: social interdependence theory and cooperative learning. Educ. Res. **38**(5), 365–379 (2009). https://doi.org/10.3102/0013189X09339057
20. Lee, Y., Schumacher, E.H.: Cognitive flexibility in and out of the laboratory: task switching, sustained attention, and mind wandering. Curr. Opin. Behav. Sci. **59**, 101434 (2024). https://doi.org/10.1016/j.cobeha.2024.101434
21. MacLeod, C.M.: Half a century of research on the stroop effect: an integrative review. Psychol. Bull. **109**(2), 163–203 (1991). https://doi.org/10.1037/0033-2909.109.2.163
22. Mathieu, J.E., Heffner, T.S., Goodwin, G.F., Salas, E., Cannon-Bowers, J.A.: The influence of shared mental models on team process and performance. J. Appl. Psychol. **85**(2), 273–283 (2000). https://doi.org/10.1037/00219010.85.2.273
23. Posner, M.I.: Orienting of attention. Q. J. Exp. Psychol. **32**(1), 3–25 (1980). https://doi.org/10.1080/00335558008248231
24. Rand, D.G., Nowak, M.A.: Human cooperation. Trends Cogn. Sci. **17**(8), 413–425 (2013). https://doi.org/10.1016/j.tics.2013.06.003
25. Stieger, S., Lewetz, D., Willinger, D.: Face-to-face more important than digital communication for mental health during the pandemic. Sci. Rep. **13**(1) (2023). https://doi.org/10.1038/s41598-023-34957-4
26. Varvel, T., Adams, S.G., Pridie, S.J., Ulloa, B.C.R.: Team effectiveness and individual myers-briggs personality dimensions. J. Manag. Eng. **20**(4), 141–146 (2004). https://doi.org/10.1061/(ASCE)0742597X(2004)
27. Webb, N.M.: The teacher's role in promoting collaborative dialogue in the classroom. Br. J. Educ. Psychol. **79**(1), 1–28 (2009). https://doi.org/10.1348/000709908X380772
28. Yeung, N., Nystrom, L.E., Aronson, J.A., Cohen, J.D.: Between-task competition and cognitive control in task switching. J. Neurosci. **26**(5), 1429–1438 (2006). https://doi.org/10.1523/jneurosci.3109-05.2006
29. Zhang, J., Yin, K., Li, S.: Leader extraversion and team performance: a moderated mediation model. PLoS ONE **17**(12), e0278769 (2022). https://doi.org/10.1371/journal.pone.0278769

Mathematics and Physics

An Investigation on Diamagnetic Levitation of Graphite Sheets

Tsin Kai Jethro Chai[1](✉), Kevin Goh Ke Ming[2], and Chloe Soh En Qi[3]

[1] Hwa Chong Institution, 661 Bukit Timah Rd, Singapore 269734, Singapore
211383n@student.hci.edu.sg

[2] NUS High School of Math and Science, 20 Clementi Ave 1, Singapore 129957, Singapore
h2110067@nushigh.edu.sg

[3] Raffles Institution, 1 Raffles Institution Ln, Singapore 575954, Singapore
25YSOHE530F@student.ri.edu.sg

Abstract. Graphite is a strongly diamagnetic material, and thin sheets of graphite can levitate stably above a 2 x 2 array of cubic permanent magnets with alternate poles facing upward. Two forms of graphite, pyrolytic graphite and isostatic graphite are observed to have different orientation in stable levitation. We hypothesize that this is due to anisotropy of pyrolytic graphite and show that it is using an analytical model. We also go on to investigate the resonant behaviour, equilibrium position and temperature dependence of pyrolytic and isotropic graphite sheets of various sizes levitating in an array of permanent magnets, making use of novel and accessible methods without the need for advanced equipment. Using these results, we characterize the magnetic susceptibility at room temperature experimentally as well as the change in susceptibility with different temperatures. We conclude that increasing temperature leads to reduced magnitude of diamagnetic susceptibility, increasing sheet size leads to decreased frequency of each mode of oscillation for both types of graphite, and the tilting and vertical modes of pyrolytic and isostatic graphite sheets have very similar frequency experimentally.

Keywords: Diamagnetism · Magnetic Levitation · Graphite Sheet

1 Introduction

While theorised to be mathematically impossible by Earnshaw [1], stable and passive magnetic levitation is in fact, possible due to the existence of materials with negative susceptibility ($\chi < 0$). This phenomenon is also known as diamagnetic levitation. When a magnetic field is applied to diamagnetic materials, a magnetic field in the opposite direction is induced, causing the material to experience repulsive force, hence granting it the ability to levitate stably in a magnetic field. Being passive, this allows for applications in sensing due to its low complexity, energy use, and advantages of physical isolation provided by levitation.

Our main objective is to analyse the behaviour of a graphite sheet levitating above an array of 2×2 magnets with alternating poles pointing up of pyrolytic graphite and isostatic graphite (Fig. 1).

H. Guo et al. (Eds.): IRC-SET 2025, PTT, pp. 385–401, 2026.
https://doi.org/10.1007/978-981-95-9630-0_33

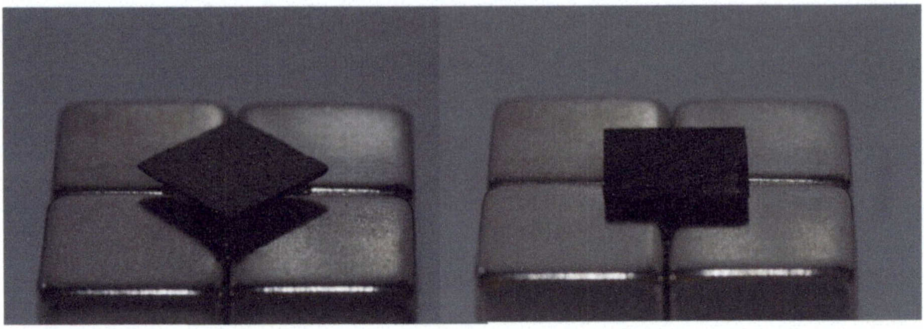

Fig. 1. Image of levitating graphite plates. Figure on the left is the pyrolytic (anisotropic) graphite sheet, levitating with $\theta_z = \frac{\pi}{4}$. . Figure on the right is the isotactic (isotropic) graphite sheet, levitating with $\theta_z = 0$.

2 Materials and Methods

2.1 Theoretical Formulations

To obtain the governing equations for the system, the gradient of magnetic equivalent of Poisson's equation is used to define the magnetic potential energy density, which is then integrated across the graphite sheet to obtain its magnetic potential energy. Fully analytical equations describing the magnetic field (Appendix A) around the magnetic array were derived and then used to calculate the magnetic potential energy and subsequently the equilibrium positions of the graphite sheet and oscillation frequencies and modes. Detailed derivation can be found in Appendix B (Fig. 2).

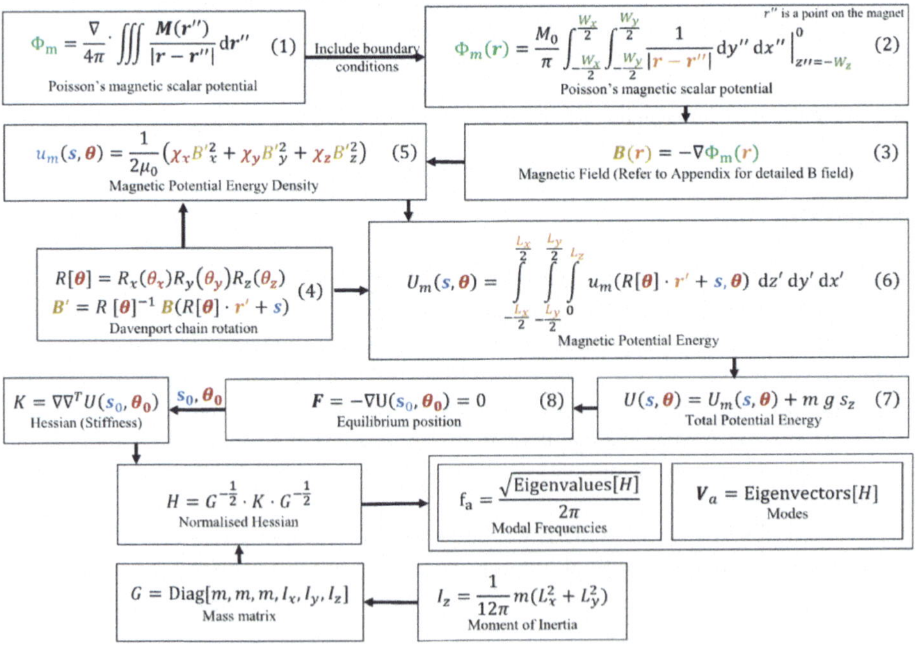

Fig. 2. Flowchart of equations used in theory model

2.2 Materials

5 sheets of 50 × 40 × 3 mm of isostatic pressed graphite sheets were obtained from Lixada mall on Lazada. 10 × 10 × 0.35 mm and 5.3 × 5.2 × 0.35 mm pyrolytic graphite sheets were obtained from Dr Chen Xianfeng (A*STAR). 10 10 × 10 × 10 mm N50 Neodymium magnets with NiCuNi plating were obtained from SG Magnetics. 30 × 30 × 4mm magnets with NiCuNi plating were obtained from Supreme Magnets®. A FLUKE Ti450 digital thermal imager, which had an error of $\pm 2° C$ [2] was used to track the temperature of the graphite during heating experiments. A high-power torchlight was used to heat the graphite sheet during levitation experiments to determine susceptibility change with temperature. Wolfram Mathematica and Octave GUI were used to visualise our mathematical models. The cameras used for these experiments are a Samsung SM-N986B (main camera) for video and a Pentax K5-IIs with DFA 100 mm F2.8 Macro (WR) for static height measurements and all other photos.

2.3 Characterization of Neodymium Magnets

To characterize the magnets used, magnetic field strength from a singular cube magnet was measured with a Gauss meter with varying distance between the probe and the magnet. The magnetization was then fit using equation (3) to the magnetic field strengths measured. Magnetisation was determined to be 1.314 T. Fit can be found in Appendix C.

2.4 Preliminary Characterization and Preparation of Graphite Sheets

Pyrolytic graphite and isostatic graphite sheets were shaped using a pen knife and thickness was controlled using sandpaper of grits 80-, 200-, 400-, 1200-. The lengths L_x, L_y of each graphite sheet were then measured using digital vernier callipers, which was also used to ensure even thickness across the sheet. Their thickness L_z was measured using a micrometre screw gauge. The mass m of the sheets were then measured with a 4 d.p. (in grams) electronic weighing scale. Resistivity was determined using the Van der Pauw method [3].

2.5 Experimental Setup

A 2 × 2 mm array of the 10 mm cube magnets with alternating North-South poles facing up was created. Non-magnetic tweezers were used to place the graphite plates in the centre where they levitate (Fig. 3).

Fig. 3. (Left) tracking tilting motion in tracker. (Right) experimental setup to determine levitation force to height.

2.6 Change in Levitation Height with Levitation Force

To determine magnetic susceptibility, the levitation height s_z of each of the sheets was measured by taking a picture of the side view of the setup using a camera and converting the pixels between the bottom of the sheet and the magnets to meters. The width of the graphite sheet is used as a reference point for focus and calibration. Magnetic susceptibility was then numerically fit. However, to improve reliability, plastic weights of known mass were added onto the graphite sheet, with the subsequent levitation heights s_z and total weight W were recorded similarly. The net vertical force was then computed for each case using equation (8) and its gradient descent minimized using simulated annealing with a, b-axis susceptibility (χ_x, χ_y) and c-axis susceptibility (χ_z) as parameters.

2.7 Change in Levitation Height with Temperature

7.5 mm isostatic graphite sheet was levitated above the magnetic array and heated with a high-power torch. The temperature of graphite sheet was quantified and recorded down using a thermal imager. The phone camera recording was used to track the motion of the sheet as it cooled.

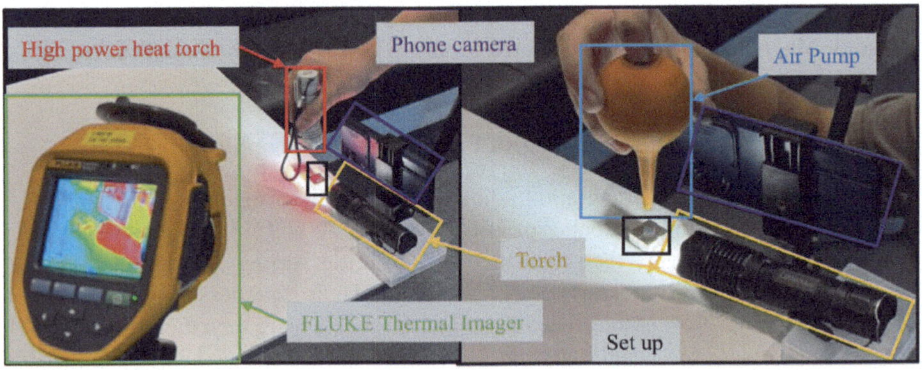

Fig. 4. (Left) Experimental setup to track levitation height against temperature change. (Right) Experimental setup to track tilting and vertical oscillations.

2.8 Motion of Graphite Sheet After Displacement

The levitating graphite sheet was displaced arbitrarily using an air pump (Fig. 4 (Right)), and its side view and top view motion was recorded separately with a phone camera. Its centre was then tracked after the oscillation stabilizes. The motion in the different degrees of freedom was then Fast Fourier Transformed (FFT) to determine frequencies.

3 Results and Discussion

3.1 Equilibrium Position of the Graphite Sheets

The equilibrium orientation for the isostatic graphite and pyrolytic graphite differs due to the anisotropy of diamagnetic susceptibility in pyrolytic graphite. Since χ_z is higher than $\chi_{x,y}$ for pyrolytic graphite ($\frac{\chi_z}{\chi_{x,y}} > 5$), it is more strongly in the regions directly above the faces of the magnets where B_z is higher, and less strongly repelled by the regions between the magnets where $B_{x,y}$ is higher, even though $|\boldsymbol{B}|$ is higher there. This can be visualized from the magnetic potential energy density plots, where the energy wells are sharply diagonal to the cube magnets for the anisotropic case (Fig. 5 (top right)).

In the case of isostatic compressed graphite, diamagnetic susceptibility is isotropic, hence Eq. (5) simplifies to only $u_m = \frac{1}{2\mu_0}\chi_{x,y,z}|\boldsymbol{B}|^2$, and the individual $\boldsymbol{B}$ field components do not matter. Hence, the sheet avoids regions with higher $|\boldsymbol{B}|$, as this would lead directly to higher magnetic potential energy density, which happen to be at the regions above the intersections between magnets (Fig. 5 (top left)). This is more clearly seen in Fig. 10 of Appendix G, where we can see that the magnetic potential energy is minimum when $\theta_z = \frac{\pi}{4}$ for the anisotropic graphite, and $\theta_z = 0$ for the isotropic graphite. Hence, the graphite would orientate itself as shown in Fig. 5 (bottom left for isotropic and bottom right for anisotropic).

For example, for a 7.5 mm long graphite sheet, the equilibrium position was mathematically determined to change from $\theta_z = 0$ to $\theta_z = \frac{\pi}{4}$ at $\frac{\chi_z}{\chi_{x,y}} > 1.203$.

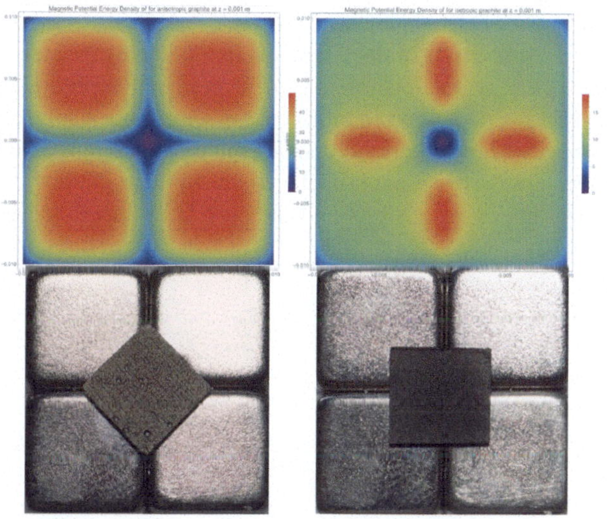

Fig. 5. Magnetic potential (top) and observed orientation (bottom) of isostatic (left) and pyrolytic (right) graphite.

3.2 Equilibrium Position of the Graphite Sheets

As can be seen from Fig. 9 of Appendix D, levitation force is higher as levitation height decreases, due to the increase in magnetic field strength closer to the magnet, leading to stronger repulsion. The susceptibility values attained from the fit: $\chi_{x,y} = -6.92 \times 10^{-5}$, $\chi_z = -4.11 \times 10^{-4}$ are within acceptable ranges provided by literature for this specific pyrolytic graphite we procured [4]. This is also the case for isotropic graphite: $\chi_{x,y,z} = -1.379 \times 10^{-4}$.

However, significant deviation can also be observed. This could be explained by the unevenness in the surfaces of the pyrolytic graphite which we were unable to smoothen due to its strong crystalline structure, as well as difference in magnetisation across the magnets led to uneven levitation of the pyrolytic graphite. This was further enhanced by the imprecision of equipment used to determine the levitation height such as the lack of focus around the edges of the graphite sheet. Moreover, larger increments in masses were needed to obtain observable changes in levitation height, leading to limited data points for the gradient descent. Hence, the fitted $\chi_{x,y}$ and χ_z may not be the most accurate one, accounting for the deviations.

3.3 Resonant Modes and Frequencies

From the FFT graphs (Fig. 6), the predominant trend is two peaks at higher frequencies (16.3 Hz,18.4 Hz) is visible in the side view but disappears in the top view. These peaks should correspond to the tip/tilt vertical modes, since the Eigenvectors of these two modes have either no (for vertical) or very low (for tip/tilt) lateral components. From the eigenvectors, we also infer that highest frequency (18.4 Hz) corresponds to the vertical mode, and the second highest, to the tip/tilt mode (16.3 Hz). There are two smaller peaks (2.2 Hz, 3.4 Hz). The peak that is visible in both views is most likely the x-y lateral mode, since it also has a vertical component, whereas the twist does not and is most likely the other peak. Due to the amplitude of the lateral oscillation being greater, it is more visible even in FFT of the z component, leading to its peak being quite prominent. Furthermore, the coupling between the modes can be more visibly seen in the displacement-time graph for the z axis because of how small the displacements are compared to the amplitudes of the lateral mode. The x component displacement-time graph, however, seems much cleaner as only two modes are visible, and they are both uncoupled with one another (Fig. 7).

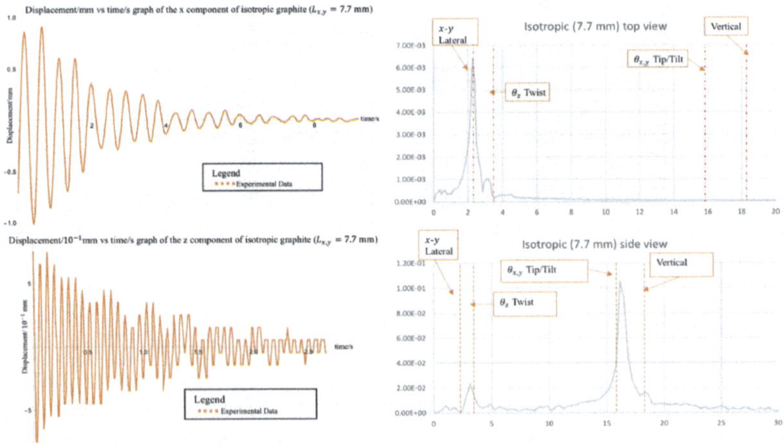

Fig. 6. (Top Left) Example of displacement vs time graph obtained from tracker for x component. (Top Right) FFT analysis of data from left, with theoretical frequency peaks marked in orange. (Bottom Left) Example of displacement vs time graph obtained from tracker for z component. (Bottom Right) FFT analysis of data from left, with theoretical frequency peaks marked in orange. Both are for the isotropic graphite at $L_{x,y} = 7.7$ mm

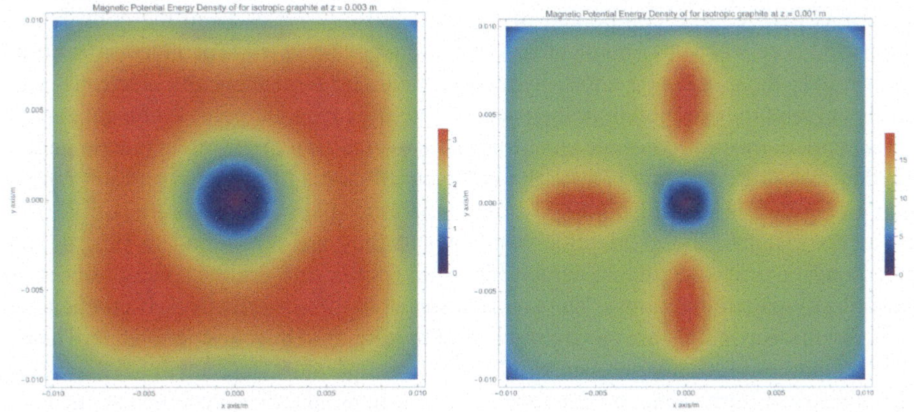

Fig. 7. (Left) and (Right) Magnetic potential energy density for isotropic (Left) and anisotropic (Right) graphite.

Generally, the observed frequencies derived through FFT correlate very closely with the expected values from theory, particularly for the isotropic case. Both theory and experimental results show both twist and tilt oscillation frequency decrease with increasing sheet size that is roughly linear (Fig. 13). However, upon further theoretical investigation, we find that the decrease in frequencies plateaus with increasing length (Fig. 16, 17). This behaviour corresponds to how the change in levitation height also decreases at a decreasing rate, similar with the change in frequencies. Comparing 7.1 and 7.2, we see that the variation in magnetic potential energy greatly decreases at higher levitation heights, hence leading to a decrease in stiffness value (since the increasing mass and moment of inertia would negate the increase in magnetic potential energy that the sheets with higher lengths receive). Hence, frequency would decrease. Past a certain length, magnetic potential energy starts to decrease with increasing x or y, which leads to the levitation height decreasing with increasing length. Since total magnetic potential energy also decreases, and mass and moment of inertia continues to increase, stiffness decreases. Hence, even in the phase where levitation height decreases with increasing length, frequency decreases. However, this trend could not be captured experimentally due to the low data range and small deviations in mass and thickness.

However, as could be seen from the Fig. 13 as well as the FFT graphs, the theory does not match well for the frequencies of pyrolytic (anisotropic) graphite. This could be due to inaccuracies in the susceptibility values fitted using the levitation force to height experiment as described in Sect. 2.2.

3.4 Effect of Temperature

Based on established literature, the magnitude of diamagnetic susceptibility of pyrolytic graphite decreases with increasing temperature, which would lead to decreasing levitation height.

From Fig. 8, we can see a clear linear relation between susceptibility and temperature, where increasing temperature leads to lower diamagnetism. This agrees with theoretical estimations [5]. A large component of the susceptibility arises from the contribution from paired core electrons and electrons in the valence band [6]. The electrons in fully filled orbitals have opposite spins, hence the magnetic field from the electrons' spin cancels out, and they do not align themselves with the magnetic field. When a magnetic field is applied, the motion of the electrons alters to create an opposing magnetic field in accordance with Lenz's law [7]. As the conduction band and valence band in graphite touch each other at certain points [7], as temperature rises, electrons transition from the valence band to the conduction band, where they are not bound to orbitals and hence contribute less to diamagnetism [8]. This results in a net decrease in observed susceptibility with temperature. The contribution from the core electrons is independent of temperature since they are too tightly bound to jump into the conduction band, hence at low temperatures, diamagnetism would be less temperature dependent [5]. However, due to logistical limitations, we are unable to investigate the diamagnetism at such low temperatures.

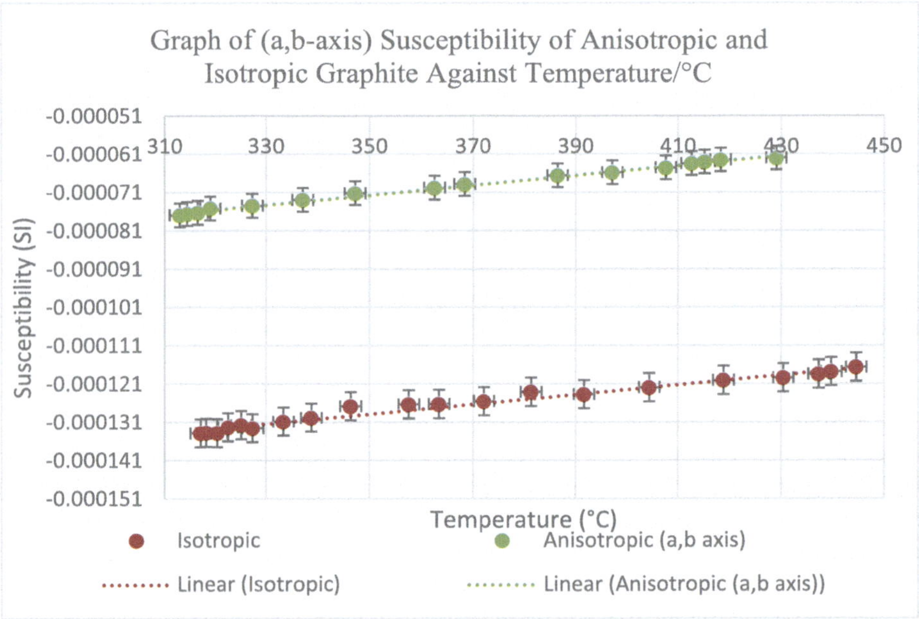

Fig. 8. Graph of susceptibility of anisotropic and isotropic graphite against temperature.

4 Conclusion

In conclusion, we have explored the resonant behaviour of isostatic graphite and pyrolytic graphite sheets levitating in a permanent magnetic trap. Initially, we explained the differing orientation of pyrolytic and isostatic graphite during stable levitation using magnetic field and magnetic potential plots obtained from our analytical model. Using FFTs, we characterize various modes of oscillation of different sizes of graphite plates and compare them to analytical theory values. Using weights to plot levitation force to levitation height for different sizes of sheet, we characterized the diamagnetic susceptibility of both types of graphite. By tracking levitation height change as heated graphite sheets cool to room temperature, we measure the change in magnetic susceptibility with temperature. Finally, we compare the changes in each mode of oscillation among the two types of graphite and differing sheet sizes, finding that tilting modes correspond well between both types of graphite, and larger sheet size leads to decreased oscillation frequency, approaching a linear relationship as sheet sizes approach the maximum size allowing stable levitation. Moreover, we demonstrate the use of accessible tools like a mobile phone camera and the Physlets Tracker app to track small scale oscillations and movement (on the scale of 0.1–1.0mm) and show that a reasonable level of precision and accuracy can be achieved, especially after processing the oscillation data with FFT analysis.

In the future, we commit to studying theoretically the damping forces, as well as the incorporation of other geometries and larger arrays to better understand its movement in different magnetic fields. The stiffness and other properties of a diamagnetic spring configured with alternating magnet and graphite sheets also present themselves as possible areas of future investigation. We also plan to study the rotational and translational motion of the graphite when a localized spot on the graphite is heated.

Appendices

Appendix A: Detailed Mathematical Descriptions

We first define the coordinate systems as in Fig. 9.

The magnetic field generated by the neodymium magnets is described by Gauss's law for magnetism, respectively:

$$\nabla \cdot \mathrm{B} = \nabla \cdot \mu_0 H = 0 \quad (1)$$

where H is the magnetic field intensity, μ_0 is the permeability of free space and B is the magnetic field vector.

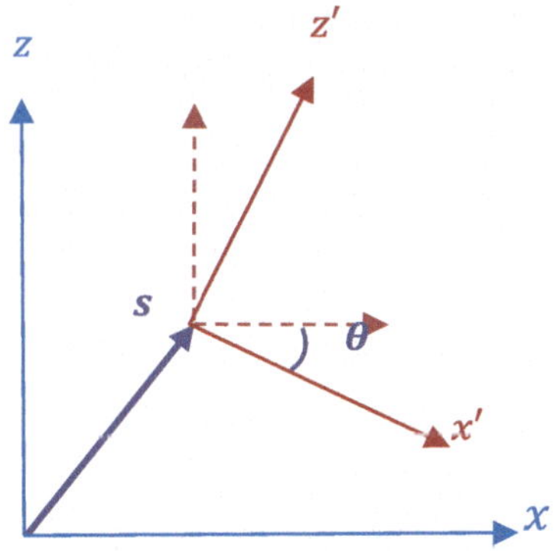

Fig. 9. Coordinate system.

The general solution to this for magnetostatics is well known:

$$\mathrm{B} = \mu_0 H = -\mu_0 \nabla \Phi_m \tag{2}$$

We can then solve it numerically within our boundary conditions as laid out in our experimental setup. However, due to the computational power required to solve such complex equations, this would be extremely and unnecessarily time consuming.

We hence utilise the magnetic equivalent of Poisson's equation:

$$\Phi_{\mathrm{m}} = \frac{\nabla}{4\pi} \cdot \iiint \frac{M\left(r''\right)}{\left|r - r''\right|} \mathrm{d}r'' \tag{3}$$

where Φ_{m} is the magnetic scalar potential, M is the magnetisation vector of the magnet, r is the observation point in Cartesian coordinates of the lab frame and r'' is a point on the magnet.

Substituting the dimensions and position of the magnet, as well as the vertical direction of the poles, Eq. (1) simplifies to

$$\Phi_m(x, y, z) = \frac{M_0}{\pi} \int_{-\frac{W_x}{2}}^{\frac{W_x}{2}} \int_{-\frac{W_y}{2}}^{\frac{W_y}{2}} \frac{1}{\left|r - r''\right|} \mathrm{d}y'' \mathrm{d}x'' \Big|_{z''=-W_z}^{0} \tag{4}$$

where W_x, $W_y and W_z$ are the respective lengths of the magnet in x, y, z coordinates. By taking the negative gradient of Φ_m as laid out in (10), we can cancel out the respective integrals and the equation for magnetic field simplifies to the equations found below:

$$B_x = -\frac{\delta \Phi_m}{\delta x} = -\frac{M_0}{4\pi} \tanh^{-1}\left(\frac{y - y''}{\sqrt{(x + x'')^2 + (y + y'')^2 + (z + z'')^2}}\right) \Big|_{x''=\frac{-W_x}{2}}^{\frac{W_x}{2}} \Big|_{y''=\frac{-W_y}{2}}^{\frac{W_y}{2}} \Big|_{z''=-W_z}^{0} \tag{5}$$

$$B_y = -\frac{\delta \Phi_m}{\delta y} = -\frac{M_0}{4\pi} \tanh^{-1}\left(\frac{x - x''}{\sqrt{(x + x'')^2 + (y + y'')^2 + (z + z'')^2}}\right) \Big|_{x''=\frac{-W_x}{2}}^{\frac{W_x}{2}} \Big|_{y''=\frac{-W_y}{2}}^{\frac{W_y}{2}} \Big|_{z''=-W_z}^{0} \tag{6}$$

$$B_z = -\frac{\delta \Phi_m}{\delta z} = -\frac{M_0}{4\pi} \tan^{-1}\left(\frac{(y - y'')(x - x'')}{(z - z'')\sqrt{(x + x'')^2 + (y + y'')^2 + (z + z'')^2}}\right) \Big|_{x''=\frac{-W_x}{2}}^{\frac{W_x}{2}} \Big|_{y''=\frac{-W_y}{2}}^{\frac{W_y}{2}} \Big|_{z''=-W_z}^{0} \tag{7}$$

where $\boldsymbol{B_x}, \boldsymbol{B_y}, \boldsymbol{B_z}$ are the $\boldsymbol{x}, \boldsymbol{y}, \boldsymbol{z}$ components of the magnetic field $\boldsymbol{B}$ respectively, from one cube magnet. The positions of each magnet are then used to translate $\boldsymbol{W_x}, \boldsymbol{W_y}, \boldsymbol{W_z}$ by that amount and the total $\boldsymbol{B}$ field is obtained by adding up the contributions of each component of each magnet respectively.

Using the definition of the magnetic potential energy density, we get:

$$u_m(x, y, z) = -\frac{\chi_v}{2\mu_0}\left(B_x^2 + B_y^2 + B_z^2\right) \tag{8}$$

To obtain total magnetic potential energy, this is integrated over the volume of the graphite sheet, and to define the Euler angles in the three dimensions, intrinsic rotations were done using the Davenport Rotation Matrices:

$$R_x = \begin{bmatrix} 1 & 0 & 0 \\ 0 & \cos\theta_x & -\sin\theta_x \\ 0 & \sin\theta_x & \cos\theta_x \end{bmatrix}$$

$$R_y = \begin{bmatrix} \cos\theta_y & 0 & \sin\theta_y \\ 0 & 1 & 0 \\ -\sin\theta_y & 0 & \cos\theta_y \end{bmatrix}$$

$$R_z = \begin{bmatrix} \cos\theta_z & -\sin\theta_z & 0 \\ \sin\theta_z & \cos\theta_z & 0 \\ 0 & 0 & 1 \end{bmatrix}$$

$$R = \begin{bmatrix} \cos\theta_y\cos\theta_z & \cos\theta_z\sin\theta_x\sin\theta_y - \cos\theta_x\sin\theta_z & \cos\theta_x\cos\theta_z\sin\theta_y + \sin\theta_x\sin\theta_z \\ \cos\theta_y\sin\theta_z & \cos\theta_x\cos\theta_z + \sin\theta_x\sin\theta_y\sin\theta_z & \cos\theta_z\sin\theta_x + \cos\theta_x\sin\theta_y\sin\theta_z \\ -\sin\theta_y & \cos\theta_y\sin\theta_x & \cos\theta_x\cos\theta_y \end{bmatrix}$$

$$R\prime = \begin{bmatrix} \cos\theta_y\cos\theta_z & \cos\theta_y\sin\theta_z & -\sin\theta_y \\ \cos\theta_z\sin\theta_x\sin\theta_y - \cos\theta_x\sin\theta_z & \cos\theta_x\cos\theta_z + \sin\theta_x\sin\theta_y\sin\theta_z & \cos\theta_y\sin\theta_x \\ \cos\theta_x\cos\theta_z\sin\theta_y + \sin\theta_x\sin\theta_z & \cos\theta_z\sin\theta_x + \cos\theta_x\sin\theta_y\sin\theta_z & \cos\theta_x\cos\theta_y \end{bmatrix}$$

With these operations, we obtain:

$$U_m\left(s_x, s_y, s_z, \theta_x, \theta_y, \theta_z\right) = \int_{-\frac{L_x}{2}}^{\frac{L_x}{2}} \int_{-\frac{L_y}{2}}^{\frac{L_y}{2}} \int_{0}^{L_z} u_m(x, y, z)\, \mathrm{d}z'\, \mathrm{d}y'\, \mathrm{d}x' \tag{9}$$

$$x = \left(z' \sin\theta_x - y' \cos\theta_x\right) \sin\theta_z + \left(x' \cos\theta_y + \left(z' \cos\theta_x + y \sin\theta_x\right) \sin\theta_y\right) \cos\theta_z + s_x$$

$$y = \cos\theta_z\left(y' \cos\theta_x - z' \sin\theta_x\right) + x' \cos\theta_y \sin\theta_z + \left(z' \cos\theta_x + y' \sin\theta_x\right) \sin\theta_y \sin\theta_z$$

$$z = \left(z' \cos\theta_x + y' \sin\theta_x\right) \cos\theta_y - x' \sin\theta_y + s_z$$

where L_x, L_y and L_z are the respective lengths of the graphite plate, s_x, s_y and s_z are the respective lateral displacements of the centre of the graphite plate and θ_x, θ_y and θ_z are the rotations in the respective axes.

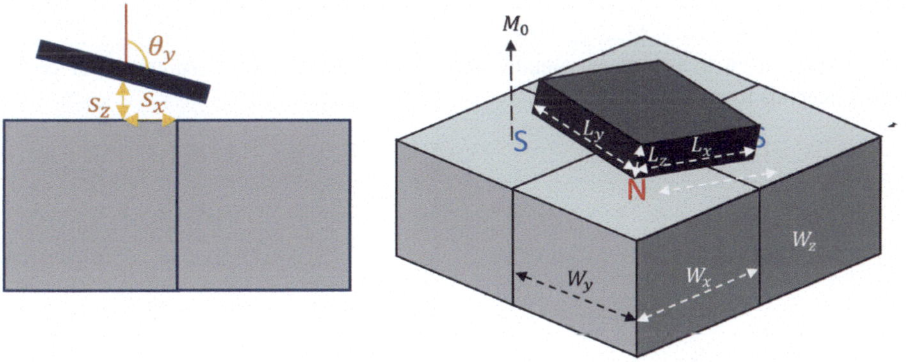

Fig. 10. Diagram relating variables to their referenced lengths.

However, to account for anisotropy, the x, y, z component of susceptibility χ_v is multiplied by the respective $\boldsymbol{B}$ field component. The components of the $\boldsymbol{B}$ field would also have to be reverse rotated back into the lab frame using the transposed rotation matrix, obtaining:

$$u_m\left(x, y, z, \theta_x, \theta_y, \theta_z\right) = -\frac{1}{2\mu_0}\left(\chi_x B_x^{2'} + \chi_y B_y^{2'} + \chi_z B_z^{2'}\right) \tag{10}$$

$$B_{x'} = B_x \cos\theta_y \cos\theta_Z - B_z \sin\theta_y + \mathrm{B_y} \cos\theta_y \sin\theta_z$$

$$B_{y'} = B_z \cos\theta_y \sin\theta_x + B_x\left(\cos\theta_z \sin\theta_x \sin\theta_y - \cos\theta_x \sin\theta_z\right) + B_y\left(\cos\theta_x \cos\theta_z + \sin\theta_x \sin\theta_y \sin\theta_z\right)$$

$$B_{z'} = B_z \cos\theta_X \cos\theta_y + B_x\left(\cos\theta_x \cos\theta_z \sin\theta_y + \sin\theta_x \sin\theta_z\right) + B_y\left(-\cos\theta_Z \sin\theta_x + \cos\theta_x \sin\theta_y \sin\theta_z\right)$$

Appendix B: Magnetic Field Fit

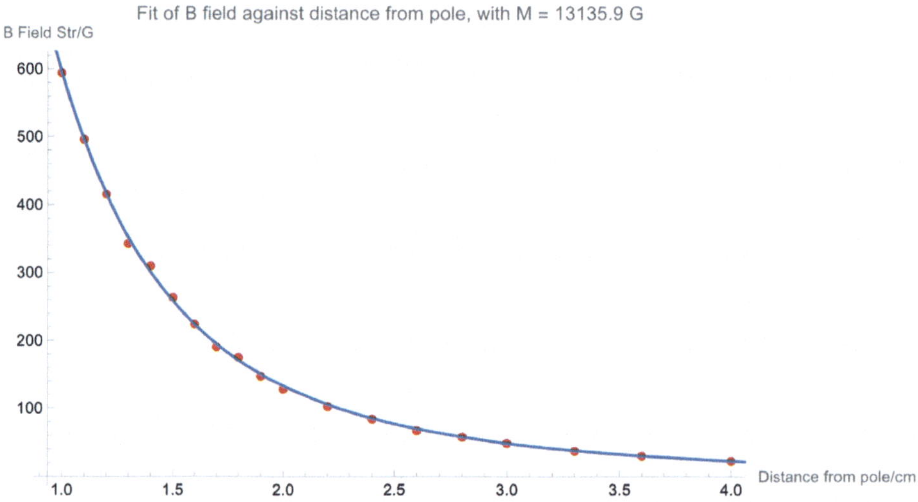

Fig. 11. Magnetic field fits experimentally derived field magnitude measurements at certain distance intervals from magnets.

Appendix C: Characteristics

Table 1. Graphite characteristics.

Index	L_x/mm	L_y/mm	L_z/mm	m/mg	s_z/mm	$\chi_{x,y}(SI)$	$\chi_z(\text{SI})$
Isostatic							
1	3.8	3.8	0.272	6.6	0.432	-1.379×10^{-4}	
2	5.3	5.3	0.361	17.3	0.482		
3	5.8	5.8	0.242	13.6	0.586		
4	7.5	7.5	0.395	37.8	0.598		
5	10.0	10.0	0.312	52.3	Unstable		
Pyrolytic							
6	5.3	5.3	0.351	19.7	0.684	-6.92×10^{-5}	-41.1×10^{-5}
7	7.5	7.5	0.458	39.2	0.875		
8	10.0	10.0	0.458	71.5	0.925		

Appendix D: Levitation Force to Levitation Height Fit

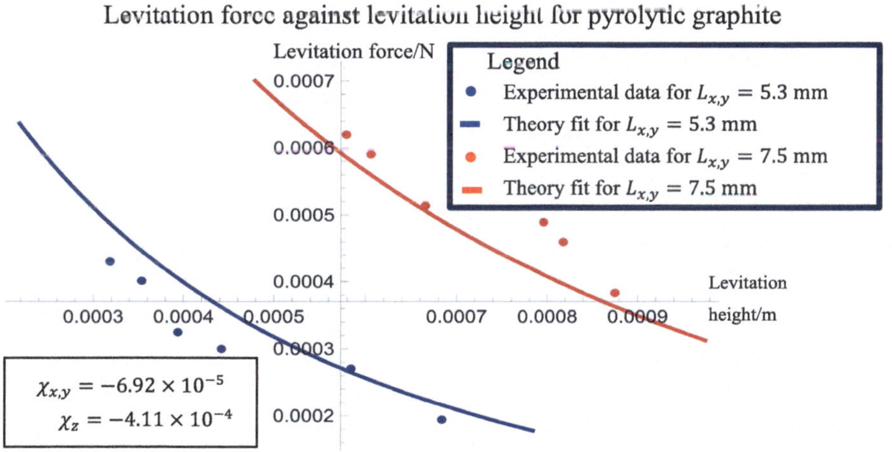

Fig. 12. Fit of levitation force against levitation height for susceptibility for anisotropic graphite.

Appendix E: Levitation Force to Levitation Height Fit.

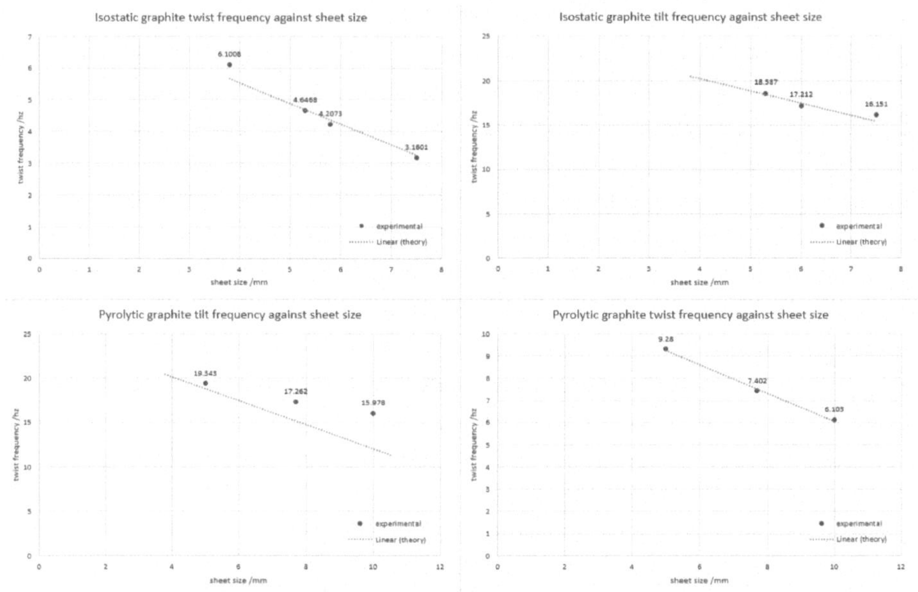

Fig. 13. Tilt and twist frequencies against sheet size for pyrolytic and isostatic graphite.

Appendix F: Magnetic Field Plots

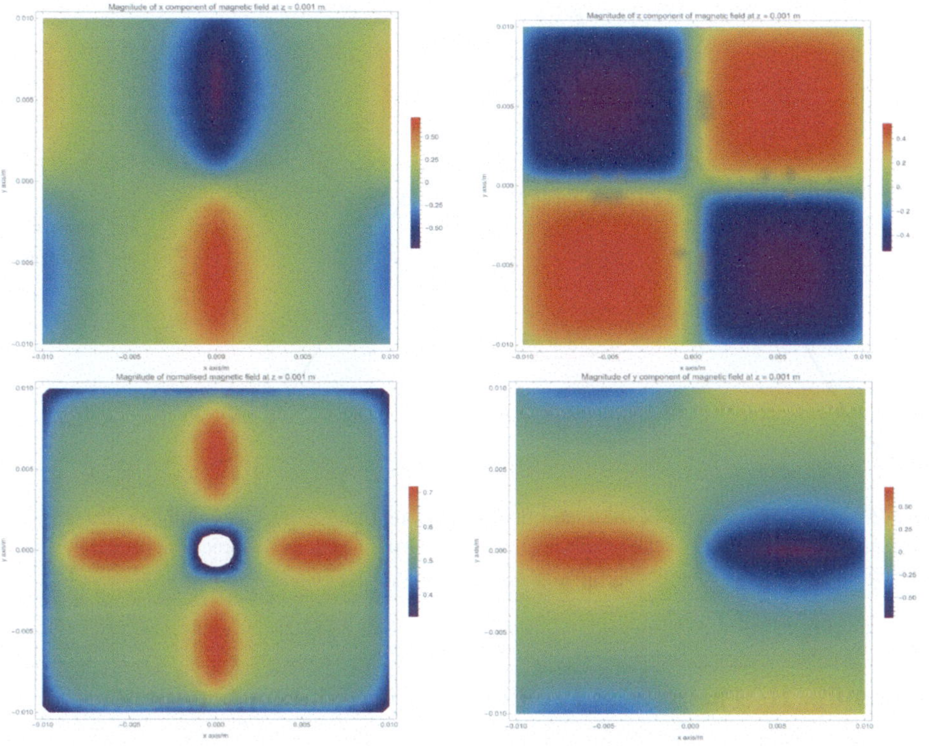

Fig. 14. Magnetic field magnitude plots. On the top left is the magnitude of the x component of the magnetic field with respect to x and y axes. On the top right is the magnitude of the z component of the magnetic field with respect to x and y axes. On the bottom right is the magnitude of the y component of the magnetic field with respect to x and y axes. On the bottom left is the normalised magnitude of the magnetic field with respect to x and y axes.

Appendix G: Stability of Twist Angles

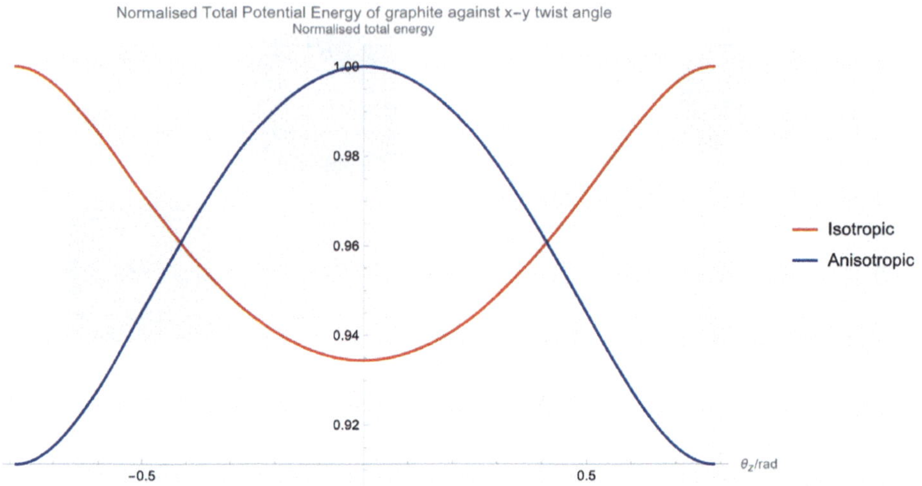

Fig. 15. Normalised graph of the theoretical total potential energy of graphite against the x-y twist angle θ_x in radians of the graphite sheet for isotropic and anisotropic graphite. The graph was normalised by dividng each graph by its max value.

Appendix H: Theoretical Prediction of Oscillation Frequencies

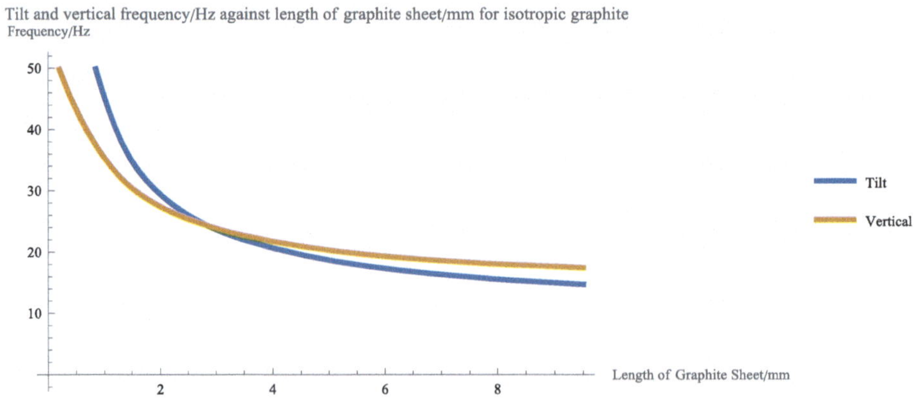

Fig. 16. Theoretical prediction of tilt and vertical frequencies against length of graphite sheet for isotropic graphite using susceptibility values as laid out in Table 1.

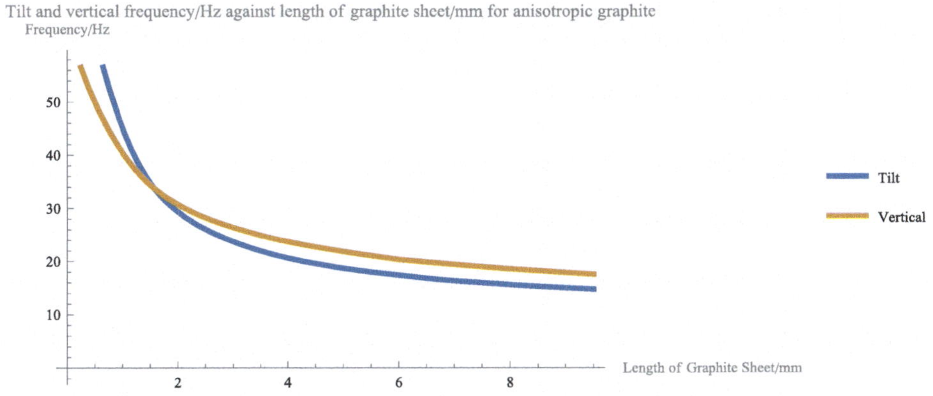

Fig. 17. Theoretical prediction of tilt and vertical frequencies against length of graphite sheet for anisotropic graphite using susceptibility values as laid out in Table 1.

References

1. Earnshaw, S.: On the nature of the molecular forces which regulate the constitution of the Luminiferous Ether. On The Nature of the Molecular Forces Which Regulate the Constitution of the Luminiferous Ether, 7, 97 (1848). https://ui.adsabs.harvard.edu/abs/1848TCaPS
2. Fluke. "Fluke Ti450 Infrared Camera | Fluke," Fluke.com. 05 Apr 2022. https://www.fluke.com/en-us/product/thermal-cameras/ti450?srsltid=AfmBOop1qrsIEqY1TjELvAFNOohD4OtrPB75WEMgH7n_81ATQil5ndc2
3. Katharina. Van-der-Pauw measurement-Linseis. Linseis (11 Dec 2024). https://www.linseis.com/en/methods/van-der-pauw-measurement/
4. Tian, S., et al.: Feedback cooling of an insulating high-Q diamagnetically levitated plate. Appl. Phys. Lett. **124**(12) (2024). https://doi.org/10.1063/5.0189219
5. Banerjee, D., Bhattacharya, R.: Temperature variation of diamagnetic susceptibility of graphite. In: Indian Association for the Cultivation of Science, Indian J. Phys. (Vol. 64, Issue 4, pp. 316–320) (1990). https://core.ac.uk/download/pdf/156953823.pdf
6. Adams, E.N.: Magnetic susceptibility of a diamagnetic electron Gas—The role of small effective electron mass. Phys. Rev. **89**(3), 633–648 (1953). https://doi.org/10.1103/physrev.89.633
7. Peres, N.M.R., Castro, E.V.: Algebraic solution of a graphene layer in transverse electric and perpendicular magnetic fields. J. Phys.: Condens. Matter **19**(40), 406231 (2007). https://doi.org/10.1088/0953-8984/19/40/406231
8. Magnetic susceptibility. Questions and Answers in MRI (n.d.). https://mriquestions.com/causes-of-susceptibility.html

Paper Folding: Extensions of Gallivan's Formula

Kabir Srivastav(✉)

NUS High School of Math and Science, Singapore, Singapore
h2210077@nushigh.edu.sg

Abstract. In 2002, Britney Gallivan derived a formula estimating the length of paper lost when it is folded repeatedly in half along the same side. This placed a theoretical upper bound on the number of times a piece of paper can be folded in half, depending on its thickness and length. In this paper, we generalize and refine Gallivan's model. We first derive the corresponding lost-length formulas for alternate-side folding and apply them to both square and rectangular paper. We then develop improved models of folding that address limitations in Gallivan's core assumption—that the fold behaves as a perfect semicircle with inner radius equal to the paper's thickness—and derive new bounds accordingly. Finally, we introduce a constraint involving only the outermost layer, leading to a sharper folding limit. Potential applications and limitations of these theoretical results are discussed.

Keywords: folding · paper · lost length · upper bound · optimisation

1 Background and Introduction

Folding materials in half is a familiar task in everyday life. It is commonly believed that a piece of paper cannot be folded in half more than 7 times. In 2002, Britney Gallivan demonstrated that this is not the case, [5] and that an upper bound for the number of folds can be determined by considering the paper's thickness t and the geometric accumulation of lost length at each fold. She derived a formula for the length of paper lost to folding after n folds:

$$L = \frac{\pi t}{6}(2^n + 4)(2^n - 1)$$

In this paper, Gallivan's formula was extended to alternate-side folding, for both square and rectangular paper. Additionally, two areas of improvement were explored to derive more accurate formulas for the maximum number of folds.

1.1 Related Work

While this paper investigates the upper limit in folding only based on the paper's folding geometry, several studies have explored the folding limit by only considering the maximum force the average person's fingers can apply [2,4,7]. Paper [4]

H. Guo et al. (Eds.): IRC-SET 2025, PTT, pp. 402–411, 2026.
https://doi.org/10.1007/978-981-95-9630-0_34

covers a general mechanical analysis of the behaviour of creased sheets, including the relationship of the folding angle with the force required. Papers [2] and [7] attempted to calculate the maximum number of folds the average person can make on a piece of paper, obtaining a result of about **6 folds** for same-side folding [2] and about **8 folds** for alternate-side folding (for A4 paper) [7] respectively. Interestingly, paper [2] suggested that the folding force required for the nth fold is independent of the dimensions of the paper (apart from thickness). As will be shown in this paper, considering the paper's folding geometry results in a stricter folding limit for common sizes of paper (e.g. A3, A4, A5), though the maximum number of folds can be made arbitrarily large with sufficiently large pieces of paper (provided sufficient force is available, say, with machinery).

2 Same-Side Paper Folding: Gallivan's Model

In Gallivan s model, each folded region is approximated as a semicircle, as shown in Fig. 1. For n folds, the cumulative loss can be expressed by evaluating the expression below.

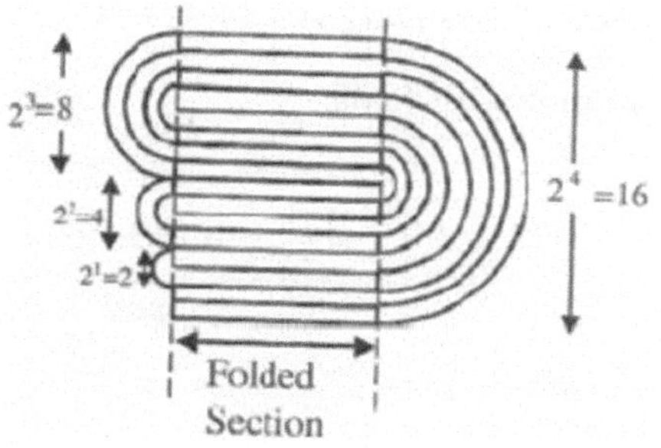

Fig. 1. Semicircular folded regions in Gallivan's Model (Source: [5])

$$L = \pi t + (\pi t + 2\pi t) + (\pi t + 2\pi t + 3\pi t + 4\pi t) + \cdots + \sum_{j=1}^{2^{n-1}} j\pi t$$

Simplifying using geometric and arithmetic progression formulas gives:

$$L = \frac{\pi t}{6}(2n + 4)(2n - 1)$$

Solving this quadratic equation (and taking the positive root) yields the maximum number of folds N for same-side folding.

$$N = \left\lfloor \log_2 \left(\frac{-3 + \sqrt{\frac{24L}{\pi t} + 25}}{2} \right) \right\rfloor$$

3 Alternate-Side Paper Folding

3.1 Length Lost to Folding

In alternate-side folding, the lost length is different for the side with odd-numbered folds and the side with even-numbered folds. Since the total lost length is greater for the side with the last fold, only the lost length for this side will limit the number of folds. However, this means that the cases for the total number of folds being even and odd will have to be considered separately.

To derive the formula for an even number of folds, only the lost length due to even-numbered folds is considered. Hence, the maximum lost length L_{even} due to folding can be expressed as follows.

$$L_{even} = (\pi t + 2\pi t) + (\pi t + 2\pi t + 3\pi t + 4\pi t + 5\pi t + 6\pi t + 7\pi t + 8\pi t) + \cdots + \sum_{j=1}^{2^n - 1} j\pi t$$

Simplifying as before,

$$L_{even} = \frac{\pi t}{15}[2(2^n) + 7][2^n - 1]$$

Similarly, for an odd number of folds:

$$L_{odd} = (\pi t) + (\pi t + 2\pi t + 3\pi t + 4\pi t) + \cdots + \sum_{j=1}^{2^n - 1} j\pi t$$

$$L_{odd} = \frac{\pi t}{15}[2(2^n) - 1][2^n + 3]$$

3.2 Maximum Number of Folds

Square Paper. With square paper, solving for the maximum number of folds is approached in the same way as with same-side folding. Here, a close approximation for the lost length will be used by taking the average lost length for even and odd number of folds.

$$L_{avg} = \frac{L_{odd} + L_{even}}{2} = \frac{\pi t}{15}[2(2^n)^2 + 5(2^n) - 5]$$

Solving this quadratic equation yields:

$$N = \lfloor \log_2 \frac{-5 + \sqrt{\frac{120 L_0}{\pi t} + 65}}{4} \rfloor$$

Rectangular Paper. For rectangular paper with side lengths $a \geq b$, to maximise the number of folds, the side of length a should have the greater lost length, so we let the last fold be along that side. Hence, at the maximum number of folds (N), the lost length for the side of length a (L_a) and the side of length b (L_b) can be written as follows.

$$L_a = \frac{\pi t}{15}[2(2^N)^2 + 5(2^N) - 5]$$

$$L_b = \frac{\pi t}{15}[2(2^{N-1})^2 + 5(2^{N-1}) - 5]$$

Hence, the following inequalities must be satisfied.

$$a \geq \frac{\pi t}{15}[2(2^N)^2 + 5(2^N) - 5]$$

$$b \geq \frac{\pi t}{30}[(2^N)^2 + 5(2^N) - 10]$$

Notice that $\frac{L_a}{L_b} \approx 4$ for large N (by taking the limit of $\frac{L_a}{L_b}$ as N approaches infinity). $\frac{L_a}{L_b} \approx 3.9$ (1dp) at $N = 6$ and $\frac{L_a}{L_b} \approx 4.0$ (1dp) for $N \geq 8$, so this approximation is a close upper bound. By using this approximation, the side of length a will have the limiting lost length for $b \leq a < 4b$ and the side of length b will have the limiting lost length for $a \geq 4b$. In both cases, the maximum number of folds is achieved by making the last fold along the side of greater length.

Hence, for $b \leq a < 4b$:

$$N = \lfloor \log_2 \frac{-5 + \sqrt{\frac{120a}{\pi t} + 65}}{4} \rfloor$$

Meanwhile, for $a \geq 4b$:

$$N = \lfloor \log_2 \frac{-5 + \sqrt{\frac{120b}{\pi t} + 65}}{2} \rfloor$$

4 Areas of Improvement in Gallivan's Formula

4.1 Radius of Fold

Gallivan's model assumes that the fold behaves as a perfect semicircle with radius equal to the paper's thickness t. This implies that the inner surface has zero length while the outer surface has length πt. Clearly, this is physically unrealistic as the ratio of outer surface length to inner surface length cannot be infinite.

There are two ways to resolve this. Case 1 is where paper does not stretch or compress. In this case, to keep the lengths of the inner and outer surface equal, the inner surface of the paper must form an inward "bulge" facilitated by delamination of fibers in paper [1,3]. This is shown in Fig. 2. Clearly, this increases the radius of fold beyond t to make space for the bulge.

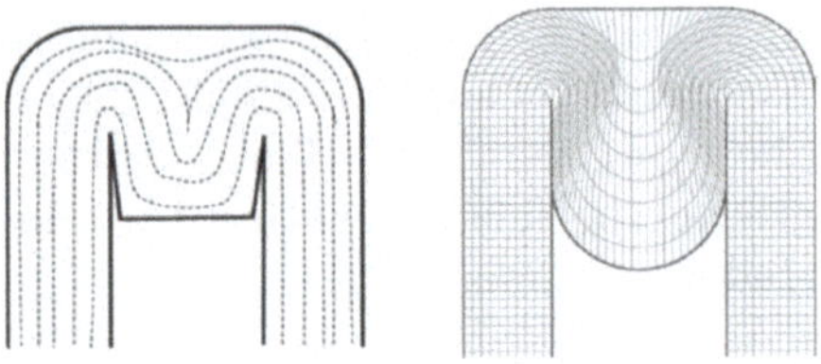

Fig. 2. Crease formation in Case 1 (Source: [1])

The second case is where one of the following occurs at the fold: linear compression of the inner surface [case 2(a)], linear stretching of the outer surface [case 2(b)] or both [case 2(c)]. Since paper cannot stretch or compress infinitely, there must be a gap between the layers of paper at the fold so that the ratio of outer surface length to inner surface length is finite (shown in Fig. 3).

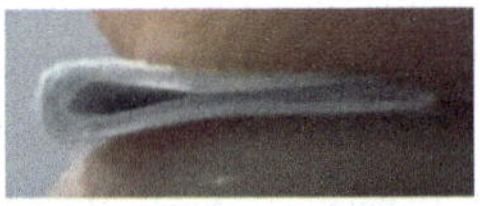

Fig. 3. Folding gap for Case 2

Modified Formulas. We now derive 3 modified formulas for the maximum number of folds (N) in same-side folding. Equation (1) holds for case 1 and case 2(a) (when the outer surface is unchanged in length). Equation (2) holds for case 2(c) (stretching and compression). Equation (3) holds for case 2(b) (stretching only).

For Eq. (1), let the minimum gap between layers of paper in the smallest fold be d. Then the minimum radius of the first (innermost) fold is $(t + \frac{d}{2})$. Note that the radius of the outermost fold increases as the number of folds increases, so the ratio of outer surface length to inner surface length (before stretching/compression) decreases. Hence, we may assume that there are no further gaps between the layers of paper outside the innermost fold (this is true for case 2 and is a close approximation for case 1 as the bulge width will be considerably smaller after the innermost fold). Now, we derive the corrected lost length L_c.

$$L_c = L + \frac{\pi d}{2}(1 + 2 + 4 + ... + 2^{n-1})$$

$$L_c = \frac{\pi}{6}[(t)(2^n)^2 + (3t + 3d)2^n - (4t + 3d)]$$

By solving this quadratic equation:

$$N = \lfloor \log_2 \frac{-(3t+3d) + \sqrt{(3t+3d)^2 + 4t(\frac{6L}{\pi} + 4t + 3d)}}{2t} \rfloor \quad (1)$$

The **neutral axis** is the region between the inner and outer surface of the material being bent where no stretching or compression occurs. For Eq. (2), the lost length for the neutral axis needs to be considered as both surfaces change in lemgth. For simplicity, we will assume the neutral axis is halfway between the outer and inner surface at bending, so the radius of the innermost fold is $(\frac{t}{2} + \frac{d}{2})$.

$$L_c = L + \frac{\pi d}{2}(1 + 2 + 4 + ... + 2^{n-1}) - \frac{\pi t}{2}(1 + 2 + 4 + ... + 2^{n-1})$$

$$L_c = \frac{\pi}{6}[(t)(2^n)^2 + (3d)2^n - (t + 3d)]$$

By solving this quadratic equation:

$$N = \lfloor \log_2 \frac{-(3d) + \sqrt{9d^2 + 4t(\frac{6L}{\pi} + t + 3d)}}{2t} \rfloor \quad (2)$$

Finally, for Eq. (3), where only the outer layer changes in length, the lost length for the inner surface at each fold needs to be considered. Here, the radius of the innermost fold is d.

$$L_c = L + \frac{\pi d}{2}(1 + 2 + 4 + ... + 2^{n-1}) - \pi t(1 + 2 + 4 + ... + 2^{n-1})$$

$$L_c = \frac{\pi}{6}[(t)(2^n)^2 + (3d - 3t)2^n - (3d - 2t)]$$

Now, before the corrected lost length L_c is equated to the length of the paper L to form a quadratic equation, consider that the outer surface of the previous fold becomes the inner surface of the current fold (for every layer). Since the outer surface at each folds stretches in this case, the final length of paper to be considered is greater than L. In particular, the relationship between L_c and L can be expressed as follows.

$$L_c = L + \pi t[1 + (1 + 2) + (1 + 2 + 3 + 4) + ... + (1 + 2 + ... + 2^{n-2})]$$

Simplifying, the length of paper L must satisfy:

$$L = \frac{\pi}{8}[t(2^n)^2 + (4d - 5t)2^n - (4d - 4t)]$$

By solving this quadratic equation:

$$N = \lfloor \log_2 \frac{-(4d - 5t) + \sqrt{(4d - 5t)^2 + 4t(\frac{8L}{\pi} - 4t + 4d)}}{2t} \rfloor \quad (3)$$

Calculating the Value of d for Case 2. For case 2 (if paper undergoes compression and/or stretching, but not inward bulging due to delamination), it is possible to express d using other variables (fundamental properties of paper). A brief description of the relevant variables is provided below.

The maximum **strain** (also known as breaking strain) an object can endure is the maximum fractional change in length ($\frac{\text{Change in Length}}{\text{Original Length}}$) it can undergo before breaking. The two types of strain explored here are tensile (during stretching) and compressive (during compression). Let the maximum tensile strain (outer surface) and compressive strain (inner surface) of paper be s_o and s_i respectively.

For case 2(a) (compression only), we may solve for the value of d as follows.

$$1 - s_i = \frac{\frac{d}{2}}{t + \frac{d}{2}}$$

$$d = 2t\frac{1 - s_i}{s_i}$$

For case 2(b) (stretching only):

$$1 + s_o = \frac{t + \frac{d}{2}}{\frac{d}{2}}$$

$$d = \frac{2t}{s_o}$$

For case 2(c) (compression and stretching), let the distance of the neutral axis from the inner surface of paper be x. Then, using the strain condition:

$$s_o = \frac{t - x}{x + \frac{d}{2}}$$

$$s_i = \frac{x}{x + \frac{d}{2}}$$

By dividing the equations, we may solve for x.

$$x = \frac{t}{\frac{s_o}{s_i} + 1}$$

By adding the first two equations and substituting in x, we may solve for d.

$$d = \frac{2t}{s_o + s_i} - \frac{2t}{\frac{s_o}{s_i} + 1} = \frac{2t}{s_o + s_i}(1 - s_i)$$

To determine the minimum gap d between the two layers of paper at a fold, 8 pieces of standard A5 notebook paper were folded in half and pressed together tightly. The widths of this stack at the folded ends and flat ends were measured to be 1.8 mm and 1.3 mm respectively, giving d = **0.06 mm** and t = **0.08 mm**. Using this value of d, **the above equations predict incredibly high values for the maximum compressive and tensile strains of paper**. For example, for case 2(b), the maximum tensile strain of paper would have to be 2.6 - much greater than is found in online literature (0.01–0.09) [6]. This suggests that case 2 is not accurate in describing the folding in paper, so case 1 is likely the dominant factor in determining the minimum folding radius.

This is further supported by a 2023 study [3] which found that a similar model to case 1 (inward bulging due to delamination) predicts accurates values for the bulge width in paperboard folding. Hence, **equation (1) from Sect.** 4.1 **is likely to give the closest estimate of the maximum number of folds.**

4.2 Additional Constraint on Lost Length

Returning to Gallivan's formula for same-side paper folding, notice that for each subsequent fold after the first one, there are two possible directions to fold the paper in. For example, if the first fold was made anticlockwise with respect to the centre, the next fold can be made in either the same direction (anticlockwise) or the opposite direction (clockwise), as shown in Fig. 4.

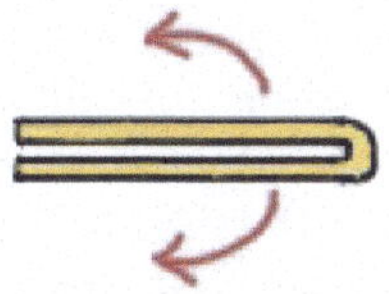

Fig. 4. Rotational direction of folding

Let us first consider same-direction folding. If the inner layer of paper is folded such that one end exactly meets the other, the outer later will not meet the other end as it has greater "lost length" than the inner layer - this is shown in Fig. 5.

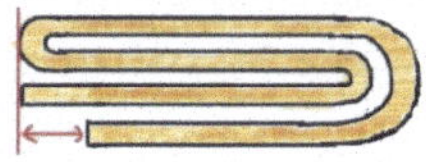

Fig. 5. Lost length of outer layer

Thus, even if the total "lost length" across all layers of paper has not exceeded the paper's original length (after some number of folds), it is possible that the outer layer can no longer fold onto the other side due to its "lost length". To ensure that this is not the case, the following inequality must be satisfied.

$$\text{Current Length of Paper} = \frac{L}{2^n} > \pi t(2 + 4 + ... + 2^{n-1}) = \text{Lost Length of Outer Layer}$$

$$L > \pi t(2^n)(2^n - 2)$$

To maximise the number of folds, alternate-direction (but same-side) paper folding is a better option. This is because the outer layer changes with every fold, so the "lost length" for the current outer layer after some number of folds only comes from the previous fold. Then, the following inequality must hold.

$$L > \frac{\pi t}{2}(2^n)^2$$

However, this is still a stricter bound than Gallivan's formula. In particular, the maximum number of folds along the same side becomes

$$N = \lfloor \log_2 \sqrt{\frac{2L}{\pi t}} \rfloor$$

.

4.3 Table for Maximum Number of Folds for Common Paper Sizes

Below, Table 1 shows how the various formulas compare to actual testing for same-side folding. All pieces of paper tested were of thickness $t = 0.08$ mm and minimum innermost gap width $d = 0.06$ mm. While the modified formula from Sect. 4.1 makes a negligible difference, **the modified formula from Sect. 4.2 makes a significant difference and seems to accurately predict the maximum number of folds.**

Table 1. Comparison of formulas with results from testing

Type of Paper	A5	A4	A3	A2	A1	A0
Max No. of Folds (by Gallivan's Formula)	6	6	6	6	7	7
Max No. of Folds (by modified formula in 4.1.1)	6	6	6	6	7	7
Max No. of Folds (by modified formula in 4.2)	5	5	5	6	6	6
Max No. of Folds (by testing)	5	5	5	Not tested	Not tested	Not tested

5 Conclusion

The mathematics explored in this paper ranges from simple extensions of Gallivan's formula (Sect. 3) to analysing more complex behaviour of paper in folding (Sect. 4). The findings of this paper can have a wide range of applications, such as when it comes to minimising the area in which foldable materials can be stored. This can potentially save costs in storage and packaging.

Limitations and Further Work:

1. The formulas derived for alternate-side paper folding and the modified formulas in Sect. 4.1 still rely on approximations, so it could be explored whether exact formulas can be derived.
2. With regards to Sect. 4.1, a formula for the minimum bulge width could not be determined as it requires complex physics beyond the scope of this paper, so it could be explored whether this is possible.
3. The formulas derived in this paper assume that the force applied in bending is not a limiting factor. In reality, however, this is not true since the required force increases exponentially with each fold. The maximum number of folds actually achievable by hand is likely lower for larger pieces of paper, so the formulas here can be taken as an upper bound instead for larger paper sizes.
4. It could be explored whether a modified version of the area of improvement under Sect. 4.2 applies to alternate-side folding.
5. It could be explored whether a formula for the maximum number of folds can be derived for other types of folding in half (e.g. folding along the diagonal).

Acknowledgments. I would like to thank my mentor, Mr Yeo Heu Ming Jeremy, for overseeing my research as part of this project, and Mr Chai Ming Huang Royce, for giving me the opportunity to carry out this research project.

Disclosure of Interests. There are no competing interests to declare that are relevant to the content of this article.

References

1. Coffins, D., Nygårds, M.: Creasing and folding. In: Advances in Pulp and Paper Research (2017)
2. Deboeuf, S.: A comparative study of crumpling and folding of thin sheets. Am. Phys. Soc. (2018)
3. Jina, W., Nagasawa, S., Yamamoto, T., Nagumo, T.: Analysis of the folding behavior of a paperboard subjected to indentation of a deviated creasing rule using the finite element method. AIMS Mater. Sci. (2023)
4. Lechenault, F., Thiria, B., Adda-Bedia, M.: The mechanical response of a creased sheet. Phys. Rev. Lett. (2014)
5. Please, M.: Say crease! folding paper in half. Fermat's Libr. **4**, 3 (2015)
6. Tensile properties of paper and paperboard (using constant rate of elongation equipment). https://www.tappi.org/content/tag/sarg/t494.pdf. Accessed 24 Sept 2010
7. Wang, X., Qiu, X.: Mechanical model analyses of a sheet under unidirectional multiple folding. Eur. J. Mech. - A/Solids **96** (2022)

Probabilistic Analysis of the Chromatic Polynomial

Ruhan Tasneem Shafa(✉) and Akash Thiagarajan

NUS High School of Mathematics and Science, 20 Clementi Ave 1,
Singapore 129957, Singapore
{h2110130,h2110006}@nushigh.edu.sg

Abstract. Network congestion is a pertinent issue in many fields, from satellite interference to transport route timings. To tackle such issues, networks are often modeled as graphs, collections of vertices joined by edges. The chromatic polynomial is a key function in graph theory, counting the number of ways to colour a graph with k colours such that no adjacent vertices have the same colour. This provides a measure of how sensitive the network is to congestion. However, in the case of dynamic networks that change with time, a probabilistic analysis is needed. Our work analyses the chromatic polynomial in the Erdős–Rényi model, where each edge appears independently at random with probability p. We provide formulas for the expectation and variance of the resulting chromatic polynomial that are more theoretically and computationally efficient. Using these formulas, we compute these polynomials explicitly for some simple classes of graphs. We also derive results on the first few terms of the expected and variance polynomials. These results allow us to make estimates on the expectation and variance, leading to "confidence interval" bounds which can prove useful in risk assessment of real networks. Our findings, supported by computational simulations, show promise to be extended to other models of random graphs, providing researchers with a versatile tool which can be used for a range of applications.

Keywords: Chromatic Polynomial · Probability · Graph Theory

1 Introduction and Background

The chromatic polynomial, introduced by George David Birkhoff in 1912 [1] as an algebraic approach to the Four Colour Problem, has since become a central concept in graph theory with deep links to combinatorics and statistical mechanics. Notably, it arises in the study of the Potts model, where it captures key features of partition functions and phase transitions.

In this work, we present a probabilistic analysis of the chromatic polynomial to model networks whose structure evolves under randomness. Graph theory provides a natural framework for such analysis, allowing us to quantify how stochastic variations in a graph affect its chromatic properties.

H. Guo et al. (Eds.): IRC-SET 2025, PTT, pp. 412–424, 2026.
https://doi.org/10.1007/978-981-95-9630-0_35

Definition 1. *A graph $G(V, E)$ (often abbreviated to just G) is defined by a set V of vertices, and a set $E \subseteq \{\{x, y\} \mid x, y \in V\}$ of edges.*

A graph can be thought of as a network of nodes and edges joining the nodes. In real world applications, we aim to avoid congestion in a given network. As such, we often allocate a key resource in a way that certain pairs of objects do not "share" a resource slot. For example, in a satelite network, satellites which have a similar orbit should not have transmit the same frequency. If we represent each possible frequency as a colour, and connect two satellites by an edge if they have a similar orbit, then no two vertices connected by an edge can have the same colour. This leads us to the next definition:

Definition 2. *Given a graph G, a proper k-colouring is a way of labelling the graph with k distinct labels (often called colours) so that no vertices which are joined by an edge have the same colour.*

If there are few ways to colour the graph, then the network is fragile, as small changes in the network could lead to a lack of a proper colouring. However, if there are a large number of ways to colour the graph, then the corresponding network has a lot more room for error as many alternative colourings are present. The chromatic polynomial gives the number of ways to properly colour the graph.

Definition 3. *For a positive integer k, the chromatic polynomial of a graph G, denoted $P_G(k)$, is the number of proper k-colourings of the graph G (with a fixed set of colours). It can be shown to be a polynomial in k.*

The following is one of the most fundamental and well-known results about chromatic polynomials:

Theorem 1. *(Deletion contraction) For any graph G and edge e of G,*

$$P_G(x) = P_{G-e}(x) - P_{G\backslash e}(x)$$

where $G - e$ denotes the graph formed by removing the edge e, while $G\backslash e$ denotes the graph formed by contracting the edge e, that is deleting the edge e and merging the two vertices that were joined by e [4].

This is enough to show that the chromatic polynomial is in fact a polynomial in k [4,9].

The following well-known formula is a key result that we use in the following section as it provides an explicit way to compute the chromatic polynomial without recursion.

Theorem 2. *For a graph G, let $E(G)$ be the set of its edges. Then*

$$P_G(x) = \sum_{S \subseteq E(G)} (-1)^{|S|} x^{c(S)}$$

where $c(S)$ is the number of connected components of the graph with S as its edges [9].

For dynamic networks, we can model them with random graphs. There are results on the chromatic number and roots [10] of random graphs, but the chromatic polynomial of random graphs has not been well studied. Through the probabilistic analysis of the chromatic polynomial, our novel approach provides insight into congestion in dynamic graphs.

2 Chromatic Polynomial of Random Graphs

2.1 Random Models of Graphs

Now that we have established some basic results, we shall move on to chromatic polynomial of random graphs. There are mutliple ways to "choose" graphs randomly, so in this report we will consider the most common model known as the Erdős-Renyi model, which we will use a variation of. Given a graph G with n vertices, we will let each of its edges appear independently at random with probability $p \in [0, 1]$. This will generate a subgraph of G, which we will call G_{random}. Typically, $G = K_n$ (complete graph on n vertices) is used, but we will work more generally. This model is often denoted as $G(n, p)$.

2.2 Expected Chromatic Polynomial in the $G(n, p)$ Model

In order to analyse the distribution of random chromatic polynomials, it helps to consider its expectation. We have come up with the following results, some of which are extensions of well-known results for the chromatic polynomial.

We denote the expected chromatic polynomial in terms of x of a graph G, with each of its edges having probability p of appearing, by $E_G(x, p)$. $E_G(x, p)$ can be found by summing over all edge subsets the product of the probability that edge subset occurs and the chromatic polynomial of the resulting subgraph. Consider a graph $G(V, E)$ with vertex set V and edge set E. The probability that $G(V, S)$ (where $S \subseteq E(G)$) occurs is given by $p^{|S|}(1-p)^{|E|-|S|}$. Summing over all possible S, we have formulated the following:

Theorem 3.

$$E_G(x, p) = \sum_{S \subseteq E(G)} p^{|S|}(1-p)^{|E|-|S|} P_{G|S}(x)$$

where $G|S$ is the graph G but with only the edges in the subset S.

The disadvantage of Theorem 3 is that it requires us to know the chromatic polynomial of all subgraphs of G. For a more direct formula, we have also extended Theorem 2 to the following result:

Theorem 4.

$$E_G(x, p) = \sum_{S \subseteq E(G)} (-p)^{|S|} x^{c(S)}$$

where $c(S)$ is the number of connected components of S.

Proof. The idea is to use linearity of expectation. Consider a fixed set $S \subseteq E(G)$. Define the indicator function $\mathbb{1}_S$ as follows:

$$\mathbb{1}_S = \begin{cases} 1, \text{if } S \subseteq E(G_{\text{random}}) \\ 0, \text{otherwise} \end{cases}$$

Then we may use Theorem 2 to write

$$P_G(x) = \sum_{S \subseteq E(G)} (-1)^{|S|} x^{c(S)} \cdot \mathbb{1}_S$$

We can see that if we were to choose a random subgraph of G using the $G(n, p)$ model and $S \subseteq E(G)$, $\mathbb{1}_S$ would have a probability $p^{|S|}$ of being 1 (since all the $|S|$ edges would need to appear). Otherwise, $\mathbb{1}_S$ would always be 0. So by linearity of expectation,

$$\begin{aligned} E_G(x, p) = \mathbb{E}\left[\sum_{S \subseteq E(G)} (-1)^{|S|} x^{c(S)} \cdot \mathbb{1}_S\right] &= \sum_{S \subseteq E(G)} \mathbb{E}\left[(-1)^{|S|} x^{c(S)} \cdot \mathbb{1}_S\right] \\ &= \sum_{S \subseteq E(G)} (-1)^{|S|} x^{c(S)} \cdot p^{|S|} \\ &= \sum_{S \subseteq E(G)} (-p)^{|S|} x^{c(S)} \end{aligned}$$

Corollary 1. *$E_G(x, p)$ is a polynomial in x with degree n where n is the number of vertices, and a polynomial in p with degree e, where e is the number of edges.*

Proof. Clearly, $E_G(x, p)$ is a polynomial in both x and p, since $c(S)$ and $|S|$ are always nonnegative integers. Also $E_G(x, p)$ has degree n as it is not possible to have more connected components than the number of vertices, hence $c(S)$ is at most n. We also know that the coefficient of x^n is non-zero as shown below. Additionally, the maximum value of $|S|$ is when all the edges are present, and hence, the degree of p is e, corresponding to the unique p^e term in Theorem 4.

Corollary 2. *For a graph G with e edges, $E_G(x, p)$ is expanded, let the coefficient of x^k be a_k. Then:*

- $a_n = 1$
- $a_{n-1} = -pe$
- $a_{n-2} = \binom{e}{2}p^2 - Tp^3$, *where T is the number of triangles.*

Proof. From Theorem 4, we know that the coefficient of x^k is given by

$$a_k = \sum_{S \subseteq S_k} (-p)^{|S|}$$

where S_k refers to the set of all edge subsets that result in a subgraph with k conected components.

- A subgraph on n vertices has n connected components if and only if it is a null graph, hence $a_n = (-p)^0 = 1$
- A subgraph on n vertices has $n-1$ connected components if and only if it has one edge, and there are a total of e edges to select. Hence, $a_{n-1} = (-p)^1 \cdot e = -pe$
- A subgraph on n vertices has $n-2$ connected components in 2 cases: when it has 2 edges, and when it has a triangle and no other edges. In the first case, we would have $\binom{e}{2}$ pairs of edges corresponding to the coefficient of $(-p)^2$. In the second case, there are T triangles corresponding to the coefficient of $(-p)^3$, as desired.

2.3 Variance of the Chromatic Polynomial in the $G(n, p)$ Model

Another quantity that is useful to consider is the variance. Let us define $V_G(x, p)$ as the variance of $P_{G_{\text{random}}}(x)$. For some event X, the standard definition of variance is:

$$\text{Var}(X) = \mathbb{E}[(X - \mathbb{E}[X])^2].$$

Hence, the variance of our chromatic polynomial can be found by using

$$V_G(x, p) = \mathbb{E}[(P_G(x) - E_G(x, p))^2], \tag{1}$$

which gives:

$$V_G(x, p) = \sum_{S \subseteq E(G)} \left[p^{|S|}(1-p)^{|E|-|S|}(P_{G|S}(x) - E_G(x, p))^2 \right], \tag{2}$$

where $G|S$ is the graph G with only the edges in S. Similar to Theorem 3, this follows by considering the probability that each subset of edges appear. We have developed the following result

Theorem 5. *Let G be a graph on n vertices with e edges. Then $V_G(x, p)$ is a polynomial in x of degree $2n-2$ and leading coefficient $ep(1-p)$.*

Proof. It is clear from (2) that $V_G(x, p)$ is a polynomial in x, as each individual term in the sum is a polynomial. By Corollary 2, both $E_G(x, p)$ and $P_{G|S}(x)$ have degree n and leading term x^n. This term cancels out in the expression $P_{G|S}(x) - E_G(x, p)$ and the highest degree term would be x^{n-1}. Since the expression $P_{G|S}(x) - E_G(x, p)$ is squared in the formula for $V_G(x, p)$, the leading term in $V_G(x, p)$ is x^{2n-2}. As for the coefficient, we may note that the second term in $E_G(x, p)$ is pe while it is $|S|$ in $P_{G|S}$ (number of edges). Hence, using (2), the leading coefficient can be written as

$$\sum_{S \subseteq E(G)} \left[p^{|S|}(1-p)^{|E|-|S|}(|S| - pe)^2 \right].$$

The key here is to note that since each edge appears with probability p, $\mathbb{E}[|S|] = pe$. Hence, the above expression is simply the variance of $|S|$, or the number of edges in our random subgraph. Also because the edges appear independently, $|S| \sim \text{Binomial}(e, p)$. Hence, the variance of $|S|$ is $ep(1-p)$ as desired.

2.4 Novel Bounds on $E_G(x, p)$ and $P_{G_{\text{random}}}(x)$

Although the formula in Theorem 4 is not explicit, we have been able to provide some novel bounds on $E_G(x, p)$. In what follows, let G be a graph with n vertices.

Theorem 6. *(Modified deletion contraction) For positive integers x,*

$$E_G(x, p) \geq E_{G-e}(x, p) - p \cdot E_{G\setminus e}(x, p)$$

Proof. We will prove this by induction on the number of edges. When there are no edges, both sides are equal to x^n. Suppose this is true for all graphs with at most $m - 1$ edges. Now, consider a graph G with m edges and consider an edge e. In G_{random}, there is a $1 - p$ probability that e does not appear, in which case the expected chromatic polynomial would be $E_{G-e}(x, p)$. If the edge e does appear, then $P_{G_{\text{random}}}(x) = P_{G_{\text{random}}-e}(x) - P_{G_{\text{random}}\setminus e}(x)$.

However, $\mathbb{E}[P_{G\setminus e}(x)] \leq E_{G_{\text{random}}\setminus e}(x, p)$. This is because if we choose the subgraph first and then contract, the edges joined to the contracted vertex (after contracting e) will have probability $> p$ of appearing, since these edges appear if and only if either one of the edges joined to the two endpoints of the edge e appear in the original graph. Hence on average, there will be more edges with a lower value of the chromatic polynomial, as more edges implies less ways to colour. Thus,

$$\begin{aligned} E_G(x, p) &= (1 - p)E_{G-e}(x, p) + p[E_{G-e}(x, p) - \mathbb{E}[P_{G\setminus e}(x)] \\ &\geq E_{G-e}(x, p) - p \cdot E_{G\setminus e}(x, p) \end{aligned}$$

as desired.

This result could be useful if we know the expected chromatic polynomial of graphs similar to G, allowing us to get a good bound on the expected chromatic polynomial of G without any computation. We can see that if this inequality was an equality, then $p^n \cdot P_G(\frac{x}{p})$ satisfies the recurrence (this follows from Theorem 2 after substituting in). Hence, we believe the following conjecture to be true:

Conjecture 1. For positive integer x,

$$E_G(x, p) \geq p^n \cdot P_G(\frac{x}{p})$$

This conjecture provides an interpretation for noninteger inputs to the chromatic polynomial as $\frac{x}{p}$ may not be an integer. These values can be used to bound the expected chromatic polynomial.

We will now analyse the deviation of $P_{G_{\text{random}}}(x)$ from $E_G(x, p)$ using Chebyshev's Inequality in conjunction with Theorem 6 on variance. Chebyshev's Inequality states that for a random variable X,

$$\Pr(|X - \mathbb{E}[x]| > a) \leq \frac{\text{Var}(X)}{a^2}$$

applying this here in the case of K_n and setting $a = \lambda x^{n-1}$.

Theorem 7. *Let x be a real number and G be a graph with n vertices and e edges. Then we have the following inequality:*

$$Pr(|P_{G_{random}}(x) - E_G(x,p)| > \lambda x^{n-1}) \leq \frac{V_G(x,p)}{\lambda^2 x^{2n-2}} = \lambda^{-2}\left[ep(1-p) + O\left(\frac{1}{x}\right)\right]$$

Proof. This follows from Chebyshev's inequality, with the second equality following from Theorem 6, as all terms with degree lower than x^{2n-2} would contribute $O(1/x)$ when divided by x^{2n-2} (Fig. 1).

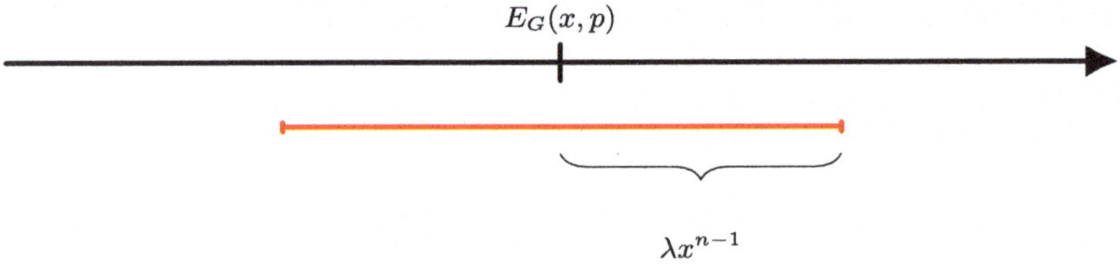

Fig. 1. Confidence interval of chromatic polynomials

Refer to the diagram above. Theorem 7 states that the probability that the value $P_{G_{\mathrm{random}}}(x)$ lies outside the red interval is at most $\lambda^{-2}ep(1-p)$ (approximately). This inequality allows us to analyse the distribution of all chromatic polynomials, and to bound the deviation from the expected $E_G(x,p)$.

In applications, the networks are often very large, and the number of colours x that are required would be large too. By varying λ, we can change the confidence level of our interval. Thus, the key result regarding confidence intervals of the chromatic polynomial is as follows:

Corollary 3. *Let $C \in (0,1)$ be a real number. Let*

$$\lambda_C = \sqrt{\frac{ep(1-p)}{1-C}}$$

Then for large x, the interval $E_G(x,p) \pm \lambda_C \cdot x^{n-1}$ will contain the value $P_{G_{random}}(x)$ with probability at least C.

Proof. We wish to have the probability lying inside the interval being C, which means the probability of being outside the interval is $1-C$. We may solve the equation

$$\lambda_C^{-2}ep(1-p) = 1-C$$

to get the desired result.

Note that we are approximating the $O(1/x)$ term to be 0 here. In reality, the value of λ_C will be slightly higher due to this term, but for large x the deviation can be ignored.

2.5 Results from Computer Simulation

One common class of graphs are complete graphs, K_n, where all pairs of nodes are connected. We were not able to get a closed form formula for $E_{K_n}(x,p)$ and its variance, but, we have written code to find these for smaller n. We were not able to continue doing so for larger n due to computational limitations as there is exponential time complexity, as will be discussed further. Also, the expressions for $E_{K_6}(x,p)$, $E_{K_7}(x,p)$, $V_{K_5}(x,p)$, $V_{K_6}(x,p)$ are rather long, hence they have been listed in the appendix instead.

$\mathbf{E_{K_n}(x,p)}$. First, we note that the chromatic polynomial function has time complexity $O(\min(2^e, 1.62^{e+n}))$ [11]. Now, there are 2 ways to compute $E_{K_n}(x,p)$ - one uses Theorem 3, and the other uses Theorem 4.

When using Theorem 3, the code iterates through all 2^e edge subsets of G, and for each edge subset, the chromatic polynomial of the subgraph is copmuted, which is itself $O(2^e)$. Hence, overall, the time complexity is $O(\min(4^e, 1.62^n \cdot 3.24^e))$. When using Theorem 4, we are doing one operation for each subset, hence the time complexity is $O(2^e)$. As a result, we opt to use Theorem 4 to compute $E_{K_n}(x,p)$. The following are some results from our code.

- $n=2$: $x^2 - px$
- $n=3$: $x^3 - 3px^2 + (-p^3+3p^2)x$
- $n=4$: $x^4 - 6px^3 + (-4p^3+15p^2)x^2 + (p^6 - 6p^5 + 15p^4 - 16p^3)x$
- $n=5$: $x^5 - 10px^4 + (-10p^3+45p^2)x^3 + (5p^6 - 30p^5 + 85p^4 - 110p^3)x^2 + (p^{10} - 10p^9 + 45p^8 - 120p^7 + 205p^6 - 222p^5 + 125p^4)x$

$\mathbf{V_{K_n}(x,p)}$. Computing the variance requires iterating over all subsets of the graph, computing the chromatic polynomial for each one (which is algebraically manipulated then). Similar to using theorem 3 to find $E_{K_n}(x,p)$, this has time complexity $O(4^e)$ (For reference: K_5 took us 19 s to compute while K_6 took 29 min).

- $n=2$: $(-p^2+p)x^2$
- $n=3$: $(-3p^2+3p)x^4 + (-6p^4+18p^3-12p^2)x^3 + (-p^6+6p^5-9p^4+p^3+3p^2)x^2$
- $n=4$: $(-6p^2+6p)x^6 + (-24p^4+84p^3-60p^2)x^5 + (12p^7 - 88p^6 + 288p^5 - 423p^4 + 196p^3 + 15p^2)x^4 + (8p^9 - 78p^8 + 300p^7 - 578p^6 + 504p^5 - 60p^4 - 96p^3)x^3 + (-p^{12} + 12p^{11} - 66p^{10} + 212p^9 - 417p^8 + 480p^7 - 257p^6 - 6p^5 + 27p^4 + 16p^3)x^2$

From these results, we can better understand the relationships between the powers of p and the powers of x in the chromatic polynomial. It also led us to develop the following conjectures.

Conjecture 2. The following results about the smallest power of p in each x^k coefficient hold:

(a) In the coefficient of x^k in $E_G(x,p)$, the smallest power of p is $n-k$
(b) In the coefficient of x^k in $V_G(x,p)$, the smallest power of p is $\lceil \frac{2n-k}{2} \rceil$

In the regime where p is small, the smallest power of p could provide a better estimate of the expected chromatic polynomial, as higher order terms can be dropped.

Lastly, we have the following conjecture about the exact form for the second term of the variance:

Conjecture 3. The second term of $V_{K_n}(x,p)$ when written down in descending powers of x is given by:

$$3\binom{n}{3}p^2(1-p)(2p-n-1)x^{2n-3}$$

3 Conclusion and Possible Extensions

3.1 Other Models and Extensions

Although our report mainly focuses on the $G(n,p)$ model, our method of proving Theorem 4 using indicator functions and linearity of expectation can be generalised to other random models, as we simply need to consider the probability of each subset S of edges appearing and weight the sum accordingly.

In general, we can apply the linearity of expectation method to obtain, in an arbitrary random model

$$\mathbb{E}[P_{G_{\text{random}}}(x)] = \sum_{S \subseteq E(G)} \Pr(S \subseteq E(G_{\text{random}})) \cdot (-1)^{|S|} x^{c(S)}$$

We may apply this in the case of the $G(n,M)$ random model, which is another common model which chooses a subset of edges of size M uniformly at random. For this model, we used a similar method to obtain

$$E_G(x,M) = \sum_{S \subseteq E(G), |S| < m} (-1)^{|S|} \binom{|E(G)| - |S|}{M - |S|} x^{c(S)}$$

which has been verified using our code. This serves as a proof of concept that our methods could be generalised to other models, allowing our results to be more applicable to a wide range of scenarios.

Our work could potentially be extended to non-simple graphs or hypergraphs, and potentially varying the vertices randomly rather than the edges, allowing us to handle even more applications. Additionally, more study could be done on $E_G(x,p)$, especially concerning the interpretation of $p \notin [0,1]$. $E_G(x,p)$ could potentially be useful as a formal polynomial of x, p, and serve another generalisation of the chromatic polynomial (similar to the Tutte polynomial). Overall, our work so far lends itself to many extensions, and the potential progress that could be made from here is exciting.

3.2 Significance of Our Results

Considering the chromatic polynomial of random graphs allows us to model graph dynamics when there is variability in the graph and conduct risk assessment for congestion of networks using "confidence interval" bounds. Our results are generalisable and applicable to a broad range of scenarios. Overall, we hope that this paper can provide more avenues for further research on $E_G(x, p)$ and $V_G(x, p)$ and that our work can potentially prove useful in practice.

Disclosure of Interests. The authors have no competing interests to declare that are relevant to the content of this article.

4 Appendix A: Code Used in the Project

We used Python code to generate expected Chromatic Polynomial and Variance of the chromatic polynomial in the $G(n, p)$ model, as well as the expected chromatic polynomial in the $G(n, M)$ model. These were used to verify our conjectures and theorems. The packages used are listed in the first 5 lines of code.

```
import numpy as np
import networkx as nx
import matplotlib.pyplot as plt
from sympy import *
from itertools import combinations
def expected_cp(graph,p,factorise=False):
  k = symbols('x')
  edges = list(graph.edges)
  expected_chromatic_poly = 0
# iterating over all subsets of edges and computing the term in the formula
  for subset_size in range(len(edges) + 1):
      for edge_subset in combinations(edges, subset_size):
          subgraph = nx.Graph()
          subgraph.add_nodes_from(graph.nodes)
          subgraph.add_edges_from(edge_subset)


          term = ((-p)**subset_size)*(k**nx.number_connected_components(subgraph))
          expected_chromatic_poly += term

  if factorise:
    return factor(expand(simplify(expected_chromatic_poly)))

  return (expand(simplify(expected_chromatic_poly)))

def format_poly(polynomial):
  poly, _ = poly_from_expr(polynomial, symbols('x'))
```

Listing 1.1. Computing the variance of the chromatic polynomial in the first model

```
def cpvar(graph, p, factorise=False):

    x = symbols('x')
    edges = list(graph.edges)
    cpsquare = 0

# iterating over all subsets of edges and computing the term in the formula
    for subset_size in range(len(edges) + 1):
        for edge_subset in combinations(edges, subset_size):
            subgraph = nx.Graph()

            subgraph.add_nodes_from(graph.nodes)
            subgraph.add_edges_from(edge_subset)

            chromatic_poly_squared = (nx.chromatic_polynomial(subgraph))**2
            prob = (p ** subset_size) * ((1 - p) ** (len(edges) - subset_size))
            cpsquare += prob * chromatic_poly_squared

    poly=cpsquare-(chromatic_polynomial_formula(graph,p))**2

    if factorise:
        return factor(expand(simplify(poly)))

    return expand(simplify(poly))
```

Listing 1.2. Computing the expected value of the chromatic polynomial in the m random edges model

```
def expected_cp2(G, m, factorise=True):
  edges = list(G.edges)
  cp = 0
  i = 0
  if m == 0 or m>len(edges):
    return 0
  for subset in combinations(edges, m):
    subgraph = nx.Graph()
    subgraph.add_nodes_from(G.nodes)
    subgraph.add_edges_from(subset)
    cp += nx.chromatic_polynomial(subgraph)
    i += 1
  if factorise:
    return factor(expand(simplify(cp/i)))
  return (expand(simplify(cp/i)))
```

Listing 1.3. Displaying the results

```
n = 6
G = nx.complete_graph(n)
p = symbols('p')

expected_poly = expected_cp(G, p)
print("Expected␣Polynomial␣(Model␣1):␣" + expected_poly)
variance = cpvar(G, p)
print("Variance:", format_poly(variance))
expected_poly = expected_cp2(G, p)
print("Expected␣Polynomial␣(Model␣2):␣" + expected_poly)
```

5 Appendix B: Long Expressions

- $E_{K_6}(x,p)$: $x^6-15px^5+(-20p^3+105p^2)x^4+(15p^6-90p^5+285p^4-435p^3)x^3+(6p^{10}-60p^9+270p^8-735p^7+1330p^6-1617p^5+1080p^4)x^2+(-p^{15}+15p^{14}-105p^{13}+455p^{12}-1365p^{11}+2997p^{10}-4945p^9+6165p^8-5700p^7+3660p^6-1296p^5)x$
- $E_{K_7}(x,p)$: $x^7-21px^6+(-35p^3+210p^2)x^5+(35p^6-210p^5+735p^4-1295p^3)x^4+(21p^{10}-210p^9+945p^8-2625p^7+5005p^6-6762p^5+5250p^4)x^3+(-7p^{15}+105p^{14}-735p^{13}+3185p^{12}-9576p^{11}+21189p^{10}-35595p^9+45990p^8-45360p^7+32417p^6-13377p^5)x^2+(-p^{21}+21p^{20}-210p^{19}+1330p^{18}-5985p^{17}+20349p^{16}-54257p^{15}+116175p^{14}-202755p^{13}+290745p^{12}-343140p^{11}+331506p^{10}-258125p^9+156555p^8-68295p^7+16807p^6)x$
- $V_{K_5}(x,p)$: $(-10p^2+10p)x^8+(-60p^4+240p^3-180p^2)x^7+(60p^7-430p^6+1520p^5-2525p^4+1330p^3+45p^2)x^6+(20p^{11}-200p^{10}+1000p^9-3390p^8+8180p^7-13510p^6+12880p^5-4320p^4-660p^3)x^5+(20p^{13}-315p^{12}+2100p^{11}-8200p^{10}+21080p^9-36715p^8+41600p^7-25495p^6+2670p^5+3145p^4+110p^3)x^4+(-10p^{16}+160p^{15}-1220p^{14}+5820p^{13}-19100p^{12}+44820p^{11}-75820p^{10}+90380p^9-70270p^8+27780p^7+860p^6-2400p^5-1000p^4)x^3+(-p^{20}+20p^{19}-190p^{18}+1140p^{17}-4835p^{16}+15344p^{15}-37540p^{14}+71680p^{13}-106555p^{12}+121020p^{11}-100535p^{10}+55490p^9-15550p^8-220p^7+25p^6+582p^5+125p^4)x^2$
- $V_{K_6}(x,p)$: $(-15p^2+15p)x^{10}+(-120p^4+540p^3-420p^2)x^9+(180p^7-1280p^6+4920p^5-9165p^4+5240p^3+105p^2)x^8+(120p^{11}-1200p^{10}+5880p^9-19560p^8+47550p^7-83460p^6+88860p^5-35580p^4-2610p^3)x^7+(-30p^{16}+450p^{15}-3150p^{14}+13890p^{13}-44715p^{12}+114780p^{11}-245310p^{10}+439240p^9-639255p^8+702240p^7-486580p^6+121620p^5+26385p^4+435p^3)x^6+(-40p^{18}+810p^{17}-7530p^{16}+43130p^{15}-172470p^{14}+516480p^{13}-1204390p^{12}+2222970p^{11}-3237480p^{10}+3621510p^9-2896770p^8+1355910p^7-113460p^6-120030p^5-8640p^4)x^5+(30p^{21}-666p^{20}+7140p^{19}-48810p^{18}+236970p^{17}-863370p^{16}+2437614p^{15}-5436285p^{14}+9664560p^{13}-13682320p^{12}+15207330p^{11}-12793035p^{10}+7468620p^9-2339310p^8-73935p^7+158590p^6+55797p^5+1080p^4)x^4+(12p^{25}-300p^{24}+3600p^{23}-27630p^{22}+152390p^{21}-642948p^{20}+2154900p^{19}-5874580p^{18}+13213260p^{17}-24695400p^{16}+38389700p^{15}-49375530p^{14}+51861330p^{13}-43392180p^{12}+27595980p^{11}-12087444p^{10}+2763300p^9+64860p^8-15120p^7-75240p^6-12960p^5)x^3+(-p^{30}+30p^{29}-435p^{28}+4060p^{27}-27405p^{26}+142494p^{25}-593475p^{24}+2032200p^{23}-5825295p^{22}+14154730p^{21}-29401401p^{20}+52465260p^{19}-80569795p^{18}+106308810p^{17}-119856345p^{16}+114246625p^{15}-90435225p^{14}+57703485p^{13}-28169165p^{12}+9484755p^{11}-1677543p^{10}+3595p^9-16455p^8+12540p^7+12660p^6+1296p^5)x^2$

References

1. Birkhoff, G.D.: A determinant formula for the number of ways of coloring a map. Ann. of Math. (2) **14**((1912/13)1-4) 42–46. MR1502436

2. Perrett, T.: Roots of the Chromatic Polynomial. Technical University of Denmark. DTU Compute PHD2016 No. 438 (2017)
3. Cameron, P.J., Morgan, K.: Algebraic properties of chromatic roots (n.d.). arXiv (Cornell University)
4. Aydelotte, A.: An Exploration of the Chromatic Polynomial (2017). Mathematics Undergraduate Theses. 7
5. Wolfram Research, Inc. (n.d.). Resistance Distance – from Wolfram MathWorld
6. Fouts, C. (n.d.). THE CHROMATIC POLYNOMIAL
7. Xu, J., Liu, Z.: The chromatic polynomial between graph & its complement—about Akiyama and Hararys' open problem. Graphs Combinatorics **11**, 337–345 (1995)
8. Farrell, E.J.: On chromatic coefficients. Discret. Math. **29**(3), 257–264 (1980)
9. Read, R.C.: An introduction to chromatic polynomials. J. Comb. Theory **4**(1), 52–71 (1968)
10. Van Bussel, F., Ehrlich, C., Fliegner, D., Stolzenberg, S., Timme, M.: Chromatic polynomials of random graphs. J. Phys. A: Math. Theor. **43**(17), 175002 (2010)
11. Wilf, H.S.: Algorithms and complexity (1994). In Lecture notes in computer science

Orbital Characterization of Hyperbolic Bolides for Interstellar Candidate Identification

Neo Xuan Che(✉)

NUS High School of Mathematics and Science, 20 Clementi Avenue 1, Singapore 129957, Singapore
h2110112@nushigh.edu.sg

Abstract. The recent discoveries of interstellar objects 'Oumuamua and 2I/Borisov have highlighted the need for systematic identification of smaller interstellar fragments entering Earth's atmosphere. We present a comprehensive multi-parameter orbital analysis framework for identifying interstellar meteor candidates from NASA's Center for Near Earth Object Studies (CNEOS) fireball database. Our approach integrates Väisälä orbit determination, Monte Carlo uncertainty quantification, gravitational focusing corrections, and Tisserand parameter analysis to discriminate between interstellar and solar system origins. Analysis of 1022 CNEOS fireballs (2014–2023) identified four candidates with excess velocities $V_\infty > 0$, representing 0.4% of the database. Synthetic validation using 20,000 test trajectories achieved 97.0% detection efficiency with 0.69% false positive rate. Bayesian analysis estimates an interstellar meteor flux of 1.60×10^{-7} events per km2 annually, though large uncertainties (95% CI: 0.51–19.93) reflect dataset limitations. While definitive identification requires additional observational confirmation, this framework provides a foundation for systematic interstellar meteor surveys and screening of existing databases.

Keywords: Meteor · Interstellar objects · Orbital mechanics · Statistical analysis

1 Introduction

1.1 Introduction

The detection of 'Oumuamua in 2017 [1], 2I/Borisov in 2019 [2] and 3I/ATLAS in 2025 [3] marked the first confirmed discoveries of macroscopic interstellar objects transiting the solar system. These discoveries suggest that interstellar objects may be relatively common but difficult to detect due to observational limitations and brief transit times through the inner solar system. Prior to these discoveries, theoretical models predicted regular incursions of interstellar material based on the local stellar density and relative motion of the solar system through the local interstellar medium [4, 5]. However, observational confirmation remained elusive until the advent of modern wide-field sky surveys capable of detecting fast-moving objects with unusual orbital characteristics.

While 'Oumuamua, 2I/Borisov and 3I/ATLAS represent the large end of the interstellar object size distribution, theoretical considerations [6] suggest that smaller fragments should be far more numerous. Interstellar meteoroids could potentially manifest

H. Guo et al. (Eds.): IRC-SET 2025, PTT, pp. 425–434, 2026.
https://doi.org/10.1007/978-981-95-9630-0_36

as meteors upon entry into earth's atmosphere, producing observable fireballs in existing detection networks. The identification of interstellar meteors presents unique challenges compared to their larger counterparts. Unlike 'Oumuamua and 2I/Borisov, which could be tracked over extended periods, meteors only provide brief observational windows of seconds to minutes. Furthermore, atmospheric entry effects, measurement uncertainties, and the vast number of solar system meteors create significant observational and analytical challenges for systematic identification.

Meteor detection networks, including the European Fireball Network [7], the Desert Fireball Network [8], and NASA's Center for Near Earth Object Studies (CNEOS), have extensive databases of bolide observations. The CNEOS database, maintained by NASA's Jet Propulsion Laboratory, represents one of the most accessible datasets for bolides, containing over 1000 fireball observations spanning the past decade. Previous attempts to identify interstellar meteors have primarily relied on single-parameter approaches, most commonly excess velocity (V_{∞}) relative to solar escape speed [9, 10]. While conceptually straightforward, this approach suffers from several limitations: measurement uncertainties can cause bound objects to appear hyperbolic, atmospheric effects can affect observed velocities, gravitational focusing effects must be properly accounted for, and solar system objects on highly eccentric orbits may exhibit similar characteristics.

To address these fundamental limitations, this work presents the first comprehensive multi-parameter framework for systematic interstellar meteor identification. Our approach integrates multiple orbital parameters, including including eccentricity, inclination, and Tisserand parameters, along with rigorous Monte Carlo uncertainty quantification. Our objectives include developing robust orbital determination methods suitable for meteor observations, integrating multiple orbital parameters for enhanced discrimination, validating the framework through synthetic testing, and applying the method to existing databases for candidate identification.

2 Methodology

2.1 Computational Methods and Implementation

All data processing and statistical analyses were performed using Python 3.11 within the Anaconda distribution environment. The core computational framework relied on NumPy 1.24.3 for numerical operations and array manipulations, while Pandas 2.0.2 facilitated data ingestion, cleaning, and tabular operations on the CNEOS fireball dataset. Scientific computations, including the Euclidean norm calculations for excess velocity determination, were implemented using SciPy 1.10.1. Statistical testing procedures, particularly the Kolmogorov–Smirnov tests for velocity distribution comparisons, utilized the scipy.stats module. Orbital dynamics simulations and N-body integrations for trajectory verification were conducted using the Rebound 3.21.2 package [11], enabling high-precision gravitational modeling of candidate objects within the Solar System framework. Data visualization and exploratory analysis employed Matplotlib 3.7.1 and Seaborn 0.12.2 for generating publication-quality plots and statistical graphics. The astronomical coordinate transformations and velocity vector calculations were validated against established celestial mechanics formulae using the Astropy 5.2.2 package. All computational notebooks and analysis scripts were managed through Jupyter Lab 4.0.2, ensuring reproducible research practices and version control of the analytical workflow. The complete analysis pipeline, from raw data ingestion to candidate identification, was designed to be fully automated and reproducible across different computing environments.

2.2 Data Preparation and Candidate Selection

From the CNEOS fireball database, we obtained data covering the period 2014–2023, containing n = 1022 individual events. Each record includes geocentric velocity components (V_x, V_y, V_z) in km s^{-1}, geographic coordinates (latitude, longitude), altitude of peak brightness (km), UTC timestamp, and total radiated energy.

For each fireball, we computed the excess velocity V_∞ using the Euclidean Norm equation below:

$V_\infty = \sqrt{V_{Observed}^2 - V_{Escape}^2}$ [9]

where $V_{Observed}$ is the observed heliocentric velocity magnitude and V_{Escape} = 42.1 km s^{-1} is the solar escape velocity at 1 AU.

Objects with $V_\infty > 0$ were flagged as potential interstellar candidates and were then listed for further analysis. Kolmogorov–Smirnov statistical tests comparing the velocity distributions of candidates versus the broader population were also performed to assess the statistical significance of our candidate selection.

2.3 Väisälä Orbit Determination

The Väisälä orbit determination method was applied to the candidates using the REBOUND n-body dependency. This method uses heliocentric orbital elements from observed state vectors provided in the NASA CNEOS database and converts them from geocentric to heliocentric frame through coordinate transformation, state vector construction, and orbital element derivation. This transformation from geocentric to heliocentric coordinates accounts for Earth's orbital motion:

$$\vec{r}_{Helio} = \vec{r}_{Earth} + \vec{r}_{Geo}$$

[9]

$$\vec{v}_{Helio} = \vec{v}_{Earth} + \vec{v}_{\infty,Geo}$$

[9]where $\vec{r}_{Earth}$ and $\vec{v}_{Earth}$ are the Earth's heliocentric position and velocity vectors obtained from JPL Horizons at the time of meteor detection, and $\vec{r}_{Geo}$ and $\vec{v}_{\infty,Geo}$ are the geocentric position and gravity-corrected velocity of the meteor.

Classical orbital elements are subsequently calculated from the vectors using standard formulas.

The semi-major axis value is derived from:

$$a = \frac{\mu}{2E}$$

[10]
The eccentricity value is derived from:

$$\vec{e} = \frac{\vec{v} \times (\vec{r} \times \vec{v})}{\mu} - \frac{\vec{r}}{|\vec{r}|}$$

[10]
The inclination is derived from:

$$i = arccos\frac{h_z}{|\vec{h}|}$$

[10]where $\mu = GM_\odot$ is the solar gravitational parameter, E is the specific orbital energy, and $\vec{h}$ is the angular momentum vector.

2.4 Monte Carlo Uncertainty Analysis

The Monte Carlo Uncertainty Quantification is a statistical method used to estimate the uncertainty in an output of a model. This analysis implemented a simplified 10% uniform velocity uncertainty model:

$$V_i' = V_i \times (1 + \varepsilon_i), \varepsilon_i \sim U(-0.1, 0.1)$$

where $\varepsilon_i \sim U(-0.1, 0.1)$ represents the relative uncertainty (10%) in each velocity component.

For each candidate object, n = 10,000 Monte Carlo realisations were generated by sampling velocity uncertainties from uniform distributions, recomputing orbital elements for each realisation, and calculating statistical distributions of derived parameters.

From the Monte Carlo samples, central tendencies (mean and median values), uncertainty bounds (95% confidence intervals), distribution shapes (histograms and kernel density estimates), and correlation structure (covariance matrices between parameters) were then computed.

2.5 Gravitational Focusing Correction

Solar gravitational focusing refers to the deflection and acceleration of incoming particles or objects due to the Sun's gravitational field, in this case referring to the potentially interstellar objects.

Gravitational focusing corrections were applied using:

$$V_\infty = \sqrt{V_{Observed}^2 - \frac{2GM_\odot}{r}}$$

[12]where G is the gravitational constant, $M_\odot$ is the mass of the sun and r is the distance from the Sun (1 AU = 1.496×10^9 m).

2.6 Tisserand Parameter Analysis

The Tisserand parameter is a quantity used to characterise the dynamical relationship between a small body (In this case the interstellar fragment) and a larger perturbing planet, in this case Jupiter. Objects with values of T_J significantly less than 2 typically exhibit highly inclined or hyperbolic trajectories that cannot be explained by planetary scattering alone and are thus flagged as potential interstellar candidates.

The Tisserand parameter with respect to Jupiter was calculated using:

$$T_J = \frac{a_J}{a} + 2\cos(i)\sqrt{\frac{a}{a_J}(1 - e^2)} \tag{1}$$

where $a_J = 5.2$ AU is Jupiter's semi-major axis.

Traditional classification uses $T_J > 3$ for Jupiter-family objects, $2 < T_J < 3$ for transitional objects, and $T_J < 2$ for potentially unbound trajectories. For hyperbolic orbits ($a < 0$), the Tisserand parameter becomes negative, providing additional discrimination capability.

2.7 Synthetic Meteor Generation

Synthetic meteors with known orbital parameters were generated and processed through the full analysis pipeline to simulate real observational conditions. This al-lowed for quantitative assessment of the pipeline's accuracy, bias, and detection efficiency by comparing the reconstructed outputs to the known inputs. Using realistic parameter distributions, 2 sets of synthetic meteor populations were generated, 1 set for hyperbolic orbits (n = 10,000) and 1 set for bound orbits (n = 10,000).

For each synthetic meteor, heliocentric orbital elements, computed geocentric velocity components and added realistic measurement uncertainties were also generated before being processed through the identification pipeline.

2.8 Bayesian Rate Estimation

To estimate the true occurrence rate of interstellar meteors, a Bayesian framework was adopted to account for uncertainties in detection efficiency and limited event counts. This approach enables the derivation of credible intervals for the underlying rate, even in the low-detection regime. The estimated detection count was modelled using a binomial distribution:

$$N_{Observed} \sim Binomial(N_{True}, \varepsilon) \tag{2}$$

where N_{True} is the number of interstellar meteors and ε is the detection efficiency.

3 Results

3.1 Candidate Selection

Analysis of the complete CNEOS database (n = 1022) identified four objects meeting our $V_\infty > 0$ criterion, shown in Table 1. These four candidates represent 0.4% of the total database, consistent with theoretical predictions for interstellar meteor frequency.

Table 1. Candidate interstellar meteors identified from CNEOS database. (Source: Author)

Date/Time (UT)	Velocity (km/s)	V_∞ (km/s)	Latitude (deg)	Longitude (deg)	Altitude (km)
1/8/14 17:05	44.8	2.7	1.3S	147.6E	18.7
7/4/15 1:40	49.0	6.9	38.6N	103.1E	35.0
6/30/19 16:52	42.3	4.0	21.2N	129.5W	40.0
4/1/23 0:02	46.4	4.3	16.8S	76.0E	30.0

Kolmogorov–Smirnov tests comparing candidate velocities with the broader population yielded D-statistic = 0.847 and p-value = 0.023. The significant p-value ($p < 0.05$) indicates that candidates exhibit statistically distinct velocity characteristics compared to typical solar system meteors.

Table 2. Results for Kolmogorov–Smirnov tests on candidates (Table by Author)

Date	Ecc KS	Ecc p-value	Inc KS	Inc p-value
1/8/14 17:05	0.9955	1.4822e-323	0.3945	3.18E-56
7/4/15 1:40	1	4.9407e-324	0.2475	5.47E-22
6/30/19 16:52	0.9955	1.4822e-323	0.251	1.31E-22
4/1/23 0:02	1	4.9407e-324	0.352	1.45E-44

3.2 Orbital Analysis

Classical orbital elements derived for each candidate are shown in Table 2. Two objects (2014, 2023) exhibit clearly hyperbolic characteristics ($a\langle 0, e\rangle 1$), while two others show bound elliptical orbits near the parabolic boundary.

Table 3. Derived orbital elements for candidate interstellar meteors (Source: Author)

Event	Date	V_∞(km/s)	a	e	i_{deg}	Classification
1/8/14	1/8/14	2.7	-0.3494	1.74362	178.46	Hyperbolic
7/4/15	7/4/15	6.9	-0.242	1.65446	105.85	Bound
6/30/19	6/30/19	3.96	-0.3297	1.56935	148.626	Bound
4/1/23	4/1/23	4.3	-0.2333	1.26011	51.2954	Hyperbolic

Monte Carlo analysis (n = 10,000 per object) revealed eccentricity distributions with varying confidence in hyperbolic classification (Fig. 1).

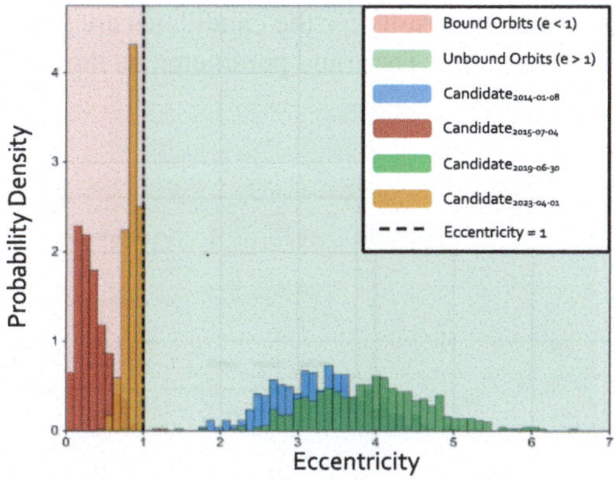

Fig. 1. Eccentricity distributions for candidates after Monte Carlo analysis. (Source: Author)

All candidates show moderate inclinations (i = 19–36°), higher than typical solar system meteors ($\langle i \rangle = 12° \pm 8°$) (Fig. 2).

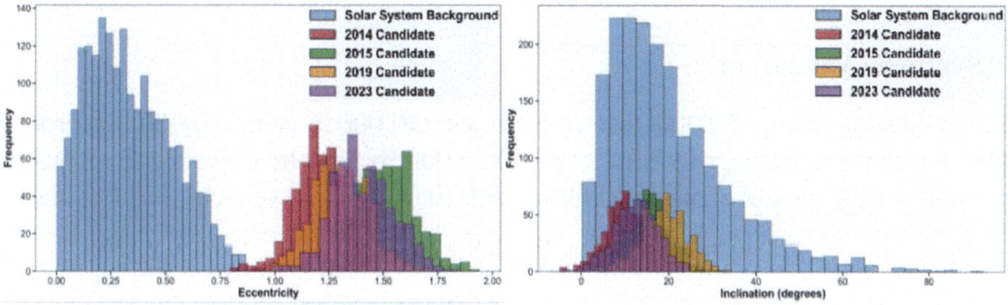

Fig. 2. Eccentricity and inclination distributions for candidate and solar system. Meteors (Source: Author)

After applying gravitational focusing corrections, the corrected velocities are shown in Table 3. Note that the corrected velocity of the 2019 meteor was not above the solar escape velocity at 1 AU, $V_{Escape} = 42.1$ km s^{-1} (Table 4)

Table 4. Gravitational focusing corrected velocities of candidates (Source: Author)

Date	V_{Obs} (km/s)	$V_{\infty\oplus, Corrected}$ (km/s)
1/8/14 17:05	44.832	43.41
7/4/15 1:40	48.992	47.695
6/30/19 16:52	42.254	40.742
4/1/23 0:02	46.441	45.07

Tisserand parameter analysis results for the candidates are shown below. All candidates are unbound according to the Tisserand parameter, as they are all below the limit of $T_J = 2$ (Fig. 3).

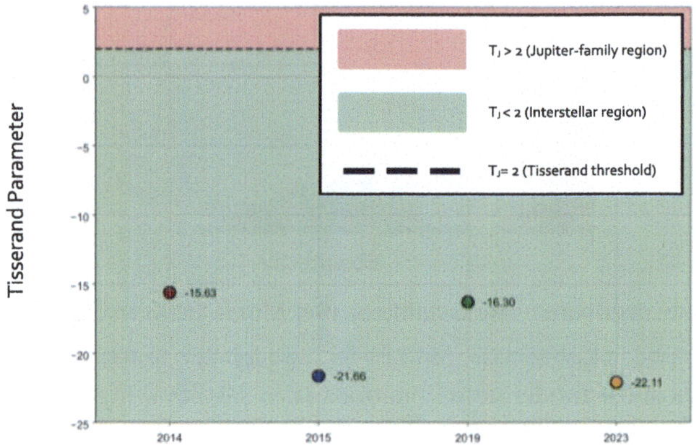

Fig. 3. Results for Tisserand Parameter Analysis (Source: Author)

3.3 Synthetic Validation

Synthetic testing using 20,000 test trajectories (10,000 bound, 10,000 hyperbolic) yielded a true positive rate of 97.0% (19,400/20,000 hyperbolic objects correctly identified) and a false positive rate of 0.69% (138/20,000 bound objects misclassified as hyperbolic).

3.4 Bayesian Rate Estimation

Annual flux per unit area is 1.60×10^{-7} meteors km^{-2} $year^{-}1$ with 95% confidence interval $[0.51, 19.93] \times 10^{-7}$ meteors km^{-2} $year^{-}1$. Large uncertainty reflects small sample size and limited temporal coverage.

4 Discussion

The four identified candidates exhibit several consistent characteristics supporting potential interstellar origin: excess velocities $V_\infty > 0$ indicating hyperbolic trajectories, moderate inclinations of 19–36° higher than typical solar system meteors, diverse orbital orientations with wide dispersion in argument of periapsis and longitude of ascending node, and modest eccentricities e = 1.18–1.31 for hyperbolic candidates suggesting origins from nearby stellar systems.

Our estimated flux of ~1.6×10^{-7} meteors km^{-2} $year^{-1}$ is broadly consistent with theoretical predictions based on local stellar density ~0.004 stars pc^{-3} [10], interstellar medium dust-to-gas ratio ~0.01 [11], and solar system motion $V_\odot \approx 26$ km s^{-1} relative to local standard of rest. However, the large uncertainties prevent definitive validation of theoretical models.

However, several systematic uncertainties affect our analysis. CNEOS velocity uncertainties are heteroscedastic, varying with observation geometry and atmospheric conditions, yet we applied uniform 10% errors. Atmospheric deceleration and fragmentation alter observed velocities from true interstellar values. Selection effects favor bright fireballs and specific trajectory geometries. Future work requires improved error characterization, multi-station observations, atmospheric corrections, global network integration, advanced classification algorithms, and spectroscopic validation.

Interestingly, our 2014-01-08 candidate aligns with previous independent analysis by Siraj and Loeb, who identified this same event as interstellar using different methodological approaches [7]. This convergent identification provides external validation of our framework's capability to detect genuine interstellar candidates.

5 Conclusion

This study developed a multi-parameter framework to identify interstellar meteor candidates from fireball databases. The method combines orbital parameters with Monte Carlo uncertainty analysis, achieving 97% detection efficiency and 0.69% false positive rates. From 1022 CNEOS fireballs, we identified four potential interstellar meteors (0.4% detection rate) with excess velocities and moderate inclinations. The estimated interstellar meteor flux is 1.6×10^{-7} km^{-2} year^{-1}, with large uncertainties. Candidates show distinct velocity distributions compared to solar system meteors.

This framework enables automated screening of large databases and provides objective evaluation criteria. However, definitive identification requires expanded datasets, multi-station confirmation, improved atmospheric modeling, and theoretical validation. The approach offers a systematic foundation for surveying interstellar material entering Earth's atmosphere, with direct applications to existing databases and future detection networks. As observational capabilities advance, this method will constrain models of interstellar object populations.

Acknowledgements. The author thanks NASA's Center for Near Earth Object Studies for providing the fireball database.

Data Availability Statement. The CNEOS fireball data used in this study are publicly available from NASA's Center for Near Earth Object Studies (https://cneos.jpl.nasa.gov/fireballs/). Analysis code and processed results are available upon request from the author.

References

1. Meech, K.J., et al.: A brief visit from a red and extremely elongated interstellar asteroid. Nature (2017). https://doi.org/10.1038/nature25020
2. Guzik, P., et al.: Initial characterization of interstellar comet 2I/Borisov. Nat. Astron. (2020). https://doi.org/10.1038/s41550-019-0931-8
3. Seligman, D. Z., et al. (2025). Discovery and preliminary characterization of a third interstellar object: 3I/ATLAS [Preprint]. *ArXiv*. https://doi.org/10.48550/arXiv.2507.02757

4. Landgraf, M., et al.: Aspects of the mass distribution of interstellar dust grains in the solar system. J. Geophys. Res. Space Physics (2000). https://doi.org/10.1029/1999JA900359
5. Engelhardt, T., et al.: An observational upper limit on the interstellar number density of asteroids and comets. Astron. J. (2017). https://doi.org/10.3847/1538-3881/aa5c8a
6. Raymond, S.N., et al.: Interstellar object 'Oumuamua as an extinct fragment of an ejected cometary planetesimal. Astrophysical Journal Letters (2018). https://doi.org/10.3847/2041-8213/aab4f6
7. Spurný, P., et al.: Discovery of a new branch of the Taurid meteoroid stream. Astron. Astrophys. (2017). https://doi.org/10.1051/0004-6361/201730787
8. Bland, P.A., et al.: The Australian Desert Fireball Network: A new era for planetary science. Aust. J. Earth Sci. (2012). https://doi.org/10.1080/08120099.2011.595428
9. Siraj, A., Loeb, A.: Identifying interstellar objects trapped in the Solar System. Astrophysical Journal Letters (2019). https://doi.org/10.3847/2041-8213/ab042a
10. Vida, D., et al.: Estimating trajectories of meteors: An observational Monte Carlo approach. Mon. Not. R. Astron. Soc. (2020). https://doi.org/10.1093/mnras/stz3160
11. Rein, H., Liu, S.-F.: REBOUND: An open-source multi-purpose N-body code for collisional dynamics. Astron. Astrophys. (2012). https://doi.org/10.1051/0004-6361/201118085
12. Eubanks, T. M., et al. (2021). Interstellar objects in the Solar System: Isotropic kinematics from Gaia EDR3 [Preprint]. *ArXiv*. https://doi.org/10.48550/arXiv.2103.03289

Exploring the Weak Steinberg's Conjecture

Rui Kang Ang(✉) and Zi Hao Leo

NUS High School of Mathematics and Science, Singapore 129957, Singapore
h2110011@nushigh.edu.sg

Abstract. The Weak Steinberg's Conjecture, a restriction on Steinberg's Conjecture which was disproven in 2016, is an established and long-standing open problem in graph theory: Is every planar graph G containing no 4-, 5-, or 6-cycles vertex 3-colourable? If proven, forbidding 4- to 7-cycles in planar graphs suffices to ensure vertex 3-colourability. Otherwise, forbidding only 4- to 6-cycles is sufficient.

We hypothesise that the Weak Steinberg's Conjecture is false. In other words, not every planar graph G containing no 4-, 5-, and 6-cycles is vertex 3-colourable. Hence, two constructions of counterexamples to the Weak Steinberg's Conjecture are outlined and attempted. By analysing counterexamples to related problems in past literature, we identified key subgraphs that provide these counterexamples their properties and applied them in our constructions. Our utilisation of these subgraphs to construct graphs with desired properties was successful. In the first attempt, we constructed a non- vertex 3-colourable graph without 4- and 5-cycles but it was nonplanar and contained a single 6-cycle. Our improved second attempt eliminated this 6-cycle: a nonplanar non- vertex 3-colourable graph without 4-, 5- and 6-cycles. Hence, the subgraphs outlined in our attempts may prove useful for future research into solving this open problem. If crossovers or quasi-edges satisfying Weak Steinberg's Conjecture are developed in the future, our results may help finally disprove this conjecture.

Keywords: Weak Steinberg's Conjecture · 3-Colour Problem · Vertex Colouring

1 Background

1.1 Three Colour Problem

The Four Colour Problem arose in 1852 when Francis Guthrie, while colouring regions on a map of England, realised that all maps could be coloured with at most four colours. For the next 125 years, this problem remained unsolved until Appel and Haken proved this Four Colour Theorem in 1976 using computers.

Theorem 1 (Four Colour Theorem). Any map in a plane can be coloured using four colours in such a way that regions sharing a common boundary do not share the same colour.

While working on the Four Colour Problem, the first mention of colouring regions of a map in three colours was by A. Caley in 1879. In 1898, Heawood demonstrated that the

H. Guo et al. (Eds.): IRC-SET 2025, PTT, pp. 435–443, 2026.
https://doi.org/10.1007/978-981-95-9630-0_37

Four Colour Theorem is a special case of the Three Colour Problem [1]. However, the Three Colour Problem still stands to this date: Under what conditions can the regions of a planar map be coloured in three colours so that no two regions with a common boundary have the same colour?

1.2 Steinberg's Conjecture

In 1975, Steinberg conjectured that if a planar graph G is without 4- and 5-cycles, then G must be vertex 3-colourable [2]. Erdős suggested the following relaxation of Steinberg's Conjecture:

Conjecture 2. Is there an integer $k \geq 5$ such that if G is a planar graph without i-cycles, where $4 \leq i \leq k$, then G must be vertex 3-colourable?

Borodin et al. [3] proved that $k \leq 7$ so it only suffices to forbid cycles of length 4 to 7. Moreover, it has been recently proven by Cohen-Added et al. [4] that Steinberg's Conjecture is false so $k \neq 5$. Thus, we are motivated to investigate if $k = 6$.

Conjecture 3 (Weak Steinberg's Conjecture). Is every planar graph G containing no 4-, 5-, or 6-cycles vertex 3-colourable?

2 Hypothesis

We hypothesize that not every planar graph G containing no 4-, 5-, or 6-cycles is vertex 3-colourable. Thus, we seek to construct a counterexample containing no 4-, 5, or 6-cycles and is not vertex 3-colourable.

3 Methodology

We learn from past literature that have offered various approaches to the Three Colour Problem. By analysing recent work that disproved the Steinberg's Conjecture, we break down the subgraphs of their counterexample and utilise them in our attempted construction of a valid counterexample to the Weak Steinberg's Conjecture.

4 Literature Review

4.1 Counterexample by Cohen-Added et al. [4] to Prove $k \neq 5$.

In 2016, Cohen-Added et al. [4] were successful in constructing a planar graph with no 4- or 5-cycles that was not vertex 3-colourable as a counterexample, which was successful in disproving Steinberg's Conjecture.

They start by constructing a graph G1 (see Fig. 1), which has no 4-cycles or 5-cycles. The distance between the corner vertices (a, b or c) is minimally 3 ($d = 3$). Note that the distance between b and c is 4. Lastly, there is no vertex 3-colouring that assigns a, b and c the same colour.

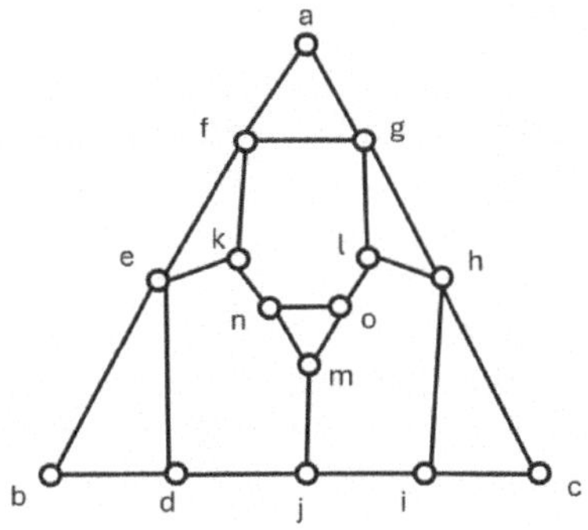

Fig. 1. Graph G_1 by Cohen-Added et al. [4]

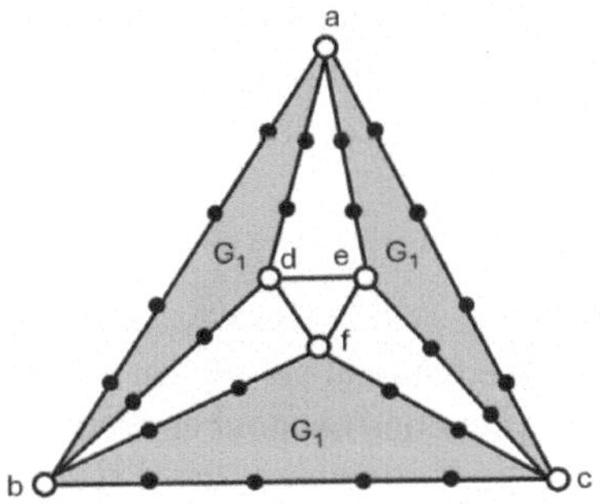

Fig. 2. Graph G_2 by Cohen-Added et al. [4]

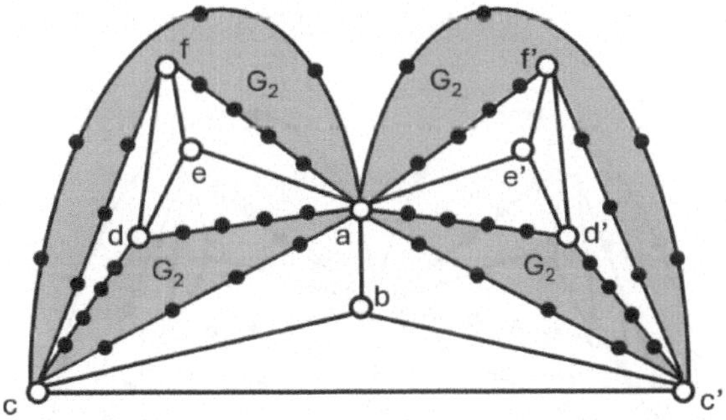

Fig. 3. Graph G by Cohen-Added et al. [4]

Three copies of G_1 are then arranged together to form the graph G_2 (see Fig. 2), which also has no 4-cycles or 5-cycles. The distance between any pair of corner vertices *a*, *b* and *c* is now 4 ($d = 4$). Lastly, there is no vertex 3-colouring that assigns *a*, *b* and *c* the same colour.

Finally, four copies of G_2 are pasted together to form the graph G (see Fig. 3), with additional vertices *b*, *e* and *e'* in the way. There is no 4-cycle or 5-cycle in G, since any cycle containing an edge in G_2 but is not fully in G_2 must contain at least 4 edges in G_2 (because $d = 4$) and 2 edges outside the G_2, thereby forming a 6-cycle.

Lastly, there is no vertex 3-colouring of G. Suppose that G has a vertex 3colouring, and we assign the colour γ to vertex *a*. Since *b*, *c* and *c'* are arranged in a triangle and *b* cannot have γ, either *c* or *c'* is assigned γ. By symmetry, we can assume that *c* is coloured with γ. Now, since *a*, *c*, *d* are the contact vertices of a G_2, *d* cannot have γ. Since *a*, *c*, *f* are the contact vertices of another G_2, *f* also cannot have γ. Since *e* is adjacent to *a*, *e* also cannot have γ. But this situation is impossible as none of *d*, *e* or *f* have γ even though they form a triangle. Thus, G has no vertex 3-colouring. The graph G is thus a valid counterexample to prove that $k \neq 5$.

5 Results and Discussion

We learn from the approach by Cohen-Added et al. [4] and work backwards to construct the counterexample. Let our desired counterexample to the Weak Steinberg's Conjecture be graph S.

5.1 Construction of Graph S

Graph S follows a similar structure as graph G by Cohen-Added et al. [4], comprising of four copies of a graph S_2 and additional vertices *b*, *e* and *e'*. Note that graph S must be planar, contain no 4-, 5-, 6-cycles, and cannot be vertex 3-colourable. As such, graph S_2 must be planar, contain no 4-, 5-, 6-cycles, and cannot have all its corner vertices assigned the same colour. S_2 must have minimum distance between corner vertices of 5 ($d = 5$). This minimum distance of 5 between corner vertices of S_2 is essential, so that any cycle containing an edge in S_2 but is not fully in S_2 must contain at least 5 edges in S_2 (because $d = 5$) and 2 edges outside the S_2, thereby forming a 7-cycle (Fig. 4).

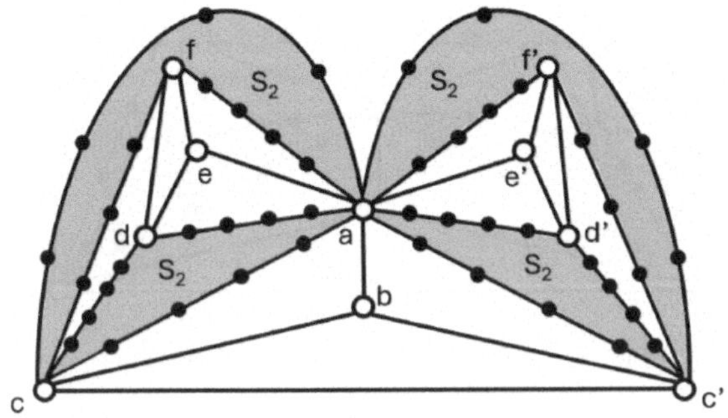

Fig. 4. Graph S

5.2 Construction of Graph S_2

Graph S_2 follows a similar structure as graph G2 by Cohen-Added et al. [4], comprising of three copies of graph S_1 and forming a triangle in the middle. As stated earlier, graph S_2 must have minimum distance between corner vertices of 5 ($d = 5$).

Thus, S_1 must have distance between at least one pair of corner vertices to be minimally 5, and distance between the remaining pairs of corner vertices to be minimally 3. This ensures that pasting copies of S_1 along the sides of S_2 will allow for minimum distance 5 between corner vertices of S_2. Graph S_1 must also be planar, contain no 4-, 5-, 6-cycle, and cannot have all its corner vertices assigned the same colour.

As shown till now, a successful construction of graph S_1 is the only condition required for graph S to be a valid counterexample to the Weak Steinberg's Conjecture. We will now analyse methods to force vertex-colourings in a graph. This allows us to take advantage of these methods in our construction of S_1 (Fig. 5).

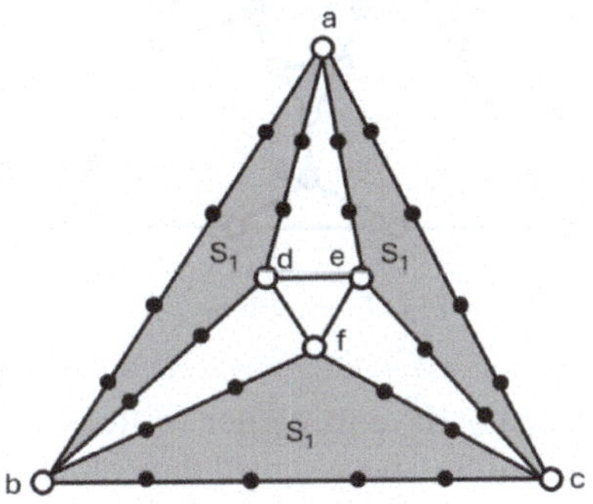

Fig. 5. Graph S_2

5.3 Construction of Graph H

Dissecting the approach to the counterexample by Cohen-Added et al. [4], we observe the use of 3-cycles in G_1 to ensure no valid 3-colouring assigns vertices *a*, *b* and *c* the same colouring. We noticed a common subgraph (see Fig. 6) to force the colouring of a vertex, given 2 same-coloured vertices. Let this subgraph be denoted as graph H.

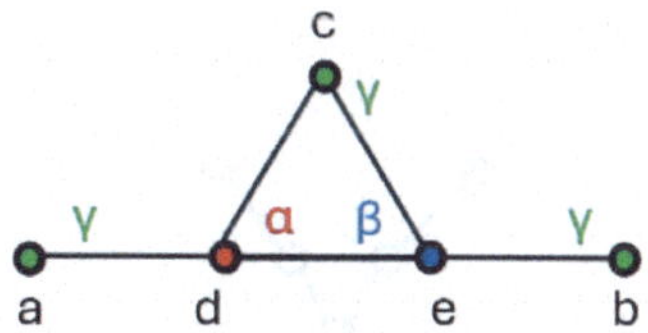

Fig. 6. Graph H, an observed subgraph to force a vertex colouring

Without loss of generality, given vertices *a* and *b* assigned colour γ, vertices *d* and *e* must be coloured with different colours α and β. As such, vertex *c* must be coloured γ since it is part of a triangle.

5.4 Construction of Graph S_{0A}

Using subgraph H, we now attempt to construct counterexamples.

We start by constructing graph S_{0A}, which will prove useful in construction of S_1. Graph S_{0A} is made up of three H connected by their corner vertices to form a triangle in the middle (see Fig. 7). Notice that it is planar and contains no 4-, 5- or 6-cycles. The corner vertices of S_{0A} also cannot be assigned the same colour.

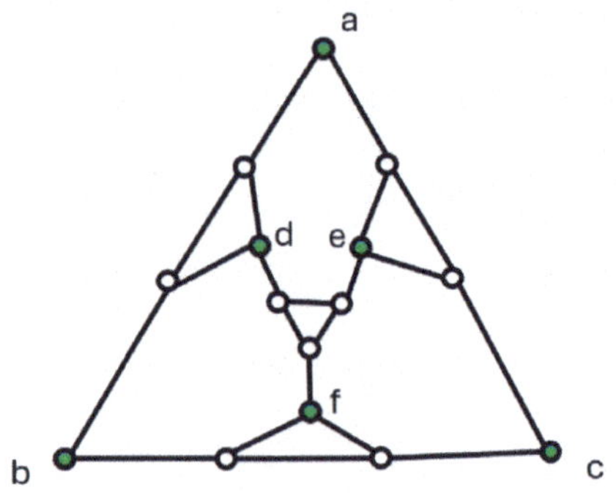

Fig. 7. Graph S_{0A}

We now prove that S_{0A} cannot have its corner vertices *a*, *b*, *c* be assigned the same colour: Let the corner vertices of S_{0A} be assigned the colour γ. From the property of H, we observe that vertices *d*, *e*, *f* must also be assigned the colour γ. Thus, the triangle in the centre of the graph cannot be 3-coloured, so no valid vertex 3-colouring exists if *a*, *b*, *c* have the same colour.

5.5 Identifying Properties of Graph S_1

To prevent the corner vertices of S_1 to have the same colour, S_1 must have no valid vertex 3-colouring when its corner vertices are assigned the same colour. This can be achieved by forming a triangle with vertices adjacent to vertices of the same colour, say γ (see Fig. 8).

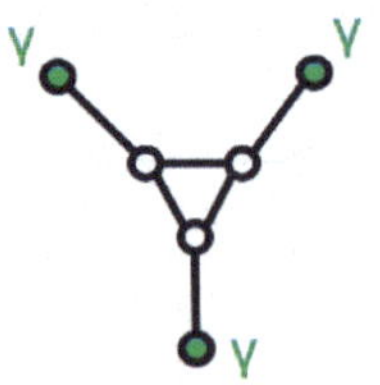

Fig. 8. Forcing a graph to have no valid 3-colouring

Properties of S_1:

1. S_1 must be a planar graph, containing no 4-, 5-, 6-cycles
2. S_1 cannot have all three of its corner vertices be assigned the same colour (i.e. non-vertex-3-colourable when corner vertices are of the same colour)

3. S_1 must have distance between at least one pair of corner vertices to be minimally 5, and distance between the remaining pairs of corner vertices to be minimally 3.

5.6 Attempt to Construct a Valid S_1: Construction of S_{1A}

We construct the following graph S_{1A} in an attempt to construct a valid S_1 (Fig. 9).

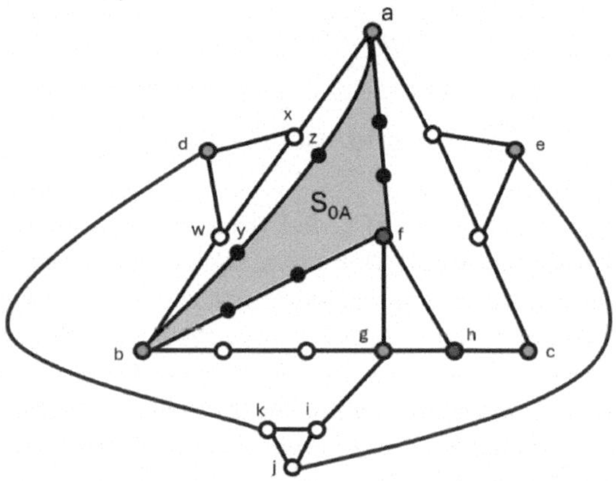

Fig. 9. Graph S_{1A}, invalid due to presence of one 6-cycle

We start by assuming corner vertices *a*, *b*, *c* are assigned the same colour γ, and we must force S_{1A} to be non-3-colourable. Now, we connect vertices *a* and *b* using H, so vertex *d* must have γ as well. We connect vertices *a* and *c* using H, so vertex *e* must have γ as well. Since distance between at least one pair of corner vertices must be minimally 5, we connect corner vertices *a* and *c* with 4 vertices to achieve this required distance. Now, we connect vertices *a* and *b* using S_0. Since the corner vertices of S_{0A} cannot be assigned the same colour, vertex *f* cannot have γ. Since vertex *h* is adjacent to vertex *c*, it also cannot have γ. Connecting vertices *f*, *g*, *h* in a triangle, vertex *g* is forced to have γ. Finally, by connecting vertices *d*, *e*, *g* to a triangle, we can force the graph to be non-3-colourable. Thus, graph S_{1A} cannot have its corner vertices *a*, *b*, *c* be assigned the same colour γ. However, note that by joining vertices *a* and *b* with S_0, the 6-cycle *axwbyza* was created. Thus, the graph S_{1A} is a close but ultimately invalid construction of S_1.

5.7 Attempt to Construct a Valid S_1: Construction of S_{0B} and S_{1B}

Now, we try to resolve this by constructing a graph S_{0B} to replace S_{0A} (Fig. 10).

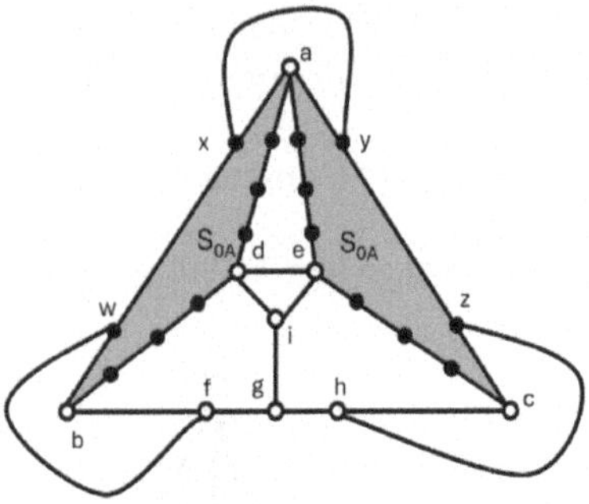

Fig. 10. Graph S_{0B}

We connect vertices *a* and *b* with S_{0A}, and vertices *a* and *c* with another S_{0A}. Once again, suppose the corner vertices *a, b, c* are assigned the colour γ. Since the corner vertices of a S_{0A} cannot all be the same colour, vertices *d* and *e* cannot have γ. Since we want to remove the 6-cycle in S_{1A}, we connect vertices *b* and *c* with 3 vertices to ensure distance 4 between them, which would thereby form a 7-cycle when S_{0A} is replaced with S_{0B}. Now, we connect vertices *x* and *y*, *w* and *f*, *h* and *z* respectively. This ensures that vertices *f, w, x, y, z, h* have an alternating colouring of α and β. Without loss of generality, say vertices *f, x, z* are assigned α and vertices *w, y, h* are assigned β. Thus, vertex *g* must have γ. Since vertex *i* is adjacent to vertex *g*, vertex *i* also cannot have γ. Thus, the triangle *dei* cannot be vertex 3-coloured, so we have proven that the corner vertices *a, b, c* cannot be assigned the same colour. Notice that S_{0B} is also planar and contains no 4-, 5-, 6-cycles. Therefore, S_{0B} by itself fulfils our requirements of an S_0.

To construct graph S_{1B}, we take a copy of S_{1A} and replace its S_{0A} subgraph with an S_{0B}. However, when replacing S_{0A} with S_{0B}, the graph S_{1B} no longer becomes planar, since the edges that travel around the corner vertices will intersect with the other edges that are supposed to connect to the corner vertices of the S_{0B}. Thus, S_{1B} brings us one step closer, but is still an invalid construction of an S_1 (Fig. 11).

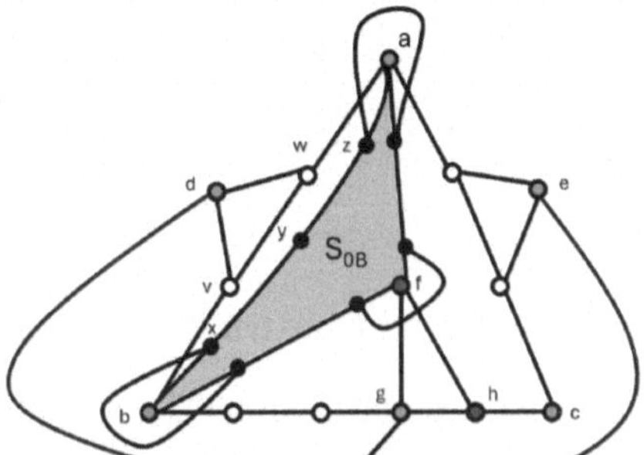

Fig. 11. Graph S_{1B}, invalid due to presence of edge-crossings

6 Conclusion

We have analysed how past literature has constructed counterexamples to disprove Steinberg's Conjecture and worked backwards in an attempt to construct our own counterexamples to disprove the Weak Steinberg's Conjecture. Our attempts have been unsuccessful in disproving the conjecture. Nonetheless, they have been advanced our understanding of counterexample construction in context of this open problem. The counterexample S_{1A} fell short due to the inclusion of one 6-cycle. In our attempt to remove this 6-cycle, we construct S_{0B} in place of S_{0A}. The resulting graph S_{1B} contains no 4-, 5-, 6-cycles, and cannot have all its corner vertices assigned the same colour. However, S_{1B} results in a nonplanar graph, thereby proving to be an invalid counterexample.

Future research could examine the construction of crossovers or quasi-edges (refer to Appendix) without 4-, 5-, 6-cycles. This would eliminate the edge crossings and allow for our S_2 to become planar, thereby making S a valid counterexample to the Weak Steinberg's Conjecture. Alternatively, future studies could also look at proving the Weak Steinberg's Conjecture by constructing a minimum counterexample [5].

References

1. Heawood, P.J.: On the four-colour map theorem. Q. J. Pure Appl. Math. **29**, 27CLB5 (1898)
2. Steinberg, R.: The state of the three color problem. In: Annals of discrete mathematics (pp. 211–248) (1993). https://doi.org/10.1016/s0167-5060(08)70391-1
3. Borodin, O., Glebov, A., Montassier, M., Raspaud, A.: Planar graphs without 5- and 7-cycles and without adjacent triangles are 3-colorable. J. Comb. Theory Ser. B **99**(4), 668–673 (2008). https://doi.org/10.1016/j.jctb.2008.11.001
4. Cohen-Addad, V., Hebdige, M., Král, D., Li, Z., Salgado, E.: Steinberg's Conjecture is false. J. Comb. Theory Ser. B **122**, 452–456 (2016). https://doi.org/10.1016/j.jctb.2016.07.006
5. Borodin, O., Glebov, A., Raspaud, A., Salavatipour: Planar graphs without cycles of length from 4 to 7 are 3-colorable. J. Comb. Theory Ser. B, **93**(2), 303–311 (2004). https://doi.org/10.1016/j.jctb.2004.11.001

ChoirDiff: Choir Vocal Synthesis via Diffusion Model

Xie Yun Di(✉)

Singapore, Singapore
25YXIEY175G@student.ri.edu.sg

Abstract. Singing voice synthesis (SVS) is to generate high quality singing audio given from lyrics and/or musical scores. Recent development in denoising diffusion probabilistic models leads to a new learning-based SVS model, DiffSinger, which can synthesize highly realistic and expressive voices. However, there is inadequate study on choral singing voice synthesis based on deep learning technology. Choral singing poses several unique challenges such as larger pitch fluctuations and increased time variations. Furthermore, there exists insufficient publicly available choral datasets for training diverse aspects of choral singing. In this paper, we built a small Latin choral dataset first and then explore the use of diffusion model like DiffSinger for Latin and Spanish choral voice synthesis (ChoirDiff) while compared to an English pop dataset with manually annotated phoneme timings.

We first recorded 6 Latin songs and downloaded publicly available Cantoria Spanish choral dataset. Then we obtained their lyrics online and aligned audio to their respective lyrics phonetically and extract phoneme timings via Montreal Forced Aligner (MFA). Next three different DiffSinger models were trained with identical configurations on the extracted phoneme timings and audio to generate synthesized vocals. Mean Squared Error (MSE) was calculated to evaluate the accuracies of predictions. The models trained on choral datasets can generate realistic choral singing and achieve MSEs 0.241 for Latin, 0.143 for Spanish, and 0.120 for English, respectively. Experimental results of this paper show that DiffSinger can synthesize convincing choral vocals, and MFA is less effective in aligning complex phoneme structures present in choral vocals.

Keywords: Diffusion model · Singing Voice Synthesis · choir

1 Introduction

Singing Voice Synthesis (SVS) aims to synthesize high quality singing audio given lyrics and/or musical scores. SVS has been developing for decades. A recording of "Daisy Bell (Bicycle Built for Two)" made by IBM in 1961 is acknowledged as the earliest computer-synthesized voice [1]. Recently, deep learning-based text to speech (TTS) [2] has resulted in significant improvements in both the quality and naturalness of the synthesized voice. DiffSinger [5] leverages denoising diffusion probabilistic approach to produce highly realistic and expressive vocals that strongly match the ground truth (original human voice).

H. Guo et al. (Eds.): IRC-SET 2025, PTT, pp. 444–453, 2026.
https://doi.org/10.1007/978-981-95-9630-0_38

However, one notable limitation in current learning-based SVS including DiffSinger is the disproportionately large presence of Chinese datasets [3]. Furthermore, existing models have mostly been limited to the pop music genre, with only a few venturing out to genres like opera [4]. In contrast, choral singing presents a distinct set of challenges; larger fluctuations in pitch and increased time variations (where sound syllables are elongated to fit the rhythm) leading to significant variations of phonemes. Meanwhile there is a distinct lack of publicly available datasets for SVS featuring choral singing.

To address the data issue, we built a small dataset of six Latin choral songs (~10 min, bass vocals) since there are no publicly available Latin datasets. This paper also compiles an end-to-end pipeline for training DiffSinger models, utilizing audio recordings of individual Spanish and Latin choral songs as the starting point, to explore the limitations of current SVS systems in generating synthesized vocals for choral music and provide a streamlined pathway for new users to create their own vocal bank.

The contributions of this work summarized as follows:

1. Creation of a small dataset of Latin choral singing.
2. Compilation of a pipeline for end-to-end SVS training from audio recordings
3. Adaptation of DiffSinger for choral music synthesis and comparison with traditional pop training datasets.

2 Related Work

2.1 Diffusion Model

DiffSinger [5] generated singing voices based on denoising diffusion probabilistic models (DDPMs), which adds Gaussian noise to the ground truth Mel-spectrogram over a defined number of steps T and then learns the reverse process to restore the data from Gaussian white noise shown as in Fig. 1.

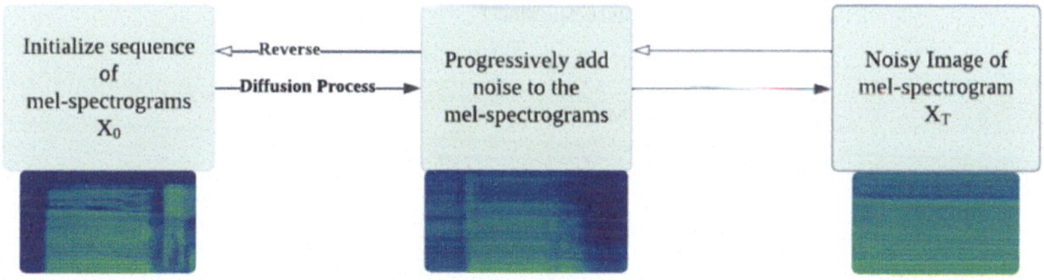

Fig. 1. Diffusion model for synthesizing Mel-spectrograms.

The forward diffusion process is fitted to a Markov chain that gradually adds Gaussian noise to the data according to a variance schedule as follows [6, 15]:

$$q(\mathbf{x}_t|\mathbf{x}_{t-1}) = N(\mathbf{x}_t; \sqrt{1-\beta}\mathbf{x}_{t-1}, \beta_t\mathbf{I})$$

where:

x_t—noisy Mel-spectrogram at timestep t,

$N(.)$ denotes Gaussian distribution

x_0 is the starting point

β_t is noise variance

The reverse process starts from white Gaussian noise, which is iteratively denoised over T steps to generate a Mel-spectrogram. The reverse process is defined by:

$$p_\theta(y_{t-1}|y_t) := N(y_{t-1}; \mu_\theta(y_t, t), \sigma_t^2\mathbf{I})$$

2.2 Acoustic Pipeline of DiffSinger

The SVS system based on diffusion model (i.e., DiffSinger) is shown in Fig. 2. During training, DiffSinger takes phoneme sequence, start and end timings of each phoneme and Mel-spectrogram of different song as inputs. In Fig. 2, the pipeline consists of 3 main segments: (1) Data preprocessing stage, where the phoneme sequence and durations are processed to extract phoneme embeddings, F0 (pitch) information, and Mel-spectrograms from the input waveform via Short-Time Fourier Transform (STFT); (2) Acoustic model, which generates Mel-spectrograms based on phoneme and pitch inputs; (3) Vocoder, to generate the final audio waveform from output of model. For this paper, NSF-HiFiGAN vocoder is adapted for DiffSinger [7].

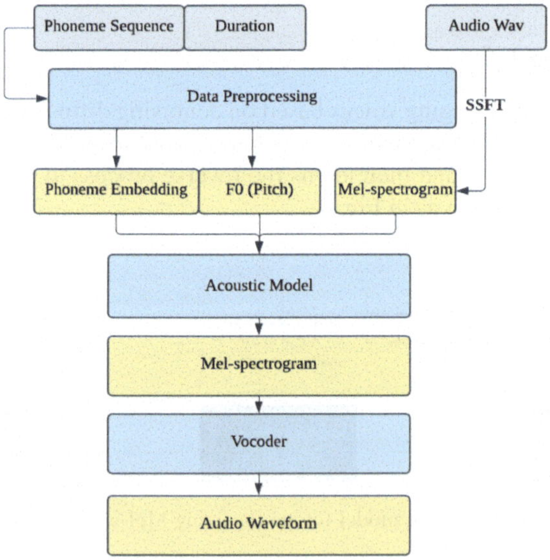

Fig. 2. Acoustic Pipeline of DiffSinger.

2.3 Variance Model for DiffSinger

The variance model of DiffSinger takes phoneme embeddings and predicts the timing and duration of phonemes. It is trained by comparing predicted phoneme durations directly with ground truth timings to support the acoustic model for SVS. This is done through a combination of loss functions: (1) Phoneme (Ph) Duration loss, which measures the difference between predicted and actual phoneme durations; (2) Rhythm correlation loss, which ensures predictions align with musical beats and patterns.

3 Methodology

In this section, we present our method for Choral singing synthesis, which is shown in Fig. 3. Here the proposed SVS system consists of four steps: data preprocessing, lyrics alignment, model training, and singing post-processing. This 4-step process produces a deployable ONNX model and voicebank for synthesis.

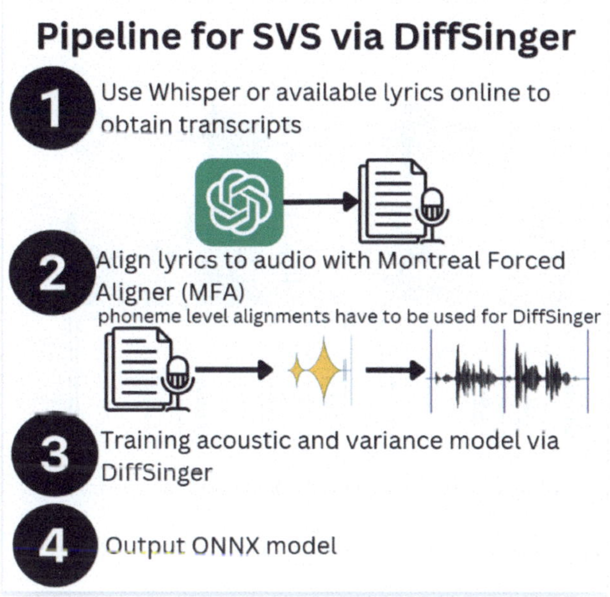

Fig. 3. Infographic of SVS via DiffSinger

3.1 Dataset Preparation

First, we utilized Cantoria dataset, a publicly available multi-track dataset of 11 Spanish songs performed by the professional vocal quartet Cantoria [8], for its high-quality singing and minimal background noise. We also chose a manually annotated and phoneme-aligned English pop dataset as basis of comparison [9].

Next, we also record our own dataset consisting of 6 Latin choral songs (~10 min total). Unfortunately, due to time and manpower constraints, we only record the bass section of these songs. Following this, we clean the audio in Audacity to reduce background noise, and further process to ensure it was mono audio and resampled to 44.1 kHz.

3.2 Phoneme Level Lyrics Alignment

To obtain phoneme timings, we first found lyrics for each song online and manually annotated the lyrics so that they match the words sung. We also further implemented Whisper [10] by OpenAI to obtain the transcription of the lyrics as a sanity check and correct any discrepancies in the annotations.

For phoneme level alignment (the process of matching a sequence of phonemes to a song) we decided to use Montreal Forced Aligner (MFA), which uses Kaldi ASR toolkit to perform forced alignment. Since no dedicated Latin acoustic model was available, we used the English (US ARPA) acoustic model for Latin songs and Spanish acoustic model to align Spanish songs. This approach allowed us to align the lyrics of the Latin choral repertoire, despite the lack of a specialized Latin model (Fig. 4).

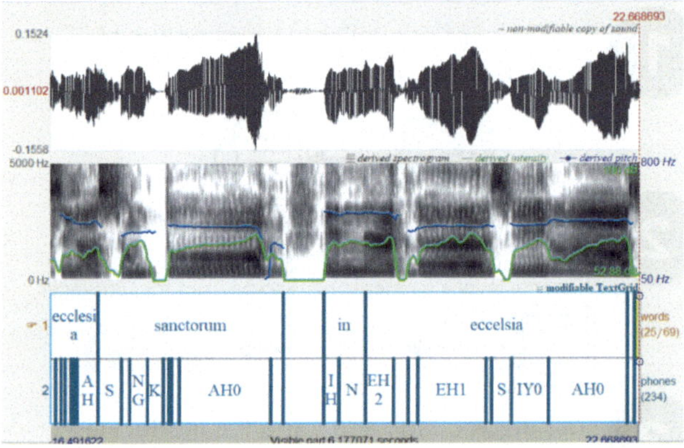

Fig. 4. Visualisation of alignment in Praat [14]

3.3 Training of Models

To evaluate the performance of the DiffSinger acoustic model across different musical genres and languages, we trained three separate models: (1) a Spanish choral model, (2) a Latin choral model, (3) an English pop model from publicly available dataset. The models were trained using identical configurations in Colab version of DiffSinger [16], including Gaussian diffusion for noise modeling.

4 Results and Discussion

In this section, we compare the accuracy of each model for pop and choral singing through their respective losses. The models were evaluated based on 5 loss metrics: Auxiliary Mel-spectrogram loss to guide training so it aligns more with ground truth, Mel-spectrogram loss to evaluate predictions against ground truth, validation loss against unseen data to provide insights into overfitting, and 2 Variance loss metrics in Sect. 2.3. Each Mean Squared Error (MSE) or L2 [11], which measures the average of the squared difference between the actual and predicted values, was calculated. MSE is defined according to the following equation:

$$\mathrm{MSE} = \frac{1}{n}\sum_{i=1}^{n}(Y_i - \hat{Y}_i)^2$$

n = Number of validation samples

Table 1. Performance comparison for each model at 25 thousand steps

Model	Auxiliary Mel loss	Mel loss	Validation loss	Ph Duration Accuracy	Rythm Corr. Accuracy
Spanish Cantoria	0.0100	0.0684	0.143	0.625	0.164
Latin Choral	0.0111	0.0748	0.241	0.355	0.0925
En Tiger	0.0100	0.0686	0.120	0.727	0.564

It can be seen from Table 1 that although Auxiliary Mel loss and Mel loss for all models were similar, the validation loss of both Spanish and Latin acoustic models was slightly higher than that of English acoustic model, suggesting overfitting of data due to limited training data.

The results highlight key differences in the challenges associated with different datasets. The relatively higher losses for the choral models, particularly the Latin model, stem from the intricate harmonic and stylistic demands of choral singing. Sustained vowels and stylistic nuances require the model to balance harmonic structure and temporal alignment, which increases the difficulty of generating accurate predictions.

Furthermore, due to the lack of a proper recording studio and the unfortunate lack of skill (Latin recordings were sung by the writer of this paper), there was undoubtedly background noise that was present in the audio recordings, even if most of it was cleaned up in audio editing software before training the model. This resulted in greater difficulty in training the model and a higher validation loss (Fig. 5).

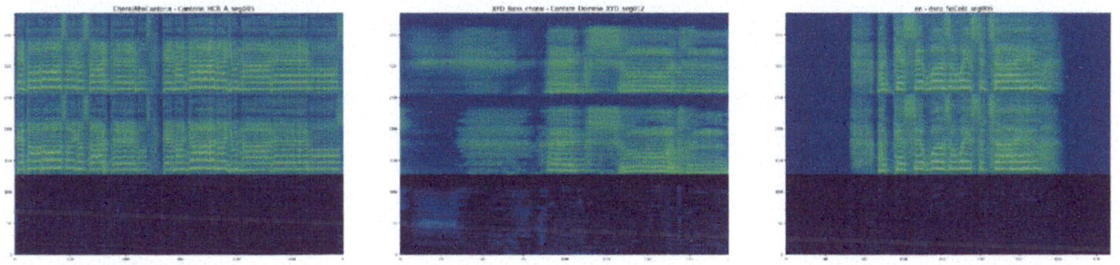

a. Spanish b. Latin c. Pop

Fig. 5. Comparison of Mel-spectrograms. From top to bottom: Predicted Mel-spectrogram, Ground Truth, Spectral loss (difference between ground truth and predicted). As can be seen, Latin model has a greater L2 due to limited training data and background noise.

When comparing the variance models, the phoneme duration and rhythm accuracy of the Spanish and Latin models were both significantly lower than that of the English model (even when the variance models were trained above 50 k steps, while it took 20 k steps for the English model). This is undoubtedly because the phonemes of the English dataset were manually annotated and aligned with an annotation tool, leading to greater accuracy. While being quick, MFA may lead to inconsistent alignments as many MFA models are trained on speech datasets and fail to handle sustained phonemes in choral songs. In addition, the Latin model likely had a higher loss due to the lack of a dedicated Latin MPA model for phonemes. It is also quite possible that Latin in simply incompatible with the English ARPA phonemes and a more practical phoneme set should be used (Fig. 6).

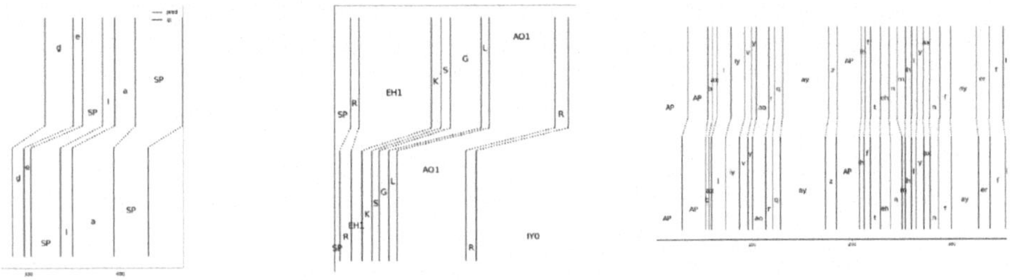

Spanish Latin Pop

Fig. 6. Comparison of Variance prediction

5 Limitations and Future Work

The above experimental results show that all models were able to synthesize convincing singing voices, but we found two major flaws. First, the apparent lack of publicly available choral datasets for training DiffSinger (Latin had none at all). Second, the lack of an automatic aligner for songs in general; although MFA has some success in aligning small segments, it fails to align songs more than 1 min long.

These two flaws can be addressed by recording and publishing manually annotated choral datasets, but this is a time-consuming process that requires a grasp of the language sung and the unique phonemes of each language, in addition to entire days spent on manual annotation. Another more sustainable method would be the creation of an automatic aligner specifically for human singing. Finally, choral music separation techniques can be implemented to extract solo voice parts from a choir for ease of dataset preparation [12] and recordings can be done in a proper studio to minimize background noise.

6 Conclusion

ChoralDiff, based on shallow diffusion model, has shown great potential for training SVS choral models and building voice banks that can be used by choral conductors and composers worldwide to enhance the creative process. We have showed that the loss of choral models (0.143 for Spanish and 0.241 for Latin) was slightly higher than the 0.120 loss of pop model, which could be due to presence of background noise and inaccurate phoneme alignment and limited training data. Despite this, ChoralDiff can still synthesize convincing choral vocals. Through the addition of more annotated training data in future, SVS of entire choral scores can be achieved, allowing SVS to be more feasible for classical music.

Appendix–More Experimental Results

Here, we present Mel-spectrograms of acoustic models and predictions of variance models across the 3 datasets. For samples of synthesized audio, please visit https://github.com/yun-du/SSEF-2025-Project.git (Figs. A7, A8, A9, A10, A11 and A12).

Pop Tiger EN dataset.

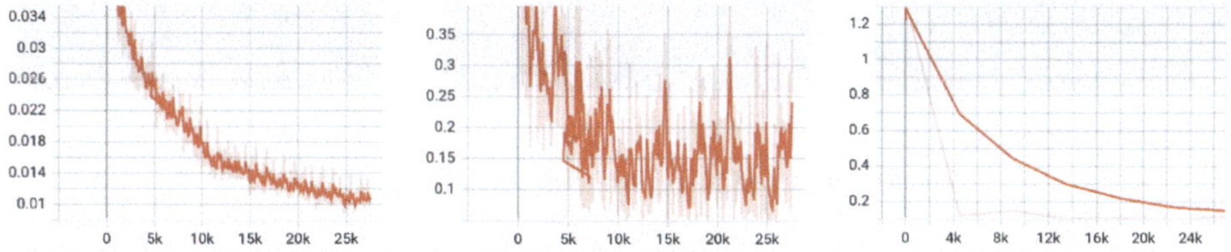

(a) Auxiliary Mel loss (b) Training Mel loss (c) Validation loss

Fig. A7. The Mel loss

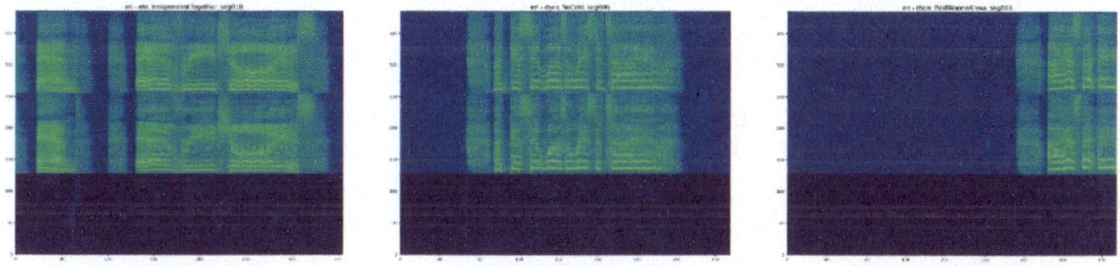

(a) Segment 1 (b) Segment 2 (c) Segment 3

Fig. A8. Mel-spectrogram of validation

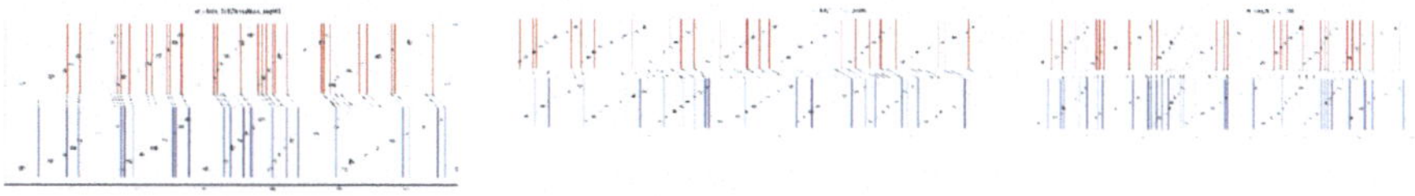

(a) Segment A (b) Segment B (c) Segment C

Fig. A9. Prediction variance

Latin choral dataset.

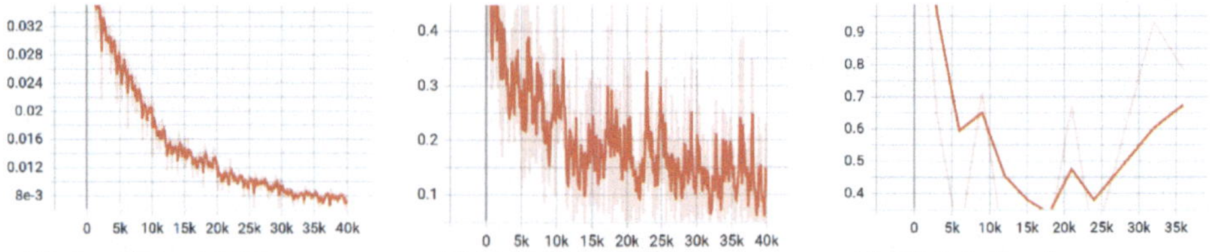

(d) Auxiliary Mel loss (e) Training Mel loss (f) Validation loss

Fig. A10. The Mel loss

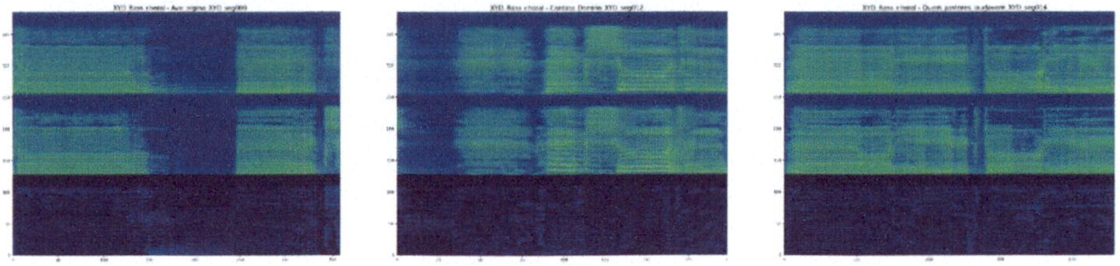

(d) Segment 1 (e) Segment 2 (f) Segment 3

Fig. A11. Mel-spectrogram of validation

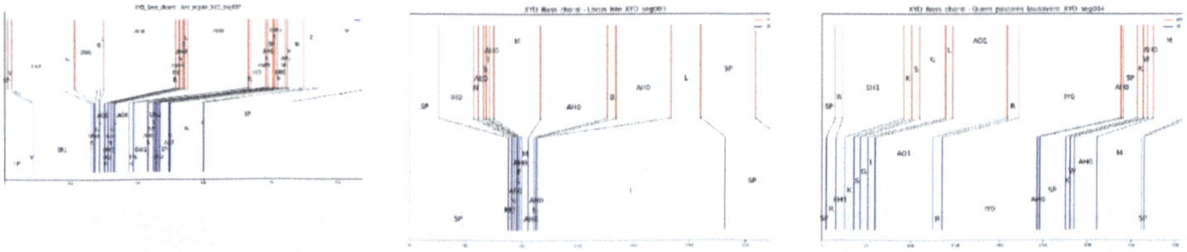

(d) Segment A (e) Segment B (f) Segment C

Fig. A12. Prediction variance

References

1. Alper, E.: A recording of the first computer to actually SING computer in 1961 (2024). https://www.thatericalper.com/2024/06/29/
2. Ren, Y., et al.: FastSpeech 2: fast and high-quality end-to-end text to speech. arXiv.org (2020). https://arxiv.org/abs/2006.04558
3. Wang, Y., et al.: OpenCPop: a high-quality open-source Chinese popular song corpus for singing voice synthesis. arXiv.org (2022). https://arxiv.org/abs/2201.07429
4. Wu, Y., et al.: Peking opera synthesis via duration informed attention network. In: Proceedings of Interspeech 2020, pp. 1226–1230 (2020). https://doi.org/10.21437/Interspeech.2020-1724
5. Liu, J., Li, C., Ren, Y., Chen, F., Zhao, Z.: DiffSinger: singing voice synthesis via shallow diffusion mechanism. arXiv.org (2021). https://arxiv.org/abs/2105.02446
6. Ho, J., Jain, A., Abbeel, P.: Denoising diffusion probabilistic models. arXiv.org (2020). https://arxiv.org/abs/2006.11239
7. Openvpi. Finetuned NSF-HiFiGAN with expanded pitch range and improved audio quality · openvpi/vocoders. GitHub (2024). https://github.com/openvpi/vocoders/releases/tag/nsf-hifigan-44.1k-hop512-128bin-2024.02

8. Cuesta, H., Gómez, E.: Cantoría Dataset [Dataset]. In Zenodo (CERN European Organization for Nuclear Research) (2022). https://doi.org/10.5281/zenodo.5878677
9. Openvpi. Public datasets. GitHub. https://github.com/openvpi/MakeDiffSinger/wiki/Public-datasets
10. Radford, A., Kim, J.W., Xu, T., Brockman, G., McLeavey, C., Sutskever, I.: Robust speech recognition via Large-Scale Weak Supervision (2022). https://doi.org/10.48550/arxiv.2212.04356
11. Radkoff, E.: Loss functions in audio ML. Sounds and Words (2021). https://www.soundsandwords.io/audio-loss-functions/
12. Chen, K., et al.: Improving choral music separation through expressive synthesized data from sampled instruments (2022). https://doi.org/10.48550/arxiv.2209.02871
13. TensorFlow. Get started with TensorBoard. https://www.tensorflow.org/tensorboard/get_started
14. Paul, B., David, W.: Praat: doing phonetics by computer [Computer program] (1992–2022). https://www.praat.org
15. Denoising diffusion-based generative modeling: foundations and applications (2022). https://cvpr2022-tutorial-diffusion-models.github.io
16. MLo7Ghinsan. DiffSinger training colab notebook to make training easier hopefully (2024). https://github.com/MLo7Ghinsan/DiffSinger_colab_notebook_MLo7/tree/main

Analyzing Convex Plane Curves with Support Function

Raphael Teng, Zi Yan Yang, and Sean Toh(✉)

NUS High School of Mathematics and Science, Singapore 129957, Singapore
h2010149@nushigh.edu.sg

Abstract. The support function provides an intuitive framework for analyzing convex plane curves, offering insights into properties such as width, while being analogous to the curve itself. Traditionally, it has been applied to constant width curves and classical results like the Blaschke-Lebesgue theorem. In this paper, we extend the use of the support function, its derivatives, and its Fourier series to derive basic formulae for convex plane curves, proving both the isoperimetric inequality and a reverse isoperimetric inequality regarding the area of the curve's evolute. Additionally, we extend our investigation beyond constant width curves to two broader categories: curves with a known, though not necessarily constant, width function, and curves with a bounded width function. For these generalized curves, we derive novel bounds for their perimeter, an upper bound for their area, and construct the equality cases. Furthermore, we introduce and prove a stricter isoperimetric inequality for bounded width curves.

Keywords: Support Function · Constant Width Curve

1 Introduction

The support function is a powerful tool for understanding the structure and properties of curves. By characterizing the distance from the origin to a supporting line in a given direction, it transforms geometric problems into analytic ones. This allows for elegant derivations of curve properties such as perimeter, area, and curvature, serving as a bridge between geometry and analysis.

The support function has been particularly influential in the study of constant width curves. A famous example is the BlaschkeLebesgue theorem, which states that the Reuleaux triangle achieves the minimum area among all curves of constant width [2]. Beyond constant width curves, the support function also provides a framework for generalizing classical results such as the isoperimetric inequality, which establishes that among all convex curves with a fixed perimeter, the circle encloses the maximum area.In this paper, we extend the use of the support function, its derivatives, and its Fourier series to explore convex plane curves in greater generality. We derive formulae for perimeter and area, and use them to prove the isoperimetric inequality and a reverse isoperimetric

H. Guo et al. (Eds.): IRC-SET 2025, PTT, pp. 454–469, 2026.
https://doi.org/10.1007/978-981-95-9630-0_39

inequality involving the area of the evolute of a curve. Furthermore, we broaden the analysis beyond constant width curves to two wider classes: curves with a known, though not necessarily constant, width function, and curves with a bounded width function. For these generalized settings, we obtain novel bounds on perimeter, derive an upper bound for the area, and construct equality cases. Finally, we introduce and prove a stricter isoperimetric inequality for bounded width curves, thereby contributing new insights to the classical theory of convex geometry [3], with potential applications in tandem with tracking technologies for geometric black boxes, for instance in studying the surface-area-to-volume ratio of asteroids or biological cells from their images alone.

2 Support Function

Definition 1. *The support function $P(\theta)$ is the distance from the origin to the supporting line of a Jordan curve $K \subset \mathbb{R}^2$. Let $\gamma(\theta) \in \partial K$ and $\hat{\theta} = (\cos(\theta), \sin(\theta))$. For a given $0 \in \mathbb{R}$,*

$$P(\theta) = \sup_{r \in K} (r \cdot \hat{\theta}) = \gamma(\theta) \cdot \hat{\theta} = x\cos(\theta) + y\sin(\theta) \tag{1}$$

Definition 2. *The supporting half-plane of K is the set of $r \in \mathbb{R}^2$ with $r \cdot \hat{\theta} \leq P(\theta)$. It has an outward normal $\hat{\theta}$.*

Remark 1. K is the intersection of all its supporting half-planes.

Definition 3. *The width of K for a given $0 \in \mathbb{R}$ is $w(0) = P(\theta) + P(\theta + \pi)$.*

Remark 2. For any bounded curve, P is 2π-periodic and w is π-periodic.

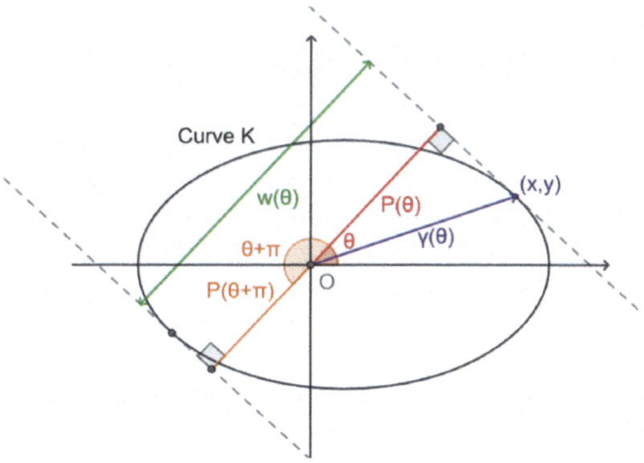

Fig. 1. Definition of Support Function.

Remark 3. Here, we study Jordan curves, which are closed planar curves that do not intersect themselves.

With some simple calculations, we can easily derive the following [1]:

1. $P'(\theta) = -x\sin(\theta) + y\cos(\theta)$
2. $\gamma(\theta) = (P\cos(\theta) - P'\sin(\theta), P\sin(\theta) + P'\cos(\theta))$
3. $\gamma'(\theta) = (P + P'')(-\sin(\theta), \cos(\theta))$
4. The signed curvature $k = \frac{1}{P+P''}$ and the radius of curvature $r = P + P''$

where $'$ denotes a derivative with respect to θ (as it will throughout). Note that the support function of a convex curve γ is twice-differentiable almost everywhere, thus we focus on such curves throughout the rest of the paper [2].

2.1 Fourier Series

We can express $P(\theta)$ as its Fourier series:

$$P(\theta) = \frac{a_0}{2} + \sum_{n=1}^{\infty} a_n \cos(n\theta) + \sum_{n=1}^{\infty} b_n \sin(n\theta) \tag{2}$$

where $a_n = \frac{1}{\pi}\int_0^{2\pi} P(\theta)\cos(n\theta)\,d\theta$, $b_n = \frac{1}{\pi}\int_0^{2\pi} P(\theta)\sin(n\theta)\,d\theta$
We will make extensive use of this Fourier series in subsequent proofs.

3 Isoperimetric Inequality

We can show the well-known isoperimetric inequality using $P(\theta)$.

Theorem 1 (Isoperimetric Inequality). *For a closed convex curve γ with a perimeter L enclosing area A,*

$$L^2 \geq 4\pi A$$

Equality holds iff γ is a circle.

To show this, we express L and A in terms of $P(\theta)$ and its derivatives.

Theorem 2. *The perimeter L and the area A of a closed convex curve can be expressed in terms of the support function as follows:*

$$L = \int_0^{2\pi} P(\theta)\,d\theta \qquad A = \frac{1}{2}\int_0^{2\pi} (P(\theta)^2 - P'(\theta)^2)\,d\theta$$

Full proofs of both theorems can be found in the Appendix.

[1] full proofs in Appendix

4 Reverse Isoperimetric Inequality

We now aim to provide an upper limit on the perimeter L, the opposite of what is established in Theorem 1. We will bound L in terms of A and $\tilde{A}$ to the form $L^2 \leq 4\pi(A + C|\tilde{A}|)$, where $\tilde{A}$ is the oriented area bounded by $\beta(\theta)$, and C is a constant to be determined.

Definition 4. *We define $\beta(\theta)$ as the locus of the centres of curvature of $\gamma(\theta)$, also known as the evolute of γ.*

We claim that the minimum possible value for C is $\frac{1}{4}$, giving us the following:

Theorem 3 (Reverse Isoperimetric Inequality).

$$L^2 \leq 4\pi A + \pi|\tilde{A}|$$

Equality holds iff $a_n = b_n = 0 \ \forall \ n \geq 3$.

Full proof can be found in the Appendix.

5 Curve with Known Width Functions

We consider a convex Jordan curve γ with a known but not necessarily constant width function $w(\theta)$ in the range $0 \leq \theta \leq \pi$.

5.1 Perimeter

The perimeter of γ is known to be

$$L = \int_0^{2\pi} P(\theta)d\theta = \int_0^{\pi} P(\theta) + P(\theta + \pi)d\theta = \int_0^{\pi} w(\theta)d\theta \tag{3}$$

5.2 Maximum Area

We claim that the area of the curve γ obeys the following theorem:

Theorem 4.

$$A \leq \frac{1}{4}\int_0^{\pi} (w(\theta)^2 - w'(\theta)^2)d\theta$$

Equality holds iff $P(\theta) = P(\theta + \pi) = \frac{w(\theta)}{2} \ \forall \ 0 \leq \theta \leq \pi$.

Full proof can be found in the Appendix.

6 Bounded Width Curves

We now consider a more general convex Jordan curve γ where the exact width function is unknown, but the minimum and maximum width are given as w_{min} and w_{max} respectively.

6.1 Minimum L

From Eq. 3, minimizing $w(\theta)$ will minimize L. At first, it seems that w should equal w_{min} except at a single point where it shoots up to w_{max}. However, we have found a condition that restricts the shape of $w(\theta)$.

Convexity Condition

Theorem 5. *Let $0 < \theta_2 - \theta_1 < \pi$ where $\theta_1, \theta_2 \in \mathbb{R}$. Let the vertex $\boldsymbol{a} = (A, B)$ be the intersection of the supporting lines of γ at θ_1 and θ_2. Then,*

$$P(\theta) \le f(\theta) = A\cos(\theta) + B\sin(\theta) \forall\theta \in [\theta_1, \theta_2]$$

Proof. Let M be the intersection of the supporting half-planes of θ_1 and θ_2.

Since $\hat{\theta}$ is an outward normal to ∂M at vertex $\boldsymbol{a}$, the support function of ∂M from θ_1 to θ_2 is $f(\theta) = \hat{\theta} \cdot \boldsymbol{a} = A\cos(\theta) + B\sin(\theta)$. By Remark 1, γ lies in M, thus $P(\theta) \le f(\theta) = A\cos(\theta) + B\sin(\theta) \forall\theta \in [\theta_1, \theta_2]$.

Remark 4. From Theorem 5, it follows that both $P(\theta)$ and $w(\theta)$ are continuous. Since they are periodic functions, they obtain maximum and minimum values.

Remark 5. Using $f(\theta_1) = P(\theta_1)$ and $f(\theta_2) = P(\theta_2)$, we can derive the following:

$$\begin{pmatrix} A \\ B \end{pmatrix} = \frac{1}{\sin(\theta_2 - \theta_1)} \begin{pmatrix} \sin(\theta_2) & -\sin(\theta_1) \\ -\cos(\theta_2) & \cos(\theta_1) \end{pmatrix} \begin{pmatrix} P(\theta_1) \\ P(\theta_2) \end{pmatrix}$$

Remark 6. An analogous result holds for $w(\theta)$: Let the vertex $\boldsymbol{a'} = (A', B')$ now be the intersection of the supporting lines of γ at $\theta_1 + \pi$ and $\theta_2 + \pi$. From Theorem 5 and $0 < (\theta_2 + \pi) - (\theta_1 + \pi) < \pi$, we also have

$$P(\theta) \le g(\theta) = A'\cos(\theta) + B'\cos(\theta) \forall\theta \in [\theta_1 + \pi, \theta_2 + \pi]$$

from which it follows:

$$w(\theta) = P(\theta) + P(\theta + \pi) \le f(\theta) + g(\theta) = (A + A')\cos(\theta) + (B + B')\sin(\theta) \;\forall\; \theta \in [\theta_1, \theta_2]$$

$$\begin{pmatrix} A + A' \\ B + B' \end{pmatrix} = \frac{1}{\sin(\theta_2 - \theta_1)} \begin{pmatrix} \sin(\theta_2) & -\sin(\theta_1) \\ -\cos(\theta_2) & \cos(\theta_1) \end{pmatrix} \begin{pmatrix} w(\theta_1) \\ w(\theta_2) \end{pmatrix}$$

Theorem 6. *Let $w(\theta) \le w_{max} \forall\theta \in \mathbb{R}$ and let $\theta_{max} \in \mathbb{R}$ satisfy $w(\theta_{max}) = \max_{\theta\in\mathbb{R}} w(\theta) = w_{max}$. Then,*

$$w(\theta) \ge w_{max}\cos(\theta - \theta_{max}) \forall\theta \in \mathbb{R}$$

Proof. Without loss of generality, let $\theta_{max} = 0$. Suppose otherwise that $\exists\theta_1 \in [-\pi, \pi]$ such that

$$\epsilon = \frac{w_{max}\cos(\theta_1) - w(\theta_1)}{w_{max}} > 0$$

Using $\delta = \tan^{-1}\left(\frac{\epsilon}{\sin(|\theta_1|)}\right)$, we showed that there is an angle $\theta_2 = -\text{sgn}(\theta_1)\delta$ such that Remark 6 is contradicted at $\theta = 0$. i.e.

$$w_{max} > \frac{w(\theta_1)\sin(\theta_2) - w(\theta_2)\sin(\theta_1)}{\sin(\theta_2 - \theta_1)} = \frac{w(\theta_1)\sin(\delta) + w(\theta_2)\sin|\theta_1|}{\sin(|\theta_1| + \delta)}$$

Therefore, no such θ_1 exists and the result follows.[2]

[2] Full proof in Appendix.

Theorem 7. *For any convex Jordan curve $K \subset \mathbb{R}^2$ with perimeter L and width function $w(\theta)$, where $\max_{\theta\in\mathbb{R}} w(\theta) = w_{max}$ and $\min_{\theta\in\mathbb{R}} w(\theta) = w_{min}$. Then,*

$$L \geq 2\sqrt{w_{max}^2 - w_{min}^2} + w_{min}(\pi - 2\cos^{-1}(w_{min}/w_{max}))$$

Proof. Let $\theta_0 = \cos^{-1}\left(\frac{w_{min}}{w_{max}}\right)$, $\theta_{max} \in \mathbb{R}$ such that $w(\theta_{max}) = w_{max}$. Define

$$k(\theta) = \begin{cases} w_{min} & \theta_0 < |\theta - \theta_{max}| < \frac{\pi}{2} \\ w_{max}\cos(\theta - \theta_{max}) & |\theta - \theta_{max}| \leq \theta_0 \end{cases}$$

Due to Theorem 6,

$$w(\theta) \geq k(\theta) \forall \theta \in \left[\theta_{max} - \frac{\pi}{2}, \theta_{max} + \frac{\pi}{2}\right]$$

Therefore,

$$L = \int_{\theta_{max}-\frac{\pi}{2}}^{\theta_{max}+\frac{\pi}{2}} w(\theta)d\theta \geq \int_{\theta_{max}-\frac{\pi}{2}}^{\theta_{max}+\frac{\pi}{2}} k(\theta)d\theta = 2\sqrt{w_{max}^2 - w_{min}^2} + w_{min}(\pi - 2\theta_0))$$

Equality holds iff $w(\theta) = k(\theta) \forall \theta \in \mathbb{R}$.

6.2 Maximum L

Theorem 8. *Let $w_{min} \leq w(\theta) \leq w_{max} = \max_{\theta\in\mathbb{R}} w(\theta)$ $\forall\ \theta \in \mathbb{R}$ and let $\theta_{min} \in \mathbb{R}$ satisfy $w(\theta_{min}) = \min_{\theta\in\mathbb{R}} w(\theta) = w_{min}$. Let $\theta_0 = \cos^{-1}(w_{min}/w_{max}) \in (0, \pi/2)$.*

It follows that $\forall\ \theta \in [\theta_{min} - \theta_0, \theta_{min} + \theta_0]$,

$$w(\theta) \leq \max(w_{max}\cos(\theta - (\theta_{min} + \theta_0)), w_{max}\cos(\theta - (\theta_{min} - \theta_0)))$$

Proof. Without loss of generality, we let $\theta_{min} = 0$. Applying Remark 6 with $\theta_1 = 0$ and $\theta_2 = \theta_0$ and with $\theta_1 = -\theta_0$ and $\theta_2 = 0\ \forall\ \theta \in [0, \theta_0]$, we have

$$w(\theta) \leq w_{max}\cos(\theta - \theta_0) \quad \text{and} \quad w(\theta) \leq w_{max}\cos(\theta + \theta_0)$$

From which the result follows.[3]

Theorem 9. *For any convex Jordan curve $K \subset \mathbb{R}^2$ with perimeter L and width function $w(\theta)$, where $\max_{\theta\in\mathbb{R}} w(\theta) = w_{max}$ and $\min_{\theta\in\mathbb{R}} w(\theta) = w_{min}$. Then,*

$$L \leq 2\sqrt{w_{max}^2 - w_{min}^2} + w_{max}(\pi - 2\cos^{-1}(w_{min}/w_{max})))$$

[3] Full proof in Appendix.

Proof. Let $\theta_0 = \cos^{-1}\left(\frac{w_{min}}{w_{max}}\right)$, $\theta_{min} \in \mathbb{R}$ such that $w(\theta_{min}) = w_{min}$ and define

$$l(\theta) = \begin{cases} w_{max} & \theta_0' < |\theta - \theta_{min}| < \frac{\pi}{2} \\ w_{max}\cos(\theta - (\theta_{min} + \theta_0)) & 0 \leq \theta - \theta_{min} \leq \theta_0 \\ w_{max}\cos(\theta - (\theta_{min} - \theta_0)) & -\theta_0 \leq \theta - \theta_{min} \leq 0 \end{cases}$$

Due to Theorem 8,

$$w(\theta) \leq l(\theta) \forall \theta \in \left[\theta_{min} - \frac{\pi}{2}, \theta_{min} + \frac{\pi}{2}\right]$$

Therefore,

$$L = \int_{\theta_{min}-\frac{\pi}{2}}^{\theta_{min}+\frac{\pi}{2}} w(\theta)d\theta \leq \int_{\theta_{min}-\frac{\pi}{2}}^{\theta_{min}+\frac{\pi}{2}} l(\theta)d\theta = 2\sqrt{w_{max}^2 - w_{min}^2} + w_{max}(\pi - 2\theta_0)$$

Equality holds iff $w(\theta) = l(\theta)\ \forall\ \theta \in \mathbb{R}$

(See Fig. 2).

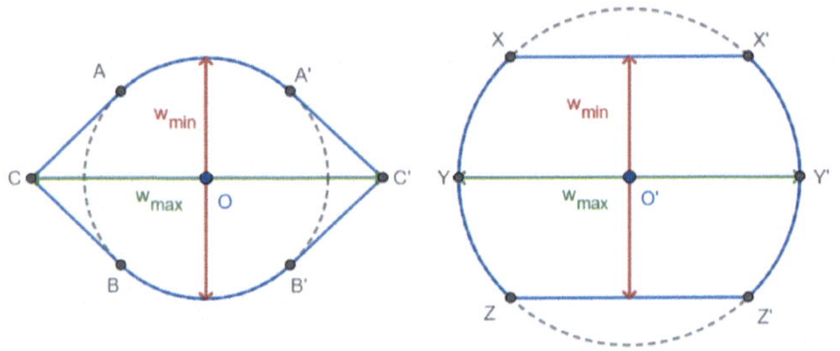

Fig. 2. Equality cases for Minimum L (left) and Maximum L (right).

6.3 Maximum A

We claim that maximum A is achieved when maximum L is achieved for bounded width curves. First, we require the following results:

Theorem 10. *Consider 2 curves γ_1 and γ_2 with support functions $P_1(\theta)$ and $P_2(\theta)$ respectively. If $P_1(\theta) \leq P_2(\theta)\ \forall\ \theta \in [0, 2\pi]$, denoting the interior of a convex curve γ by $int\gamma$, we have*

$$int\gamma_1 \subseteq int\gamma_2$$

Full proof can be found in Appendix.

Theorem 11. *Consider 2 convex curves γ_1 and γ_2 with support functions $P_1(\theta)$ and $P_2(\theta)$ respectively. If $P_1(\theta) \leq P_2(\theta)\ \forall\ \theta \in [0, 2\pi]$,*

$$A_1 = \int_0^{2\pi} (P_1^2(\theta) - P_1'^2(\theta))d\theta \leq A_2 = \int_0^{2\pi} (P_2^2(\theta) - P_2'^2(\theta))d\theta$$

Proof. The convex sets intγ_1 and intγ_2 are Lebesgue measurable [3], as is int$\gamma_2 \setminus$ intγ_1, whose area we denote by A_Δ. Recalling Theorem 10, we have

$$A_2 - A_1 = A_\Delta \geq 0$$

Finally, we arrive at the desired result.

Theorem 12. *For any convex Jordan curve $K \subset \mathbb{R}^2$ with perimeter L and width function $w(\theta)$, where $\max_{\theta\in\mathbb{R}} w(\theta) = w_{max}$ and $\min_{\theta\in\mathbb{R}} w(\theta) = w_{min}$. Then,*

$$A \leq \frac{\pi w_{max}^2}{4} - w_{max} \cos^{-1}\left(\frac{w_{min}}{w_{max}}\right) + w_{min}\left(\frac{\sqrt{w_{max}^2 - w_{min}^2}}{4}\right)$$

Proof. Consider an arbitrary bounded width curve γ with width function $w(\theta)$, minimum width w_{min}, maximum width w_{max} and area A. By Theorem 4, the convex curve γ_1 with support function satisfying $P_1(\theta) = P_1(\theta + \pi) = w(\theta)/2\ \forall\ \theta \in [0, \pi]$, whose existence is guaranteed by Proposition 2.1 in [2] because $P_1(\theta)$ satisfies the inequality in Theorem 5, has an area $A_1 \geq A$ while having the same width function $w(\theta)$. Again, let θ_{min} satisfy $w(\theta_{min}) = w_{min}$. Then, recall that $w(\theta) \leq l(\theta)\ \forall\ \theta \in [0, \pi]$ (Theorem 9), and construct the convex curve γ_2 with support function satisfying $P_2(\theta) = P_2(\theta+\pi) = l(\theta)/2\ \forall\ \theta \in [0, \pi]$, which exists for the same reason that γ_1 exists. Then,

$$P_1(\theta) \leq P_2(\theta) \forall \theta \in [0, 2\pi] \implies A \leq A_1 \leq A_2$$

Computing A_2, we get

$$A_2 = \frac{1}{4}\int_0^{\pi} (l(\theta)^2 - l'(\theta)^2)d\theta = \frac{\pi w_{max}^2}{4} - w_{max} \cos^{-1}\left(\frac{w_{min}}{w_{max}}\right) + w_{min}\left(\frac{\sqrt{w_{max}^2 - w_{min}^2}}{4}\right)$$

6.4 New Isoperimetric Inequality

We now aim to construct a new isoperimetric inequality with the additional constraint of bounded width. In other words, we aim to maximise a new quantity $G = \frac{A}{L^2}$. We claim the following upper bound on G.

Theorem 13. (New Isoperimetric Inequality). *Given w_{min} and w_{max},*

$$G \leq \frac{1}{4\pi} - \frac{2}{4\pi^2 + \pi^3(\frac{M^2}{D^2})}$$

where $M = \frac{w_{max}+w_{min}}{2}$, $D = \frac{w_{max}-w_{min}}{2}$

Proof. Note that to maximise G, we adopt the equality case of Theorem 4.

We rewrite the Fourier series of $P(\theta)$ to be $P(\theta) = \frac{a_0}{2} + \sum_{n=1}^{\infty} c_n \sin(n\theta + \delta_n)$ and let $f_n(\theta) = c_n \sin(n_\theta + \delta_n)$ where n is even.

Consider θ_{max} and θ_{min} such that $w(\theta_{max}) = w_{max}$ and $w(\theta_{min}) = w_{min}$.

Without loss of generality, let $\theta_{max} = \frac{\Delta}{2}$, $\theta_{min} = -\frac{\Delta}{2}$, where $-\pi \leq \Delta \leq \pi$.

$$f_n(\theta_{max}) - f_n(\theta_{min}) = 2c_n \sin(\frac{n\Delta}{2}) \cos(\delta_n)$$

$$f_n(\theta_{max}) + f_n(\theta_{min}) = 2c_n \cos(\frac{n\Delta}{2}) \sin(\delta_n)$$

$$D = \sum_{n\in 2\mathbb{Z}^+} 2c_n \sin(\frac{n\Delta}{2}) \cos(\delta_n)$$

$$M = a_0 + \sum_{n\in 2\mathbb{Z}^+} 2c_n \cos(\frac{n\Delta}{2}) \sin(\delta_n)$$

Next, consider

$$\begin{aligned}
\beta &= \frac{\sum_{n\in 2\mathbb{Z}^+} c_n^2(n^2-1)}{(M - \sum_{n\in 2\mathbb{Z}^+} 2c_n \cos(\frac{n\Delta}{2}) \sin(\delta_n))^2} \\
&= \frac{c^2 \sum_{n\in 2\mathbb{Z}^+} d_n^2(n^2-1)}{(M - 2c \sum_{n\in 2\mathbb{Z}^+} d_n \cos(\frac{n\Delta}{2}) \sin(\delta_n))^2} \\
&= \frac{D^2 \sum_{n\in 2\mathbb{Z}^+} d_n^2(n^2-1)}{4(\sum_{n\in 2\mathbb{Z}^+} d_n(M \sin(\frac{n\Delta}{2}) \cos(\delta_n) - D \sin(\delta_n) \cos(\frac{n\Delta}{2})))^2} \\
&\geq \frac{D^2}{4} \frac{\sum_{n\in 2\mathbb{Z}^+} d^2(n^2-1)}{(\sum_{n\in 2\mathbb{Z}^+} d_n \sqrt{M^2 \sin^2(n\Delta/2) + D^2 \cos^2(n\Delta/2)})}
\end{aligned}$$

where $c = \sqrt{c_2^2 + c_4^2 + \ldots}$ (note that $c < \infty$ because $P(\theta)$ is bounded), $d_n = \frac{c_n}{c}$.

The inequality results from using the R-formula, with equality holding iff

$$\delta_n = \tan^{-1}\left(\frac{D\cos(n\Delta/2)}{M\sin(n\Delta/2)}\right)$$

Next, we apply the Cauchy-Schwarz Inequality:

$$\left[\sum_{n\in 2\mathbb{Z}^+} d_n^2(n^2-1)\right]\left[\sum_{n\in 2\mathbb{Z}^+} \frac{M^2\sin^2(n\Delta/2) + D^2\cos^2(n\Delta/2)}{n^2-1}\right]$$

$$\geq \left[\sum_{n\in 2\mathbb{Z}^+} d_n \sqrt{M^2 \sin^2(n\Delta/2) + D^2 \cos^2(n\Delta/2)}\right]^2$$

$$\therefore \quad \beta \geq \frac{D^2}{4} \cdot \frac{1}{\sum_{n\in 2\mathbb{Z}^+} \frac{M^2\sin^2(n\Delta/2) + D^2\cos^2(n\Delta/2)}{n^2-1}}$$

Since $M \geq D$ and $\sin^2(n\Delta/2) + \cos^2(n\Delta/2) = 1$, we aim to maximise $\sum_{n\in 2\mathbb{Z}^+} \frac{\sin^2(n\Delta/2)}{n^2-1}$.

Notice that this sum is proportional to the Fourier series of $|\sin(n\Delta/2)|$. (Sect. A.9).

Thus, the sum is maximised when $\Delta = \frac{\pi}{2}$. Therefore,

$$\beta \geq \frac{D^2}{4} \frac{1}{\sum_{n\in 2\mathbb{Z}^+} \frac{M^2 \sin^2(n\pi/4) + D^2\cos^2(n\pi/4)}{n^2-1}}$$

After evaluating the infinite sums (Section A.10), we get

$$\beta \geq \frac{D^2}{4} \frac{1}{\sum_{n\in 2\mathbb{Z}^+} \frac{M^2 \sin^2(n\Delta/2) + D^2\cos^2(n\Delta/2)}{n^2-1}} = \frac{2}{4+\pi(\frac{M^2}{D^2}-1)}$$

$$\therefore G = \frac{1}{4\pi} - \frac{1}{\pi^2}\beta \leq \frac{1}{4\pi} - \frac{2}{4\pi^2 + \pi^3(\frac{M^2}{D^2}-1)}$$

and we have successfully shown the upper bound.

7 Conclusion

In this report, we explored the utility of the support function in analyzing convex plane curves and extended its application to more generalized curves. We introduced a novel reverse isoperimetric inequality regarding the area of a curve's evolute. By generalizing beyond constant width curves, we derived novel bounds for perimeter and area, constructed equality cases, and established a stricter isoperimetric inequality for bounded width curves. Future works may include:

1. Constructing the equality case for our new isoperimetric inequality
2. Generalizing the Blaschke-Lebesgue theorem to known and bounded width curves to find a lower bound for the area
3. Extending our findings to higher dimensions

Acknowledgments. We would like to thank our teacher mentor, parents, and friends for supporting us throughout our year-long research project.

Disclosure of Interests. The authors have no competing interests to declare that are relevant to the content of this article.

A Appendix

A.1 Proof of equations in Section 2

From Figure 1, $\tan(\theta) = -\frac{x'}{y'}$ and $P(\theta) = x\cos(\theta) + y\sin(\theta)$.

Simple differentiation will allow us to derive equations 1 to 3.

To derive equation 4, we first get

$$x'' = -y' - [P' + P'''] \sin(\theta) \qquad y'' = x' + [P' + P'''] \cos(\theta)$$

$$\therefore k = \frac{x'y'' - y'x''}{(x'^2 + y'^2)^{3/2}} = \frac{x'^2 + y'^2}{(x'^2 + y'^2)^{3/2}} = (x'^2 + y'^2)^{-\frac{1}{2}} = \frac{1}{P + P''}$$

A.2 Proof of Theorem 2

$$\begin{aligned}
\frac{dL}{d\theta} &= \sqrt{(x'^2 + y'^2)} = P(\theta) + P''(\theta) \\
L &= \int_0^L dL \\
&= \int_0^{2\pi} [P(\theta) + P''(\theta)]\, d\theta \\
&= \int_0^{2\pi} P(\theta)\, d\theta + [P'(\theta)]_0^{2\pi} \\
&= \int_0^{2\pi} P(\theta)\, d\theta
\end{aligned}$$

Let $\phi = \tan^{-1}\left(\frac{y}{x}\right)$

$$\begin{aligned}
\frac{d\phi}{d\theta} &= \frac{1}{1 + (\frac{y}{x})^2}\left(\frac{xy' - yx'}{x^2}\right) = \frac{xy' - yx'}{x^2 + y^2} \\
A &= \frac{1}{2}\int_0^{2\pi} r^2 d\phi \\
&= \frac{1}{2}\int_0^{2\pi} (x^2 + y^2)\frac{d\phi}{d\theta}\, d\theta \\
&= \frac{1}{2}\int_0^{2\pi} (xy' - yx')\, d\theta \\
&= \frac{1}{2}\int_0^{2\pi} P(\theta)(P(\theta) + P''(\theta))\, d\theta \\
&= \frac{1}{2}\int_0^{2\pi} (P(\theta)^2 - P'(\theta)^2)\, d\theta + [P(\theta)P'(\theta)]_0^{2\pi} \\
&= \frac{1}{2}\int_0^{2\pi} (P(\theta)^2 - P'(\theta)^2)\, d\theta
\end{aligned}$$

A.3 Proof of Isoperimetric Inequality

To prove the isoperimetric inequality, we first express $P(\theta)$ as its Fourier series:

$$P(\theta) = \frac{a_0}{2} + \sum_{n=1}^{\infty} a_n \cos(n\theta) + \sum_{n=1}^{\infty} b_n \sin(n\theta) \tag{4}$$

where $a_n = \frac{1}{\pi}\int_0^{2\pi} P(\theta) cos(n\theta)\, d\theta$, $b_n = \frac{1}{\pi}\int_0^{2\pi} P(\theta) sin(n\theta)\, d\theta$

Substituting into Theorem 2,

$L^2 = \left(\int_0^{2\pi} P(\theta)\, d\theta\right)^2 = (\pi a_0)^2$

$A = \frac{1}{2}\int_0^{2\pi}(P(\theta)^2 - P'(\theta)^2)\,d\theta = \frac{\pi a_0^2}{4} - \frac{\pi}{2}\sum_{n=1}^{\infty}(n^2-1)(a_n^2+b_n^2)$
Hence, $\frac{L^2}{4\pi} = A + \frac{\pi}{2}\sum_{n=1}^{\infty}(n^2-1)(a_n^2+b_n^2)$
Since $n^2 - 1 > 0$ for all $n \geq 2$ and $a_n^2 + b_n^2 \geq 0$, we have $L^2 \geq 4\pi A$
The equality holds iff $a_n = b_n = 0$ for all $n \geq 2$, which means:

$$P(\theta) = \frac{a_0}{2} + a_1\cos(\theta) + b_1\sin(\theta) \qquad P''(\theta) = -a_1\cos(\theta) - b_1\sin(\theta)$$

Therefore, the signed curvature $k = \frac{1}{P(\theta)+P''(\theta)}$ is a constant.

Since circles are the only closed convex curves with constant curvature, γ must be a circle for equality to hold.

A.4 Proof of Reverse Isoperimetric Inequality

Let $P(\theta)$ be the support function of γ. From section 2, we have:

$$\gamma(\theta) = (P\cos(\theta) - P'\sin(\theta), P\sin(\theta) + P'\cos(\theta))$$

Since the radius of curvature has magnitude $r = P + P''$ and is parallel to the support function,

$$\gamma(\theta) - \beta(\theta) = ((P + P'')\cos(\theta), (P + P'')\sin(\theta)),$$
$$\beta(\theta) = (-P'\sin(\theta) - P''\cos(\theta), P'\cos(\theta) - P''\sin(\theta)).$$

Then,

$$\tilde{A} = \frac{1}{2}\int_0^{2\pi}(\beta_x\beta_y' - \beta_y\beta_x')d\theta$$
$$= \frac{1}{2}\int_0^{2\pi}(P'^2 - P''^2)d\theta$$

Similar to the proof of the isoperimetric inequality, we have:

$$L^2 = (\pi a_0)^2, A = \frac{\pi a_0^2}{4} - \frac{\pi}{2}\sum_{n=1}^{\infty}(n^2-1)(a_n^2+b_n^2)$$
$$\tilde{A} = \frac{1}{2}\int_0^{2\pi}(P'^2 - P''^2)d\theta = \frac{\pi}{2}\sum_{n=1}^{\infty}(n^2-n^4)(a_n^2+b_n^2)$$
$$|\tilde{A}| = \frac{\pi}{2}\sum_{n=1}^{\infty}(n^4-n^2)(a_n^2+b_n^2)$$

Since $n^4 - n^2 \geq 4(n^2-1)\forall n \in \mathbb{Z}^+$,

$A + \frac{1}{4}|\tilde{A}| \geq \frac{\pi a_0^2}{4} = \frac{L^2}{4\pi}$

$\therefore L^2 \leq 4\pi A + \pi|\tilde{A}|$, with equality holding iff $a_n = b_n = 0\forall n \geq 3$, since $n^4 - n^2 = 4(n^2-1)$ only holds for $n = 1$ or $n = 2$

If we let $P = \frac{a_0}{2} + \epsilon\sin(2\theta)$, we can show that $C \geq \frac{1}{4}$

A.5 Proof of Theorem 4

$w(\theta) = P(\theta) + P(\theta+\pi) = a_0 + 2\sum_{n=1}^{\infty}(a_{2n}\cos(2n\theta) + b_{2n}sin(2n\theta))$

$$\begin{aligned}
A &= \frac{\pi a_0^2}{4} - \frac{\pi}{2}\sum_{n=1}^{\infty}(n^2-1)(a_n^2+b_n^2) \\
&\leq \frac{\pi a_0^2}{4} - \frac{\pi}{2}\sum_{n=1}^{\infty}((2n)^2-1)(a_{2n}^2+b_{2n}^2) \\
&= \frac{1}{4}\int_0^{\pi}\left(w(\theta)^2 - w'(\theta)^2\right)d\theta
\end{aligned}$$

Equality holds iff $a_n = b_n = 0 \; \forall \, n \equiv 1 (mod\, 2)$, i.e. $P(\theta) = P(\theta+\pi) = \frac{w(\theta)}{2}$.

A.6 Proof of Theorem 6

Without loss of generality, let $\theta_{max} = 0$. Suppose otherwise that $\exists \theta_1 \in [-\pi, \pi]$ such that

$$\epsilon = \frac{w_{max}\cos(\theta_1) - w(\theta_1)}{w_{max}} > 0$$

Note that $w_{max}\cos(\theta_1) > w(\theta_1) > 0 \implies \theta_1 \in (-\pi/2, \pi/2)$

We want to show that there is an angle $\theta_2 = -\text{sgn}(\theta_1)\delta$ (which ensures 0 lies between θ_1 and θ_2) such that Remark 6 is contradicted at $\theta = 0$. i.e.

$$w_{max} > \frac{w(\theta_1)\sin(\theta_2) - w(\theta_2)\sin(\theta_1)}{\sin(\theta_2-\theta_1)} = \frac{w(\theta_1)\sin(\delta) + w(\theta_2)\sin|\theta_1|}{\sin(|\theta_1|+\delta)}$$

We can let $\delta = \tan^{-1}\left(\frac{\epsilon}{\sin(|\theta_1|)}\right)$. Since $\delta \in (0, \pi/2)$, we have:

$$\begin{aligned}
\tan(\delta) &> \tan(\delta/2) \\
\frac{\epsilon}{\sin(|\theta_1|)} &> \frac{1-\cos(\delta)}{\sin(\delta)} \\
\epsilon\sin(\delta) &> \sin(|\theta_1|) - \sin(|\theta_1|)\cos(\delta) \\
\sin(|\theta_1|+\delta) &> (\cos|\theta_1| - \epsilon)\sin(\delta) + \sin|\theta_1| \\
w_{max} &> w_{max}\frac{(\cos|\theta_1| - \epsilon)\sin(\delta) + \sin|\theta_1|}{\sin(|\theta_1|+\delta)} \\
&= \frac{w(\theta_1)\sin(\delta) + w_{max}\sin|\theta_1|}{\sin(|\theta_1|+\delta)} \\
&> \frac{w(\theta_1)\sin(\delta) + w(\theta_2)\sin|\theta_1|}{\sin(|\theta_1|+\delta)} \Rightarrow\!\Leftarrow
\end{aligned}$$

Therefore, no such θ_1 exists and the result follows.

A.7 Proof of Theorem 8

Without loss of generality, let $\theta_{min} = 0$. Applying Remark 6 with $\theta_1 = 0$ and $\theta_2 = \theta_0$, $\forall\, \theta \in [0, \theta_0]$,

$$\begin{aligned} w(\theta) &\leq \frac{w_{min}\sin(\theta_0) - w(\theta_0)\sin(0)}{\sin(\theta_0 - 0)}\cos(\theta) + \frac{-w_{min}\cos(\theta_0) + w(\theta_0)\cos(0)}{\sin(\theta_0 - 0)}\sin(\theta) \\ &= w_{min}\cos(\theta) + \frac{-w_{min}\cos(\theta_0) + w(\theta_0)}{\sin(\theta_0)}\sin(\theta) \\ &\leq w_{max}\cos(\theta_0)\cos(\theta) + \frac{w_{max}(1 - \cos^2(\theta_0))}{\sin(\theta_0)}\sin(\theta) \\ &= w_{max}\cos(\theta - \theta_0) \end{aligned}$$

Similarly, applying Remark 6 with $\theta_1 = -\theta_0$ and $\theta_2 = 0$, $\forall \theta \in [-\theta_0, 0]$,

$$\begin{aligned} w(\theta) &\leq \frac{w(-\theta_0)\sin(0) - w_{min}\sin(-\theta_0)}{\sin(0 - (-\theta_0))}\cos(\theta) + \frac{-w(-\theta_0)\cos(0) + w_{min}\cos(-\theta_0)}{\sin(0 - (-\theta_0))}\sin(\theta) \\ &= w_{min}\cos(\theta) + \frac{-w(-\theta_0) + w_{min}\cos(\theta_0)}{\sin(\theta_0)}\sin(\theta) \\ &\leq w_{max}\cos(\theta_0)\cos(\theta) + \frac{w_{max}(\cos^2(\theta_0) - 1)}{\sin(\theta_0)}\sin(\theta) \\ &= w_{max}\cos(\theta + \theta_0) \end{aligned}$$

Therefore, $w(\theta) \leq \max(w_{max}\cos(\theta - (\theta_{min} + \theta_0)), w_{max}\cos(\theta - (\theta_{min} - \theta_0)))\forall\theta \in [\theta_{min} - \theta_0, \theta_{min} + \theta_0]$

A.8 Proof of Theorem 10

Every convex set in $\mathbb{R}^2$ is the intersection of all its supporting half-planes.[1] Thus, defining

$$\begin{aligned} H_1(\theta) &= \{r \in \mathbb{R}^2 | r \cdot \hat{\theta} \leq P_1(\theta)\} \\ H_2(\theta) &= \{r \in \mathbb{R}^2 | r \cdot \hat{\theta} \leq P_2(\theta)\} \end{aligned}$$

We have

$$H_1(\theta) \subseteq H_2(\theta) \forall\theta \in [0, 2\pi]$$

such that

$$\left(x \in \bigcap_{\theta\in[0,2\pi]} H_1(\theta) \implies x \in \bigcap_{\theta\in[0,2\pi]} H_2(\theta) \right) \implies \gamma_1 = \bigcap_{\theta\in[0,2\pi]} H_1(\theta) \subseteq \bigcap_{\theta\in[0,2\pi]} H_2(\theta) = \gamma_2$$

A.9 Fourier Series of $|\sin(x)|$

The function $f(x) = |\sin(x)|$ is even and π-periodic. Therefore, its Fourier series is given by:

$$f(x) = \frac{a_0}{2} + \sum_{k=1}^{\infty} a_k \cos(2kx)$$

where $a_k = \frac{2}{\pi}\int_0^{\pi} f(x)\cos(2kx)\,dx = \frac{2}{\pi}\int_0^{\pi} \sin(x)\cos(2kx)\,dx$

It follows that

$$\begin{aligned} a_k &= \frac{1}{\pi}\int_0^{\pi} [\sin((1+2k)x) + \sin((1-2k)x)]\,dx \\ &= \frac{1}{\pi}\left[\frac{-\cos((1+2k)x)}{1+2k} + \frac{\cos((1-2k)x)}{1-2k}\right]_0^{\pi} \\ &= \frac{2}{\pi}\left(\frac{1}{2k+1} - \frac{1}{2k-1}\right) = -\frac{4}{\pi(4k^2-1)} \end{aligned}$$

Thus, the Fourier series for $|\sin(x)|$ is:

$$|\sin(x)| = \frac{2}{\pi} - \frac{4}{\pi}\sum_{k=1}^{\infty}\frac{\cos(2kx)}{4k^2-1}.$$

A.10 Infinite sum evaluations

$$\begin{aligned} \sum_{n\in 2\mathbb{Z}^+}\frac{\sin^2(n\pi/4)}{n^2-1} &= \sum_{n=0}^{\infty}\left[\frac{1}{(8n+2)^2-1} + \frac{1}{(8n+6)^2-1}\right] \\ &= \frac{1}{2}\sum_{n=0}^{\infty}\left[\frac{1}{8n+1} - \frac{1}{8n+3} + \frac{1}{8n+5} - \frac{1}{8n+7}\right] \\ &= \frac{1}{2}\sum_{n=0}^{\infty}\frac{(-1)^n}{2n+1} \\ &= \frac{\pi}{8} \\ \sum_{n\in 2\mathbb{Z}^+}\frac{\cos^2(n\pi/4)}{n^2-1} &= \sum_{n\in 2\mathbb{Z}^+}\frac{1}{n^2-1} - \sum_{n\in 2\mathbb{Z}^+}\frac{\sin^2(n\pi/4)}{n^2-1} \\ &= \frac{1}{2}\sum_{n\in 2\mathbb{Z}^+}\left[\frac{1}{n-1} - \frac{1}{n+1}\right] - \frac{\pi}{8} \\ &= \frac{1}{2} - \frac{\pi}{8} \end{aligned}$$

References

1. Boyd, S.P., Vandenberghe, L.: Convex Optimization. Cambridge University Press, Cambridge (2004). https://web.stanford.edu/~boyd/cvxbook/
2. Hynd, R.: The blaschke-lebesgue theorem revisited. Expo. Math. **42**(6), 125617 (2024). https://doi.org/10.1016/j.exmath.2024.125617
3. Lang, R.: A note on the measurability of convex sets. Arch. Math. **47**, 90–92 (1986). https://doi.org/10.1007/BF01202504

Author Index

H. Guo et al. (Eds.): IRC-SET 2025, PTT, pp. 471–472, 2026.
https://doi.org/10.1007/978-981-95-9630-0

The manufacturer's authorised representative in the EU is Springer Nature Customer Service Centre GmbH, Europaplatz 3, 69115 Heidelberg, Germany. If you have any concerns regarding our products, please contact ProductSafety@springernature.com

Printed and bound by CPI Group (UK) Ltd, Croydon, CR0 4YY
20/10/2026
02236465-0016